Commission of the European Communities

environm
and quality

Climatic change and impacts: A general introduction

Proceedings of the European School of Climatology and Natural Hazards course, held in Florence from 11 to 18 September 1988

Edited by:

R. Fantechi, [1] G. Maracchi, [2] M. E. Almeida-Teixeira [1]

[1] Commission of the European Communities
200, Rue de la Loi
B-1049 Brussels

[2] Università di Firenze
Istituto di analisi ambientale e telerilevamento applicati all'agricoltura – Consiglio nazionale delle ricerche
I–50100 Firenze

Directorate-General
Science, Research and Development

1991

EUR 11943 EN

Published by the
COMMISSION OF THE EUROPEAN COMMUNITIES
Directorate-General
Telecommunications, Information Industries and Innovation
L-2920 Luxembourg

LEGAL NOTICE

Neither the Commission of the European Communities nor any person acting on behalf of the Commission is responsible for the use which might be made of the following information

Cataloguing data can be found at the end of this publication

Luxembourg: Office for Official Publications of the European Communities, 1991

ISBN 92-826-0564-7 Catalogue number: CD-NA-11943-EN-C

Printed in France

EUROPEAN SCHOOL OF CLIMATOLOGY AND NATURAL HAZARDS

Foreword

The training of new scientific and technical staff and the development of highly qualified scientific specialists are and always have been among the prime concerns of the Commission of the European Communities, which regards this task as a major requirement for the implementation of a common policy in science and technology.

The European School of Climatology and Natural Hazards is a part of the training and education activities of EPOCH (European Programme on Climatology and Natural Hazards).

It annually organizes courses open to graduating, graduate or post graduate students in appropriate fields of Climatology, Natural Hazards and closely related fields.

The courses are organized in cooperation with European institutions involved in the carrying out of the Community's R&D programmes on Climatology and Natural Hazards, and are aimed at allowing students to attend formal lectures and to participate in informal discussions with leading research workers. The opportunity for demonstrations, case studies or presentation of posters is given to the students attending the courses.

The teachers are selected among European Scientists who are leading authorities in their respective fields.

The present volume contains the lessons delivered at the course held in Florence from 11th to 18th September 1988 on the subject **"Climatic Change and Impacts: A General Introduction"**, together with short presentations from the students of their own research activities and interests.

It is hoped to constitute a valuable permanent record of the course, and to stimulate further interest on the part of teachers and students alike.

R. FANTECHI
Head of Service

TABLE OF CONTENTS

Foreword III

List of teachers VII

List of students XI

CLIMATIC CHANGE AND IMPACTS: A GENERAL INTRODUCTION 1

1. Introduction

P. MOREL – An Overview of the Climatic System 3

2. Past Climate Changes 11

J.C. DUPLESSY – The Last Climatic Cycle and Past Ocean Changes 13

A. BERGER – Milankovitch Effects on Long-Term Climatic Changes 29

3. Climate Processes and Climate Modelling 47

A. BERGER – Basic Concepts of Climate Modelling 49

A. BERGER and Ch. TRICOT – Simple Climate Models 77

H.J. BOLLE – Radiative Transfer and Greenhouse Effect 119

G. BRASSEUR – Atmospheric Chemistry and Climate 133

J.F.B. MITCHELL – Simulation of Climate 157

H. CATTLE – Ocean Models and Ocean-Atmosphere Coupling 177

J. OERLEMANS – Land Ice and Climatic Change – An Introduction 187

H.J. BOLLE – Land-Surface – Vegetation – Climate Feedbacks .. 207

G. DUGDALE – Satellite Data for Climate Studies 223

4. The Greenhouse Gas Induced Climate Change 237

U. SIEGENTHALER – Carbon Dioxide: Its Natural Cycle and Anthropogenic Perturbation 239

J.F.B. MITCHELL – Simulation of Climate Change 289

5. Climatic Impacts 319

M.L. PARRY and T.R. CARTER - Climatic Changes and Future Land Use Potential in Europe 321

G. MARACCHI - Impacts of Climatic Change on Crops 343

H. LIETH - Interaction Between Biosphere and Climate 351

R. A. WARRICK - Sea Level Change and the Climate Connection: Past and Future 375

STUDENTS' PAPERS 391

M. FORSTREUTER - Dry Matter Accumulation, CO_2 Gas Exchange, and Water Uptake of Plants and Plant Communities Under Increased Atmospheric CO_2 Concentration 393

D. GREGORY - Inclusion of Downdraught Processes in the UK Meteorological Office Mass Flux Correction Scheme 403

J. GUIOT, J.L. de BEAULIEU, A. PONS and M. REILLE - A 140,000 - Year Climatic Reconstruction of the Continental Climate from Two European Pollen Records 405

J.E. HOSSELL - Hydrometeorological Assessment Program in East Africa 409

W.J. INGRAM - Cloud Radiative Properties and Climate Sensitivity 415

E. LOPEZ-BAEZA - Determination of Surface Albedo for Agroclimatological Uses 419

A.J.L. MANNING - Modelling Radiative Transfer in the Middle Atmosphere 425

I. MARSIAT, Th. FICHEFET, H. GALLEE, Ch. TRICOT and A. BERGER - Modelling the Eurasian Ice Sheet Over the Last 122,000 Years 431

J.H. PORTER - A Comparison of Two Agroclimatic Indices for Use in Investigating the Effects of Climatic Change on Agriculture in Europe 437

C. SENIOR - The Role of Sea - Ice in the Simulated Antarctic Winter Climate 445

Rosamund A. SPENCE - The Effects of Climatic Change and Increasing CO_2 on European Crop Production - Initial Studies 449

List of Teachers

Prof. André BERGER, Université Catholique de Louvain, B

Doctor of Sciences, Université Catholique de Louvain, where he is professor of meteorology and head of the Institute of Astronomy and Geophysics. He is the president of the International Commission of Climate of IUGG and of the Paleoclimate Commission of INQUA. His main research is on modelling the astronomical theory and the man's impact on climate.

Prof. Dr. Hans-Jürgen BOLLE, Freie Universität Berlin, D

Prof. Bolle works at the Institute für Meteorology, Free University of Berlin. His main scientific interests are: radiative transfer in the atmosphere, radiation properties of atmospheric trace gases, soils and vegetation, evaluation of satellite measurements for climate studies.

Dr. Guy BRASSEUR, Institut d'Aéronomie Spatiale de Belgique, B

Leader, Atmospheric Chemical Modelling Section, National Center for Atmospheric Research, Boulder Colorado. Lecturer, Free University of Brussels (lectures on "External Geophysics" and "Geochemistry of the atmosphere"). Activities: development of atmospheric models of the stratosphere and troposphere including chemical, radiative and dynamical processes. Study of the ozone layer and its perturbations, climatic impacts of man-made perturbations, etc.

Dr. Howard CATTLE, Meteorological Office, UK

Dr. Cattle joined the United Kingdom Meteorological Office in 1973, having carried out doctoral and postdoctoral studies of the observed structure of the boundary layer at Imperial College, London. He then worked on problems of four dimensional data assimilation before taking up the post of Deputy Principal of the Meteorological Office College in 1976. In 1980 he joined the Dynamical Climatology Branch of the Meteorological Office as leader of a new group with the task of the development of global ocean models for climate studies. In 1986 he took charge of the group concerned with the parametrization of physical processes in atmospheric models prior to being appointed Assistant Director and head of the Dynamical Climatology Branch in 1987. The interests of the branch lie in the development and use atmosphere, ocean and coupled ocean-atmosphere models for studies of climate and of climate change, including, in particular, the impact of man's activities in these areas.

Dr. George DUGDALE, University of Reading, UK

Department of Meteorology, University of Reading, UK. Research in surface energy exchange and satellite meteorology with particular reference to applications in tropical areas.

Dr. Jean-Claude DUPLESSY, C.N.R.S. - Gif-sur-Yvette, F

Director of Research at Centre National de Recherche Scientifique (CNRS), France. Director, Centre des Faibles Radioactivités, Joint Laboratory of CNRS and CEA (Atomic Energy Commission). Chairman of the Committee of Climatic Change and the Ocean (CCCO), Paleoclimatology panel. Physicist working in geochemistry in order to reconstruct the evolution of past climates from marine sediment data.

Prof. Helmut LIETH, University of Osnabrück, D

University of Osnabrück, FRG.
Scientific activities: modelling vegetation/climate relations, modelling vegetation/soil/hydrology relations, interdisciplinary modelling of ecology/economy/sociology relations, geoecology, remote sensing.
Prof. Helmut Lieth has a vaste scientific career both in Germany and the United States, which also includes field work carried out in Europe, Africa, North and South America. He also works for quite a few international scientific societies.

Prof. Giampiero MARACCHI, University of Florence, I

Professor of Agrometeorology and Climatology, University of Florence. Director IATA - CNR, and CESIA. He does research in the field of agrometeorology, crop modelling and remote sensing. Principal investigator and co-investigator of several international programmes of EEC, WHO and FAO. Author of about 200 scientific publications.

Dr. John MITCHELL, Meteorological Office, UK

Scientist, UK Meteorological Office. His main scientific interests are: numerical modelling of climate and climate change (particulary that due to man's activities). Over the last 10 years, has investigated the climatic effects of increased CO_2. Has also studied the effects of changing the earth's orbital parameter, the climatic effects of nuclear war, and the dependence of simulated climate sensitivity on the representation of physical processes.

Prof. Pierre MOREL, Organisation Météorologique Mondiale, CH

P. Morel has established the "Laboratoire de Météorologie Dynamique", National Center of Scientific Research, Paris, France and initiated the METEOSAT and ARGOS satellite systems. He has also led the French space science programmes as Deputy Director General of the French space agency, CNES. He joined the International Science Management Community in 1982 as Director of the World Climate Research Programme of the World Meteorology Organisation (WMO) and the International Council of Scientific Unions (ICSU). He is a Professor of Physics at the University of Paris since 1964.

Prof. Hans OERLEMANS, University of Utrecht, NL

Professor at Faculty of Physics and Astronomy, University of Utrecht. His research interests are: mass balance and dynamics of ice sheets and glaciers and mesoscale meteorology.

Dr. Martin PARRY, University of Birmingham, UK

Reader in Resource Management, University of Birmingham, Department of Geography.
1982-1985 Project leader, IIASA Climate Impacts Project (which investigates possible effects of climate change on agriculture). Currently directing projects for EC on impacts on agriculture in Europe and for UNEP on impacts of climate change in SE Asia.

Dr. Dominique RAYNAUD, C.N.R.S., Saint-Martin d'Hérès, F

Directeur de Recherche, CNRS. His scientific activities lie on climatic and environmental record of the last 150.000 years from ice core studies.
He works with the Centre National de la Recherche Scientifique, mostly at the Laboratoire de Glaciologie et de Géophysique de l'Environnement.
Current interests: interaction between atmospheric composition and climate and interaction between climate and ice-sheets.

Dr. Ulrich SIEGENTHALER, University of Bern, CH

University of Bern. His research activities concentrate upon modelling of the global carbon cycle and its perturbations, using isotopes and studies of paleoclimate using stable isotopes.

Dr. Richard WARRICK, University of East Anglia, UK

Climate Research Unit, East Anglia University, UK. Current research activities include: the greenhouse effect and sea level changes, impacts on the UK and Europe; CO_2, climatic change and agricultural effects; development of models for probabilistic estimates of climatic change for decision-making.

Prof. Tom WIGLEY, University of East Anglia, UK

Director of the Climatic Research Unit of East Anglia University - UK. Prof. Wigley works on various aspects of climatology including: monitoring of recent climatic change; paleoclimatology, especially the last 20.000 years; mechanisms of climate change; modelling and the greenhouse effect.

List of Students

BARTHELMIE Rebecca	Norwich - UK
BINDI Marco	Firenze - Italy
COSTA Rita	Lisboa - Portugal
DUPRAT Josette	Bordeaux - France
FERRINI Francesco	Firenze - Italy
FORSTREUTER Manfred	Osnabrück - FR Germany
GREGORY David	Bracknell - UK
GUIOT Joël	St. Savournin - France
HOSSELL Joanne	Reading - UK
HOTCHKISS Dawn	Reading - UK
INGRAM William	Bracknell - UK
LAMARQUE Jean-François	Braine-le-Comte - Belgium
LOPEZ-BAEZA Ernest	Burjasot - Spain
LOUTRE Marie-France	Tournai - Belgium
MADEIRA Ana	Lisboa - Portugal
MAITRE Florence	Paris - France
MANNING Amanda	Oxford - UK
MARSIAT Isabelle	Court-St. Etienne - Belgium
MÜLLER Kerstin	Hamburg - FR Germany
MURPHY James	Wokingham - UK
PIERROS Frangiskos	Athens - Greece
PORTER Julia	Birmingham - UK
RIEWENHEM Sabine	Osnabrück - FR Germany
SENIOR Catherine	Wokingham - UK
SPENCE Rosamund	Reading - UK
SWEET Nina	Yiewsley - UK
TALAIA Mario	Agueda - Portugal

EUROPEAN SCHOOL OF CLIMATOLOGY AND NATURAL HAZARDS

Course on

"Climatic Change and Impacts: A general Introduction"

1. INTRODUCTION

AN OVERVIEW OF THE CLIMATE SYSTEM

P. MOREL
World Climate Research Programme
ICSU/WMO

Summary

Although real climate changes are easily masked by all kinds of meteorological perturbations or vacillations of the atmospheric circulation, and correspondingly hard to detect, paleoclimatology and even the instrumental record in the last hundred years give solid evidence of significant climatic variations. The most spectacular change in the last 20,000 years is the transition from a glacial age to the present benign climate. The analysis of variations in the global pattern of sea surface temperature allows separating the most significant low-frequency modes of atmospheric (and oceanic) variability: quasi-biennial changes in the mean strength of winds, the El Niño/Southern Oscillation and longer-period oscillation of the global atmosphere/global ocean system. Climate is also highly sensitive to the multiple roles of water in all three phases. In particular, atmospheric water vapour and clouds have profound effect on the radiative balance of the earth. In-depth study of all components of the climate system, comprised of the atmosphere, ocean, land- and sea-ice, terrestrial vegetation and surface water, is necessary to understand and predict global climate change which may result from the impact of man's activities on the environment as well as natural causes.

1. INTRODUCTION

Climate may be defined as the cumulative effect of weather, summed or averaged over a suitable period of time. In effect, one could say that climate is "made of weather" since variability exist on all time scales, from days to centuries. Day-to-day variations, up to 10 or even 20 days in some cases, are associated with individual meteorological events such as the development and transit of specific baroclinic waves and cyclonic disturbances in the region of polar air front. Large-scale waves may be predictable up to about 10 days in most cases, on the basis of the information provided by daily global meteorological observations.

Longer period variations are considered as anomalies in the normal seasonal or annual regime of the atmosphere. For example, tropical atmospheric circulation anomalies associated with changes in tropical Pacific ocean currents and thermal stratification, are well documented and known as the Southern (atmospheric) Oscillation and El Niño (oceanic circulation) events. Such phenomena, which may last a few months to one year, and recur every 3-5 years, are in fact the main source of interannual variability of the sea surface temperature in tropical oceans and seasonal disturbances of the tropical atmospheric regime, including considerable shifts in the tropical storm tracks, serious drought in some locations and excessive rainfall elsewhere, as well as anomalous strong westerly winds instead of the usual easterly Trades. On somewhat longer time-scales, the dry period of the north American "dust-bowl" lasted for about ten years and the characteristic period of successive drought episodes in the African Sahel is 5-10 years. Furthermore, we know from the instrumental records that the mean global surface temperature of the earth, ocean and continents alike, has been generally rising, by some 0.6 K during the last hundred years.

So, it is a good question to ask what distinguishes climatic variations from normal weather variability, and the answer can be nothing but the arbitrary definition of a low-pass time filter, which traditionally has been taken to be a 30 year average. Alternatively, one may define as climatic variability (transient fluctuations and long-term trends), all variations beyond the range of deterministic predictions, which does not normally exceed 10 days or so, i.e. when memory information contained in the observed initial state of the atmospheric circulation is lost. The latter definition is that adopted by the World Climate Research Programme (WCRP) and includes a good deal of spontaneous atmospheric variability or vacillations such as the random development of stationary blocks which transform the fast westerly circulation into a relatively slow, meandering flow. Other vacillations are associated with the propagation of wave-like perturbations along the equator, e.g., the 30-50 oscillation which controls the break- and active periods of the Indian monsoon or the stratospheric Quasi-Biennial Oscillation (QBO). The signature of the QBO is now being detected in several tropospheric phenomena.

It should be clear from this (brief) enumeration that changes from year to year or decade to decade, most likely to be noticed by an individual, are not long-term trends at all but manifestations of essentially random transient fluctuations of natural origin, caused by vacillations in the dynamics of the earth's system, constituted by the atmosphere, ocean, ice and continental land surface and soil. Real "climate change" over long periods like 30 years are far more difficult to document, so much so that one may legitimately ask: "does climate change?"

2. NATURAL CLIMATE VARIATIONS

Meteorological statistics show that climate change over a period of 30 years, which would be attributed to a long-term trend, is currently so much smaller than transient fluctuations that climate is still taken as constant in usual engineering practice. Possible climate change is generally not considered a significant factor in planning new investments, meaning that design specifications for civil engineering and other environment dependent activities are simply based on climatological statistics of the last decade or two. It should be pointed out that this is a perfectly valid approach as long as substantial climate change is not expected within the time span of a particular installation, a few decades for an industrial plant to a century for civil engineering works. Looking back into past history, one would take comfort in the apparent stability of climate conditions: for example, the map of Atlantic trade winds established by Halley in the seventeenth century could still be used with effect by modern sailors!

And yet, ample proof of past climate change can be obtained from plain observation or not so simple investigations of paleoclimatological markers. The history of Viking nautical voyages across the north Atlantic and the foundering of Norse settlements in Greenland are clear evidence of relatively benign conditions in the high Middle Ages and the subsequent cooling, known as the "Little Ice Age", which began in the 14th century and caused big storms and storm surges in the North Sea as well as a marked extension of land glaciers and sea-ice. On the other hand, the well-documented retreat of alpine glaciers during the last 200 years is an indication of recovery from the previous cold period. The retreat of snow and glaciers stands out when 19th century drawings or water colours of alpine landscapes are compared with the present scenery.

The mapping and dating of submerged fossil coral terraces* around the tropical ocean shores tell us a fantastic story: the mean level of the ocean went down everywhere by as much as 130 metres below present global mean sea level, during glacial climate episodes which seems to be recurring every 100,000 years or so. That much water was then sequestered in massive north American, Eurasian and Antarctic ice sheets, some 3 to 5000 m thick.

The analysis of deep ocean sediments provides us with even more detailed information on global sea surface temperature patterns in past ages. Indeed, the uppermost euphotic layer of the ocean was then teaming with precisely the same species of phyto- and zooplankton as we find now, only in different proportions depending upon the upper ocean temperature. These microscopic sea creatures live their short lifes, die and sink down to the bottom of the ocean to become part of abyssal sediments. Thus, by analysing the populations of microfossils recognizable in deep ocean sediment cores taken now, and dating each layer of the core, one can reconstitute the mix of planktonic species which were striving 10, 20 or 30,000 years ago at one particular location, and therefore estimate the upper ocean temperature at that time. Project CLIMAP collated this type of information from many sediment cores to reconstitute successive maps of the global sea surface temperature across the ages, showing graphically the strong extra-tropical cooling associated with maximum glaciation 18,000 years ago, and remarkably enough, almost constant temperature conditions near the equator.

Even more impressive, by the extreme sensitivity of geochemical techniques which are brought to bear, are the fine details of the climatic record obtained from the chemical and isotopic analysis of Greenland or Antarctic deep ice cores. In particular, a 2000 metres long core taken by the USSR Antarctic mission at Vostok, and analysed by French glaciologists and geochemists, yielded a detailed continuous record of isotopic ratio anomalies (D/H or O^{18}/O^{16}) which correlate rather well with mean atmospheric temperature at high latitudes, and also changes in atmospheric composition (especially the concentration of carbon dioxide) through the last 165,000 years. The striking relation between temperature and the amount of carbon dioxide is a clear indication of the central role played by this absorbing gas in controlling the atmosphere greenhouse effect and the earth temperature.

To answer the question raised in the introduction, one can state with confidence that the earth climate did change on geological time scales of 10,000 to 100,000 years, with major effects on the environment, temperature, moisture, aridity and vegetation. It is reasonable to

---

* Corals can only live at a very shallow depth

believe that similar natural phenomena are occurring now, and in addition, that man's activities such as the worldwide development of agriculture and industry, deforestation, the rapid combustion of fossil fuels, have reached such a high rate that global climatic impacts must be expected.

3. THE ROLE OF THE OCEAN IN THE CLIMATE SYSTEM

If we define the useful climate signal (for application to human endeavour) as change on periods longer than the characteristic time scales of atmospheric vacillations (like the 30-50 day oscillation or the QBO), yet not so long as to involve major redistributions of ice sheets and terrestrial vegetation), then natural climatic variability and response to external forcing must be determined by the coupled dynamical behaviour of the two fluid components of the earth system: the atmosphere and the ocean.

Now, the atmosphere acts on the ocean in a variety of ways: pure mechanical forcing of the ocean surface by wind stress, and gravitational (density) forcing by transfer of heat or fresh water (changing oceanic temperature or salinity). On the other hand, the ocean reacts on the atmospheric circulation through heat fluxes which are determined by only one parameter, the sea surface temperature. Thus, a diagnostic of large-scale atmosphere-ocean interaction is possible by analysing the time and space dependence of the annual or seasonal mean sea surface temperature (SST) field, which fortunately is know over almost the whole ocean from ship measurements for the last 80-100 years. This information may be presented in the form of global mean or zonal mean temperature trends, but more usefully, in the form of a modal expansion, based on the analysis of interannual temperature variability in terms of empirical orthogonal functions (EOF).

The time history of synchronous SST variations associated with the first mode (EOF-1) of this expansion is undistinguishable from that of the mean global temperature. It shows substantial variations on quite short time-scales, 2 years or so, and also a much slower and possibly "irreversible" trend (at least for the near future). The fast variations must necessarily be associated with quasi-biennial fluctuations in the mean strength of the atmospheric circulation or "index cycle", with concurrent changes in the depth of the thermocline or the amount of mixing of relatively warm superficial waters with deep (cold) oceanic waters, without substantially affecting the total (annual mean) heat content of the ocean. The low-frequency variations are an altogether different matter (they involve net oceanic heat loss or storage) and are not fully understood at present. Suffice it to say that the general warming trend observed since 1880 could be explained by the computed excess heat flux caused by the increased greenhouse effect of carbon dioxide, if the mean global thermal inertia of the ocean was that of an homogeneous mixed layer 200-300 m thick, without any obvious change in long distance heat transport by oceanic currents.

The distinctive spatial patterns and time history of the second EOF leads us to recognize a major mode of interannual variability in the coupled ocean-atmosphere system, the importance of which has, however, been recognized only fairly recently. This mode is known as the El Niño-Southern Oscillation (ENSO). As can be seen plainly by examining EOF-2, the dynamically significant response of the ocean, involving substantial long-range transport of heat, occurs essentially in the tropical Pacific, while relatively weak, spatially homogeneous and synchronous changes in the tropical Atlantic and Indian oceans, as well as in the whole Pacific basin, are being driven by the associated changes in the global atmospheric circulation. The ENSO is not a significant component of European climate as the atmospheric circulation over the Eurasian continental mass is, to a very large degree, decoupled from tropical SST anomalies. On the other hand, the ENSO has serious climatic impacts in the tropics, especially on the western shore of South America, Australia, Indonesia and other Pacific basin countries. A major activity in the WCRP is the Tropical Ocean and Global Atmosphere (TOGA) study, undertaken in 1985, to understand the physics of the tropical ocean-atmosphere processes and develop means to predict the outcome of incipient ENSO events.

The third most significant mode, EOF-3, has yet different characteristics: its spatial pattern is, essentially, a seesaw between the northern and southern oceans and its time-dependence involves mostly longer periods (10-50 years). The remarkable correlation between the time variation of the EOF-3 ocean temperature pattern and that of rainfall over the African Sahel region is definite proof that the phenomenon is indeed governed by coupled ocean-atmosphere dynamics on the global scale. The nature of the oceanic circulation change associated with this third mode is not known at present, but it is quite obvious that it does involve the global ocean as well as the global atmosphere. Maybe what we are witnessing is a vacillation of the earth system on a grand scale between the present state of oceanic circulation with substantial convective overturning (deep water formation) in the North Atlantic, and strong meridional heat transport, and a more stable stratified state, with much reduced convection and meridional heat transport, which we believe existed during periods of glacial climate in the past. Obviously, much improved knowledge of global ocean dynamics and thermodynamics, is needed to address this problem and to separate natural variability on low frequencies (time-scales of years to decades) from the true response to persistent change in the composition of the atmosphere and other climate factors. A major oceanographic programme, the World Ocean Circulation Experiment is planned under the WCRP, to begin in 1990-91 with the launching of a new series of oceanographic satellites (the US-French TOPEX/POSEIDON altimetric mission and the European Space Agency ERS-1 satellite) and the implementation of a vast hydrographic programme to span the whole ocean.

4. ROLE OF ATMOSPHERIC WATER

Meteorologists and research scientists alike have long recognized the need to envisage the behaviour of the earth atmosphere as a global fluid dynamics problem. As long as 115 years ago, the International Meteorological Organization (now WMO) was established to collect worldwide meteorological observations as quickly as possible and produce, in nearly real-time, a synoptic (instantaneous) picture of the atmospheric circulation. Hundred and fifteen years later, with the development of a worldwide ground-based observing network spanning 156 countries and several operational meteorological satellite systems covering the whole earth, that goal has essentially been achieved, allowing global weather prediction over a range of 5-10 days.

Yet, our quantitative knowledge of global atmospheric thermodynamics is still inadequate to pin down the relatively small difference between the large incoming flux of solar energy and outgoing infrared radiation, which controls climate. The root of the problem lies in the complicated role of water in the earth atmosphere and, in general, the extreme space- and time-variability of the atmospheric, land or ocean surface processes which determine the global hydrological cycle.

In the first place, water vapour itself is a strong absorber of the upwelling infrared radiation emitted by the earth surface. As we well know, radiative cooling at night can be fierce when the atmosphere is very dry, e.g. on mountain tops or in deserts. On the contrary, surface air temperature remains warm throughout the night in hot, moist climates, Indeed, atmospheric water vapour is the single most important constituent of the atmosphere contributing to the overall greenhouse effect and could even, under more extreme conditions as was found on primeval Venus, lead to a "run-away greenhouse" causing excessive warming and the eventual loss of atmospheric water through photodissociation. This phenomenon was a manifestation of the unstable (positive) feedback loop between surface temperature, atmospheric water content and the greenhouse effect.

Furthermore, water condenses in the atmosphere to form all kinds of clouds, low and high, which interfere with radiation transfer. It is known that low clouds, being both relatively dense and warm, reflect a lot of solar radiation back to space (accounting for the high albedo of the planet) and emit about as much infrared radiation as the underlying surface: their net effect is to <u>reduce</u> the net radiation input to the surface and therefore to cool the earth. On the other hand, high clouds (ice cirrus) are relatively translucent and cold. They do scatter incoming solar radiation but with only little loss to space, and they contain a substantial amount of water vapour which cuts down significantly the outgoing infrared radiation: their net effect is to <u>increase</u> the net radiation input to the surface and therefore to warm up the earth. As far as our still quite imperfect observing techniques allow us to judge, it appears that the present global distribution of low, intermediate and high clouds has a slight net negative effect (cooling) on the earth radiation balance, but only a modest change in the low cloud/high cloud ratio would make the net radiative effect neutral or even positive (warming).

We have, of course, undertaken the task of assembling detailed global cloud distribution statistics, deduced with considerable difficulty and remaining uncertainty from cloud images received from all geostationary and polar orbiting meteorological satellites (WCRP International Satellite Cloud Climatology Project). We are comparing the outgoing solar and infrared radiation fluxes stemming from cloudy and clear air patches of the atmosphere with the help of the medium resolution scanning radiometers of the Earth Radiation Budget Experiment. And yet, this is not enough because what we really want is to relate (predict) the cloud distribution to the large-scale atmospheric temperature and moisture parameters which are likely to change under different climatic conditions. We need to understand, for example, how cloud would change when the atmosphere becomes loaded with twice the pre-industrial amount of carbon dioxide.

This, in turn, means that we want to understand all factors which finely control the water vapour content and relative humidity of the atmosphere, i.e., the various aspects of the global hydrological cycle, a very tall order indeed. For the hydrological cycle involves not only the atmospheric transport of water vapour and transient rain-making convection, but also the extremely complex interaction between terrestrial vegetation or bare soil and the atmospheric boundary layer which determines evapotranspiration of water. We have good reason to believe that this complexity cannot effectively be tackled with existing global observing systems, designed mainly for the needs of operational weather forecasting, and that more powerful global remote sensing techniques, using a new generation of ambitious orbital platforms, are needed to collect the required information and document global environmental change. In the field of climate research, this ambition is borne by the plan for a Global Energy and Water Cycle Experiment (GEWEX) which would begin at the turn of the century.

5. THE PROBLEM OF GLOBAL CHANGE

We are all aware of the increasing stress caused by man's activities on the natural environment. The occurrences of significant toxic pollution of air, water and soil are commonplace in the news. The damages caused by acid rain to forests and by artificial chlorinated products to stratospheric ozone are constantly being brought to public attention. And yet it is not generally recognized that the most portentous impact of man on the planet is the slow and unspectacular alteration of the earth atmosphere composition through the release of carbon monoxide and dioxide, methane, chlorofluorocarbons (CFCs), nitrogen compounds, etc., and the equally hidden consequences on atmospheric climate and the oceanic circulation regime. We have observational evidence showing that the concentration of atmospheric carbon dioxide rose from 280 parts per million (ppm) in the 18th century to about 350 ppm at present and that the rate of increase is still accelerating. We can see that the concentration of methane is closely correlated to the total population of the earth (as it is directly linked to the expansion of rice crops and livestock). Nitrogen compounds are released in the atmosphere as a result of various chemical or industrial processes, but also, to a large extent, by the increasing use of fertilizer subsidies. And, as is well known, CFCs are products of the chemical industry with various applications from refrigeration to the production of polymer foams. Altogether, the cumulative effect of these "greenhouse gases" will approximate than of doubling the pre-industrial concentration of CO_2 about the middle of next century, an unprecented change on account of both rapidity and scale: the fastest variations of paleoclimate had a time-span of 500 years or so, and CO_2 varied "only" from about 200 to 300 ppm throughout the last glacial cycle. In effect, we are now in the process of conducting a giant "experiment" with our planet, for which there is no obvious precedent in paleohistory. It is quite a pressing problem to predict the likely outcome of this experiment of we are to provide human societies with a fair warning of global change.

EUROPEAN SCHOOL OF CLIMATOLOGY AND NATURAL HAZARDS

Course on

"Climatic Change and Impacts: A General Introduction"

2. PAST CLIMATE CHANGES

THE LAST CLIMATIC CYCLE AND PAST OCEAN CHANGES

J.C. DUPLESSY

Centre des Faibles Radioactivités
Laboratoire mixte CNRS-CEA
F-91198 Gif Sur Yvette Cédex

ABSTRACT

During the last 150,000 years, the earth's climate alternated from warm conditions to cold conditions. The peak of the last glaciation occurred about 18,000 years ago and a deglaciation led to the present climatic pattern.

In order to reconstruct past climatic and past oceanographic conditions, geologists must define proxy-data, which describe various climatic parameters in the past and determine the time scale of the records of these proxy-data.

The age of the sediment is determined by absolute dating either by the carbon-14 method (for the last 40,000 years) or by uranium series disequilibrium (for the last 300,000 years). A good stratigraphy for marine sediments can also be derived from the oxygen isotope record of planktonic and benthic foraminifera, which partly reflects the variations of the volume of ice stored over the high-latitude continents.

Conditions prevailing at the sea surface may be derived through transfer functions linking the fossil fauna to the water temperature or salinity. A similar approach can be used to derive temperature and humidity at the surface of the continents from the pollen content of lake sediments or peat bogs. Conditions prevailing in the deep ocean waters can be estimated from the oxygen and carbon isotopic composition of benthic foraminifera, which provides information on both the deep water temperature and the isotopic composition of the total dissolved carbon dioxide.

A synthesis of the proxy-data obtained for the last climatic cycle shows:

1. Conditions as warm as those of today occurred during the last 10,000 years and during about 10,000 years around 125,000 years B.P.
2. The volume of ice frozen over the continents increased progressively from about 115,000 years B.P. to 18,000 years B.P., but this increase was marked by sharp pulses of ice volume increase or decrease.

3. Conditions prevailing during the last glacial maximum at the Earth's surface were described by CLIMAP (1976,1981) and were noticeably colder than today at high latitudes. The cooling was much smaller at low latitudes. The mean sea surface temperature drop was estimated close to 2°C, but the mean global temperature of the Earth's atmosphere was 4-6°C lower than the present one.
4. The global ocean deep water circulation during the last glacial maximum was characterized by well-oxygenated intermediate waters, whereas deep and bottom waters were poorly oxygenated, a configuration which helped to sequester CO_2 in the deep ocean and thus to lower the atmospheric CO_2 content (see the ice record).
5. Carbon-14 dating performed by Accelerator Mass Spectrometry (AMS) on monospecific samples of planktonic and benthic foraminifera indicates that the climatic changes associated with the last deglaciation have been rather rapid (10°C variations in less than 4 centuries in the Atlantic Ocean off Europe).
6. A comparison of AMS carbon-14 ages measured on planktonic and benthic foraminifera at the same sediment level in deep sea cores can be used to estimate the "residence time" or "ventilation age" of the deep ocean. Results indicate that the ventilation age of the Pacific ocean was about 2,100 years during the last glaciation (to be compared with the present value of 1,500 years), but decreased sharply to about 1,000 years during the deglaciation.

INTRODUCTION

Geologists have for long been interested in past climate, and there are many known climatic indicators in the geological record. Evidence such as moraines and other features of glacial conditions have been observed in northern Europe and northern America. However, such isolated pieces of evidence are not sufficient for understanding climatic change and its causes. For this one requires measurements of a particular climatic variable, either as a function of time at one location or as a function of geographical location at one time. For the last million years or so, this is most easily achieved by using radioactive isotopes for dating, and stable isotope techniques and transfer functions calibrating micropaleontological data in climatic terms.

In the last 25 years, a great deal of information about long-term climatic change in the last million years or more have been obtained from the study of marine and continental sediments. A reasonably consistent picture emerges, showing alternations between glacial and interglacial conditions, manifested by waning and waxing of large continental ice sheets coinciding with temperature, salinity and circulation changes in the ocean, temperature and humidity changes over the continents. In this paper, we shall concentrate on the reconstruction of the last major oscillation of the Earth's climate, generally referred to as the last climatic cycle, which occurred over the last 135,000 years. A general review on climatic variations and variability has been given by Berger (1981).

METHODS

Cores from the ocean floor or from the bottom of lakes or peat bogs contain material deposited at a fairly constant rate, generally in the order of a few centimeters per thousand years in the ocean and of a few centimeters per century in the continent. In order to derive a useful information on past climates, sediment must first be dated. Then its fossil content (foraminifera, coccoliths or radiolaria in deep sea sediments, pollen or diatoms in continental deposits) are translated into quantitative estimates of those climatic variables that physical model use and simulate.

Dating

Among the simpler methods used by geologists, counting of annual deposits is one of the most precise. Yearly growth rings are prominent in the stems of trees that grow in the seasonal climates of middle and higher latitudes. Microscopic examination of sections from sufficiently sensitive types of tree can thus reveal and date the succession of periods within individual seasons which are more or less favorable for growth (Fritts et al., 1971). Varves in lake sediments are produced by regularly alternating sequences of color, and mineral and organic composition due to seasonal differences in the flux of organic and mineral particles deposited at the bottom of the lake.

The longest tree ring chronology has been established for very long-living trees from the western U.S.A. and extends back over 7,000 years. Several chronologies have been established in other parts of the world. For some of them, the starting date is not known (floating chronologies) and they may extend to about 9,000 years. The varve chronology is essentially restricted to lake deposits from continental high latitudes and extends back to about 12,000 years, but a continuous series covering the whole range of this time span has yet to be found.

Physical methods of dating have a lower precision than the methods previously described, but provide a time scale for much older deposits. These methods depend on the measurement of the residual radioactivity in sediment in which this activity has been decaying since some definable time of interest in the past.

Radiocarbon dating is the principal technique by which paleoclimatologists construct their calendar of the past 40,000 years. The method relies on the fact that a radioactive carbon isotope, carbon-14, is assimilated into the molecular structure of living organisms. Because a population of radioactive atoms decreases at a regular rate, such atoms can serve as a clock by which to measure how long ago a plant or animal died (Broecker and Peng, 1982; Delibrias, 1985).

The rate at which a population of radioactive atoms decays is expressed in terms of a half-life. The half-life of ^{14}C is 5,730 years. This means that a sample consisting of 10,000 ^{14}C atoms would contain only half this amount, i.e. 5,000 atoms after 5730 years (the other 5,000 having transformed into nitrogen-14). After another 5,730 years, only 2,500 ^{14}C atoms would remain, and so on. Therefore, if one knows the original amount of ^{14}C in one sample, one can easily calculate the age of this sample by measuring the residual number of ^{14}C atoms today.

Fortunately the original percentage of ^{14}C among all the carbon atoms in any organic substances is fairly well known. ^{14}C is continuously produced in the upper atmosphere by the interaction of cosmic rays with nitrogen atoms. The ^{14}C then combines with oxygen to form radioactive carbon dioxide, which mixes with the atmospheric carbon dioxide and that dissolved in sea water. ^{14}C enters the biosphere mainly through plant respiration and passed up through the food chain to higher organisms. It is also sequestered in carbonate shells deposited by aquatic organisms from the bicarbonate dissolved in lake and sea water.

Until recently, the $^{14}C/^{12}C$ ratio of a sample was determined by measuring the very weak radioactivity of the sample. The major drawback of this conventional method was the large quantity of carbon (about 1g of pure carbon), which was necessary to detect a statistically significant number of disintegrations in a reasonable amount of time. This problem has now been solved by measuring the ^{14}C atoms with mass spectrometers coupled to a small electrostatic accelerator (Accelerator Mass Spectrometry, A.M.S.). ^{14}C atoms are here characterized by their mass. They can be counted with an accuracy of 1% within one to two hours in samples as small as 1 mg of pure carbon. (Duplessy et Arnold, 1985; Hedges and Gowlett, 1986).

Uranium series disequilibria are used to estimate ages in the range 5,000 to 350,000 years. The various naturally occurring isotopes of uranium (^{235}U, ^{238}U,...) decay ultimately to lead, by passing through characteristic successions of radioactive elements with widely differing half-lives. Among them are Thorium-230 (half-life 75,200 years) and Protactinium-231 (half-life 32,500 years). Both are highly insoluble in sea water. As uranium is by contrast very soluble in sea water, the ^{230}Th and ^{231}Pa isotopes produced by the radioactive decay of uranium are quickly absorbed on small particles and fall to the sediment. The decreasing abundance of these two isotopes with depth in a deep sea core can thus be used to estimate the rate at which the particulate matter accumulated over the ocean floor.

The ^{230}Th and ^{231}Pa content of carbonate shells can also be used to estimate the time elapsed since the death of the animals which grew these shells. This method relies on the fact that calcium and uranyle ions have similar ionic radii, which are fairly different from those of thorium and protactinium. As a consequence, uranium is included in small amounts within calcitic shells, whereas thorium and protactinium are not absorbed. The shell of a recently dead animal therefore contains uranium, but neither ^{230}Th nor ^{231}Pa. However the radioactive decay of uranium produces both isotopes, so that the radioactivity of ^{230}Th and that of ^{231}Pa increase progressively as a function of time. After an infinite time (in practice five half-lives of these isotopes), the radioactivity of ^{230}Th is equal to that of ^{238}U and the radioactivity of ^{231}Pa is equal to that of ^{235}U. The measurement of the $^{230}Th/^{238}U$

ratio can thus be used to date samples formed during the last 350,000 years, while the $^{231}Pa/^{235}U$ method can be applied for samples younger than 150,000 years (Ivanovich, 1984; Lalou, 1987).

Isotope geochemistry

Most elements occur in nature as a mixture of stable isotopes. These isotopes of an element differ slightly in their chemical and physical properties, as a consequence of the differences in their thermodynamic properties. Therefore, many natural processes entail slight separation of stable isotopes. For example, the evaporation and precipitation of water give rise to variations in the $^{18}O/^{16}O$ ratio, which reach an extreme in the difference between the isotopic composition of ice stored in the Antarctic ice sheet, and that of water in the ocean. Chemical processes also fractionate the isotopes, so that the $^{18}O/^{16}O$ ratio in calcite shells laid down by organisms is slightly greater than that of the water they inhabit. In this case, the fractionation factor is particularly interesting, because it is temperature dependent. This means that it may be possible to estimate past ocean temperatures by measuring the $^{18}O/^{16}O$ ratio of calcitic fossils (Emiliani, 1955; Duplessy, 1978; 1981; Shackleton, 1981).

As these isotopic variations are small, they are generally expressed as:

$$\partial^{18}O = \frac{(^{18}O/^{16}O)_{sample}}{(^{18}O/^{16}O)_{PDB}} - 1$$ which is expressed in per mil.

PDB is a reference standard.

A precise interpretation of the variations of the $^{18}O/^{16}O$ ratio of the calcitic shells of foraminifera in deep sea sediments is however difficult, because both the sea water temperature and $^{18}O/^{16}O$ ratio vary with the climate: since continental ice sheets are depleted in ^{18}O, as the amount of ice present on the continent increases during a glaciation, the water remaining in the ocean becomes richer in ^{18}O and its $^{18}O/^{16}O$ ratio increases. This isotopic enrichment is entirely reflected in the calcitic test of foraminifera.

As the temperature at which the shell formed decreases, the $^{18}O/^{16}O$ ratio of the calcite increases. Thus heavy isotopic ratios indicate cold climate. Conversely, temperature increases and ice cap melting will be reflected by light $^{18}O/^{16}O$ ratio. It has thus been necessary to estimate separately the variations of the two variables, sea water temperature and $^{18}O/^{16}O$ ratio. Recently, Labeyrie et al. (1987) have derived a $\partial^{18}O$ record which shows only the mean sea water $\partial^{18}O$ variations during the last 135,000 years due to continental ice volume changes (Fig. 1). As deep waters are well mixed, this record is common to all benthic foraminiferal records. We can thus reconstruct the changes in deep water temperature for this period in the different oceanic basins by subtracting this common sea water $\partial^{18}O$ record from each benthic foraminiferal record. As the sea water $\partial^{18}O$ record constitutes 60 to 100% of these benthic records (depending on their location and depth), the $\partial^{18}O$ variations in a deep sea core can be used to determine unambiguously the sediment deposited during the major events of the last climatic cycle. This is the basis of the isotope stratigraphy (Shackleton and Opdyke, 1973).

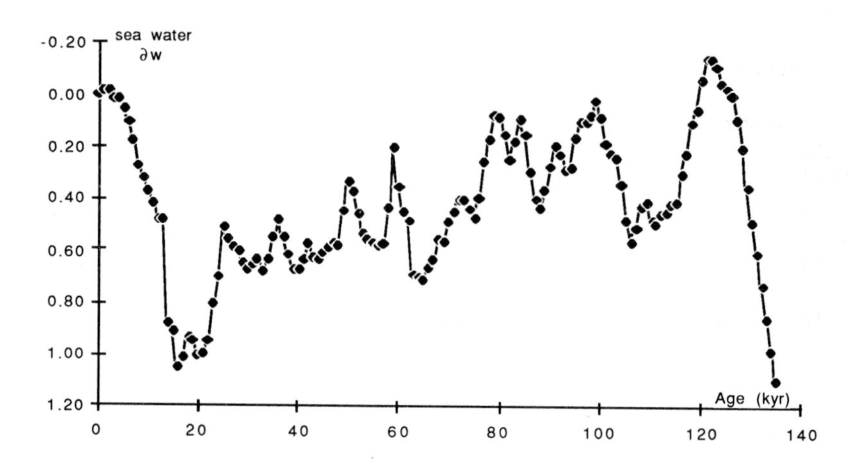

Figure 1: Variations of the $^{18}O/^{16}O$ ratio of the global mean oceanic water during the last 135,000 years. The value 0 indicates that the isotopic composition of the ocean is identical to that of the standard, which is the mean modern sea water isotopic composition (SMOW). Positive ∂w values (reported downward and expressed in per mil) indicate that the $^{18}O/^{16}O$ ratio of the global mean oceanic water was higher than the modern one, in relation with the waxing of continental ice sheets strongly depleted in ^{18}O.

Some species of benthic foraminifera (Genus *Cibicides*) closely record the modern $\partial^{13}C$ distribution in the world ocean, providing a good proxy for the reconstruction of the past $\partial^{13}C$ gradients in the ocean (Duplessy et al., 1984). Kroopnick (1985) has demonstrated that the distribution of $^{13}C/^{12}C$ ratio of the total CO_2 (ΣCO_2) dissolved in sea water delineates the general distribution of water masses and that modern $\partial^{13}C$ gradients in the deep ocean follow the net flow of the deep waters. The distribution of the $\partial^{13}C$ value of the ΣCO_2 dissolved in intermediate, deep and bottom waters of the ocean can be used as a tracer of the abyssal circulation under the following conditions: (i) within a deep water mass, the $\partial^{13}C$ value decreases with increasing oxidation of organic matter, such that the $\partial^{13}C$ decrease is related both to the surface productivity and the time elapsed since the water mass was isolated from the atmosphere; (ii) when two water masses are compared, the $\partial^{13}C$ value in each depends not only on their residence time at depth, but also on the ratio between the CO_2 produced from organic matter oxidation and the bicarbonate and carbonate ions derived from carbonate dissolution (note that this ratio is poorly controlled under past conditions); (iii) mixing between two water masses results in $\partial^{13}C$ values intermediate between the two original components (see a detailed discussion in Duplessy and Shackleton, 1985).

Transfer functions

Transfer functions are empirically derived equations for calculating quantitative estimates of past atmospheric and oceanic conditions from paleontological data. Ideally, this problem requires the formulation of a model that explains the biological responses in terms of the multitude of factors that comprise the observed ecological system. The basic assumptions are:

1. No significant changes occur to the interactions among organisms over the time scale studied and within the ecosystems sampled.
2. The modern observations provide sufficient information for analyzing the fossil observations, and further that a snapshot of modern spatial patterns is a sufficient basis for interpreting changes through time.
3. The biological responses are systematically related to the physical attributes of the biotic environment. This assumption implies that the relationships can be represented by mathematical expressions.

Transfer functions techniques, which use multivariate statistical procedures, were first applied to marine plankton data by Imbrie and Kipp (1971), to terrestrial pollen data by Webb and Bryson (1972), to tree ring data by Fritts et al. (1971). They provide typically temperature estimates either for the atmosphere (pollen) or for the sea water (foraminifer), with an accuracy of ± 1.5°C. Estimates for continental humidity and surface water salinity are more difficult to derive from the same fossil populations. Changes in diatom flora have been interpreted in terms of variations of the chemical composition of lake water.

THE LAST CLIMATIC CYCLE

Ice volume and sea level

Figure 1 shows the variations of the global mean oxygen isotopic ratio of ocean during the last 135,000 years. These variations reflect those of the volume of ice frozen over the high latitude continents. After a rapid deglaciation which occurred about 128,000 years ago, the Earth's climate was characterized by a volume of continental ice smaller than the present one. As a consequence, the sea level was higher than today by about 6 meters. This interglacial called Eemian lasted for about 10,000 years and was followed by a first phase of glaciation, which had culminated by about 107,000 years B.P. (Before Present). This ice advance was followed by several phases of ice disintegration, but the volume of ice was always larger than today: The sea level was estimated to be about -18 m during the two high levels around 80,000 and 99,000 years B.P.. Several phases of continental ice accumulation developed until the peak of the glaciation 18,000 years ago, when about 50,000,000 km^3 of ice were frozen, mainly over Canada and northern Europe. The final deglaciation began about 15,000 years ago and the ice melted in two separate phases, each lasting about 2,000 - 3,000 years. During these phases, the sea level increased at a rate of about 2 meters per century.

Temperatures

During the last interglacial, most low and mid-latitude oceanic areas experienced climatic conditions very similar to the present ones. Departures from the modern conditions have been recorded only at high latitudes, and only as barely significant deviations, taking into account of the standard error on the transfer function estimates (CLIMAP, 1984). Over the continents, quaternary deposits (i.e. formed during the last 1.8 M. years) are rare and, on a world-wide scale, only the broad features of the Eemian climate can be reconstructed. All the pollen data indicate that the climate was warmer and moister than today in northern Eurasia and northern America. A similar warming is suggested by the southern limit of permafrost in Eurasia (Flohn and Fantechi, 1984).

By the end of the Eemian, sea surface temperatures generally decreased. Close to the american continent in the Atlantic, the temperature drop was in phase with the ice volume increase. On the european side of the Atlantic, warm temperatures persisted for several thousand years, so that the glaciation of northern Europe began after that of northern America. However, pollen data indicate glacial conditions over northern Europe by about 107,000 B.P..

The pattern of temperature change during the last climatic cycle is only well known in the ocean. It strongly depends on the location and varies with latitude as expected from variations in solar radiation (see A. Berger, this volume). For example, the high latitudes exhibit a strong response in the 41,000 year band, an indication that the insolation budget primarily depends on the obliquity of the Earth's axis. At lower latitude, the temperature show sharp variations with a frequency close to 23,000 years, characteristics of the precession.

The surface of the ice-age Earth at the last glacial maximum (LGM) has been reconstructed by CLIMAP (1976). The northern hemisphere differed markedly from today by the presence of huge land-based ice sheets, which were as much as 3 km in thickness, and by a significant increase in the extent of sea ice and marine-based ice sheets. In the southern hemisphere, the most striking contrast was the greater extent of sea ice during LGM. The sea level was about 100 m below the modern level. On the continents, grasslands, steppes, sandy outwash plains, and deserts spread at the expense of forests and the extent of snow-covered land was significantly greater. The global average sea surface temperature change associated with ice-age cooling was close to 2°C, but the magnitude of the cooling depended strongly on the geographic location: it was as much as 10°C in the north Atlantic, but almost zero in the central gyres of the Atlantic, Indian and Pacific oceans. This pattern resulted in a marked steepening of thermal gradients along frontal systems. As the cooling was much stronger over the continents than over the oceans, the estimate of the global average surface air temperature cooling is about 4-5°C, as calculated by General Circulation models of the atmosphere.

The deglaciation is a climatic event, abrupt on the geological time scale. The chronological resolution of the dating of paleoclimatic signals appears to be the most critical factor in quantifying the rate of temperature change of sea surface temperatures. Detailed analysis of deep sea cores with high sedimentation rates enables one to minimize the effects of bioturbation (the activity of benthic organisms, which permanently mix the upper centimeters of sediment). The dating of these cores by Accelerator Mass spectrometry demonstrated that, in the North Atlantic, the polar front retreated at a mean rate higher than 2 km/year. Off Portugal, this retreat resulted in a 10°C sea surface temperature increase in less than 400 years (Figure 2). This first warming, which led to temperatures similar to those of today, was followed by a sharp an abrupt cooling called the Younger Dryas which occurred in the whole high latitude northern hemisphere. The occurrence of a similar cooling in the southern hemisphere is still a matter of debate and the Younger Dryas is still poorly understood. The Younger Dryas lasted less than a millenia and was followed by a warming, less abrupt than the preceding one, leading to the Holocene conditions.

Deep water circulation

Oxygen isotope analysis of benthic foraminifera from the various basins of the ocean indicates that the deep waters had cooled by several degrees by the end of the Eemian, probably in response to the high latitude surface water cooling (Oppo and Fairbanks, 1987). By contrast, the temperature of intermediate waters did not change significantly (Kallel et al., 1988).

The $\partial^{13}C$ distribution of the total CO_2 dissolved in the deep ocean during the last interglaciation shows that the deep water sources were active in both hemispheres, but that the production of Antarctic Bottom Water (AABW) was stronger than that of North Atlantic deep Water (NADW) (Duplessy et al., 1984).

The global deep water circulation is very sensitive to the climate of high latitudes. For example, the production of NADW ceased during the peak of the melting of the ice sheets during the penultimate deglaciation. By contrast, this production increased during the inception of the glaciation (Duplessy and Shackleton, 1985).

During the last glacial maximum, the northern hemisphere source of deep water was weak in the North Atlantic and most bottom and deep water in the whole

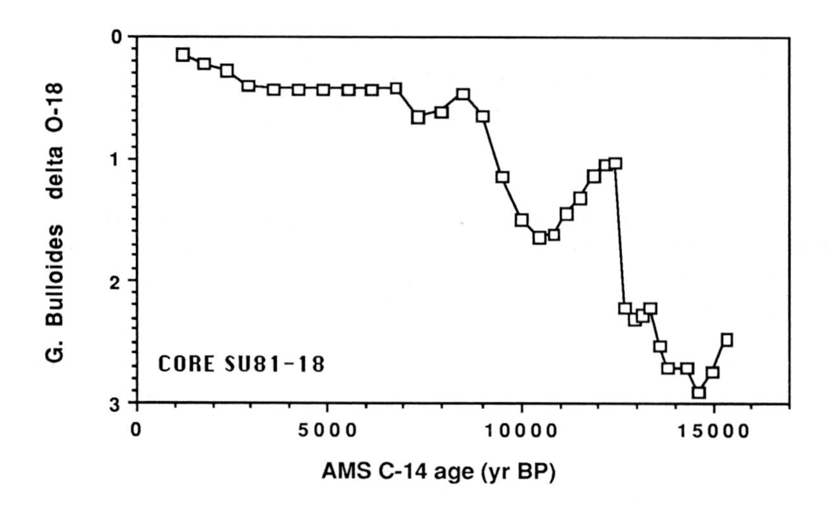

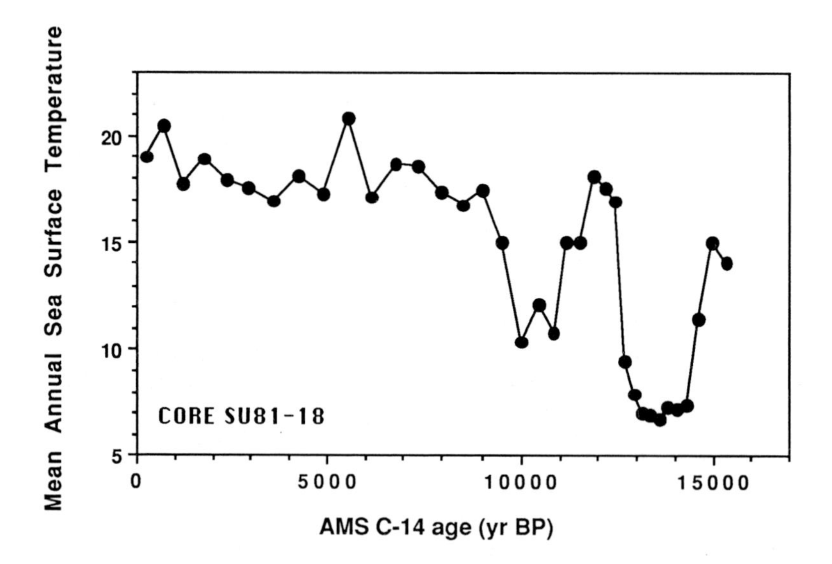

Figure 2: Paleoclimatic record of the last deglaciation in core SU 81-14 off Portugal. C-14 ages have been obtained by measuring the C-14 content of the shells of the planktonic foraminifera *Globigerina bulloides* at different levels in the core by Accelerator Mass Spectrometry. The upper curve shows the variations of the $^{18}O/^{16}O$ ratio of the planktonic foraminifera *Globigerina bulloides* . The lower curve displays the sea surface temperature estimates deduced from the variations of the foraminiferal fauna in the core

ocean originated from poorly-oxygenated surface water, sinking in the high latitudes of the Southern Ocean (Shackleton et al., 1983; Duplessy et al, 1988).

By contrast, Intermediate waters extended somewhat deeper than under modern conditions and were much more oxygenated than today in the Atlantic, Indian and Pacific oceans. We do not have enough data to interpret in detail the hydrology of the Intermediate waters in the Atlantic Ocean. In the Indian and Pacific oceans, intermediate waters were separated from the deep water by a well-developed deep thermocline, resulting in a strong stratification of the deep water realm. This resulted in strong changes in ocean chemistry which caused the decrease of the atmospheric CO_2 content (Broecker, 1982; Boyle and Keigwin, 1985).

The global circulation of deep and intermediate waters is strongly dependent on the earth's climate. Accelerator Mass Spectrometry (AMS) radiocarbon determination has made it possible to measure changes in the radiocarbon age of the deep ocean. Initial attempts to achieve this were thwarted by the difficulties imposed by bioturbation, which permanently mixes the upper ten or more centimeters of sediment. Shackleton et al (in press) have overcome this difficulty by making measurements in a core with a sufficiently high accumulation rate ($\geq$ 10 cm/kyr) to make the effects of bioturbation negligible. These authors demonstrated that the uncorrected age difference between the planktonic foraminifera *N. dutertrei* and the benthic foraminifera *Uvigerina* has changed during the last 30,000 years. During the last glaciation, it was close to 2,100 years, indicating that the ventilation age of the deep Pacific was about 500 years greater than it is today.

Duplessy et al. (in press) followed the same strategy in order to determine the ventilation age of the Pacific Intermediate waters and compared the uncorrected ^{14}C ages of planktonic and benthic foraminifera in core CH 84-14 (41°44' N, 142°33' E, 978 m). They dated only the peaks of abundance of planktonic and benthic foraminifera, in order to minimize the effects of bioturbation, because the abundance of foraminifera vary by one to two orders of magnitude in response to local climatic variations. The development of benthic organisms is directly related to the surface productivity because food is brought to the abyss by sinking of particulate matter formed in the surface water. As a consequence, the peaks of abundance of planktonic and benthic foraminifera are generally found in the same sediment levels.

Carbon-14 determinations indicate that the mean difference is always lower than 1,000 years in this core before 10,000 year B.P., indicating that the ventilation of the intermediate water was much more important during the deglaciation than today (1250 years). During the Holocene, dissolution is high in this core, as in most north Pacific sediment cores, and planktonic foraminifera are not abundant enough to pursue this study.

In order to compare the variations of the ventilation time of the intermediate and deep waters of the Pacific oceans during the deglaciation, we show in Figure 3 the age difference between benthic and planktonic foraminifera in core CH 84-14 and in core TR 163-31B (3°37' S, 83°58' W, 3210 m) analyzed by Shackleton et al. (in press). The ventilation rate of intermediate and deep pacific waters exhibits a similar pattern

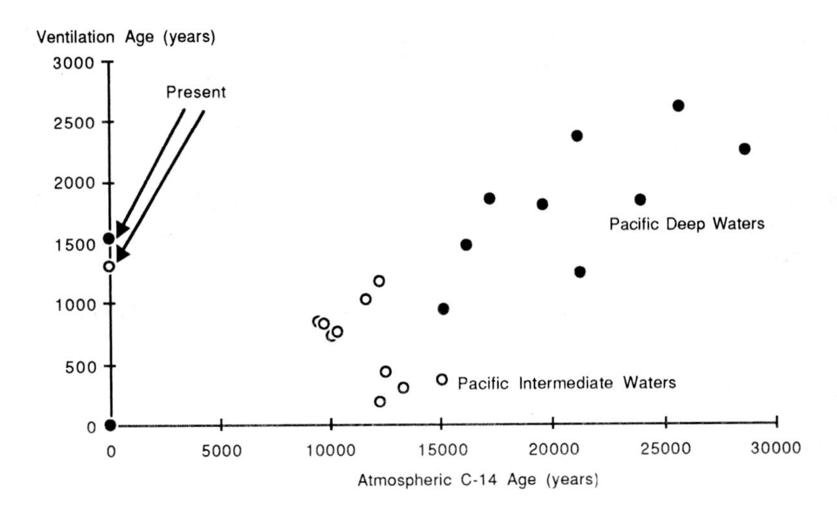

Figure 3: Variations of the ventilation rate of the Pacific deep and intermediate waters oceanic water during the last 30,000 years, deduced from the C-14 age difference between planktonic and benthic foraminifera found at the same sediment level in two deep sea cores.

of variation and was smaller than under the modern conditions. An active deep and intermediate water circulation was an efficient factor diluting the water resulting from the melting of the ice sheets within the ocean water, which was salty and therefore denser than the meltwater.

ACKNOWLEDGEMENTS

Thanks are due to J.F.B. Mitchell for a careful review of the manuscript and to M. Arnold and L.D. Labeyrie for stimulating discussions.

This is C.F.R. Contribution N° 985.

BIBLIOGRAPHY

Bard, E., Arnold, M. , Duprat, J., Moyes, J., and Duplessy, J.C., Reconstruction of the last deglaciation: deconvolved records of $\partial^{18}O$ profiles, micropaleontological variations and accelerator mass spectrometry ^{14}C dating, Clim. Dyn., 1, 101-112, 1987.

Bard, E., Arnold, M. , Maurice, P., Duprat, J., Moyes, J., and Duplessy, J.C., Retreat velocity of theNorth Atlantic Polar Front during the last deglaciation determined by ^{14}C Accelerator Mass Spectrometry, Nature, 328, 791-794, 1987.

Berger, A. Ed., Climatic Variations and Variability: Facts and Theories, NATO A.S.I. Series, D. Reidel, 1981.

Boyle, E. A., and Keigwin L. D., Comparison of Atlantic and Pacific paleo-chemical records for the last 215,000 years: Changes in deep ocean circulation and chemical inventories, Earth Planet. Sci. Lett., 76, 135–150, 1985.

Broecker, W. S., Glacial to interglacial changes in ocean chemistry, Prog. Oceanogr., 11, 151–197, 1982.

Broecker, W.S. and Peng, T.H., Tracers in the Sea, Eldigio Press, 1982.

CLIMAP Project Members, The surface of the Ice Age Earth, Science, 191, 1131–1137, 1976.

CLIMAP Project Members, The last interglacial ocean, Quat. Res., 21, 123-224, 1984.

Delibrias, G., Le Carbone 14. In "Méthodes de datation par les phénomènes nucléaires naturels", E. Roth et B. Poty Eds, Masson, 421-458, 1985.

Duplessy, J. C., Isotope Studies. In Climatic Change, Ed. by J. Gribbin, Cambridge University Press, 46-67, 1978.

Duplessy, J. C. et Arnold, M., La mesure du carbone 14 en spectrométrie de masse par accélérateur: Premières applications. In "Méthodes de datation par les phénomènes nucléaires naturels", E. Roth et B. Poty Eds, Masson, 459-473, 1985.

Duplessy, J. C., and Shackleton N. J., Response of global deep-water circulation to the Earth's climatic change 135,000–107,000 years ago, Nature, 316, 500–507, 1985.

Duplessy, J. C., Shackleton, N. J., Matthews, R. K., Prell, W., Ruddiman, W. F., Caralp, M., and Hendy, C. H., ^{13}C record of benthic foraminifera in the last interglacial ocean: Implications for the carbon cycle and the global deep water circulation, Quat. Res., 21, 225–243, 1984.

Duplessy, J. C., Shackleton, N. J., Fairbanks, R. G., Labeyrie, L. D., Oppo, D. and Kallel, N., Deep water source variation during the last climatic cycle and their impact on the global deep water circulation, Paleoceanography, in the press, 1988.

J.C. Duplessy, Arnold, M., Bard, E., Juillet-Leclerc, A., Kallel, N., and Labeyrie, L. D., AMS ^{14}C study of transient events and of the ventilation rate of the Pacific Intermediate Water during the last deglaciation, submitted to Radiocarbon.

Emiliani, C., Pleistocene temperatures, J. Geol., 63, 538-578, 1955.

Flohn, H. and Fantechi, R. Eds, The climate of Europe: Past, Present and Future, D. Reidel, 1984.

Fritts, H. C., Blasing, T. J., Hayden, B. P., and Kutzbach, J. E., Multivariate technique for specifying tree-growth and climate relationship and for reconstructing anomalies in paleoclimate. J. Applied Meteorol., 10, 845-864, 1971.

Hedges, R. E. M. and Gowlett, J. A. J., Radiocarbon dating by Accelerator Mass Spectrometry, Scientific American, 82-89, 1986.

Imbrie, J. and Kipp, N. G., A new micropaleontological method for quantitative paleoclimatology: Application to a late Pleistocene Caribbean core. In "The Late Cenozoic Glacial Ages", Ed. by K.K. Turekian, 77-181, Yale Univ. Press., 1971.

Ivanovich, M., Uranium series disequilibria applications in geochronology. In "Uranium series disequilibrium: Applications to environmental problems", Ed by M. Ivanovich and R.S. Harmon, Oxford Science Publications, 56-78, 1984.

Kallel, N., Labeyrie, L. D., Juillet-Leclerc, A. and Duplessy, J. C., A deep hydrological front between intermediate and deep-water masses in the Glacial Indian Ocean, Nature, 651-655, 1988.

Kroopnick, P., The distribution of ^{13}C of ΣCO_2 in the world oceans, Deep Sea Res., 32, 57–84, 1985.

Labeyrie, L. D., Duplessy, J. C., and Blanc, P. L., Variations in mode of formation and temperature of oceanic deep waters over the past 125,000 years, Nature, 327, 477–482, 1987.

Lalou, C., Déséquilibres radioactifs dans la famille de l'uranium, in "Géologie de la Préhistoire", Ed. by J. C. Miskovsky, 1073-1085, Masson, Paris, 1987

Mix, A., and Fairbanks, R. G., North Atlantic surface-ocean control of Pleistocene deep ocean circulation, Earth Planet. Sci. Lett., 73, 231–243, 1985.

Oppo, D. W., and Fairbanks, R. G., Variability in the deep and intermediate water circulation of the Atlantic during the past 25,000 years: Northern hemisphere modulation of the Southern Ocean, Earth Planet. Sci. Lett., 86, 1–15, 1987.

Shackleton, N. J., Imbrie, J., and Hall,M., Oxygen and Carbon isotope record of east Pacific core V 19-30: Implications for the formation of deep water in the Late Pleistocene North Atlantic, Earth Planet. Sci. Lett., 65, 233–244, 1983.

Shackleton, N.J. and Opdyke, N. D., Oxygen isotope and paleomagnetic stratigraphy of equatorial Pacific core V28-238: oxygen isotope temperatures and ice volumes on a 10^5 year and 10^6 year scale, Quat. Res., 3, 39-55, 1973.

Shackleton, N. J., Duplessy, J. C., Arnold, M., Maurice, P., Hall, M., and Cartlidge, J., Radiocarbon age of Last Glacial Deep water, Nature, in press.

Webb, T., and Bryson, R. A., Late and post glacial climatic change in the northern Midwest, USA: quantitative estimates derived from fossil pollen spectra by multivariate statistical analysis, Quat. Res., 2, 70-115, 1972.

MILANKOVITCH EFFECTS ON LONG-TERM CLIMATIC CHANGES

A.L. BERGER
Catholic University of Louvain
Institute of Astronomy and Geophysics G. Lemaître
2 Chemin du Cyclotron
B-1348 Louvain-la-Neuve, Belgium

Abstract

In the 19th century, Croll (1875) and Pilgrim stressed the importance of severe winters as a cause of Quaternary ice ages. Later on, mainly during the first half of the 20th century, Köppen, Spitaler and Milankovitch regarded high winter and low summer insolation as favoring glaciation. After Köppen and Wegener related the Milankovitch new radiation curve to Penck and Brückner's subdivision of the Quaternary, there was a long lasting debate whether or not such changes in the insolation can explain the Quaternary glacial-interglacial cycles. In the 1970s, with the improvements of radiometric dating, and of acquiring and interpreting the long continuous geological records, with the advent of computers and with the development of more sophisticated astronomical and climate models, the astronomical theory and in particular its Milankovitch version were revived. Over the last 5 years, most geological, astronomical and climatological obstacles have been overcome and the causal role of the orbital variations in long-term variation of climate is taken almost universaly for granted.

1. MILANKOVITCH ERA AND DEBATE

It is only during the first decades of the 20th century that Spitaler (1921) rejected Croll's theory that the conjunction of a long cold winter and a short hot summer provides the most favorable conditions for glaciation. He adopted the opposite view, first put forward by Murphy already in 1876, that a long cool summer and short mild winter are the most favorable. The diminution of heat income during the summer half-year has also been claimed by Brückner, Köppen and Wegener (1925) as the decisive factor in glaciation. Milankovitch (1920, 1941), however, was the first to present a complete astronomical theory of Pleistocene ice ages where computing the changes of orbital elements over time and linking the changes in insolation to climate (Imbrie and Imbrie, 1979; Berger, 1988).

Milutin Milankovitch was a Yugoslavian astronomer born in Dalj on 28 May 1879 who died in Beograd on the 12 of December 1958. He was a contemporary of Alfred Wegener [1880-1930] with whom he became acquainted through Vladimir Köppen [1846-1940], Wegener's father-in-law (Schwarzbach, 1985; Milankovitch V., 1988; Berger and Andjelic, 1988).

It is roughly between 1915 and 1940 that Milutin Milankovitch put the astronomical theory of the Pleistocene ice ages on a firm mathematical basis. He calculated how the intensity of radiation striking the top of the atmosphere during the caloric summer and winter half years varied as a function of latitude and the orbital parameters e, ε, $e \sin \omega$. He then emphasized the importance of the summertime insolation at 65°N as a controling factor of the Northern Hemisphere glaciations, with its dominant obliquity-driven periodicity of 41,000 years. Finally, he estimated the magnitude of ice-age departures from the present-day air surface temperatures, estimating the radiation balance at the Earth's surface.

The essential product of the Milankovitch theory is his curve that shows how the intensity of summer sunlight varied over the past 600,000 years, on which he identified certain low points with four European ice ages reconstructed 15 years earlier by Albrecht Penck and Eduard Brückner (1909) and from which he concluded that these geological data constituted a verification of his theory.

If we consider this curve, however, we are left in no doubt that Milankovitch's success was only an apparent one, because the Quaternary has had many more glacial periods than was claimed during the first part of the 20th century (e.g., Shackleton and Opdyke, 1973, 1976; Morley and Hays, 1981) and because the ice volume record is dominated by a 100 kyr rather than by a 41 kyr cycle. Moreover, during the late 1960s, modern detailed studies of Alpine terraces showed that the climatic reconstruction of Penck and Brückner was wrong : the time scale was grossly in error and the terraces are tectonic rather than climatic features (Kukla, 1975a). Although it is now clear that the empirical argument used by Milankovitch to support his theory was misinterpreted, the modern evidence now strongly supports its essential concept, namely that the orbital variations exert a significant influence on climate.

From roughly 1950's to 1970's, the Milankovitch theory was largely disputed, with discussions based on fragmentary geological records supported by incomplete and frequently incorrect radiometric data. The accuracy of the astronomical parameters and of the related insolation fields was not known, and the climate was considered too resilient to react to "such small changes" as observed in the summer half-year caloric insolation of Milankovitch. At the end of the 1960's, climatologists have attacked the problem theoretically by adjusting the boundary conditions of energy-balance models (EBM) and observed that the magnitude of the calculated response was indeed very small (e.g. Budyko, 1969; Sellers, 1970). However if these early numerical experiments are viewed narrowly as a test of the astronomical theory, they are open to question because the models used are far from being complete and contain untested parameterization of important physical processes.

2. MILANKOVITCH REVIVAL

In the late 1960's, judicious use of radioactive dating and paleomagnetic techniques gradually clarified the Pleistocene time scale (Broecker et al., 1968). Better instrumental methods came on the scene by applying oxygen isotope in ice-age foraminifera relies (Emiliani, 1966; Shackleton and Opdyke, 1973; Duplessy, 1978), ecological methods of core interpretation were perfected (Imbrie and Kipp, 1971), global climates in the past were reconstructed (CLIMAP, 1976, 1981) and atmospheric general circulation models (Smagorinsky, 1963) and climate models became available (Alyea, 1972). With these improvements of dating and interpretating of geological data in terms of paleoclimates, with the advent of computers and the development of astronomical and climate models, a more critical and deeper investigation became necessary of all four main steps of any astronomical theory of paleoclimates, namely of :

1. the computation of the astronomical elements (Berger, 1976a)
2. the computation of the appropriate insolation parameters (Berger, 1978a,b)
3. the development of suitable climate models (Kutzbach, 1985; Crowley, 1988)
4. the analysis of geological data in the time and frequency domains designed to investigate the physical mechanisms, and calibrate and validate the climate models (Berger et al., 1984).

It is this systematic approach with modern powerful techniques which has brought, mainly since 1975, the following major discoveries supporting progressively the astronomical theory.

2.1. Bipartition of the precessional period

In 1976, Hays, Imbrie and Shackleton demonstrated from spectral analysis of the climate sensitive indicators in selected deep-sea records that the astronomical frequencies (corresponding to the 100, 41, 23 and 19 kyr periods) are significantly superimposed upon a general red noise spectrum. It is the geological observation of the bipartition of the precessional peak (23 and 19 kyr were found instead of the usual 21 kyr), confirmed in the astronomical computations made independently by Berger (1977), which was one of the first most delicate and impressive of all tests for the Milankovitch theory.

2.2. Monthly insolations

The long-term deviations from today-values of the caloric half year insolation introduced by Milankovitch (1941), used in particular by Bernard (1962) and Kukla (1972) and revised by Sharaf and Budnikova (1969), Vernekar (1972) and Berger (1978b) amount up to 3 to 4 per cent at the maximum. However, if the monthly insolation values (Berger, 1978a) are used instead, important fluctuations masked by the half year averaging method become easily recognizable. For example, 125 kyr BP, during the Eemian interglacial, all latitudes were overinsolated in July with respect to the present, particularly in the northern polar regions, where the positive anomaly reached up to 12 %; and the same is for 10 kyr BP. This is especially significant when a delay of some thousands of years, required for the ice sheets to melt, is taken into account in climate modeling.

Moreover, a detailed treatment of the seasonal cycle is much more meaningful and indeed required for explaining climate variations (Kukla, 1975b) in realistic climate models. Thus the computation of insolation variations over a complete seasonal cycle introduced in the early 1970's by Berger (1976b) and their change in time are significant. The well-known high sea level stands of the Barbados III (∿124 kyr BP) and II (∿103 kyr BP) marine terraces (Broecker et al., 1968) clearly correspond in time to high summer insolation anomaly which amounts to some 10 % of current values particularly in the high latitudes. An abortive glaciation at 115 kyr BP which separated these two warm intervals was successfully simulated by Royer et al. (1983) using the associated insolation minimum as the only external forcing. The main glacial transition between stages 5 and 4, at 72 kyr BP, was underinsolated and, more important, this drop in insolation was not compensated by any significant increase during the whole Würm glaciation phase. On the contrary, yet another important decrease at 25 kyr BP augers the 18 kyr BP maximum extent of ice in the Northern Hemisphere. Indeed, between 83 kyr and 18 kyr BP there was an overall solar energy deficiency of 2.5×10^{25} calories north of 45°N, sufficient to compensate for the latent heat liberated in the atmosphere during the formation of snow required by the buildup of the huge 18 kyr BP ice sheets (Mason, 1976).

2.3. Astronomical frequencies documented in diverse geological records

Since 1976, spectral analysis of climatic records of the past 800,000 years or so, has provided substantial evidence that, at least near the frequencies of variation in obliquity and precession, a considerable fraction of the climatic variance is driven in some way by insolation changes accompanying the perturbations of the Earth's orbit (Imbrie and Imbrie, 1980; Berger, 1987). For example, if some long deep-sea cores are used and the uncertainty in the geological time scale and in the spectral analysis is accounted for, Berger and Pestiaux (1987) have shown that, the following peaks are significantly present in the astronomical bands:

103,000 with a standard deviation of 24,000
42,000 with a standard deviation of 8,000
23,000 with a standard deviation of 4,000

However, the interpretation of the results is not always as clear. The 100 kyr cycle, so dominant a feature of the late Pleistocene record, does not exhibit a constant amplitude over the past 2-3 million years as displayed on a 3-dimensional time evolutionary spectrum of deep-sea core V28-238 (Pestiaux and Berger, 1984). Clearly, this periodicity disappears before 10^{6} years ago, at a time the ice sheets were much less developed over the Earth, reinforcing the idea that the growth of the major ice sheets may have played a role in the modulation of the 100 kyr cycle.

Moreover, the shape of the spectrum depends also upon the location of the core and the nature of the climatic parameter analysed (Hays et al., 1976). For example in core V30-97, the 41 kyr cycle is not seen at all whereas the 23 kyr cycle is dominant in Atlantic summer sea-surface temperatures of the last 250,000 years (Ruddiman and McIntyre, 1981; Imbrie, 1982). This is not too surprising as these spectra depend upon the way the climate system reacts to the insolation forcing and upon which type of insolation it is sensitive too. Indeed, contrary to the well-received Milankovitch idea that the high polar latitudes must record the obliquity signal (as shown in the Vostok core, for example, Jouzel et al., 1987) whereas low latitudes record only the precessional one, the latitudinal dependence of the insolation parameters is more complex. Clearly the mid-month high-latitude summer insolation displays a stronger signal in the precession band than in the obliquity one (Berger and Pestiaux, 1984).

2.4. Phase Coherency

As already shown by Hays et al. (1976), the variance components centered near a 100 kyr cycle which dominates most climatic records, seem to be in phase with the eccentricity cycle (high eccentricity at low ice volume). The exceptional strength of this cycle calls, however, for a stochastic (Hasselman, 1976; Kominz and Pisias, 1979) amplification of the insolation forcing, or for a non-linear amplification through the deep ocean circulation, the carbon dioxide, the ice sheet related mechanisms and feedbacks, the isostatic rebound of the elastic lithosphere and of the viscous mantle and the ocean-ice interactions. The 100 kyr climatic cycle can indeed be explained both (i) from the eccentricity signal directly, provided an amplification mechanism can be found (as in the double potential theory of Nicolis 1980, 1982 and Benzi et al. 1982); and/or (ii) by a beat between the two main precessional components as shown by Wigley (1976) from non-linear climate theory. It must be stressed here that the same arises for the 100 kyr eccentricity cycle in celestial mechanics : the frequency corresponding to the second period of the eccentricity (94945 years in Table 3 of Berger (1977)) is obtained from precession frequencies number 3 and 1 in Table 2 of the same reference (1/18976 - 1/23716). The other ones (number 1 and 3 to 6) are coming respectively from the following combinations : 2-1, 3-2, 4-1, 4-2 and 3-4 (more details are available in Berger et al., 1987).

It is worth here to remember that Milankovitch requested a high eccentricity for an ice age to occur - which is just the reverse of the correlation claimed at the beginning of this section. He viewed indeed the effect of the eccentricity through the precessional parameter alone. However, if a higher degree of accuracy is used in the insolation computations, $(1-e^2)^{-3/2}$ appears as a factor in all insolation parameters at all latitudinal bands, reflecting the full equation of the elliptical motion and the variation of the Earth-Sun mean distance in terms of the invariable semi-major axis of the Earth orbit (Berger, 1978b). Although its absolute effect is relatively small (1 % at the most), this term increases the annual global insolation totals at high eccentricity times (for example, at e = 0.075, the total solar energy received by the Earth is increased by 0.8 %) and decreases it during low eccentricity, a result seemingly coherent with the recent findings. It can, therefore, not be ignored any more, especially since it reinforces the impact on climate in the (i) variant of the 100 kyr cycle explanation.

However, it must be stressed that Imbrie and his collaborators (1984) have made it clear that the coherency of orbital and climatic variables in the 100 kyr band is enhanced significantly when the geologic record is tuned precisely to the obliquity and precession bands. Using the so-called orbitally-tuned SPECMAP time scale (refined by Martinson et al., 1987, for the last 300,000 years), they indeed found a coherency in the astronomical bands significant at more than 99% level of confidence. This supports the second mechanism (beat) rather than the first !

Such a fairly coherent phase relationship was also reasonably well defined between insolation and ice volume in Kominz and Pisias (1979) where obliquity consistently lead the ^{18}O record by about 10,000 years, whereas precession seemed to be in phase with the 23 kyr geological signal. However, the recent results obtained by SPECMAP (Imbrie et al., 1988) show that these leads and lags are more complicated.

CLIMAP (1976) and more recently SPECMAP (Imbrie et al., 1984) teams have indeed shown that the phase lags in the climate response to orbital forcing depend upon the nature of the climatic parameters themselves and upon their geographical location. For example, in their data the sea-surface temperature of the southern oceans seems to lead the response of the northern hemisphere ice sheets by roughly 3,000 years.

2.5. Other astronomical frequencies

Ruddiman et al. (1986) succeeded in finding in the geological records one of the secondary astronomical periods that was predicted in 1976, the 54 kyr one (Berger, 1977). A similar period of 58,000 year was found in a 400 kyr record of the paleomagnetic field from Summer Lake in South-central Oregon (Negrini et al., 1988).

Although, the most important term of a 412 kyr period in the eccentricity was already predicted by Berger in 1976, it is only two years later that appropriate statistical techniques and geological time series, long enough to discover such a periodicity, became available. Shackleton suggested this periodicity in 1978 and Briskin and Harrell (1980) using 2-million year sediment cores from both the Atlantic and the Pacific found it in the oxygen isotopic record of planktonic foraminifers, in coarse sediment fraction and in the magnetic inclination (which lead them to propose that a relationship may exist between the eccentricity and the Earth's core modulation of the magnetic field). Kominz et al. (1979) reported it also from two spectra covering the 730,000-year long records.

The investigation by Moore et al. (1982) of calcium carbonate concentrations in equatorial Pacific sediments (core RC11-209), have shown that the Pacific carbonate spectrum has been dominated for the past 2 million years by variance in the 400 kyr band, with more modest contributions in the 100 kyr and 41 kyr bands (matching the variations of respectively the eccentricity and obliquity).

2.6. Data from the Pliocene and late Miocene

Very interestingly, with the existence of a few deep-sea cores extending through the first appearance of abundant ice rafted debris in North Atlantic, 2.5 million years ago, this most important eccentricity term of a period of 412 kyr, was confirmed in geological record. Using cross-spectral analysis, Moore et al. (1982) discovered that the 400 kyr eccentricity and sedimentary cycles were in phase over the last 8 million years, but the 100 kyr cycle which dominated climatic variability during the Pleistocene ice ages (29% of the total variability) had only a minor effect 5 to 8.5×10^6 years ago, where it accounted for 6 times less variability than as today at the DSDP site 158. Prell (1982) also failed to find evidence of a strong 100 kyr cycle in pre-Pleistocene sediments at the DSDP site 502 in the western Caribbean and at the site 503 in the eastern equatorial Pacific. He did apparently find evidence of the 41 kyr cycle in a 7 million-year sediment record, as well as the evidence of a 250 kyr cycle.

Spectral analysis of DSDP Hole 552A reveals also such a dominant quasi-periodicity associated with obliquity-induced temperature variations in surface water and weaker peaks at the eccentricity and precession periods (Backman et al., 1986). In the Mediterranean Pliocene, rhythmic lithological variations in the Trubi and Narbone Formations of Sicily and Calabria show cycles that could be related to precession and eccentricity (Hilgen, 1987). In particular, the precession cycle corresponds well with the mean duration of the deposition of the basic rhythmites, the eccentricity cycle of about 400 kyr would match the average duration of the carbonate units and the 100 kyr cycle, the arrangement of the sapropelitic intercalations.

This late Miocene-Pliocene time scale, requires thus our most urgent attention. The upper Pliocene is indeed at the limit of validity of the astronomical calculations as far as the time domain is concerned.

However, there is still a large confidence in the value of the astronomical frequencies, in such a way that conclusions from a comparison between geological and astronomical data may still be convincingly drawn in the frequency domain. Moreover, a new astronomical solution valid over the last 10×10^6 years is now underway (Berger, Loutre and Laskar, 1988). It is thus challenging to see to which extent may the methodology developed for the Mid- and Upper-Pleistocene (Berger, 1987) be extended to Early Pleistocene, Pliocene and Miocene.

2.7. Pre-Cenozoic evidence of astronomical signal

There is also evidence that the orbital variations were linked to climate at periods shorter than 100 kyr during the past few hundred million years. This appeared at times when major ice masses were probably absent. Walsh power spectra of the Blue Lias Formation (basal Jurassic) show two cycles with duration less than 93 kyr which may record changes in orbital precession and obliquity (Weedon, 1985/86). Carbonate production in pelagic mid-Cretaceous sediments, quantified by calcium carbonate and optical densitometry time series, reflects the orbital eccentricity and precessional cycles (Herbert and Fisher, 1986). Fourier analysis of long sections of the Late Triassic Lockatong and Passaic formation of the Newark Basin show periods in thickness corresponding roughly to the astronomical periodicities (Olsen, 1986). All these interesting results encourage research of the stability of the solar system in order to determine to which extent the changing Earth-Moon distance, for example, influenced the length of the main astronomical periods. Recent astronomic computations (Berger et al., 1987) show (Table 1) that the precession and obliquity cycles should indeed be reduced drastically prior to 2 billions years ago with the obliquity cycle starting to approach the precessional ones even if we take into account this varying Earth-Moon distance only. Already at 250 Myr ago, changes were not negligible : the main precessional periods were 18.4 kyr and 22.1 kyr and the obliquity one was 38.2 kyr.

TABLE I. Changes in the main periods of precession and obliquity due to changes in the Earth-Moon distance only.

Epoch (billion years BP)	Precession		Obliquity
0	19,000	23,000	41,000
0.5	17,600	21,000	35,100
1	16,900	20,000	32,300
2	15,500	18,100	27,600
2.5	14,700	16,900	25,000

If we accept the astronomical theory of paleoclimates as a fundamental principle, a time will come when geology will provide astronomers with periodicities which will allow to test the theories of the planetary system and of its stability over much of the Earth's history.

2.8. Combination tones

Pestiaux et al. (1988) used cores with a high sedimentation rate covering the last glacial-interglacial cycle to resolve the higher freqency part of the spectrum. Besides the 19 kyr precessional peak, three other periods were detected at significant levels :

10,300 with a standard deviation of 2,200
4,700 with a standard deviation of 800
2,500 with a standard deviation of 500

These preferential frequency bands of climatic variability outside the direct orbital forcing band are still too broad to allow for a definite physical explanation. A tentative interpretation, though, may be given in terms of the climatic system's nonlinear response to variations in the insolation available at the top of the atmosphere. The 10.3, 4.7 and 2.5 kyr near-periodicities are indeed rough combinations tones of the 41, 23 and 19 kyr peaks found in the main insolation perturbations. Moreover, Pestiaux et al. (1988) succeeded to predict these shorter periods by using the Ghil - Le Treut - Kallen non-linear oscillator climate model (Ghil and Le Treut, 1981; Le Treut and Ghil, 1983).

On the other hand, a tendency to obtain free oscillations at periods up to several tens of thousands of years in complex climate models has also been mentioned by Sergin (1979), Kallen et al. (1979) and Ghil (1980).

2.9. Ice sheets modeling and the 100,000 year periodicity

Hays et al. (1976) were among the first to suggest that the enhancement of the 100 kyr cycle may be due to non-linearities in the climate response. Wigley (1976) showed that it may indeed be a beat effect from the two main precession periodicities.

Imbrie and Imbrie (1980) have developed a simplified glacial dynamics model especially designed for the explicit purpose of reproducing the Pleistocene ice volume record from orbital forcing. The rate of climatic change is made inversely proportional to a time constant that assumes one of two specified values, depending on whether the climate is warming or cooling. Such a model tuned over the last 150,000 years, is forced with orbital input corresponding to an irradiation curve for July 65N, with a mean time constant of 17,000 years and a rate of 4 : 1 between the time constants of glacial growth and melting. Model's simulation of the isotopic record of ice volume over the past 150,000 years is reasonably good but results for earlier times are mixed and parametric adjustments do little or nothing to improve the matter.

The models based on a beat or on a simple form of non-linearity (like the asymmetrical response of Imbrie and Imbrie, 1980, in which the time constant governing ice decay is smaller than the time constant governing the ice growth), could hide a second problem : it is difficult to introduce substantial 100 kyr power into the climate response without also introducing power reflecting the 413 kyr eccentricity cycle in amounts that are much greater than have been detected in most Late Pleistocene climatic records (Kukla et al., 1981). However, it is worth mentioning that the fit is much better in the results obtained by Moore et al. (1982) where the whole Pleistocene is considered as seen in section 2.6 !

The role if ice sheets in determining the long-period climate response, namely in the 100 kyr range, can possibly be clarified with some realistic parametric modelling of the ice sheet dynamics. Following Birchfield and Weertman (1978), solar radiation variations seem to be large enough to account for the ice age cycles when glacier mechanics is properly taken into account. Adding calving of icebergs into a marginal lake in his ice sheet model, Pollard (1982, 1984) did create a sharp 100,000-year cycle that stayed roughly in step with the geological record of ice volume as far back as 600,000 years ago. He found also that the eccentricity cycle could impose itself only indirectly through its accentuation of the precession cycle and that a sharp termination resulted when an eccentricity-strengthened precession cycle coincided with a large ice sheet in existence.

Another mechanism for amplifying the effects of orbital variations, and namely of the eccentricity peak, is the interaction between the ice sheet and the underlying bedrock through isostatic rebound (Oerlemans, 1980; Birchfield et al., 1981). High latitude topography (Birchfield et al., 1982) and the role of ice albedo-temperature feedback (Birchfield and Weertman, 1982) are also included in the models.

It must also be pointed out that recent ice sheet models show that the 100 kyr cycle can be simulated with (Ghil and Le Treut, 1981; Saltzman et al., 1984) or without (Lindzen, 1986) internal free oscillations related to resonances when astronomically forced. It is significantly reinforced when isostatic rebound (Hyde and Peltier, 1985) and iceberg calving are taken into account (Pollard, 1982).

2.10. Equilibrium 3-D climate models astronomically controled

Another suggestion that has generated considerable interest is that geography may help explain climate's sensitivity (North et al., 1983; Mitchell et al., 1987). So, when orbital variations are used that favor increasingly cooler summers (as at the transition between 125 kyr and 115 kyr BP and at the last glacial maximum 18 kyr BP), models with realistic distribution of continents and oceans generate the largest ice cap over northern Canada and Scandinavia. These are obviously the most sensitive spots of the climate system to orbital influences (Royer et al., 1983).

Glacial ice has generally received most of the attention but evidence exists that orbital variations also influence the behaviour of the North Atlantic deep ocean water and atmospheric features such as the intensity of the westerly winds and of the Indian monsoon. For example, changing the orbital configurations to that of 9 kyr BP, when insolation seasonality was 14 percent higher than today, leads to an intensified southwest monsoon (Kutzbach, 1981).

In fact, a simulation over the last 150,000 years (Prell and Kutzbach, 1987) has shown that under glacial conditions, the simulated monsoon is weakened in Southern Asia but precipitation is increased in the equatorial west Indian Ocean and equatorial North Africa. Moreover, the monsoon is strongly tied to the precession parameter (their maxima coincide) as it is also the case for the variations in tropical (Bernard, 1962; Rossignol-Strick, 1983; Short and Mengel, 1986) and equatorial climate (Pokras and Mix, 1987).

2.11. Transient response of the climate system to the orbital forcing and change in the seasonal pattern of insolation

In addition to the calculation of the Earth's climate which is in equilibrium with a particular insolation pattern and other boundary conditions (like the ice-sheets, for example), the simulation of the transient response of a realistic climate system to orbital variations must allow better understanding of the physical mechanisms linking astronomical forcing with climate. Berger suggested earlier (1979) that the long term astronomical variation of the latitudinal distribution of the seasonal pattern of insolation is the key factor driving the climate system, and the complex interactions among its different parts. A 2.5-dimension time-dependent physical climate model, which takes into account the feedbacks between the atmosphere, the upper ocean, the sea-ice, the ice-sheets and the lithosphere, thought to be the most important at the astronomical time scales, strongly supports such hypothesis (Berger et al., 1988). In this model, the simulated long-term variations of the global ice volume over the past 125,000 years agree remarkably well with the reconstructed sea-level curves of Chappell and Shackleton (1986) and Labeyrie et al. (1987).

Using climate models of various complexity, astronomically forced, researchers have shown that the dynamic behaviour of the climate over the last 400,000 years reproduced fairly well (Imbrie and Imbrie, 1980; Berger, 1980; Berger et al., 1988). Extrapolation thus begins to be allowed at least for a period over which we can assume there is sufficient predictability (Nicolis and Nicolis, 1986) : assuming no human interference at the astronomical scale, orbital forcing predicts that the general cooling that began 6 kyr BP will continue with a first moderate cold peak around 5 kyr AP, a major cooling about 23 kyr AP and full ice age conditions 60 kyr AP (Berger, 1980 and 1988).

3. CONCLUSIONS

Recent evidence seems thus to have laid to rest the argument that orbital variations might cause minor climate fluctuations but not major climatic changes as the Pleistocene glacials and interglacials. Among the number of competing theories to explain the coming and going of the Pleistocene ice sheets and other climatic variations of the past, only the astronomical theory (of which Milankovitch theory is a particular version) has been supported so far by substantial physical evidence.

This evidence, both in the frequency and time domains that orbital influences are felt by the climate system imply that the astronomical theory of climates has the following advantages :

1. It provides an absolute clock with which to date Quaternary sediments with a precision several times greater than is otherwise possible.
2. It provides the boundary conditions necessary for a better understanding of the climatic system and the interactions between the atmosphere, hydrosphere, cryosphere, biosphere, and lithosphere, which, at the astronomical time scales, all play a role.
3. It allows a better understanding of the seasonal cycle and can be used to test the performance of the climate models over a broad spectrum of climatic regimes.
4. It allows a better understanding of the other forcings, in particular the CO_2 cycle (Pisias and Shackleton, 1984; Barnola et al., 1987), by extracting the astronomical signal from the climate variability.
5. It predicts gross natural climate changes to be expected at the geological time scale in the next 100,000 years, an approximate decay period of radioactive wastes.
6. It allows a better understanding of the sensitivity of our present-day interglacial climate and of the possible superinterglacial that could be generated by human activities within the next 50 years or so.

7. It may provide data for astronomers with which to test the stability of the planetary system in pre-Quaternary times.

8. It enables accurate computation of the insolation changes at the decadal time scale due to change in orbital elements, in relation to the satellite measurements of the solar constant and its variations.

9. It allows a better understanding of the planetary system and the climatic variations of the planets (Ward, 1974; Ward et al., 1979; Pollack, 1979).

10. It allows us to transfer theoretical knowledge (spectral analysis, numerical schemes, etc.) and technologies (deep-sea drilling, satellites, supercomputers, etc.) to society at large.

4. ACKNOWLEDGMENT

I thank Dr. George Kukla from Lamont Doherty Geological Observatory, Columbia University, New York, for his careful reading of the manuscript and for fruitful discussions, and Mrs Materne-Depoorter N. for typing the manuscript.

5. REFERENCES

Alyea, F.N. (1972). Numerical simulation of an ice-age paleoclimate, Atmos. Sci. Paper n°193, Colorado St. University, 120pp.

Backman, J., Pestiaux, P., Zimmerman, H., Hermelin, O. (1986). Paleoclimatic and palaeoceanographic development in the Pliocene North Atlantic : Discoaster accumulation and coarse fraction data. In : C.P. Summerhayes and N.J. Shackleton (Eds), "North Atlantic Palaeoceanography", Geological Society, Special Publ. n°21, 231-242.

Barnola, J.M., Raynaud, D., Korotkevitch, Y.S., and Lorius, C. (1987). Vostok ice Core : a 160,000 year record of atmospheric CO_2, Nature, 329 (6138), 408-414.

Benzi, R., Parisi, G., Sutera, A. and Vulpiani, A. (1982). Stochastic resonance in climatic change, Tellus, 34, 10-16.

Berger, A. (1976a). Obliquity and precession for the last 5,000,000 years, Astronomy and Astrophysics, 51, 127-135.

Berger, A. (1976b). Long-term variations of daily and monthly insolation during the Last Ice Age, EOS, 57(4), 254.

Berger, A. (1977). Support for the astronomical theory of climatic change, Nature, 268, 44-45.

Berger, A. (1978a). Long-term variations of daily insolation and Quaternary climatic changes, J. Atmospheric Sciences, 35(12), 2362-2367.

Berger, A. (1978b). Long-term variations of caloric insolation resulting from the Earth's orbital elements, Quaternary Research, 9, 139-167.

Berger, A. (1979). Insolation signatures of Quaternary climatic changes, Il Nuovo Cimento, 2C(1), 63-87.

Berger, A. (1980). Milankovic astronomical theory of paleoclimates : a modern review, Vistas in Astronomy, 24(2), 103-122.

Berger, A. (1987). Pleistocene climatic variability at astronomical frequencies. In : H. Faure and N. Rutter (Eds), "Global Change", International Quaternary Association, Ottawa, Ont.

Berger, A. (1988). Milankovitch Theory and Climate, Review of Geophysics, 26(4), 624-657.

Berger, A. (1989). The spectral characteristics of Pre-Quaternary climatic records, an example of the relationship between the astronomical theory and Geo-Sciences. In : A. Berger, S. Schneider, J.Cl. Duplessy (Eds), "Climate and Geo-Sciences, a Challenge for Sciences and Society in the 21st Century", Kluwer Academic Publishers, Dordrecht, Holland (in press).

Berger, A., and Pestiaux, P. (1984). Accuracy and stability of the Quaternary terrestrial insolation. In : A. Berger, J. Imbrie, J. Hays, G. Kukla, B. Saltzman (Eds), "Milankovitch and Climate", 83-112, D. Reidel Publ. Company, Dordrecht, Holland.

Berger, A., Imbrie, J., Hays, J., Kukla, G., and Saltzman, B. (Eds) (1984). Milankovitch and Climate. Understanding the Response to orbital Forcing, D. Reidel Publ. Company, Dordrecht, Holland, 895p.

Berger, A., Dehant, V., Loutre, M.F. (1987). Origin and stability of the frequencies in the astronomical theory of paleoclimates, Scientific Report 1987/5, Institut d'Astronomie et de Géophysique G. Lemaître, Université Catholique de Louvain-la-Neuve.

Berger, A., and Pestiaux, P. (1987). Astronomical frequencies in Paleoclimatic data. In : Ye Duzheng, Fu Congbin, Chao Jiping, M. Yoshino (Eds), "The Climate of China and Global Climate", 106-114, China Ocean Press, Springer Verlag.

Berger, A., Gallée, H., Fichefet, Th., Marsiat, I., Tricot, Ch. (1988). Transient response of the climate system to the astronomical forcing. In : A.H. Cottenie and A. Teller (Eds), "Global Change IGBP", 10-24, Scope Belgium, Académie Royale des Sciences, des Lettres et des Beaux-Arts de Belgique.

Berger, A., Loutre, M.F., Laskar, J. (1988). Insolation values for the climate of the last 10 millions years, Scientific Report 1988/13, Institute of Astronomy and Geophysics Georges Lemaître, Catholic University of Louvain-la-Neuve.

Berger, A., Andjelic, T.P. (1988). Milutin Milankovitch, Père de la théorie astronomique des paléoclimats, Histoire et Mesure, Editions du CNRS, Paris.

Bernard, E.A. (1962). Théorie astronomique des pluviaux et interpluviaux du Quaternaire Africain, Memoire in 8*, nouvelle série, Acad. Roy. Sc. Outre-Mer, Brussels, Classe Sci. Tech., tome 12(1), 232pp.

Birchfield, G.E., and Weertman, J. (1978). A note on the spectral response of a model continental ice sheet, J. of Geophysical Research, 83 n°C8, 4123-4125.

Birchfield, G.E., Weertman, J., and Lunde, A.T. (1981). A paleoclimate model of Northern Hemisphere Ice Sheets, Quaternary Research, 15 n°2, 126-142.

Birchfield, G.E., Weertman, J. (1982). A model study of the role of variable ice albedo in the climate response of the earth to orbital variations, ICARUS, 50, 462-472.

Birchfield, G.E., Wertman, J., and Lunde, A.T. (1982). A model study of the role of high-latitude topography in the climatic response to orbital insolation anomalies, J. of Atmospheric Sciences, 39, 71-87.

Briskin, M., Harrell, J. (1980). Time series analysis of the Pleistocene deep-sea paleoclimatic record using periodic regression, Marine Geology, 36, 1-22.

Broecker, W.S., Thurber, D.L., Goddard, J., Ku, T., Matthews, R.K., Mesolella, K.J. (1968). Milankovitch hypothesis supported by precise dating of coral reefs and deep sea sediments, Science, 159, 297-300.

Brückner, Ed., Köppen, W., Wegener, A. (1925). Uber die Klimate der geologischen Vorzeit. Zeitschrift für Gletscherkunde, vol. 14.

Budyko, M.I. (1969). Effect of solar radiation variations on the climate of Earth, Tellus, 21 n°5, 611-620.
Chappell, J., and Shackleton, N.J. (1986). Oxygen isotopes and sea level, Nature, 324, 137-140.
CLIMAP Project Members (1976). The surface of the Ice-Age Earth, Science, 191, 1131-1137.
CLIMAP Project Members (1981). Seasonal reconstruction of the Earth's surface at the Last Glacial Maximum, McIntyre A. and Cline R. (Eds), Geological Society of America, Map and Chart Series MC-36, Boulder, 1-18.
Croll, J. (1875). Climate and time in their geological relations. Appleton, New York.
Crowley, Th.J. (1988). Paleoclimate Modelling. In : M. Schlesinger (Ed.), "Physically-Based Modelling and Simulation of Climate and Climatic Change", 883-949, Kluwer Academic Publishers, Dordrecht, Holland.
Duplessy, J.Cl. (1978). Isotope studies. In : J. Gribbin (Ed.), "Climatic Change", 46-68, Cambridge University Press, Cambridge.
Emiliani, C. (1966). Isotopic paleotemperatures, Science, 154(3751), 851-857.
Ghil, M. (1980). Internal climatic mechanisms participating in glaciation cycles. In : A. Berger (Ed.), "Climatic Variations and Variability : Facts and Theories", 539-557, D. Reidel Publ. Company, Dordrecht, Holland.
Ghil, M. and Le Treut, H. (1981). A climate model with Cryodynamics and Geodynamics, J. of Geophysical Research, 86, 5262-5270.
Hasselman, K. (1976). Stochastic climate models, part I., Tellus, 28, 473.
Hays, J.D., Imbrie, J., and Shackleton, N.J. (1976). Variations in the Earth's orbit : pacemaker of the Ice Ages, Science, 194, 1121-1132.
Herbert, T.D., and Fisher, A.G. (1986). Milankovitch climatic origin of mid-Cretaceous black shale rhythms, Central Italy, Nature, 321 (6072), 739-743.
Hilgen, F.J. (1987). Sedimentary rhythms and high resolution chronostratigraphic correlations in the Mediterranean Pliocene, Newsl. Stratigr., 17(2), 109-127.
Hyde, W.T., Peltier, W.R. (1985). Sensitivity experiments with a model of the Ice Age cycle. The response to harmonic forcing, J. of Atmospheric Sciences, 42(20), 2170-2188.
Imbrie, J. (1982). Astronomical theory of the Pleistocene Ice Ages : a brief historical, Icarus, 50, 408-422.
Imbrie, J., and Imbrie, K.P. (1979). Ice Ages, Solving the Mystery, Enslow Publishers, New Jersey.
Imbrie, J., and Imbrie, J.Z. (1980). Modelling the climatic response to orbital variations, Science, 207, 943-953.
Imbrie, J., and Kipp, N.G. (1971). New micropaleontological method for quantitative paleoclimatology : application to a Late Pleistocene Caribbaen Core. In : Turekian K.K. (Ed.), Late Canezoic Glacial Ages, 71-181, Yale University Press, New Haven.
Imbrie, J., Hays, J., Martinson, D.G., McIntyre, A., Mix, A.C., Morley, J.J., Pisias, N.G., Prell, W.L., Shackleton, N.J. (1984). The orbital theory of Pleistocene climate : support from a revised chronology of the marine ^{18}O record. In : A. Berger, J. Imbrie, J. Hays, G. Kukla, B. Saltzman (Eds), "Milankovitch and Climate", 269-305, D. Reidel Publ. Company, Dordrecht, Holland.

Imbrie, J., McIntyre, A., Mix, A. (1988). Oceanic response to orbital forcing in the Late Quaternary : observational and experimental strategies. In : A. Berger, S. Schneider and J.Cl. Duplessy (Eds), "Climate and Geo-Sciences", D. Reidel Publ. Company, Dordrecht, Holland.

Jouzel, J., Lorius, Cl., Petit, J.R., Genthon, C., Barkov, N.I., Kotlyakov, V.M. (1987). Petrov Vin, Vostok ice core : a continuous isotope temperature record over the last climatic cycle, Nature, 329 (6138), 403-408.

Kallen, E., Crafoord, C., and Ghil, M. (1979). Free oscillations in a coupled atmosphere-hydrosphere-cryosphere system, J. of Atmospheric Science, 36, 2292-2302.

Kominz, M.A. and Pisias, N.G. (1979). Pleistocene climate : deterministic or stochastic ? Science, 204, 171-173.

Kominz, M.A., Heath, G.R., Ku, T.L., Pisias, N.G. (1979). Brunhes time scales and the interpretation of climatic changes, Earth Planet Sci. Letters, 45, 394-410.

Kukla, G. (1972). Insolation and glacials, Boreas, 1(1), 63-96.

Kukla, G. (1975a). Loess stratigraphy of Central Europe. In : K.W. Butzer and G.U. Isaac (Eds), "After the Australopithecines", 99-188, Monton Publishers, The Hague, Holland.

Kukla, G. (1975b). Missing link between Milankovitch and climate, Nature, 253, 600-603.

Kukla, G., Berger, A., Lotti, R., Brown, J. (1981). Orbital signature of interglacials, Nature, 290(5804), 295-300.

Kutzbach, J.E. (1981). Monsoon climate of the early Holocene : climate experiment with the earth's orbital parameters for 9,000 years ago, Science, 214, 59-61.

Kutzbach, J.E. (1985). Modeling of paleoclimates, Adv. Geophys., 28A, 159-196.

Labeyrie, L.D., Duplessy, J.Cl., Blanc, P.L. (1987). Deep water formation and temperature variation over the last 125,000 years, Tellus, 327, 477-482.

Le Treut, H., and Ghil, M. (1983). Orbital forcing, climatic interactions and glaciation cycles, J. of Geophysical Research, 88 C9, 5167-5190.

Lindzen, R.S. (1986). A simple model for 100K-year oscillations in glaciation, J. Atmospheric Sciences, 43(10), 986-996.

Martinson, D.G., Pisias, N.G., Hays, J.D., Imbrie, J., Moore, T.C., and Shackleton, N.J. (1987). Age dating and the orbital theory of the ice ages : Development of a high-resolution 0 to 300,000-year chronostratigraphy, Quat. Res., 27(1), 1-29.

Mason, B.J. (1976). Towards the understanding and prediction of climatic variations.,Q.J.R. Meteorol. Soc., 102, 473-499.

Milankovitch, M.M. (1920). Théorie mathématique des phénomènes thermiques produits par la radiation solaire, Académie Yougoslave des Sciences et des Arts de Zagreb, Gauthier-Villars.

Milankovitch, M. (1941). Kanon der Erdbestrahlung. Royal Serbian Academy, Spec. publ. 132, section of Mathematical and Natural Sciences, vol. 33 (published in English by Israel Program for Scientific Translation, for the U.S. Department of Commerce and the National Science Foundation, Washington D.C., 1969).

Milankovitch V. (1988). My Father : Milutin Milankovitch, 1988, in press.

Mitchell, J.F.B., Gahame, N.S., and Needham, K.J. (1987). Climate simulations for 9,000 years before present : seasonal variations and the effect of the Laurentide ice sheet. Dynamical Meteorol. Rep. DCTN57, Meteorol. Office, Bracknell, England.

Moore, T.C., Pisias, N.G., Dunn, D.A. (1982). Carbonate time series of the Quaternary and late Miocene sediments in the Pacific ocean : a spectral comparison, Marine Geol., 46, 217-233.

Morley, J.J. and Hays, J.D. (1981). Towards a high-resolution, global, deep-sea chronology for the last 750,000 years, Earth and Planetary Science Letters, 53, 279-295.

Murphy, J.J. (1876). The glacial climate and the polar ice-cap, Quarterly J. Geological Society, 32, 400-406.

Negrini, R.M., Verosub, K.L., Davis, J.O. (1988). The middle to late Pleistocene geomagnetic field recorded in fine-grained sediments from Summer Lake, Oregon and Double Hot Springs, Nevada, USA, Earth and Planetary Science Letters, 87, 173-192.

Nicolis, C. (1980). Response of the Earth-atmosphere system to a fluctuating solar input. In : "Sun and Climate", 385-396, CNES-CRS-DGRST, Toulouse, October 1980.

Nicolis, C. (1982). Stochastic aspects of climatic transitions-response to a periodic forcing, Tellus, 34, 1-9.

Nicolis, C., and Nicolis, G. (1986). Reconstruction of the dynamics of the climate system from time series data, Proc. Natl. Acad. Sci. Geophysics, 83, 536-540.

North, G.R., Mengel, J.G., and Short, D.A. (1983). Simple energy balance model resolving the seasons and the continents. Application to the Astronomical Theory of the Ice Ages, J. of Geophysical Research, 88 C11, 6576-6586.

Oerlemans, J. (1980). Model experiments on the 100,000-yr glacial cycle, Nature, 287(2), 430-432.

Olsen, P.E. (1986). A 40-million-year lake record of Early Mesozoic orbital climatic forcing, Science, 234, 842-848.

Penck, A. and Brückner, E. (1909). Die Alpen im Eiszeitalter. Tauchnitz, Leipzig.

Pestiaux, P., and Berger, A. (1984). An optimal approach to the spectral characteristics of deep-sea climatic records. In : "Milankovitch and Climate", A. Berger, J. Imbrie, J. Hays, G. Kukla, B. Saltzman (Eds), 417-446, D. Reidel Publ. Company, Dordrecht, Holland.

Pestiaux, P., Duplessy, J.Cl., van der Mersch, I., and Berger, A. (1988). Paleoclimatic variability at frequencies ranging from 1 cycle per 10,000 years to 1 cycle per 1,000 years : evidence for nonlinear behavior of the climate system, Climatic Change, 12(1), 9-37.

Pisias, N.G., and Shackleton, N.J. (1984). Modelling the global climate response to orbital forcing and atmospheric carbon dioxide changes, Nature, 310, 757-759.

Pokras, E.M., and Mix, A.C. (1987). Earth's precession cycle and Quaternary climatic change in equatorial Africa : tropical Africa, Nature, 326(6112), 486-487.

Pollack, J.B. (1979). Climatic change on the terrestrial planets, Icarus, 37, 479-553.

Pollard, D. (1982). A simple ice sheet model yields realistic 100 kyr glacial cycles, Nature, 296, 334-338.

Pollard, D. (1984). Some ice-age aspects of a calving ice-sheet model. In : "Milankovitch and Climate", A. Berger, J. Imbrie, J. Hays, G. Kukla, B. Saltzman (Eds), 541-564, D. Reidel Publ. Company, Dordrecht, Holland.

Prell, W.L. (1982). Oxygeen and carbon isotope stratigraphy for the Quaternary of Hole 502B : evidence for two modes of isotopic variability. In : "Initial reports of the Deep Sea Drilling Project", 455-464, W.L. Prell and J.V. Gardner et al. (Eds), vol. LXVIII, Washington D.C.

Prell, W.L., and Kutzbach, J.E. (1987). Monsoon variability over the past 150,000 years, J. of Geophysical Research, 92, D7, 8411-8425.

Rossignol-Strick, M. (1983). African monsoons, an immediate climate response to orbital insolation, Nature, 303, 46-49.

Royer, J.F., Deque, M., and Pestiaux, P. (1983). Orbital forcing of the inception of the Laurentide ice sheet, Nature, 304, 43-46.

Ruddiman, W.E., Shackleton, N.J. and McIntyre, A. (1986). North Atlantic sea-surface temperatures for the last 1.1 million years. From Summerhayes C.P. and Shackleton N.J. (Eds), North Atlantic Palaeoceanography Geological Society Special Publication, 21, pp. 155-173.

Ruddiman, W.F., and McIntyre, A. (1981). Oceanic mechanisms for amplification of the 23,000-year ice-volume cycle, Science, 212, 617-627.

Saltzman, B., Hansen, A.R., and Maasch, K.A. (1984). The Late Quaternary glaciations as the response of a three-component feedback system to earth-orbital forcing, J. of Atmospheric Sciences, 41 nr 23, 3380-3389.

Schwarzbach, M. (1985). Wegener, le Père de la Dérive des Continents, Bélin, Paris.

Sellers, W.D. (1970). The effect of changes in the earth's obliquity on the distribution of mean annual sea-level temperatures, J. of Applied Meteorology, 9, 960-961.

Sergin, V.Ya. (1979). Numerical modeling of the glaciers-ocean-atmosphere global system, J. Geophysical Research, 84, 3191-3204.

Shackleton, N.J., and Opdyke, N.D. (1973). Oxygen isotope and paleomagnetic stratigraphy of equatorial Pacific core V28-238 : Oxygen isotope temperatures and ice volumes on a 10^5 and 10^6 year scale, Quaternary Research, 3, 39-55.

Shackleton, N.J., and Opdyke, N.D. (1976). Oxygen isotope and paleomagnetic stratigraphy of Pacific core V28-239 late Pliocene to latest Pleistocene, Geological Society of America, memoir 145, pp. 449-464.

Sharaf, G.S., and Budnikova, N.A. (1969). Secular perturbations in the elements of the Earth's orbit and the astronomical theory of climate variations (in Russian), Tr. Inst. Teor. Astron., 14, 48-85.

Short, D.A., Mengel, J.G. (1986). Tropical climate phase lags and Earth's precession cycle, Nature, 322, 48-50.

Smagorinsky, J. (1963). General circulation experiments with the primitive equations. I. The basic experiment, Mon. Weather Rev., 91, 99-165.

Spitaler, R. (1921). Das Klima des Eiszeitalters, Prag. (lithographed).

Vernekar, A.D. (1972). Long-period global variations of incoming solar radiation, Meterol. Monogr., 12(34), 130pp.

Ward, W.R. (1974). Climatic variations on Mars, 1, Astronomical theory of insolation, J. Geophys. Res., 79, 3375-3387.

Ward, W.R., Burns, J.A., and Toon, O.B. (1979). Past obliquity oscillations of Mars : the role of the Tharsis uplift, J. Geophys. Res., 84, 243-259.
Weedon, G.P. (1985/86). Hemipelagic shelf sedimentation and climatic cycles : the basal Jurassic (Blue Lias) of South Britain, Earth and Planetary Science Letters, 76, 321-335.
Wigley, T.M.L. (1976). Spectral analysis : astronomical theory of climatic change, Nature, 264, 629-631.

EUROPEAN SCHOOL OF CLIMATOLOGY AND NATURAL HAZARDS

Course on

"Climatic Change and Impacts: A General Introduction"

3. CLIMATE PROCESSES AND CLIMATE MODELLING

BASIC CONCEPTS OF CLIMATE MODELLING

A. BERGER
Université Catholique de Louvain
Institut d'Astronomie et de Géophysique G. Lemaître
2 Chemin du Cyclotron
1348 Louvain-la-Neuve, Belgique

Summary
The climate system consists of the atmosphere, the hydrosphere, the cryosphere, the lithosphere and the biosphere. The difficulty of modelling this system and its variations is related not only to the complexity of each of these physical entities, but also to their interactions. Moreover the large difference between their characteristic time prevents the climate to ever reach an equilibrium. For simulating the climate variations at a specific time scale, the slow parts of that system are thus considered as prescribed boundary conditions to the more "mobile" ones (for example, the dynamics of the ice sheets are not taken into account for the interannual climate variability).

Changes in one of the sub-systems generally affect the behaviour of other parts and set in motion a change of effects which may either reinforce or cancel the original change (feedback mechanisms). The number of possible mechanisms causing changes of climate is therefore rather large and the observed variations of past climate represent the attempts of the system to reach an equilibrium. The climate and its variations are the result of such processes and the climate therefore may be changing in different ways and at different rates in various parts of the world at the same time.

Climate being an extraordinary complex physical problem, the study of the many processes responsible for the climate structure and variations requires to construct mathematical models in order to represent quantitatively the various interacting elements. The proper representation of effects on very different scales of space and time in a climate model provides an otherwise advantage of the use of climate models, namely the possibility of considering many interacting processes at the same time and of studying the response of the climate system to external forcings and internal perturbations. The basic physical laws which govern the behaviour of the components of the climate system and the physics of the various interactive processes linking the components together are relatively well-known. For the atmosphere for example, these laws are expressed by the equations which describe the conservation of momentum, energy and mass (including moisture).

1. INTRODUCTION

The climate is now recognized as being continuously variable and can change in several ways. The most obvious is through a secular trend which produces a change in the average, but another very important factor is the variability of climate, in which the mean may remain the same, the oscillations about that mean becoming larger or smaller.

The climate coupled system seems capable of undergoing variations on all time scales ranging from the longest observable climatic changes (million years time scales) to climatic fluctuations and climatic variability of interannual and intra-annual character. In consideration of the enormous complexity of the behaviour of the climatic system, it is not surprising that a large variety of theories has been advanced to account for these climatic variations. The components of the climatic system being non-linearly coupled by physical, biological and chemical processes, it is probable that most factors postulated have had some bearing on the past climate.

To understand change on the Earth requires that we establish plausible hypotheses of its cause, identify the processes involved, and ascertain the limits of its variability. This knowledge must then be integrated into a conceptual framework that makes the observations meaningful.

The climate of the Earth is the result of a balance between many different and interrelated physical, chemical and biochemical processes which take place in the enormously complex "climate system" encompassing the atmosphere, the oceans, ice, soil and vegetation. Understanding the role of these processes in shaping past and present climates is based on the ability to account quantitatively for their effects and to incorporate them in climate simulations. The non-linear interactions in the climate system require the numerical integration of a set of many prognostic and diagnostic equations which constitute a mathematical model of the climate system. Such models are used for simulating the variability of climate and may eventually be used to predict climate variations. Such models will not only help to reveal the scientifically important questions but will also provide useful guidance for the evolution of the increasingly effective observational program.

Modelling the Earth Climate System requires that we go beyond simplicity, with the ultimate aim of modelling all Earth processes over all timescales (Ad Hoc Planning Group on Global Change, 1986; Earth System Sciences Committee, 1988; IGBP, 1988). When a basic understanding of the climatic system processes has been achieved, and modeling has advanced to the stage of quantitative simulations, we may then exploit the great range of conditions contained in the geological and paleoclimatological records to provide a variety of cases for verification studies. We also hope to determine whether the climatic system will achieve equilibrium, or rather will tend to oscillate between quasi-stable states, with dramatic episodes of global change accompanying the transitions between states.

Different degrees of parameterization may be used for the purpose of modelling the climate system, ranging from one-dimensional energy balance models to high resolution three-dimensional simulations of the general circulation of the atmosphere and ocean. The simpler models cannot account for the effect of inhomogeneous boundary forcing such as the effect of vegetation or the contrast between oceanic and continental areas, nor can they provide indications on regional climate anomalies. However, highly parameterized models are convenient to give first order estimates of the sensitivity of climate to planetary-scale influences.

The purpose of this paper is to summarize the basic concepts used in climate modelling and to provide some insights for developing the so-called simple models. Our task was greatly made easier by the existence of excellent review papers (Gates, 1981; Meehl, 1984; Peixoto and Oort, 1984; Dickinson, 1986), proceedings (Berger, 1981; Hansen and Takahashi, 1984; Hougton, 1984; Hecht, 1985; Manabe, 1985; Schlesinger, 1988) and books (Lamb, 1972; Budyko, 1977; Monin, 1986; Washington and Parkinson, 1986; Henderson-Sellers and McGuffie, 1987) on this subject.

2. CLIMATE AND WEATHER

There is a clear distinction between weather and climate. Weather is concerned with detailed instantaneous states of the atmosphere, and the day-to-day evolution of individual synoptic systems. The study of its evolution requires solving an initial value problem on a finite number of grid points. Because of the random errors in meaning and analysing the meteorological fields and of the discretization problem (leading to truncation errors and errors related to the parameterizations of the sub-grid scale processes) there will unavoidably be a limited predictability of weather systems ranging from a few days to 2-3 weeks (Lorenz, 1969; Lieth, 1978).

Climate may be understood as the sum total of the weather experienced over a specified period of time long enough to establish its statistical properties (like average - conveniently called a climatic state - variance, and probability of extreme events). If we consider the atmosphere alone, climatic states will be defined for any interval of time which exceeds, at least, the average life span of the synoptic weather systems. The traditional 30-year averaging interval (of the World Meteorological Organisation) is a particular case which refers to the so-called "normal climate". In general terms, we can thus define climate as the ensemble averages of climate states of the internal system, together with some measure of its variability for a specified interval of time and with a description of the interactions between the internal and external systems. This variability could in principle be separated into a deterministic part, the signal, and a random part associated with weather fluctuations, the noise (for example, one of the largest difficulties in the CO_2 problem, resides precisely in the determination of the signal to noise ratio).

Weather and climate must also be understood in terms of a spatial domain, the size of which, in some senses, determines the temporal separation to be used to distinguish between weather and climate. Atmospheric disturbances occur on many time and space scales, from weather systems to planetary scale circulations, including the Southern Oscillation, which have aspects of both weather and climate (Earth System Sciences Committee, 1988). This wide range of phenomena, characteristic of the climate and Earth system, makes the demarcation of appropriate space scales very difficult. Acknowledging implicitely the bounds of present scientific understanding and the limited abilities to predict regional climate, climate studies usually refer to regions with scales of tens to hundreds of kilometers, with each region assumed to have relatively homogeneous conditions. In fact, fundamental which promises to provide a deeper understanding of the interactions which bind the

Earth climate's components into a unified dynamical system is related to a set of interacting processes operating on a wide range of spatial and temporal scales, rather than as a collection of individual components. Important interactions connect many of these processes and thus bridge widely separated spatial and temporal regions. Once change is introduced, it can propagate through the entire Earth system.

3. CLIMATE SYSTEM

The climate system can be defined within the thermodynamical framework as a finite region in a space specified by a set of physical extensive (additive) variables, such as the volume, the internal energy, the mass and the angular momentum (Peixoto and Oort, 1984). The amount of such a variable transferred from a subsystem to an adjacent subsystem during an interval of time represents a thermodynamic process which makes these systems coupled or interactive.

The state of the climate system is also specified by intensive properties such as temperature, pressure, density, velocity, etc. These are defined at a given point of the system and at a given instant and may change in time.

The climate system

(i) may be regarded as closed for the exchange of matter;

(ii) is a composite system formed by the conjonction of five spatially disjoint physical interactive systems (Fig. 1) : the atmosphere, the hydrosphere (oceans), the cryosphere, the lithosphere and the biosphere. Any attempt to understand the climate, therefore, requires consideration of not just the potential state of the atmosphere but that of the oceans, of the land, of the ice on sea and land and of the living environment (biota).

(iii) is also a cascading system as these subsystems are open with a non uniform distribution of their intensive properties and are dynamically connected by flows of energy occuring in a variety of forms, such as heat, potential energy, kinetic energy, chemical energy, short- and long-wave radiation, momentum and matter (Fig. 2).

The climate system is subject to two external energy inputs that condition its global behaviour : solar radiation and gravity. As the solar radiation provides almost all energy that drives the climate system, orbital, rotational and shape characteristics of the Earth must be considered.

4. CLIMATE COMPONENTS

The atmosphere is formed by several layers which differ with regard to composition and nature of the energy processes involved. It is the component of the climate system which is the most variable in time and space, with a response time on the order of days to weeks. It shows a broad general circulation with less organized eddy motions, such as weather systems in middle latitudes, and turbulent motions mainly in the planetary boundary layers. The atmosphere is a heat engine with heat flowing from the tropics to the polar and upper regions. The work performed is used to maintain the kinetic energy of the circulations against the continuous drain by friction.

The hydrosphere consists of oceans, seas, lakes, rivers, and other water reservoirs but the most important for climate studies are the oceans. Due to their large mass and specific heat, they constitute an enormous reservoir to store energy and to regulate the temperature (the heat capacity of the atmosphere is equivalent to only about a 2.5 m depth of water; thus its thermal inertia, or resistance to temperature change, is relatively small compared with that of the ocean, which is 4.5 km deep. Response time of processes in the atmosphere can therefore be quite short). They have a large mechanical inertia and a pronounced stratification, with a mixed surface layer and a thermocline. They show much slower circulation than the atmosphere, forming large quasi-horizontal circulation gyres with the ocean currents and slow thermohaline overturnings, but also small scale eddies. The relaxation time for the oceans varies within a wide range that stretches from weeks or months in the upper layers to centuries or millenia in the deep oceans. On the other hand, the main sources of water vapor for the atmosphere are observed to be in the subtropics, predominently over the oceans. The water vapor in the atmosphere is transported from the surface source to the sink regions by the general circulation, leading to the atmospheric branch of the hydrological cycle. The vapor will condense and precipitate in the equatorial convergence zone and in the middle to high latitude zones associated with the polar front. The cycle is closed through runoff by the terrestrial branch which is a crucial factor in the global climate and influence the climate on a regional or local scale.

The cryosphere includes the ice sheets, continental glaciers, snow fields and sea ice. Continental snow cover and sea ice produce large intra-annual and interannual variations : the continental ice sheets play a major role in climatic changes on scale of tens of thousand of years such as the glacial-interglacial periods that have occured during the Pleistocene.

The lithosphere include the continents with their orography and the bottom of the oceans. It has the longest response time if one exclude the surface active layer.

The biosphere comprises the vegetation cover, the continental fauna and the flora and fauna of the oceans. The biosphere influences many surface processes (as albedo, roughness, evaporation) and the carbon cycle (photosynthesis and respiration). It is very sensitive to change in the atmospheric climate. Anthropogenic interaction with the climatic system through human activities such as agriculture, urbanization, industry and pollution must be mentionned here.

A global view of the entire Earth-system thus includes the solid, the fluid and the biological Earth. By contrast with the solid Earth, changes in the fluid and biological Earth are highly sensitive to the Earth's external environment, being driven almost entirely by the energy of solar radiation. Diurnal and annual variations in insolation play a central role and changes in the Earth's orbital parameters have important long-term climatic effects. To the complexity of resulting motions in the atmosphere and the oceans must be added the extraordinary richness and variety of the biosphere which has profoundly affected Earth evolution since the origin of life, more than three billion years ago.

5. CLIMATIC VARIATIONS AND THEIR CAUSES

The study of the past climate conditions (paleoclimate) is particularly important to our understanding of the more recent and future changes in the fluid and biological Earth and to the testing of climate models.

Differences between climatic states of the same kind over monthly, seasonal, annual, or decadal time scales are referred to as a climatic variation. A climatic variation will also include, in general, a change in the statistics, as well as a change of the time means. In some cases, a change in the variance may in fact be a more important aspect of a climatic variation than a change of the average. We may also introduce the concept of a climatic anomaly (defined as the departure of a particular climatic state from the average of a number of climatic states, such as the climatic anomaly of a particular January) and the concept of climatic variability (defined as the variance among a number of climatic states of the same kind, such as the variability of January climates). Climatic variability is usually considered to occur as a result of causes that are not yet completely understood. Climate change, however, refers to shifts in the normal climate, usually a consequence of perturbations of some known (or potentially knowable) factor, lasting over many years. Thus, year-to-year events, such as several years of extremely wet or dry conditions, are climatic variations if they are within the expected statistical deviations around the normal climate, whereas the gradual melting of sea ice due to hemispheric warming may result from a climate change.

The large variability in time for the atmosphere can be illustrated by the wind speed and temperature spectra. The kinetic energy spectrum illustrates the relative importance of the various atmospheric motions for periods between seconds and several years. Most of this kinetic energy is concentrated in the low frequencies related to the daily and annual cycles and associated with large scale disturbances that occur in middle latitudes along the polar front. The idealized variance spectrum of the atmospheric temperature (Fig. 3) created by Mitchell (1976) shows a background level of variability, driving from internal stochastic mechanisms and corresponding to a low degree of predictability, which appears to increase in amplitude towards the longer time scales, superimposed on which is a band-limited variability due to external forcing processes and correspondingly to a much higher degree of predictability (narrow spikes and broader peaks). The spikes are astronomically dictated; they are strictly periodic components of climate variation, such as the diurnal and annual variations and their harmonics, whereas the peaks represent variations that are, according to Mitchell, either quasiperiodic or aperiodic - but, with a preferred time scale of energization. The peak at 3-7 days is associated with the synoptic disturbances mainly at middle latitudes. The peak near 2,500 years is, perhaps, due to variations like the cooling observed after the so-called Postglacial "Climatic Optimum" (also named the Atlantic interstade or hypsithermal in the american literature) which predominated during the great ancient civilizations about 5,000 years ago, and might be related to the characteristic time scale of the ocean (Pestiaux et al., 1987). The next three peaks are related to the ice ages (the glacial-interglacial stages of the Quaternary). They are also related to deterministic astronomical variations in the orbital parameters of the Earth (Milankovitch, 1941; Berger, 1988) - the cycle at around 21,000 years to the axial precession; the period of about 41,000 years with the change in the obliquity of the ecliptic, or the axial tilt; and the cycle at about 100,000 years to the eccentricity of the orbit of the Earth. Finally, the peaks near 45 and 350 million years may be related to glaciations due to orogenic and tectonic effects and to the "drift" of the continents (plate tectonics).

Five distinct temporal bands can be used to define the major time scales involved in global change (Crowley, 1983; Bradley, 1985). These bands are

(i) Millions to billions of years : since the formation of the Earth (4.5 10^9 years ago), a metallic core (responsible for the magnetic field) has remained largely isolated from the overlying convective mantle and moving lithosphere. The characteristic time scales of these subsystems are millions of years; the evolution of life and the associated development of the chemical composition of the atmosphere occured on similar time scales.

(ii) Thousands of years : the oscillations between glacial and interglacial periods with associated variations in atmospheric chemical composition, the development of soils, and the distribution of biological species occured largely in response to changes in the Earth's orbit around the Sun that recur in cycles of tens of thousands of years. The astronomical theory focuses on this very well known deterministic forcing mechanisms of the climate system and on the characteristic frequencies of the climatic response (Berger et al., 1984).

(iii) Decades to centuries: changes in climate, chemical composition of the atmosphere, pattern of surface aridity or acidity, and in terrestrial and marine biological systems occur at 10 to 100-year time scales. Biological processes take place on all time scales but those occuring in this band are of paramount importance to the concerns and planning of human societies.

(iv) Days to seasons: weather phenomena, eddies in ocean currents, seasonal growth and melting of the sea-ice covers, surface runoff and weathering, and the annual cycle of plant growth are all confined to time scales regulated by the annual cycle of insolation. A large part of the feedbacks from the biogeochemical cycles occurs through the alteration of the radiative processes that supply energy to the major subsystem.

(v) Seconds to hours: the flux of mass, momentum, and energy among the land, the ocean, the ice, the atmosphere and the biota are all dominated by processes with time scale shorter than one day. Over the land and the ocean, these exchanges occur through the medium of turbulent transport that are themselves responsive in part to diurnal heating cycles.

In consideration of the enormous complexity of the behaviour of the climatic system, it is not surprising that a large variety of theories has been advanced to account for these climatic variations (Fig. 4). Many possible causes of climatic change have been postulated over the years (Berger, 1979; Wigley, 1981). Some operate only on a geological time scale (plate tectonics), while others may be important from year to year (sea surface temperature). Climatic change mechanisms can also be subdivided into those which are of an external (solar radiation) or an internal nature (snow and ice cover) or which are natural or man-made. These are not exclusive categories, however, since changes in carbon dioxide for instance may have a natural or man-made source.

The climate system being extremely complex, it is highly probable that most factors postulated have had some bearings on the past climate and all causes are complicated by feedback links which may enhance or cancel an original perturbation.

Moreover any discussion of causal mechanisms must take note of the possibility that the global climate may be in a continual state of transient adjustment, or is not a unique state for any particular set of climatic forcing functions and boundary conditions (Lorenz, 1976). Non-deterministic factors could thus be wholly or partly responsible for long-period fluctuations of the climatic system.

Let us assume (Fig. 5) that two different states, A and B, of a climatic system are possible at a time t=0, and let us consider that A is the climatic state that would normally be "expected" under the given constant boundary condition. In a completely transitive system, the climatic state B would progressively approach the state A with the passage of time. In a completely intransitive system, on the other hand, the climatic state B would remain unchanged, and two possible solutions would exist. There would, in this case, be no way in which we could continue to identify the state A as the "normal" solution, as state B would presumably furnish an equally acceptable set of climatic statistic. A third behavior, however, is perhaps the most interesting of all, and is displayed by an almost-intransitive system. In this case, the system in state B may behave for a while as though it were intransitive, and then shift toward an alternate climatic state A, where it might remain for a further period of time. The system might then return to the original climatic state B, where it could remain or enter into further excursions. The climate exhibited by an almost-intransitive system would consist of two (or more) quasi-stable states, together with periods of transition between them. For longer periods of time the system might have stable statistics, but for shorter periods it would appear to be intransitive.

6. THE RESEARCH APPROACH OF CLIMATE SYSTEM SCIENCE

The reality of global change shown in the previous section stimulates the understanding of its causes and the determination of the limits of the variability that arises through interactions among the components of the climate system. To reach such goals, the observations must be assembled into a conceptual framework that permits quantitative simulations to be developed. A key component of this research strategy is the use of global and processed observations to create models both of the subsystems and of the climate system itself, and the use of these models to refine the observing system.

Global Earth observations are basic to achieving such an understanding. It is only through measurements that we can record changes in the climate system in a quantitative fashion, thus constructing the data base required to specify its present state and to test theories of climate evolution and global change.

Through data analysis and interpretation (descriptive models or qualitative knowledge), we then seek to discern patterns in the data (in the time and space domains using spectral and principal components analysis, for example) that can be explained in terms of processes (i.e. associations of phenomena governed by physical, chemical and biological laws).

Mathematical and numerical models (quantative understanding) of the climate system processes carry the analysis further, incorporating numerical algorithms for processes that permit quantitative links with other processes and hence the incorporation of interactions among them.

Models that provide complete and explicit specification of the rates of change of system variables as functions of present values of the variable, external forcing effects and other parameters are often referred to as dynamical systems. Although this definition implies that the system evolution is deterministic, we know from experience with non linear subsystems of the climate system that the solutions are strongly dependent upon initial conditions. Since these are rarely known precisely, the most we can expect from this procedure, is a correct representation of the statistical character of the solution, including information about major cyclic variations and the expected ranges of system variables. Modeling provides thus a framework for assembling data and knowledge, but also supplies the only responsible means for conducting numerical simulations designed to show how the climate system would respond to different input scenarios, such as variations in external forcing, disturbances by natural events or human activities.

The results of modeling effort must, at least, reproduce the present state of the climate and explain its past. Verification allows to test the model, frequently exposing their inadequacies while at the same time suggesting both new models and new observations.

The processes of global climate change on the time scales of years to millenia may be grouped into two basic classes : the physical climate system and the biogeochemical cycles, woven together by the ubiquitous presence of global moisture in the forms of vapor, liquid water and ice.

7. PROCESSES IN AND INTERACTIONS BETWEEN THE CLIMATE SUBSYSTEMS

The operation of the physical climate system is driven largely by the variation in solar heating with latitude due to the sphericity of the Earth, the orbital motion and the tilt of the Earth's axis (Berger, 1978; Fig. 6a). More solar radiation reaches and is absorbed in the intertropical regions than at polar latitudes. Due to the observed range of temperature between the equator and the poles, the decrease of emitted terrestrial radiation with latitude is much less pronounced than the decrease in absorbed solar radiation, leading to a net excess of energy in the tropics and a net deficit poleward of 40° latitude (Fig. 6b). In order to keep the poles from getting colder and the tropics from getting warmer, energy therefore must be transferred meridionally from lower to higher latitudes (Fig. 7). This horizontal heat exchange is carried out mainly by the poleward transfer of sensible heat by the atmospheric circulation and ocean currents and by the release of latent heat through condensation of water vapor carried poleward in the atmosphere.

This latitudinal thermal imbalance produces thus a global circulation of the atmosphere. The zonal circulation shows the existence of two maxima of eastwards winds in the upper troposphere at wind-latitudes (jet stream). In the tropical regions the lower-level winds are mainly from the east forming the intertropical convergence zone. The meridional circulation (Fig. 8) revealed a three-cell regime in each hemisphere with two thermally direct cells (the Hadley and polar cells) and in between the thermally indirect Ferrel cell. This global circulation leads to great wind systems that are powerful engines for the global redistribution of heat, momentum, and material substances raised from the Earth's surface by convective currents. Passing over the oceans, these winds apply a stress to the upper ocean layer that helps to shape and drive global ocean current systems, as well as mixing this biologically productive surface region. Winds and ocean currents are thus tying the fluid and biological Earth, producing both balance and change at the Earth's surface (Table 1).

Table 1. Annual mean radiation and energy balances on horizontal surfaces in Wm^{-2}.

	middle high lat. 60-70 N	Tropics 10-20 N	Earth
TOP ATMOSPHERE	220	400	340
SURFACE			
net radiative balance (1)	27	138	93
latent heat (evaporation)	-26	-105	-76
sensible heat	-13	- 21	-17
oceanic currents	12	- 12	0
ATMOSPHERE			
net radiative balance	-104	- 97	-93
latent heat (precipitation)	34	88	76
sensible heat	13	21	87
atmospheric currents	57	- 12	0
SURFACE+ATMOSPHERE			
net radiative balance	- 77	41	0(2)
latent heat	8	- 17	0(2)
sensible heat	0	0	0(2)
oceanic+atmospheric currents	69	- 24	0(2)

(1) absorbed solar minus emitted long-wave radiation; (2) taken over the globe as a whole, observations show that the system looses about the same energy through infrared radiation as it gains from the incoming solar radiation. However, small disturbances could occur for short and long periods.

Carbon, nitrogen, sulfur and oxygen play primary roles, cycling in various forms through the atmosphere, hydrosphere and lithosphere and interacting with other essential elements. Each cycle is marked by particular pathways and time scales, but they are all interrelated by biological processes. In addition to sustaining life, the biogeochemical cycles also play a role in determining the atmospheric concentration of the greenhouse gases that influence the Earth's energy budget (carbon dioxide, methane, nitrous oxide and chlorofluorocarbons). These gases are emitted in the course of biological or human activity on land and in the oceans. Once released, they circulate through the atmosphere and the oceans until their ultimate destruction or deposition through a variety of chemical reactions. In the atmosphere, such reactions play a role in determining the extent of global air pollution and the chemistry of the Earth's protective ozone layer. In the oceans, some of these gases or their oxidation products, are taken up by living organisms and thus eventually brought to the ocean floor with organic sediments.

The existence of abundant water in all three phases is a primary difference between the Earth and the other planets in the solar system, and is critical to the maintenance of life. The global circulation of water in all forms plays a fundamental role in interactions of the Earth's surface with the atmosphere, particularly those governing the physical climate system and the biogeochemical cycles. The distribution of rainfall, snow, evaporation and runoff affects the extent and distribution of biomass and biological productivity; these can, in turn, affect hydrological processes on both local and global scales.

Because they have not yet been adequately studied and because of their crucial importance for an understanding of global change in the climate system, the measurements of global moisture of land and ocean biota, atmospheric chemistry and composition, and ocean sediments must be a primary objective within the years ahead.

8. EQUILIBRIUM IN THE CLIMATE SYSTEM

The atmosphere and oceans are strongly coupled through the exchange of energy, matter and momentum on many scales in space and time. Moreover the study of the variability of climate during the past ages requires also the cryosphere to be included together with the biosphere mainly because of the varying nature of the atmospheric composition (e.g. Lorius et al., 1988).

Because of the different ranges of relaxation times for the various components of the climatic system, the whole climate system must be regarded as continuously evolving with parts of the system leading and others lagging in time. The highly nonlinear interactions between the subsystems are governed by fundamental physical laws that drive the system toward a global and local equilibrium (or balance) of energy and momentum. Equilibrium, however, is never quite achieved, because the daily, seasonal, annual and longer term changes of solar insolation and factors such as atmospheric composition are constantly forcing the atmosphere to seek an equilibrium different from that which it was seeking earlier. Therefore the subsystems are not always in equilibrium with each other and not even in internal equilibrium. There are also feedbacks which may amplify or attenuate a perturbation (for example, the albedo-temperature positive feedback and the temperature-infrared negative feedback). Although we expect a compensation between positive and negative feedbacks in the mean, there is geological evidence for some catastrophic changes in the climate state that could involve some runaway process in which a change to a new and different state occured (Berger and Labeyrie, 1987).

9. BASIC LAWS FOR MODELING

The climatic system is highly complex because

(i) of the many scales of atmospheric motion

(ii) of the difference in physical characteristics of the climatic subsystems which are nonlinearly coupled on many different time and space scales

(iii) many classes of perturbations are unstable.

In view of these difficulties in dealing with the Earth climate system, we may select certain combinations of its components and define a hierarchy of internal systems, considering the remaining components as the external system.

Because the atmosphere is the most responsive component, it is common practice to consider it first, although it is now recognized that ocean and atmosphere constitute a more complete internal system.

The physical laws which govern the evolution of weather and of climate are basically the same. However, in climate studies it is necessary to average the equations in time and to consider the complex interactions between the atmosphere and its external system. For weather prediction, the atmosphere behaves almost inertially, so that slowly acting boundary conditions can be ignored, although these changes can gradually become important.

9.1. Basic equations

The governing equations of the atmosphere express the principle of conservation of mass, momentum and energy (fundamental principles are, for example, in Atkinson, 1981; Gill, 1982; Holton, 1979; Houghton, 1986; Morel, 1986; Wells, 1986). It is also assumes that the atmosphere behaves as a homogeneous ideal gaseous system when unsaturated and that the water substance is conserved in the various phases of the entire climatic system. Radiation laws are also taking into account short-wave absorption, reflection, scattering and infrared radiative transfer.

The basic hydrodynamic and thermodynamic laws are then represented by the following equations

$$\frac{d\rho}{dt} = -\rho\ \nabla.\vec{V} \qquad \text{conservation of mass} \qquad (1)$$

$$\frac{d\vec{V}}{dt} = -2\vec{\Omega}x\vec{V} - \frac{\nabla p}{\rho} + \vec{g} + \vec{F} \qquad \text{conservation of momentum} \qquad (2)$$

$$C_p\ \frac{dT}{dt} = Q + \frac{1}{\rho}\frac{dp}{dt} \qquad \text{1st law of thermodynamics} \qquad (3)$$

$$p = R\ \rho\ T \qquad \text{ideal gas law} \qquad (4)$$

$$\frac{dq}{dt} = f(q) \qquad (5)$$

- $\vec{V}$ is the velocity vector in (x,y,z) : (u,v,w), is the specific mass, $\vec{\Omega}$ is the Earth's angular velocity, p the atmospheric pressure, $\vec{g}$ the gravity, C_p the specific heat at constant pressure p, $\vec{F}$ the frictional force, R the gas constant for dry air, q the specific humidity.

- in (2), F can be written as the divergence of the stress tensor τ
  $(F = - \nabla.\tau/\rho)$

- in (3), Q is the net heating rate per unit mass. It includes various diabatic effects, namely radiative heating, latent heating, frictionnal and turbulent heating and boundary layer heating. In terms of entropy, s, it can be written in the form :

  $$\frac{ds}{dt} = \frac{Q}{T}$$

  As the solar radiation originates in a source with a temperaturee on the order of 6000 K, whereas the terrestrial radiation is emitted at a temperature of about 250 K, the gross generation of entropy for all internal processes of the climatic system is 20-30 times larger than the amount of imported entropy.

- (4) shows that the atmosphere is usually baroclinic. If we assume that the atmosphere behaves as an ideal gas, we can use the Poisson equation for the adiabatic expansion and compression :

$$T\, p^{-R/C_p} = \text{const.}$$

which leads to the concept of potential temperature θ allowing to discuss the static stability of the atmosphere ($\frac{T}{\theta}\frac{\partial\theta}{\partial z}$) and to relate the specific entropy to $\theta(s = C_p \ln \theta)$.

- in (5), $f(q)$ represents the sources and sinks of water vapor. If we assume that water vapor falls out as precipitation as soon as it condenses, equivalent temperature may be defined which takes into account the latent heat released : $T_e = T\,(1 + Lq/C_p T)$.

This set of 5 equations forms a closed system of 1 diagnostic and 4 prognostic equations. With appropriate boundary conditions it is thus possible to describe the evolution of the system starting with a given initial state. However, due to the complexity of the problem, these equations must not only be discretezised but also particular class of motions which are not desired must be filtered out using scale analysis.

The equation of motion (2) can be rewritten in spherical coordinates (λ, ϕ, z)

$$\frac{du}{dt} = \frac{\tan\phi}{a} uv - \frac{uw}{a} + fv - f'w - \frac{1}{\rho}\frac{\partial p}{a\cos\phi\,\partial\lambda} + F_\lambda \qquad (6.1)$$

$$\frac{dv}{dt} = -\frac{\tan\phi}{a} u^2 - \frac{vw}{a} - fu \quad - \frac{1}{\rho}\frac{\partial}{a\partial\phi} \quad + F_\phi \qquad (6.2)$$

$$\frac{dw}{dt} = \frac{u^2}{a} + \frac{v^2}{a} \quad + f'u - \frac{1}{\rho}\frac{\partial p}{\partial z} \quad + F_z \qquad (6.3)$$

where $u = a\cos\phi\,\frac{d\lambda}{dt}$, $\quad v = a\,\frac{d\phi}{dt}$, $\quad w = \frac{dz}{dt}$

$f = 2\,\Omega\,\sin\phi$, $\quad f' = 2\,\Omega\,\cos\phi$

a = mean radius of the Earth

In these scalar equations, the curvature (ζ_i/a) terms reflect the influence of the matrics in the expression of the motion field. These terms do not perform work as $\vec{V}.\vec{\zeta}/a = 0$ (the Coriolis terms satisfy a similar invariance relationship).

Because the vertical extent of the atmosphere is small compared with its horizontal dimensions, there is a tendancy for the circulation to be predominently horizontal and in quasihydrostatic equilibrium. As the primitive equations contain all scales of motion up to planetary waves, some of these scales (sound and gravity waves, in particular) may produce a noise level during the integration obscuring the real meteorological signal and must be filtered. For synoptic-scale motions above the planetary boundary layer, such a filtering of the third equation of motion (6.3) leads to :

$$\frac{\partial p}{\partial z} = - \rho g \tag{7}$$

which is the condition of hydrostatic equilibrium (the equation links pressure and height and allows to use pressure as a vertical coordinate leading to the isobaric coordinates (x,y,p,t)).

Applying a filter of 10^{-3} m s^{-2} to the horizontal equations leads to an approximate balance which defines the geostrophic wind. The Rossby number $R_0 = (dv/dt)/fu = u/fL$ (L being the horizontal length scale) which sets a criterion for the validity of this geostrophic approximation, is equal to 10^{-1} in the atmosphere and 10^{-3} in the oceans, implying a very strong geostrophic contraint for the ocean circulation.

If we use a smaller filter of 10^{-4} m s^{-2}, for example, the horizontal equations can be written

$$\frac{du}{dt} = f\,v - \frac{\partial p}{\rho\ a \cos\phi\ \partial\lambda}$$

$$\frac{dv}{dt} = -fu - \frac{\partial p}{\rho\ a\ \partial\phi} \tag{8}$$

and become prognostic equations again by introducing the ageostrophic components of the wind ($\vec{U}_{ag}$). The horizontal divergence of mass is, for instance, almost completely due to $\vec{U}_{ag}$.

We are now left with a closed system of partial differential equations which is a initial value-boundary conditions problem which will be solved in wave number space or in the actual space using a grid-point network.

If n is the number of variables necessary to define the state of the atmosphere at each point, S the area of the globe, Δ the average horizontal grid size, K the number of levels in the vertical, nKS/Δ^2 gives the number of degrees of freedom for the system. From this number, the number of operations required at each grid points and the number of time steps to cover the simulation period, it is obvious that such models require very powerful computers to perform the calculations.

9.2. Balance Equations

These basic equations are highly non linear in nature and no general analytical methods are available to solve them. Then the partial differential equations have to be replaced by equivalent finite difference equations. The utilization of numerical methods brings also problems of a mathematical nature, such as the convergence of the solutions and their stability (it is why the increment of time Δt and Δx fixed in agreement with the nature of the problem to be studied, must obey the Courant-Friedrichs-Lewy condition : $c\ \Delta t/\Delta x \leq 1$ where c is the speed of the fastest wave). In such discretum, subgrid scale phenomena, cannot be resolved explicitely and therefore must be parameterized (expressed in terms of the resolved macroscale parameters).

If spherical harmonics are used to solve these equations and only a few components are included, they give a low resolution of the physical phenomena involved and require that the eddies be described by a small number of spectral components.

As, in addition, some physical processes are not very well known, there must be a control of the physical reality of these solutions. For any time step of integration of the system (1) to (5), the balance equations for angular momentum, mass and energy must thus be satisfied, as they are all at the basis of the equations (1) to (5).

In our rotating coordinate system, the time rate change of total angular momentum for a unit volume, $M = \Omega a^2 \cos^2 \phi + u a \cos \phi$, is given by :

$$\rho \frac{dM}{dt} = - \frac{\partial p}{a \cos \phi \, \partial \lambda} a \cos \phi + F_\lambda a \cos \phi \qquad (9)$$

Therefore, for the atmosphere and the oceans, only pressure and friction torques are important to generate absolute angular momentum, and the sources and the sinks are to be found near the Earth's surface. Since the surface stress τ_0 is directed against the surface winds (τ_0 is counted positive when the winds are from the east), the tropics act as the main source of absolute angular momentum for the atmosphere, and the mid latitudes, where the westerlies predominate, as the principle sink.

In the atmosphere, the meridional transport occurs mainly in the upper troposphere with a strong divergence from the tropics. In fact, it is the transient(*) waves in the upper atmosphere which dominate by far the transport (process of negative viscosity of Starr), the meridional circulations being the dominant mechanism in the lower part of the atmosphere only. As shown in Fig. 9, in the northern hemisphere for example, the maximum northward angular momentum transport by turbulent eddies $[\overline{u'v'}]$ is indeed located just south of the maximum zonal angular velocity about the polar axis $u/\cos \phi$. To close the cycle, it is necessary that the angular momentum return from middle to low latitudes within the oceans and/or the continents. Over land, not only surface friction but also mountain torques due to pressure differences across mountain ranges play an important role for transfering angular momentum between the land and the atmosphere. Over the oceans, the angular momentum exchanges with the atmosphere generate the main ocean surface currents. Finally it is shown that the seasonal balance of the total climatic system relative angular momentum has to be maintained by changes in the rotation rate of the Earth (there is indeed a superrotation of the atmosphere during northern winter with a corresponding lengthening of the day by about 0.7 ms going from July to January (Rosen and Salstein, 1983)).

The main source of energy for the climatic system being the Sun, the forms of energy that determine the energetics of the climatic system are the solar and terrestrial radiant energy, the potential energy Φ, the internal energy I, the latent heat LH, and the kinetic energy K per unit mass :

(*) We will see in the next chapter that the total meridional transport of the relative angular momentum can be broken down into transient eddy, stationary (stanting) eddy and mean meridional circulation : $[uv] = [\overline{u'v'}] + [\overline{v}^* \overline{u}^*] + [\overline{u}][\overline{v}]$.

$$\Phi = g\,z \qquad I = C_v\,T, \qquad H = Lq$$

$$K = \frac{1}{2}\,\vec{V}.\vec{V} \tag{10}$$

$$E = K + \Phi + I + H$$

E is the total energy and the total potential energy $\Phi + I$ is introduced as in an atmosphere in hydrostatic equilibrium the internal energy for a whole unit-area column is proportional to its potential energy :

$$\langle I \rangle = \langle \Phi \rangle \frac{C_v}{R} \qquad \text{(in the absence of topography)}$$

$$\text{and} \quad \langle I \rangle + \langle \Phi \rangle \quad \langle H \rangle \qquad \langle H \rangle = \langle \Phi \rangle \frac{C_p}{R} \tag{11}$$

According to the fundamental set of equations, the corresponding balance equation is then :

$$\frac{d}{dt}\,(\Phi + I) = gw + Q - p\,\alpha\,\nabla.\vec{V} \tag{12}$$

$$\text{or} \qquad = gw + Q - \alpha\,\vec{V}.\nabla p - \alpha\,\nabla.(p\vec{V})$$

For the kinetic energy :

$$\frac{dK}{dt} = -gw - \alpha\,\vec{V}.\nabla p - \alpha\,\nabla.(\nabla.\tau) \tag{13}$$

$$\text{or} \qquad = -\,gw + p\,\alpha\,\nabla.\vec{V} - \alpha\,\nabla.(p\vec{V} + \tau.\vec{V}) + \alpha\,\tau \odot (\nabla\vec{V})$$

where the different terms on the right hand side of this last equation stand for, respectively, the rate of work done against gravity and performed by expansion against pressure, the work at the boundaries done by pressure and friction and the frictional heating.

For the latent heat we have

$$L\,\frac{dq}{dt} = L\,(e-r) \tag{14}$$

Fig. 10 illustrates the interpretation of the energy balance equations showing the meaning of the various source, conversion and sink terms of energy in the system (10 to 14).

For the total energy E, we obtain :

$$\frac{dE}{dt} = -\alpha\,\nabla.p\vec{V} - \alpha\,\vec{V}.(\nabla.\tau) + L(e-r) + Q \tag{15}$$

where e is the rate of evaporation and r the rate of precipitation.

$$\text{Noting that} \quad Q = -\alpha\,\nabla.\vec{F}_{rad} - L(e-r) - \alpha\,\tau \odot (\nabla\vec{V}) \tag{16}$$

which gives respectively the radiational, latent and frictional heatings, (15) becomes :

$$\frac{dE}{dt} = -\alpha \ \nabla . p\vec{V} - \alpha \ \nabla . \vec{F}_{rad} - \alpha(\nabla . \tau) . \vec{V} \quad (17)$$

showing that the net rate of energy production by unit mass derives from the work done at the boundaries by the pressure and the friction forces and from radiative heating. However, we usually do not study the properties of a fluid following the motion of the individual particles according to a Lagrangian scheme. Instead, the Eulerian approach is used to study the behaviour of the fluid at a fixed point in space and time. This is accomplished through the use of the following expression for a climatic variable A :

$$\frac{dA}{dt} = \frac{\partial A}{\partial t} + (\vec{V} . \nabla) A \quad (18)$$

which combined with the equation of continuity leads to the general balance equation :

$$\frac{\partial \rho A}{\partial t} = \rho \frac{dA}{dt} - \nabla . (\rho A \vec{V}) \quad (19)$$

(19) shows that the local rate of change of A per unit volume results from the generation or destruction of A, and the net inflow across the boundaries into the volume.

10. CONCLUSIONS

We have seen how the different parts of the climate system are connected in complex ways. The physics of climate is studied using some of the basic laws which are a consequence of the conservation principles of angular momentum, energy and mass. Averaging these equations in space and time will now lead to the climate equations which are an essential tool for conducting experiments on the climate system. Mathematical models (Berger and Tricot, 1989) will therefore allow to assess the sensitivity of climate to variations in external forcing and to understand better the nonlinear feedback processes operating in the climatic system.

There remains however many unresolved issues related to the climate processes and to the coupling of the subsystems. Such problems are : the parameterizations of subgrid-scale phenomena into the equations, the climate response to clouds and changes in clouds amount, types, distribution and properties, the role of various chemical substances released into the atmosphere (their cycle and their chemical reactions), the representation of the land surface processes and their interactions with the hydrological cycle, the snow and ice budgets and the long-term variations in the solar forcing. But the most important unknown component of the climate system is the ocean. Large inter-annual anomalies in their heat capacity and their poleward transport of heat must be measured with a higher resolution and a better accuracy. Worldwide teleconnections must still be understood, such as in the case of the El Nino-Southern Oscillation phenomena which might also affect the weather in middle latitudes. Coupled atmosphere-ocean models are now being developed but more complete models taking also into account the cryosphere, the lithosphere and the biosphere are still in a primitive stage of development and will require a huge effort of observation and modelling in the next decades (IGBP, 1988).

11. REFERENCES

Ad Hoc Planning Group on Global Change (1986). The International Geosphere-Biosphere Programme : A study of Global Change. ICSU, Paris.

Atkinson B.W. (1981). Dynamical Meteorology. Methuen, London.

Bach W. (1984). Our Treatened Climate. Reidel Publ. Company, Dordrecht, Holland.

Berger A. (1979). Spectrum of climatic variations and their causal mechanisms. Geophysical Surveys, 3, pp. 351-402.

Berger A. (Ed.) (1981). Climatic Variations and Variability : Facts and Theories. NATO ASI Series C vol. 72, D. Reidel Publ. Company, Dordrecht, Holland.

Berger A. (1988). Milankovitch Theory and Climate. Review of Geophysics.

Berger A., Imbrie J., Hays J., Kukla G., and Saltzman B. (Eds) (1984). Milankovitch and Climate. NATO ASI Series C vol. 126, D. Reidel Publ. Company, Dordrecht, Holland.

Berger A., and Tricot Ch. (1989). Simple models for climate simulation. This volume.

Berger W.H., and Labeyrie L.D. (1987). Abrupt Climatic Change, Evidence and Implications. D. Reidel Publ. Company, Dordrecht, Holland.

Bradley R.S. (1985). Quaternary Paleoclimatology. Methods of Paleoclimatic Reconstruction. Allen and Unwin, Boston, 472pp.

Budyko M.I. (1977). Climatic Changes. American Geophysical Union, Washington D.C.

Crowley T.J. (1983). The geologic record of climatic change. Reviews of Geophysic and Space Physics, 21 n°4, pp. 828-877.

Dickinson R. (1986). How will climate change ? In : The Greenhouse Effect, Climatic Change and Ecosystems. B. Bolin, B. Döös, J. Jaeger, R. Warricck (Eds), SCOPE 29, pp. 207-270, John Wiley, New York.

Earth System Sciences Committee (1988). Earth System Science, A Closer View. NASA Advisory Council, NOAA, Washington D.C.

Gates W.L. (1981). The climate system and its portrayal by climate models : a review of basic principles. I. Physical basis of climate. II. Modeling of climatic changes. In : Climatic Variations and Variability : Facts and Theories, A. Berger (Ed.), pp. 3-39 and 435-459, D. Reidel Publishing Company, Dordrecht, Holland.

Gill A. (1982). Atmosphere-Ocean Dynamics. International Geophysics Series, vol. 30, Academic Press, New York.

Hansen J.E., and Takahashi T. (Eds) (1984). Climate Processes and Climate Sensitivity. Geophys. Monogr. Series, vol. 29, AGU, Washington D.C.

Hecht A. (Ed.) (1985). Paleoclimate Analysis and Modeling. John Wiley and Sons Limited, New York.

Henderson-Sellers A., and McGuffie K. (1987). A Climate Modelling Primer. J. Wiley and Sons, Chichester.

Holton J.R. (1979). An Introduction to Dynamic Meteorology. International Geophysics Series, vol. 23, Academic Press, New York.

Houghton J.T. (Ed.) (1984). Global Climate. Cambridge University Press, Cambridge.

Houghton J.T. (1986). The Physics of Atmosphere. Cambridge University Press, Cambridge.

International Geosphere Biosphere Programme (1988). Plan for Action. Report n°4, IGBP Secretariat, Suède.

Lamb H.H. (1972, 1977). Climate : Present, Past and Future. Vol. 1 - Fundamentals and Climate Now; Vol. II - Climatic History and the Future, Methuen and Co Ltd, London.

Lieth C.E. (1978). Predictability of Climate. Nature, 276, pp. 352-355.

Lorenz E.N. (1968). Climatic Determinism. Meteorological Monograph, 8(30), pp. 1-3.

Lorenz E.D. (1976). Nondeterministic theories of climatic change. Quaternary Research, 6 n°4, pp. 495-507.

Lorius Cl., Barkov N.I., Jouzel J., Korotkevitch Y.S., Kotlyakov V.M. and Raynaud D. (1988). Antarctic ice core : CO_2 and climatic change over the last climatic cycle. EOS, 69 n°26, pp. 681, 683-684.

Manabe S. (Ed.) (1985). Issues in Atmospheric and Oceanic Modeling. Part A : Climate Dynamics; Part B : Weather Dynamics. Advances in Geophysics, vol. 28, Academic Press, Inc., Orlando, Florida.

Meehl G.A. (1984). Modeling the Earth's climate. Climatic Change, 6(3), pp. 259-286.

Milankovic M. (1941). Kanon der Erdbestrahlung und Seine Anwendung auf das Eiszeitenproblem. Koniglich Serbische Akademie, Beograd. (English translation by Israel progr. for Scientific Translation and published for the US Department of Commerce and the National Science Foundation : Canon of Insolation and the Ice Age Problem, 1969).

Mitchell J.M.Jr. (1976). An overview of climatic variability and its causal mechanisms. Quaternary Research, 6 n°4, pp. 481-495.

Monin A.S. (1986). A Introduction to the Theory of Climate. D. Reidel Publishing Company, Dordrecht, Holland, 272pp.

Morel P. (1986). Introduction à la Dynamique de l'Atmosphère et des Océans. Oceanus, 12 (fasciule hors-série).

Peixoto J.P., and Oort A.H. (1984). Physics of Climate. Reviews of Modern Physics, 56(3), pp. 365-429.

Pestiaux P., van der Meersch I., Berger A., and Duplessy J.Cl. (1988). Paleoclimatic variability at frequencies ranging from 1 cycle per 10 000 years to 1 cycle per 1 000 years : evidence for nonlinear behaviour of the climate systems. Climatic Change, 12 n°1, pp. 9-37.

Rosen R.D., and Salstein D.A. (1983). Variations in atmospheric angular momentum on global and regional scales, and the length of the day. J. Geophys. Res., 88, p. 5451.

Schlesinger M. (Ed.) (1988). Physically-based Modelling and Simulation of Climate and Climatic Change. NATO ASI Series C, vol. 243, Kluwer Academic Publishers, Dordrecht, Holland.

Starr V.P. (1968). Physics of Negative Viscosity Phenomena. McGraw-Hill Book Company, New York.

Washington W.M., and Parkinson C.L. (1986). An introduction to Three-Dimensional Climate Modeling. Univ. Sc. Books, Mill Valey, Calif., and Oxford Univ. Press, N.Y., 422p.

Wells N. (1986). The Atmosphere and Ocean. Taylor and Francis, London.

Wigley T.M.L. (1981). Climate and paleoclimate : what we can learn about solar luminosity variations. Solar Phys., 74, pp. 435-471.

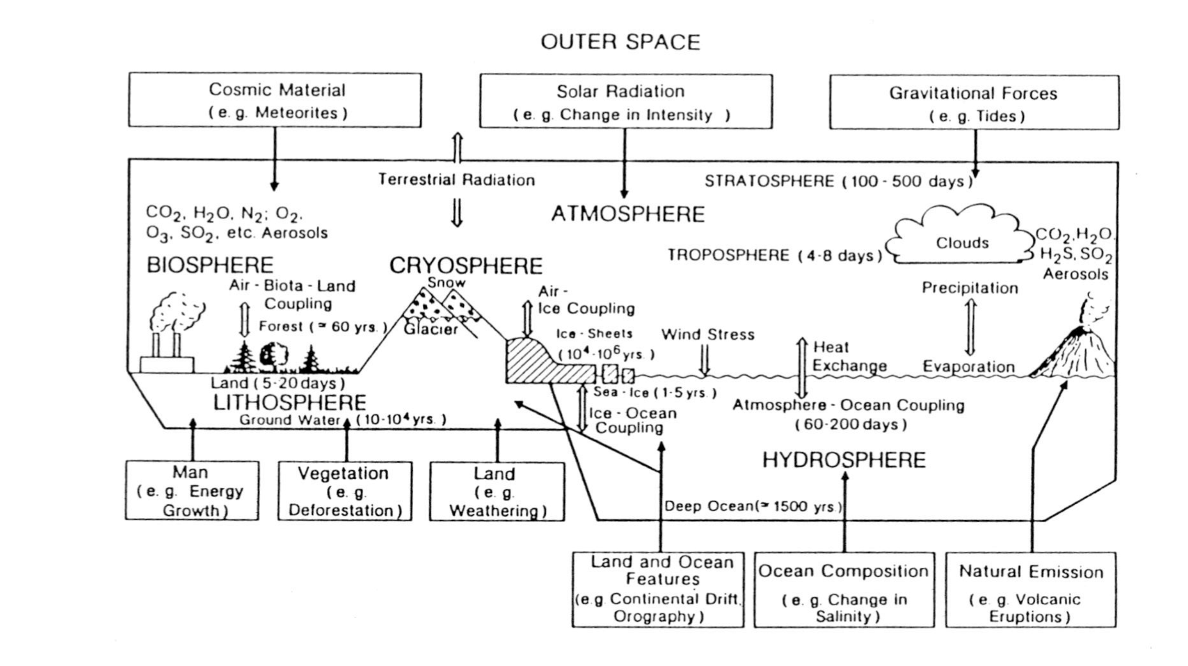

Figure 1. The climate system, its components and their characteristic time (Bach, 1984). (reprinted from Our Threatened Climate, by W. Bach, 1984, with permission from the author and Reidel Publishing Company and Verlag C.F. Müller GmbH Karlsruhe, first published in Gefahr für unser Klima, Bach W., 1982, Verlag Mueller).

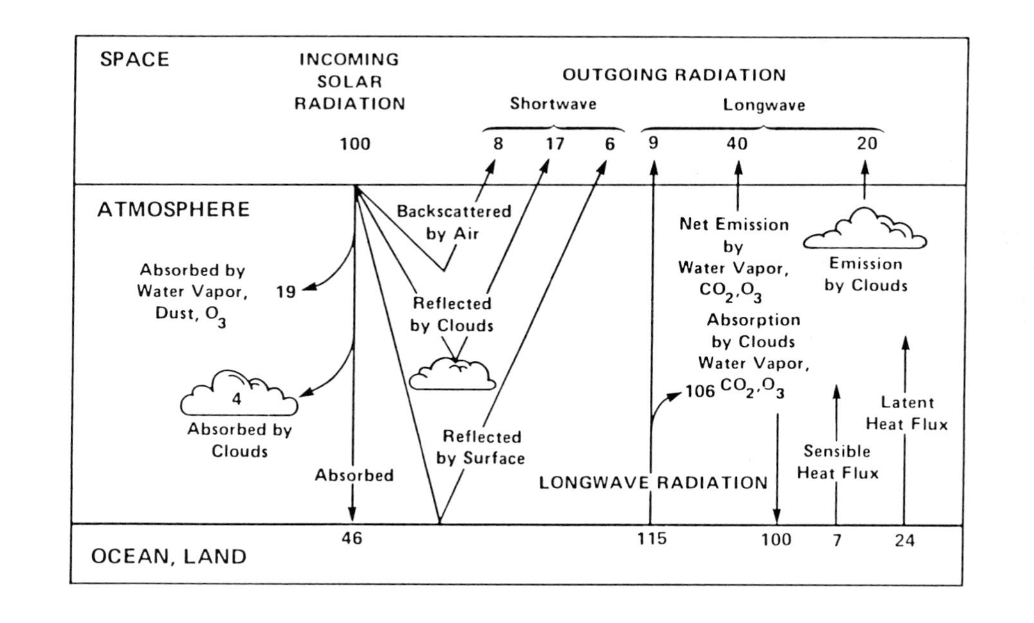

Figure 2. Global average components of the Earth's energy balance. The incoming solar radiation (100%) is roughly equal to 340 Wm^{-2} and the planetary albedo 31 %. The internal plus potential energy amount 69 % (solar and infrared radiation absorbed by the atmosphere, 23 and 15 % respectively, latent and sensible heat, 7 and 24 %). It equals the infrared radiation lost to space by the atmosphere-surface system.

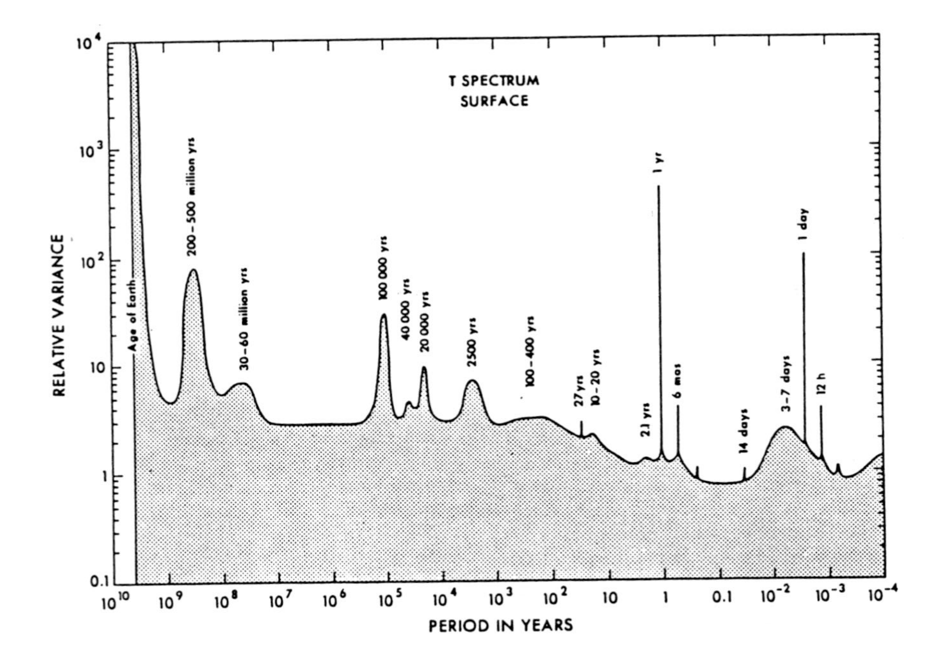

Figure 3. Idealized schematic spectrum of atmospheric temperature between 10^{-4} and 10^{10} years (Mitchell, 1976).

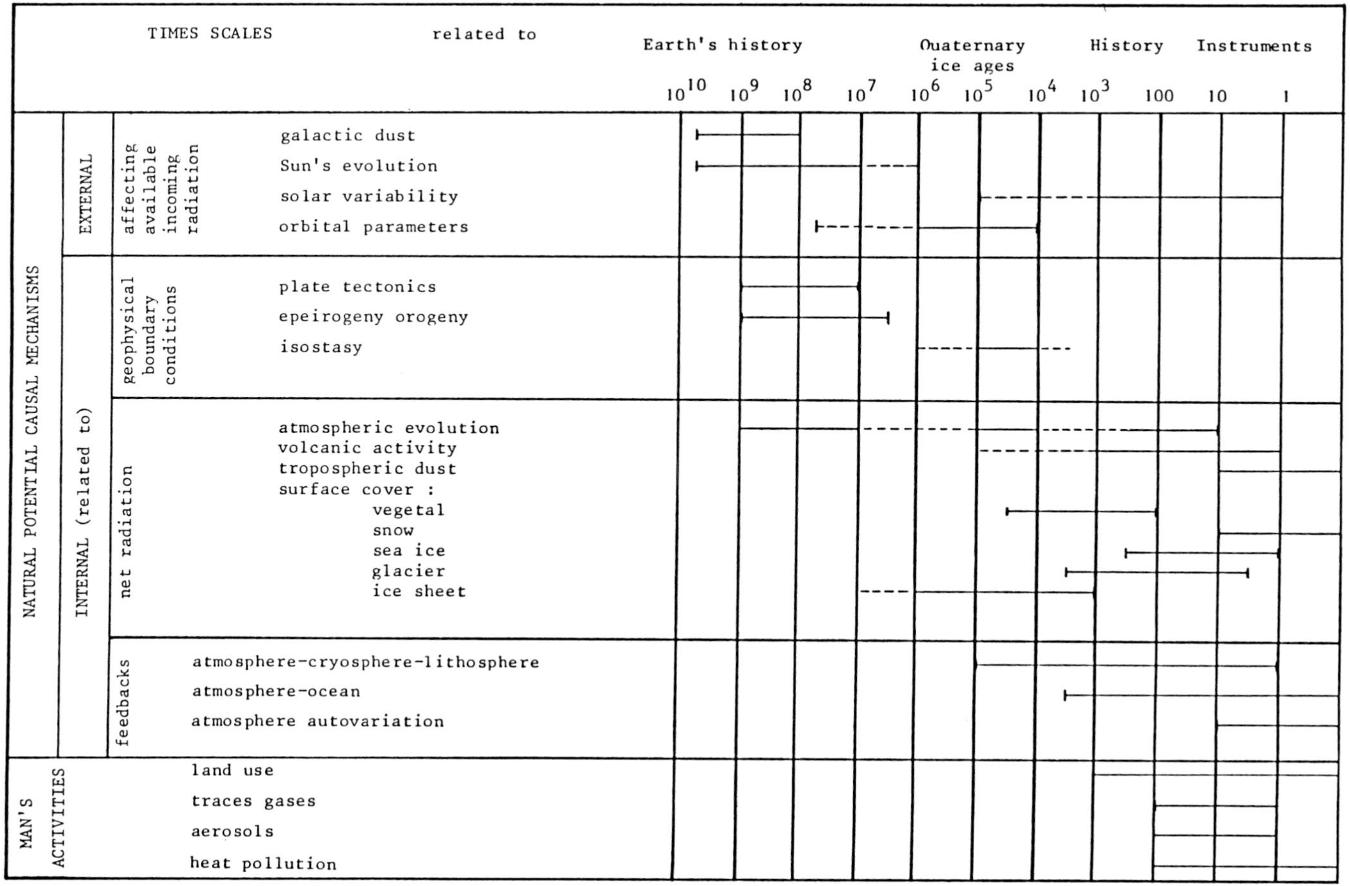

TIMES SCALES related to: Earth's history — Quaternary ice ages — History — Instruments

10^{10} 10^{9} 10^{8} 10^{7} 10^{6} 10^{5} 10^{4} 10^{3} 100 10 1

NATURAL POTENTIAL CAUSAL MECHANISMS

EXTERNAL — affecting available incoming radiation
- galactic dust
- Sun's evolution
- solar variability
- orbital parameters

INTERNAL (related to) — geophysical boundary conditions
- plate tectonics
- epeirogeny orogeny
- isostasy

INTERNAL (related to) — net radiation
- atmospheric evolution
- volcanic activity
- tropospheric dust
- surface cover :
  - vegetal
  - snow
  - sea ice
  - glacier
  - ice sheet

INTERNAL (related to) — feedbacks
- atmosphere-cryosphere-lithosphere
- atmosphere-ocean
- atmosphere autovariation

MAN'S ACTIVITIES
- land use
- traces gases
- aerosols
- heat pollution

Figure 4. Causes of climatic variations (Berger, 1981).

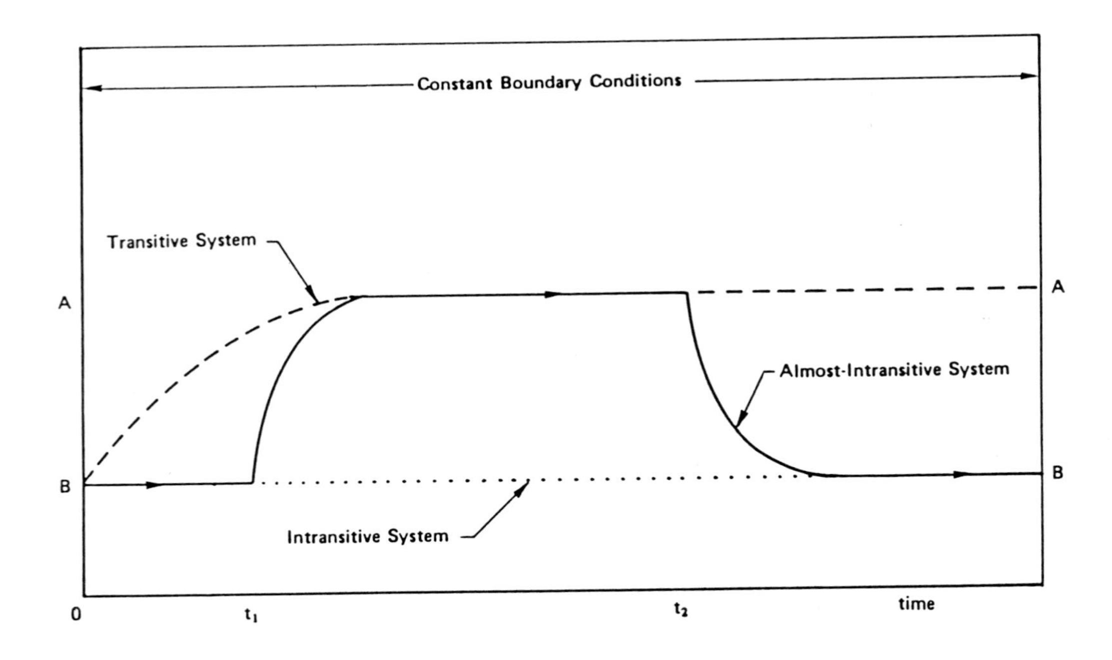

Figure 5. Illustration of the behaviour of transitive, intransitive and almost-intransitive climatic systems with respect to an initial climatic state B. The climatic state A is an alternative bistable state under the same boundary conditions (Lorenz, 1969 and 1976).

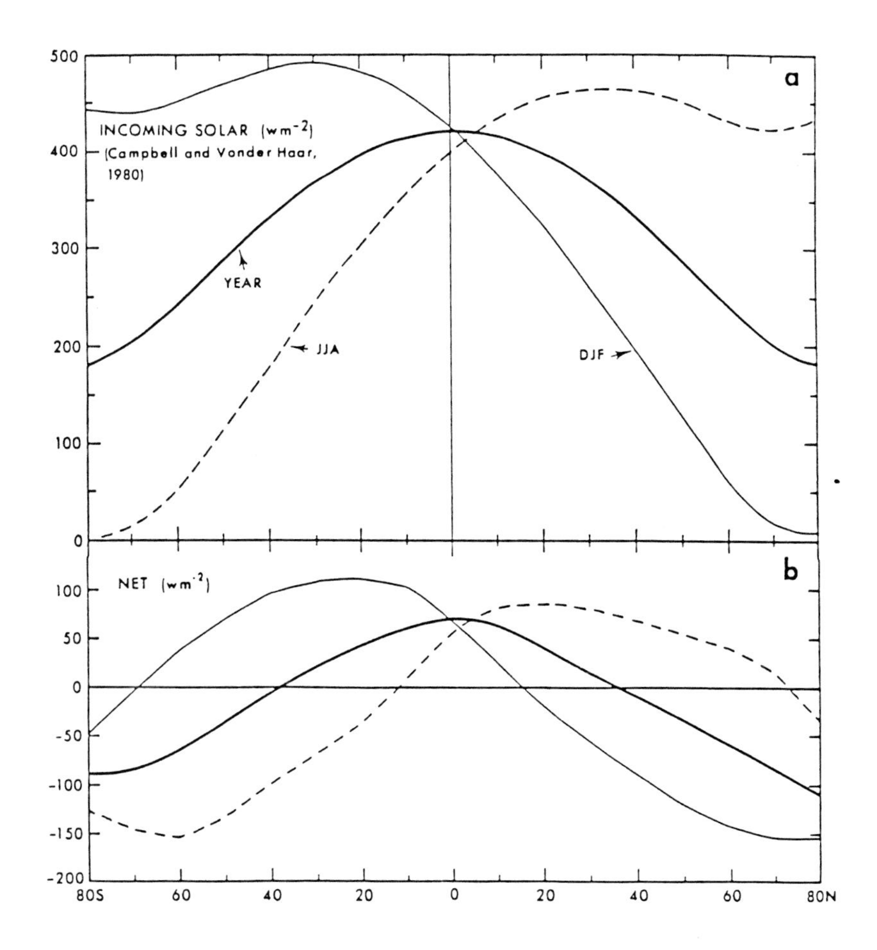

Figure 6. Latitudinal profile of incoming solar radiation (a) and net incoming radiation (b) (mean annual and seasonal). Units are Wm^{-2}. (reproduced by permission of the authors and American Physical Society from Peixoto and Oort, 1984).

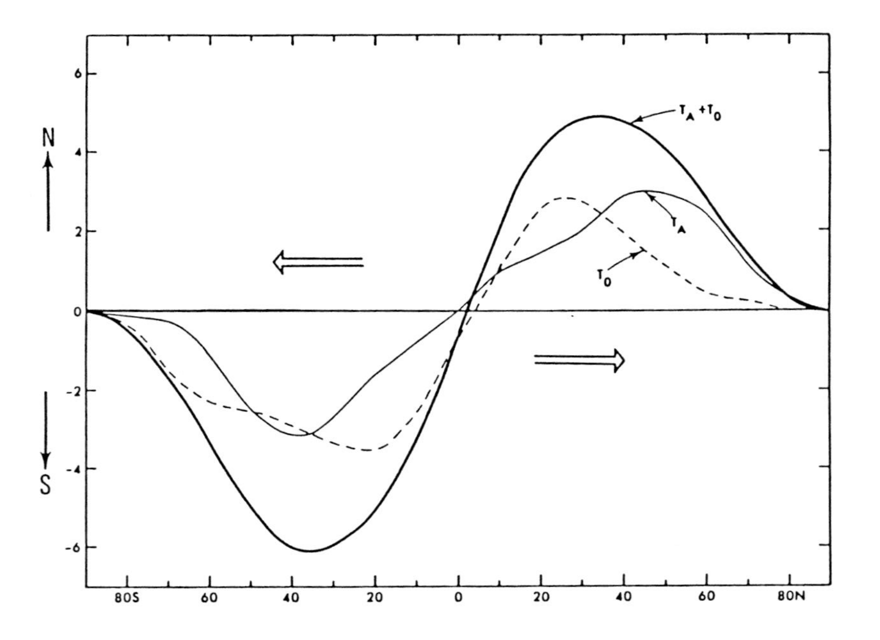

Figure 7. Annual zonal mean northward transports of energy in the atmosphere-ocean system (T_O + T_s) based on radiation measurements in the atmosphere (T_A). T_O is inferred as a residual. Units in 10^{15} W. (reproduced by permission of the authors and American Physical Society from Peixoto and Oort, 1984).

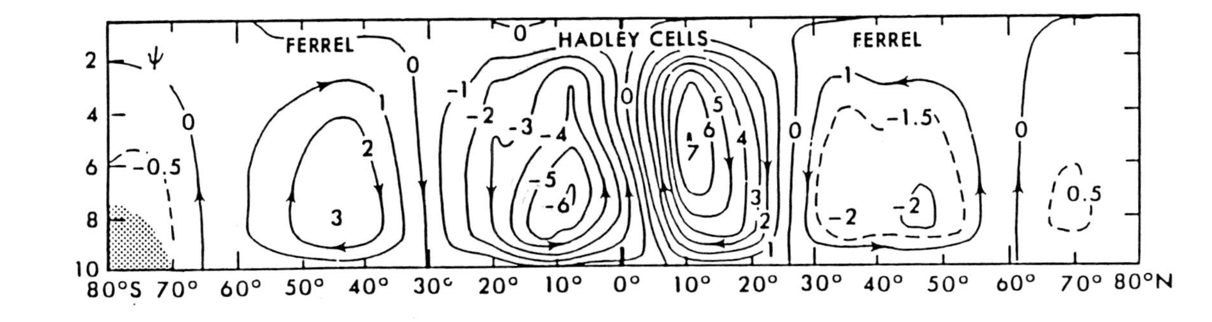

Figure 8. Zonal mean cross-section of flow of atmospheric mass in 10^{10} kg s^{-1} for annual mean and zonally averaged conditions (Peixoto and Oort, 1984).

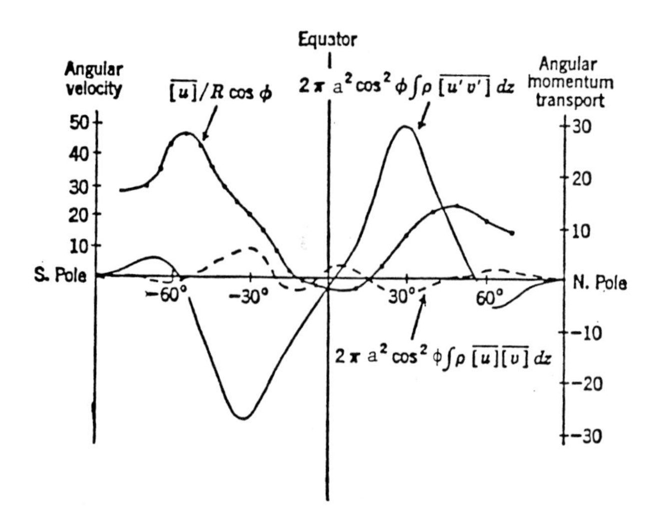

Figure 9. Relative mean angular velocity about the polar axis ($[u]/a\cos\phi$), the northward angular-momentum transport by turbulent eddies : $2\pi a^2 \cos^2\phi \int \rho\ \overline{[u'v']}\ dz$, and the angular momentum transported by the mean meridional circulation : $2\pi a^2 \cos^2\phi \int \rho\ [\overline{u}][\overline{v}]\ dz$. The values are calculated over the entire mass of the atmosphere (Starr, 1968).

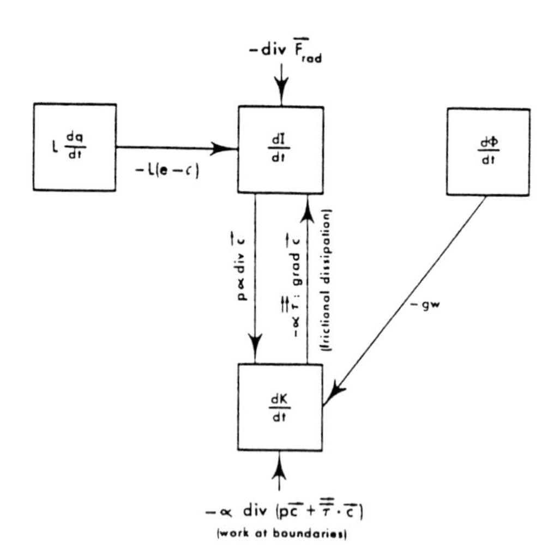

Figure 10. Box diagram showing the connections between the various forms of energy in the atmosphere. (reproduced by permission of the authors and American Physical Society from Peixoto and Oort, 1984).

SIMPLE CLIMATE MODELS

A. BERGER and Ch. TRICOT
Université Catholique de Louvain
Institut d'Astronomie et de Géophysique G. Lemaître
2 Chemin du Cyclotron
B-1348 Louvain-la-Neuve, Belgium

Summary

Climate being an extraordinary complex physical problem, the study of the main processes responsible for the climate structure and for the climatic variations requires to construct mathematical models in order to represent quantitatively the various interacting elements. These climate models can be subdivided into different categories according to different specific criteria like the dimensionality, the main physics involved and the time dependency of their simulation. So there are snapshot models providing the climatic equilibrium under some particular boundary conditions and models providing the transient response of the climate system to forcing; there are also general circulation and statistical-dynamical models; and finally 0-D to 3-D models involving only one component of the climate system (as AGCM) or considering some or all components and their coupling.

In this view, "simple models" means those where some physical processes have been parameterized in order to speed up the computation and reduce the complexity of all the interactions.

1. INTRODUCTION

It is already in the latter half of the nineteenth century that the first applications of very simple climate models were made. Tyndall (1863) and Arrhenius (1896) suggested indeed that a climate change may be induced by a change of CO_2 concentration in the atmosphere (Manabe, 1983). Their calculations were based on the radiative heat budget at the Earth's surface, as did later Callendar (1938) and Plass (1956), in particular. In the early part of the twentieth century, Milankovitch (1920) had developed a full study of a grey atmosphere forced by the astronomical insolations to test his hypothesis that cool summers in high northern hemisphere might be at the origin of the Quaternary glacial-interglacial cycles (Berger, 1988; Berger and Andjelic, 1988).

It is in the late 1960s, with the advent of modern computers, that the first atmospheric general circulation models were being developed. They were derived from numerical models of the atmosphere designed for weather forecasting (Smagorinsky, 1963 and 1983). Concurrently the first radiative-convective models were developed for studying the thermal equilibrium of the Earth-atmosphere system (Manabe and Stricker, 1964).

During the same time, other global climate models were also designed to simulate the climatic system. Adem (1964, 1988) described the first low-resolution thermodynamic model which includes in a highly parameterized way, many dynamic, radiative and surface features and feedback effects.

Budyko (1969) and Sellers (1969) built two very similar energy balance models which simple parameterization schemes have allowed long time scale changes to be simulated and stability of alternate climatic states to be investigated.

The statistical dynamical models (Saltzman, 1964, 1978) then appeared to account for the observed state of averaged atmospheric motion, temperature and moisture on time scales shorter than seasonal but longer than those characteristic of weather systems. Theoretical studies of large scale atmospheric eddies and their transfer properties combined with observational work (Peixoto and Oort, 1984) led to the parameterizations employed in two-dimensional climate models (North et al., 1983; Stone and Mao-Sung Yao, 1987; Berger et al., 1988).

By 1980 most of the computational power used by climate modellers was being consumed by general circulation models (Washington and Parkinson, 1986; Mitchell, this volume), although "simple" models must still be used to try to isolate essential processes responsible for the results which are observed from these more comprehensive, highly non-linear and highly complex models (Henderson-Sellers and McGuffie, 1987).

A recent desire to better understand global changes make in the climate system (IGBP, Special Committee, 1988) has led to the building of general climate models where all the five components, atmosphere, hydrosphere, cryosphere, lithosphere and biosphere, will be included, taking into account their individual complex behaviour but also the non-linear interactions between them.

Simple zero- and one-dimensional models will be presented in this review as an illustration of how to approach the highly complex and non-linear climate system.

2. CLIMATE MODELING

2.1. Averaging the balance equations

Climate is assumed to represent a statistical, quasi steady state of the atmosphere that depends only on the boundary conditions imposed by the external subsystems. Thus the thermodynamical equations have to be prepared through adequate averaging in time and space, leading to the climate equations.

The fields characterizing the state of the atmosphere being quite variable in time, it is important to determine the mean state of the atmosphere to describe the climate. Introducing a time average operator A and its departure A', the balance equations can be rewritten in the following way :

1. For the angular momentum (see eq. 9 in Berger, this volume)

$$\overline{\frac{\partial \rho M}{\partial t}} = - \nabla . \, \bar{M} \, \overline{\rho \vec{V}} - \nabla . \overline{M'(\rho \vec{V})'} - \frac{\partial \bar{p}}{\partial \lambda} - R \cos \phi \, \frac{\partial \bar{\tau}_{zx}}{\partial z} \qquad (1)$$

2. For potential energy ($\frac{d\Phi}{dt} = g\, w$)

$$\overline{\frac{\partial \rho \Phi}{\partial t}} = -\nabla . \bar{\Phi} \, \overline{\rho \vec{V}} + g \, \overline{\rho w} \quad (2)$$

3. For internal energy ($\overline{\frac{dI}{dt} = Q - p\alpha \, \nabla . \vec{V}}$)

$$\overline{\frac{\partial \rho I}{\partial t}} = \nabla . \bar{I} \, \overline{\rho \vec{V}} - \nabla . \overline{I'(\rho \vec{V})'} + \overline{\rho Q} - \overline{p \nabla . \vec{V}} \quad (3)$$

4. For kinetic energy (eq. 13, Berger, 1989, this volume)

$$\overline{\frac{\partial \rho K}{\partial t}} = -\nabla . \bar{K} \, \overline{\rho \vec{V}} - \nabla . \overline{K'(\rho \vec{V})'} - \overline{\vec{V} . \nabla p} - \overline{\vec{V} . \nabla \tau} - g \overline{\rho w} \quad (4)$$

5. For total energy (eq. 15, Berger, 1989, this volume)

$$\overline{\frac{\partial \rho E}{\partial t}} = -\nabla . \bar{E} \, \overline{\rho \vec{V}} - \nabla . \overline{E'(\rho \vec{V})'} - \nabla . \bar{J}_q \quad (5)$$

6. and for water vapor (eq. 14, Berger, 1989, this volume)

$$\overline{\frac{\partial \rho q}{\partial t}} = -\nabla . \bar{q} \, \overline{\rho \vec{V}} - \nabla . \overline{q'(\rho \vec{V})'} + \overline{(e-r)} \quad (6)$$

where

$$J_q = F_{rad} + p\vec{V} + \tau . \vec{V} \quad (7)$$

is the sum of the radiation flux and the mechanical energy fluxes by pressure work and frictional stresses. The terms involving products of departures are the transient eddy terms.

The predominance of zonal symmetry of the various climatic quantities with respect to the rotation axis of the Earth made it desirable to introduce the zonal averaging operator [A] and the departure A^*. These departures from zonal symmetry are due not only to anomalies in the thermal and mechanical surface boundary conditions but also to inherent hydrodynamic instabilities in the atmosphere associated with the Earth's rotation rate and the equator-to-pole heating gradient. The zonally-averaged set of equations involving stationary or standing eddies can be obtained from the time mean equations by an analogous procedure (Van Mieghem, 1973). This breakdown involving the zonal and time mean conditions is useful to elucidate the dominant mechanisms responsible for achieving the balances.

2.2. Cycle of energy in the atmosphere

The atmosphere is a heat engine which performs work to maintain the kinetic energy of the circulation against frictional dissipation. We have seen (Berger, 1989, this volume) that the source for kinetic energy is the total potential energy. However, only a fraction of this total potential energy is available to be converted into kinetic energy. This fraction is the available potential energy, P, which can be defined as the difference between the total potential energy of a closed system and the minimum total potential energy which could result from an adiabatic redistribution of mass. Lorenz (1955, 1967) was able to derive an approximate formula for P, involving the variance of temperature on a constant pressure level and the mean static stability.

Because of the zonal character of the general circulation it is useful to partition the kinetic energy into the kinetic energy of the mean zonal flow K_M and the kinetic energy of the perturbations K_E. We will call these components mean and eddy kinetic energy respectively. Similarly, the available potential energy may be partitioned into the components P_M and P_E.

The balance equations for the four basic forms of energy can be derived from equations 3 and 6 of Berger (1989, this volume) and can be written in a symbolic form, where G(x) indicates the rate of generation of x, C(x,y) the rate of conversion from x into y, and D(y) the rate of dissipation of y :

$$\frac{\partial P_M}{\partial t} = G(P_M) - C(P_m,K_M) - C(P_M,P_E) \quad (8)$$

$$\frac{\partial P_E}{\partial t} = G(P_E) - C(P_E,K_E) + C(P_M,P_E) \quad (9)$$

$$\frac{\partial K_M}{\partial t} = C(P_M,K_M) + C(K_E,K_M) - D(K_M) \quad (10)$$

$$\frac{\partial K_E}{\partial t} = C(P_E,K_E) - C(K_E,K_M) - D(K_E) \quad (11)$$

With this set of equations, we can describe the mechanisms by which the incoming solar radiation maintains the kinetic energy of the atmosphere against dissipation :

G(P) is proportional to the covariance of the diabatic heating. Therefore when regions of high temperature are heated and/or regions of low temperature are cooled, available potential energy (APE) is generated since the center of gravity of the atmosphere is raised. The radiative balance tends to generate an almost zonally symmetric distribution of temperature ($G(P_M)$). However, meridional transport of enthalpy by the eddies deforms this ideal zonal distribution of temperature leading to $G(P_E)$.

D(K) dissipation of K by turbulence and friction

$C(P_M,P_E)$ represents the effect of meridional heat transport. In fact, the transport of heat by the eddies from the warmer tropics to the cooler high latitudes reduces the mean north-south temperature gradient resulting in a decrease of APE_M.

$C(P_E,K_E)$ baroclinic instability (cyclogenesis). Baroclinic instability is an amplification process of the adiabatic redistribution of mass between a warm dilated air column located southward and a cold compressed one located northward, when the meridional atmospheric temperature gradient is greater than a critical value.

$C(K_E,K_M)$ in transporting angular momentum poleward, the eddies tend to intensify the mean zonal current. This is what Starr (1966) has called negative viscosity phenomena.

$C(K_M,P_M)$ the indirect circulation in midlatitudes consumes mean zonal kinetic energy at a rate slightly exceeding the production in the

Hadley cell and therefore converts some of the mean zonal kinetic energy back into mean zonal potential energy.

Schematically, the energy cycle in the atmosphere (Figs 1 and 2) proceeds from P_M to K_M through the following scheme

$$P_M \rightarrow P_E \rightarrow K_E \rightarrow K_M \rightarrow P_M$$

It therefore appears that the eddies play a crucial role in regulating the general circulation of the atmosphere.

The efficiency of the atmosphere can now be taken as the ratio of the K dissipated by friction (2 Wm^{-2}) and the mean incoming solar radiation (238 Wm^{-2}, 70% of 340 Wm^{-2}) which leads to a value of 0.8% (more realistic than the Carnot efficiency of roughly 10%).

2.3. Climate Equations

A climate model is thus a set of specialized thermo-hydrodynamic equations with prescribed boundary and initial conditions, certain given values of the physical constants and specified schemes of parameterization of the subgrid-scale fluxes of matter, momentum and energy. In studying climate, we must consider the behaviour of each individual component of the climatic system and the strong non-linear interactions between the components. As the characteristic phenomena in the various components have very different internal time scales (from 10^{-3} to 10^5 years, Fig. 1 of Berger, 1989, this volume) the integration of a fully coupled model including the atmosphere, ocean, land and cryosphere with a similar complexity as described here for the atmosphere alone, poses almost insurmountable difficulties in reaching a final solution.

Even if all interacting processes were completely understood, the model should indeed ideally be run at time steps applicable to the fastest component of the system, i.e. the atmosphere. To circumvent this difficulty, special techniques for integration of coupled systems have been used : (i) asynchronous coupling (e.g., Manabe and Bryan, 1969); (ii) regarding the influence of the rapidly varying atmosphere component on the slowly moving ones (e.g., ocean) as a stochastic process (Hasselman, 1976), or (iii) using statistical-dynamical models (it means for the atmosphere-ocean system, models where atmospheric large-scale eddy processes are parameterized and coupled to a more complete ocean model).

2.4. Hierarchy of Climate Models

To make the set of equations more tractable, a variety of interdependent assumptions and approximations (often referred to as parameterizations) must be made. The choice of assumptions selected is usually tailored to the type of problem and the resources available (Figs 3 and 4).

One class of assumptions concern the components of the climate system included explicitely in the model. In some cases, for example, only the atmosphere is treated with the ocean, land and cryosphere temperatures being parameterized.

Another class of assumptions concerns the set of equations treated by the models, it means that the hierarchy is now described in terms of the degree of their resolution and complexity. Two main classes are considered : stochastic and deterministic models.

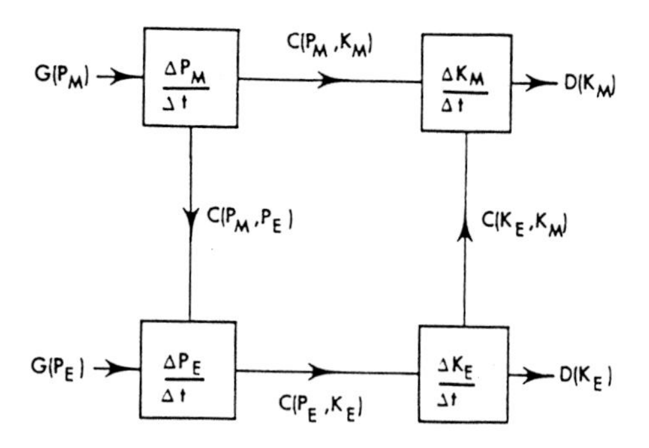

Figure 1. Box diagram of the energy cycle in the atmosphere giving the generation G, conversion C and dissipation D of available potential energy P and kinetic energy K for the mean zonal flow M and the perturbations E. (reproduced by permission of the authors and American Physical Society from Peixoto and Oort, 1984).

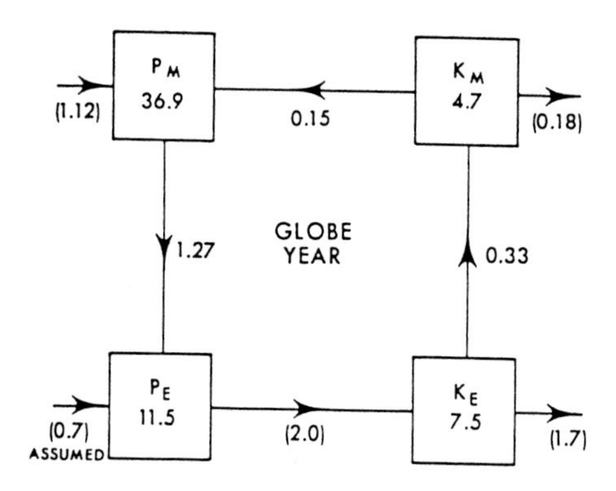

Figure 2. Observed energy cycle for the atmosphere according to Fig. 1. Energy amounts are given in 10^5 Jm^{-2} and rates in Wm^{-2}. Not directly measured terms are in parentheses (reproduced by permission of the authors and Academic Press from Oort and Peixoto, 1983).

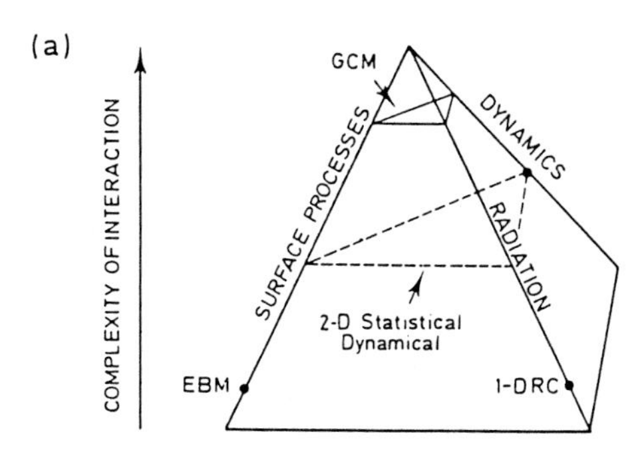

Figure 3. The climate-modelling pyramid. Energy balance models (EBM), radiative-convective models (RC) and general circulation models (GCM) are located according to the complexity with which radiation, dynamics and surface processes interact (reproduced by permission of the author and of the Royal Meteorological Society from Shine and Henderson-Sellers, 1983; also Henderson-Sellers and McGuffie, 1987).

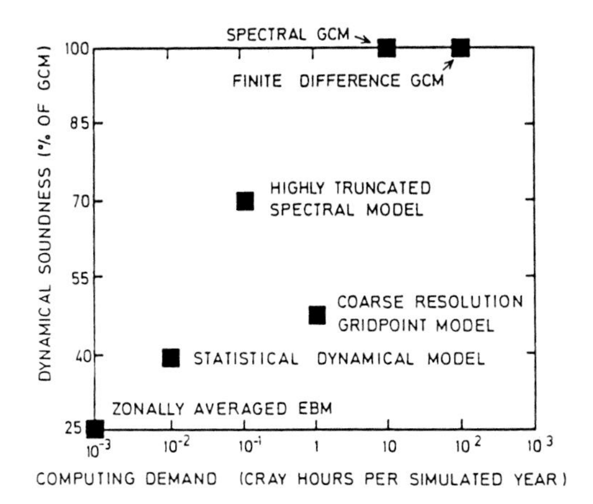

Figure 4. Computing demand versus dynamical soundness of atmospheric models (reproduced by permission of the author and American Meteorological Society from Semtner, 1984).

2.4.1. Stochastic models

As already mentioned, the stochastic models (Lemke, 1977) were developed because the physical processes in some of the subsystems cannot be modeled or parameterized in a deterministic way due to, among other factors, the large difference in internal time scales. Any perturbation leads to very different response times in each of the subsystems due to the coupling between the fast and slow subsystems. The evolution of the slowly changing variables can be expressed in terms of the faster variables using stochastic differential equations, such as the Fokker-Planck equation.

In the atmosphere-ocean internal system, the atmosphere acts as a white-noise signal generator (bounded by the prescribed variance of the fluctuations according to observations) which leads to a low-frequency integral response in the oceans.

2.4.2. Deterministic models

The important components to be discussed in constructing a deterministic model of the climatic system are at the present : radiation, dynamics and surface processes which describe the interchanges between the subsystems (Fig. 5). The relative importance of these processes in different types of models can be discussed using the Henderson-Sellers climate-modelling pyramid (Fig. 3). This figure illustrates also the relative performance against the computational effort expended (Fig. 4).

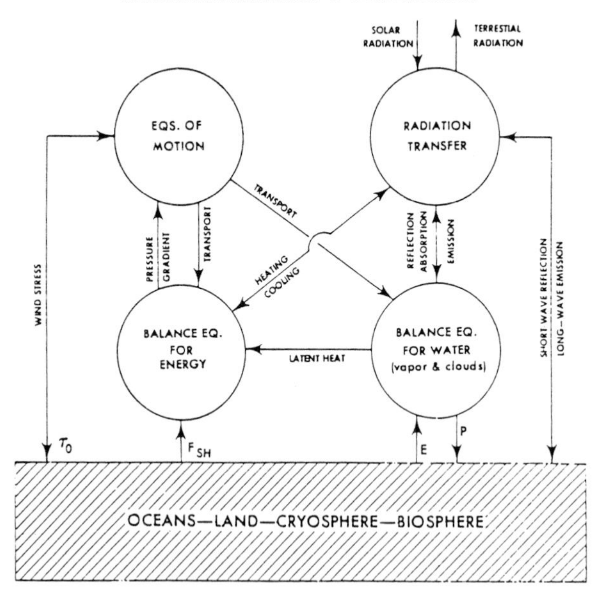

Figure 5. Major components of a mathematical climate model (reproduced by permission of the authors and American Physical Society from Peixoto and Oort, 1984).

1. General Circulation Models.

The general circulation models (GCM) are three dimensional time-dependent models that explicitely includes the weather equations. The evolution of the large scale eddy circulation is followed on a day-to-day basis. Although the GCMs treat all three dimensions of the atmosphere (Gilchrist, 1979) and/or oceans (Robinson et Baker, 1979; Semtner and Chervin, 1988), typically with a resolution of a few degrees in latitude and longitude and with numerous layers in the vertical, and attempt to incorporate all relevant physical processes that are known to be important in the real atmosphere, uncertainties in the observed initial conditions bring theoretical limit for locally accurate forecasts (a few days) and for prediction of major global features (a few weeks). However, many aspects of the space and time statistics of the model's atmosphere compare well with observed statistics, if the model is "properly constructed and suitably complete" through a verification-calibration adjustment (MacCracken and Luther, 1985).

In most climate studies, the model simulations of the present climate are referred to as the control simulations. The initial approach to studying climatic effects is then to change a parameter and to determine the new equilibrium state, refered to as the perturbed state.

The difference between the perturbed and control simulations is a measure of the climate sensitivity to the selected perturbation (Washington and Parkinson, 1986; Schlesinger and Mitchell, 1987). These arbitrary perturbations are usually chosen to be large enough so that a statistically significant difference (signal) can be seen in the model's results, particularly in GCMs, which have an internal or natural climatic variability (noise) introduced by the weather the real atmosphere (Chervin, 1981; Rind et al., 1989).

Because GCMs require extensive computer time, they have been used primarily for sensitivity studies involving arbitrary, step function changes (i.e., discontinuous in time) in various parameters. In making these simulations, the GCMs calculate the transient changes in response to step function forcing. Unless a succession of very small increments is treated, however, this transient response does not necessarily represent the manner in which the real atmosphere would respond e.g., to a continuously changing concentration because different climate system components are responding at different rates than they normally would, and adjustments (or approximations) are often made in the models to achieve a more rapid approach to climatic equilibrium (Schlesinger and Mitchell, 1987) than would actually be the case.

Instead of studying the evolution of the model climate to time-dependent forcing, interpolations of various types (typically logarithmic when dealing with CO_2 concentration) are usually made. The difficulty with such interpolations, however, is that the climate system has many time constants, and simple interpolations, especially of seasonally dependent effects, may result in misleading estimates.

2. Statistical-dynamical models

They are based on averaged equations and the effects of large eddies are parameterized in terms of mean quantities (Schneider and Dickinson. 1974). These models can be divided into thermodynamic or energy balance models, EBM (North et al., 1981), which are based solely on the first law of thermodynamics, and into momentum models which are based on the same first law completed with the parameterized meridional transport of momentum and heat.

More classes of climate models can then be obtained by considering the modes of averaging in space. Starting with 3-dimensional climate fields, one might :

- average in the vertical : 2 - (λ ,ϕ) models (Adem, 1964; North et al., 1983)

- average in the zonal direction : 2 - (ϕ,z) axially symetric models (Ohring and Adler, 1978; Berger et al., 1988). These can be associated to 2 - (ϕ,z) energy balance and radiative-convective models (Held and Suarez, 1974; Peng et al., 1982; Wang et al., 1984)

- average once more : 1 - λ longitudinal model
  1 - ϕ latitudinal model (Sellers, 1969; Budyko, 1969)
  1 - z radiative-convective model, RCM (Ramanathan and Coakley, 1978; Ramanathan et al., 1987)

- integrate these last models with respect to the remaining dimension leading to global mean or zero-dimensional models (Ghil, 1981; 1985).

As the more averaging operations are performed, the lower the resolution needed, but the more eddy components have to be parameterized, GCMs and 0-D climate models require respectively the least and the most amount of parameterization.

EBMs tend to treat the Earth globally or to divide it into latitudinal bands but do not treat vertical variations. RCMs divide the atmosphere vertically assuming buoyantly stable stratification, but do not treat horizontal variations. They are particularly well suited to detailed study of the radiative effects of trace gases and of chemical interactions among species.

Analyses of the results of sensitivity studies, particularly with EBMs and other simple models, have improved the understanding of many climatic interrelationships and interactions. One of the interesting results of model studies has been the identification of important feedbacks, in which a change in one variable changes another (or a chain of other) variables, which in turn either amplifies or moderates the change in the first parameter, thereby causing either a positive or negative feedback (Cess, 1976; Dickinson, 1986). An other important advantage of simpler models is that because they do not attempt to predict the weather, their signal-to-noise ratio is substantially higher than for studies with GCMs, and hence it is somewhat easier to identify the effects of smaller or shorter term perturbations in simpler models, even though climatic processes may not be as accurately represented.

3. ENERGY BALANCE MODEL

3.1. Energy balance diagram

The simplest method of considering the climate system of the Earth and of any planet, is in terms of its global energy balance. Taken over the Globe as a whole, observations show (Fig. 2 of Berger, 1989, this volume) that the system loses about the same amount of energy through long wave (also called thermal, infrared or terrestrial) radiation as it gains from the incoming (shortwave) solar radiation (236 W m^{-2}) (however, small imbalances can occur for short and long periods).

The 0-D climate model considers the Earth as a simple point in space having a global mean effective temperature T_e defined from the time averaged solar energy input over the whole Earth. As this input is equal to S/4 = 340 W m^{-2} (or ∿ 0.5 cal cm^{-2} min^{-1} ∿ 720 cal cm^{-2} day^{-1} ∿ 263 kcal cm^{-2} $year^{-1}$) T_e is given by :

$$(1 - \alpha) \frac{S}{4} = \sigma T_{eff}^4 \tag{12a}$$

where α is the planetary albedo
S is the solar constant (1370 W m^{-2})
σ is the Stefan-Boltzman constant (5.67 10^{-8} W m^{-2} K^{-4}).

which leads to T_{eff} = 255 K, smaller than the present-day surface temperature by roughly 33 K.

The fluxes of energy within the atmosphere-surface system can be illustrated using an energy balance diagram (Fig. 2 of Berger, 1989, this volume). Although many measurements have been made at the surface and from satellites, there are still uncertainties of 10-20% in the values of some of the fluxes because of the difficulty of making representative global measurements. The atmosphere absorbs approximetely 23% of the incoming solar radiation, mostly in the troposphere. The atmosphere is driven primarily by the energy transferred to it from the land and ocean via infrared radiation (equivalent, somewhat surprisingly, to about 115% of incoming solar radiation), heat released when evaporated water vapor condenses (about 24% of incoming solar radiation), and the direct transfer of heat, often called sensible heat (about 7% of incoming solar radiation). The atmosphere loses heat almost exclusively by infrared emission to space and to the surface (about 60% and 100% of incoming solar radiation at the top of the atmosphere, respectively).

It might seem that significantly altering the climate would required substantial changes in these fluxes. This turns out not to be the case. For example, doubling the CO_2 concentration without allowing the climate system to readjust leads to reduction of the loss of infrared energy to space by less than 1% of the incoming solar radiation (Fig. 6a). That small amount of heat, and an additional increment as a result of a vertical readjustment of the infrared flux, causes the atmosphere to warm enough to again radiate to space as much energy as was being radiated before the CO_2 concentration was increased (assuming no change in planetary reflectivity) (Fig. 6b). Because the infrared radiation emitted from the atmosphere to the surface is a very large fraction of the infrared radiation emitted by the surface, a doubling of the CO_2 concentration can significantly raise the global average surface temperature (however, calculating how much this warming will be requires interactive consideration of many processes; e.g., clouds, sea ice, water vapor).

3.2. Greenhouse theory

If the atmosphere of a planet contains gases which absorb long wave radiation, and if most of the solar radiation is absorbed at the surface, then the surface temperature, T_S, will be greater than T_e.

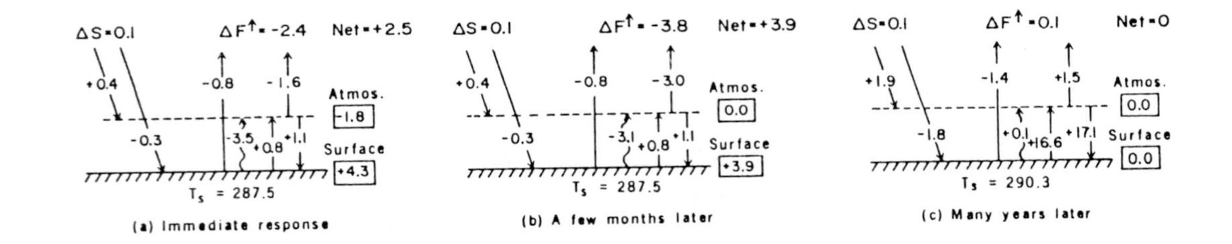

Figure 6. Changes in radiative fluxes calculated by a 1-D radiative convective model when atmospheric CO_2 is doubled from 300 ppmv to 600 ppmv. Radiative fluxes are represented by straight lines; ΔS : change in solar radiation absorbed by the atmosphere and surface; ΔF↑ : change in outward infrared radiation at top of the atmosphere. Convective flux is represented by a wavy line. Panel(a) gives the immediate response before the flux at the stratosphere is changed; (b) gives the situation a few months later before the surface-oceans temperature has changed; (c) gives the new equilibrium state with a surface temperature increase of 2.8°C (reproduced by permission of the authors and Science, copyright @ 1981 by the AAAS, from Hansen et al., vol. 213, 28 August 1981, page 962).

For the Earth, about 70 % of the energy which drives the climatic system is absorbed at the surface (from Fig. 2 of Berger, 1989, this volume : 46/70 = 0.66). The mean surface temperature being 288 K, the surface emits about 390 W m^{-2} of terrestrial radiation but due to atmospheric absorption (leading to the greenhouse effect), only 236 W m^{-2} of terrestrial radiation leaves the top of the Earth's atmosphere, so that the entire Earth-atmosphere system is in radiative balance. This balance can be described as

$$S/4\ (1 - \alpha) = \varepsilon \sigma T_S^4 \qquad (12b)$$

ε can be understood as a kind of emissivity of the Earth-atmosphere system ($\varepsilon \simeq 0.60$).

Without an atmosphere ($\varepsilon = 1$) the equilibrium temperature at the Earth's surface would be some 30 to 35 K lower than in reality. However, the atmosphere contains a number of gases which absorb infrared radiation and partly reradiate it towards the Earth's surface. About 90% of this atmospheric absorption is from H_2O, CO_2 and clouds, and the remaining 10% is due to atmospheric O_3, CH_4 and N_2O (Fig. 7). CO_2 absorbs and emits mainly in a band around 15 μ m, and in a great number of "weak" bands. A particularly important part of the atmospheric absorption spectrum for the greenhouse effect is the atmospheric window region from 7-13 μm. This is the spectral region through which approximately 70-90 % of the surface and cloud emitted terrestrial radiation escapes to space. Therefore, if there is a sizeable increase of greenhouse gases concentration which absorb in the atmospheric window region, this will tend to increase the greenhouse effect (although this is not the only process involved). An increase in the CO_2 content leads to a decrease in the emissivity : the Earth-atmosphere system has become a less effective black emitter, so that the temperature has to increase to reach a new equilibrium. The question is how large this temperature increase should be, to lead to a new radiative equilibrium with the same 236 W m^{-2} leaving the Earth-atmosphere system.

3.3. Climatic feedbacks

Energy balance models predict the change in temperature from the requirement that the change (ΔN) in the net energy flux, N, will be equal to zero (equilibrium state).

For example, at the top of the atmosphere N will be defined as the difference between the solar energy input absorbed by the climatic system, R_i, and the infrared output, R_0,

$$N = R_i - R_0$$

N is a function of quantities, E_i, which are external to the climatic system, that is quantities whose changes can lead to a change in climate, but which are independent of climate, and of quantities, I_j, which are internal to the climatic system, that is quantities that can change as the climate changes (the surface air temperature, T_a, for example) and, in so doing, feed back to modify the climate change.

Therefore : $N = N(E_i, I_j, T_a)$

$$\text{and : } \Delta N = \sum_i \frac{\partial N}{\partial E_i} \Delta E_i + \left(\frac{\partial N}{\partial T_a} + \sum_j \frac{\partial N}{\partial I_j} \frac{dI_j}{dT_a} \right) \Delta T_a \qquad (13)$$

More conveniently : $$\Delta N = \Delta Q_T - \lambda \Delta T_a \tag{14}$$

where $$\Delta Q_T = \sum_i \frac{\partial N}{\partial E_i} \Delta E_i \tag{15}$$

$$\lambda = - \frac{dN}{dT_a} \tag{16}$$

ΔQ_T is the change of N due to a change in one or more external quantities (i.e., before any temperature change).
λ describes the total change in N resulting from a temperature change, T_a. λ may also be described in terms of the gain, G_f, of the system (Schlesinger, 1985) :

$$\lambda = G_f^{-1} = G_0^{-1} - F \tag{17}$$

where G_0 is climate system gain in the absence of feedbacks and F represents the feedbacks :

$$G_0^{-1} = - \frac{\partial N}{\partial T_a} \equiv \lambda_0 \tag{18a}$$

$$F = \sum_j \frac{\partial N}{\partial T_j} \frac{dI_j}{dT_a} \tag{18b}$$

The equilibrium requirement, $\Delta N = 0$, for the energy balance (14) gives :

$$\Delta T_{ae} = \frac{\Delta Q_T}{\lambda} = \frac{G_0}{1 - G_0 F} \Delta Q_T \tag{19}$$

If N is independent of I_j or I_j are independent of T_a, then F = 0 and the input to the system Q_T is directly transferred to the output with in this case

$$\Delta T_{ae} = \frac{\Delta Q_T}{\lambda_0} \tag{20}$$

For the planetary radiative budget (12a) or (12b), the zero-feedback parameter λ_0 is given by (18a) from

$$N_0 = \frac{1-\alpha}{4} S_0 - \varepsilon \sigma T_a^4 \tag{21}$$

therefore : $$\lambda_0 = 4 \varepsilon \sigma T_a^3 \tag{22}$$

Taking T_a as being approximately equal to 288 K and the effective planetary emissivity $\varepsilon \sim 0.61$, gives

$$\lambda_0 = 3.33 \text{ Wm}^{-2} \text{ K}^{-1} \text{ or } G_0 \sim 0.3°\text{C (Wm}^{-2})^{-1} \tag{23}$$

The assumption behind Eq.(19) is that the surface-troposphere system responds as a single coupled thermodynamic system if λ is independent of the type and magnitude of the forcing. Therefore the λ value would depend only on the climate model.

For climatic change induced by changes in the atmospheric concentration of greenhouse gases, Eq.(22) is thought to give a good first-order estimate of the equilibrium surface temperature response because, for such climatic perturbations, the surface and the troposphere will most probably remain strongly coupled by radiative and non-radiative processes. Therefore, we can assume the surface warming to be linked to the greenhouse gas-induced radiative perturbation at the tropopause, ΔQ_T (see annex 1). We now have to specify the change in the heat balance (ΔQ_T) due to change in the considered external parameter (e.g., CO_2 concentration) and the climate system change in response to a given ΔQ_T (λ value).

For a CO_2 increase, the radiative schemes give a change at the tropopause which can be estimated by :

$$\Delta Q_T = H \ln [p_{CO_2} / p_{CO_2}(0)] \quad (24)$$

with $H = 6\ Wm^{-2}$; p_{CO_2} is the partial pressure of CO_2, $p_{CO_2}(0)$ being the initial value. From (24), for a CO_2 doubling, the perturbation at the tropopause will be 4.2 Wm^{-2} with an uncertainty of 0.3 Wm^{-2}

$$Q_T\ (2 \times CO_2) = 4.2 \pm 0.3\ Wm^{-2} \quad (25)$$

On the other hand, from semi-empirical and various model studies, it can be concluded that the most reasonable range of values for λ is :

$$\lambda = 1.8 \pm 1.0 \quad W\ m^{-2}\ K^{-1} \quad (26)$$

A convenient reference value for λ is λ_0 which gives immediately :

$$(\Delta T_{ae})_0 = 4.2 \times 0.3 = 1.26°C \quad (27)$$

On the ground of this simple model, one has to expect that a doubling of the atmospheric CO_2 content leads to an air temperature increase at the Earth's surface of well over 1°C, because the following feedbacks are expected :

i. water vapour feedback : surface temperature increase increases evaporation and therefore the atmospheric water vapour content, which in turns absorbs the infrared radiation, even more effectively than CO_2; this causes the counter-radiation to increase and therefore the air temperature.

ii. ice albedo feedback : if the albedo decreases at high latitudes due to the melting of ice and snow, the absorption of solar radiation by the surface will increase and the temperature increases further.

iii. Clouds feedback : Clouds influence both the planetary albedo and the infrared radiation because they are composed of water and water vapour. Temperature change can influence the cloud amount and cloud height in a very complex way and vice versa, leading to either a positive or negative feedback mechanism (or both). Thus even the direction of these feedbacks is difficult to establish. For example, it has been suggested that for low and middle-level clouds, the albedo effect would dominate over the greenhouse effect, so that increased cloudiness would result in an overall cooling. On the other hand, cirrus clouds having a smaller impact upon the albedo and a temperature much lower than the surface would tend to warm the system by enhancing the greenhouse effect (this will be discussed later with the radiative-convective models).

The overall climate system sensitivity parameter λ is therefore composed of all contributing feedback factors such as :

$$\lambda = \lambda_0 + \lambda_{water\ vapour} + \lambda_{ice\ albedo} + \lambda_{clouds} \tag{28}$$

where various estimates give (Henderson-Sellers and McGuffie, 1987)

$$\lambda_{wv} = -\ 1.7\ W\ m^{-2}\ K^{-1}$$
$$\lambda_{ice} = -\ 0.6\ W\ m^{-2}\ K^{-1} \tag{29}$$

the estimates range for λ_{clouds} being quite large depending upon the physical and radiative characteristics of the clouds themselves (Schlesinger, 1988).

Using this range of radiative perturbations and the λ values given by (26), the following range of estimates is obtained through (19) for the equilibrium surface air temperature response to a CO_2 doubling :

$$\Delta T_{ae}\ (2xCO_2) = 3.5 \pm 2.1\ K \tag{30}$$

i.e., a temperature increase ranging from 1.4 to 5.6 K.

3.4. 0-D atmosphere-surface EBM

This equilibrium formulation is a particular case of the following time-dependent model with the rate of change of temperature with time given by

$$m\ C\ \frac{dT}{dt} = R\ s \tag{31}$$

where m is the mass of the system

C is the specific heat capacity ($J\ kg^{-1}\ K^{-1}$)

s is the area of the Earth ($510\ 10^6\ km^2$)

R is the net radiative balance (difference between the net incoming and the net outgoing radiative fluxes per unit area).

On the Earth, the value of C is largely determined by the oceans : the specific heat for water ($4200\ J\ kg^{-1}\ K^{-1}$) is 4 times that of the air ($1004\ J\ K^{-1}\ kg^{-1}$) and the mass of the ocean is much greater than that of the atmosphere ($1.35\ 10^{21}$ kg and $5.3\ 10^{18}$ kg respectively; solid Earth : $5.98\ 10^{24}$ kg). If we assume that the energy is absorbed by a top mixed layer of 70 m covering 70 % of the Earth's surface, the total heat capacity m C is equal to (ρ_w, C_w and d are the density, specific heat capacity and depth of the mixed layer) :

$$C^* = \rho_w\ C_w\ d\ s = 1.05 \times 10^{23}\ J\ K^{-1} \tag{32}$$

so that (31) becomes

$$\frac{dT}{dt} = \frac{s}{C^*} \{S'(1-\alpha) - \varepsilon\sigma T_S^4\} \qquad \text{with } S' = S/4 \qquad (33)$$

The equilibrium state is then ascertained by setting $dT/dt = 0$ which gives $T_s = 288$ K for $S = 1370$ W m^{-2}, $\alpha = 0.3$ and $\varepsilon = 0.62$. An alternative use of (33) is the calculation of the rate of temperature change following an absorption by the atmosphere of, for example, 58 Wm^{-2} corresponding to an absorption of 17% of 340 Wm^{-2} : 0.5°C day^{-1}.

Finally, it is worth noting that gases that absorb solar radiation may also add to the effective greenhouse effect if most of this absorption takes place in the lower half of the troposphere due to the efficient thermal coupling of the lower troposphere-surface system. If however, most of the solar absorption takes place in the stratosphere, a surface cooling should result since this is reducing the amount of solar energy that is available to be absorbed by the surface-troposphere system (Table 1).

Table 1. The nuclear war scenario (Turco et al., 1984). The numbers are given as percentage of incoming solar radiation.

	Normal case	Nuclear case
Solar radiation reflected by atmosphere	25	20
Solar radiation absorbed by atmosphere	26	75
Solar radiation reflected by surface	6	0
Solar radiation absorbed by surface (or net energy flux from surface)	43	5
Outgoing infrared radiation	69	80
Solar radiation reflected by surface-atmosphere	31	20

EBMs derive from the work of Budyko (1969) and Sellers (1969) and has been investigated extensively since (Ghil, 1985), in particular to discuss the stability of annually averaged temperature, T, of the climatic system to perturbations :

$$c\,\frac{dT}{dt} = \mu S'(1 - \alpha(T)) - g\,\sigma\,T^4 \equiv \mu\ R_i - R_0 \qquad (34)$$

μR_i is the energy input to the system which may vary following a parameter μ and R_0 is the infrared output, $c = \frac{mC}{s}$.

In principle, α and g (grayness of the system) are functions of T and the dependence of α on T should express the change in both cloud and surface albedo with T. But the dependence of cloud albedo on T is still not well understood. On the other hand, surface albedo varies most strongly with the presence or absence of snow and ice. Hence is taken to decrease linearly with T, as ice cover decreases with latitudes, and to be constant for all T for which the Earth would be either entirely ice covered, $T < T_{ic}$, or ice free, $T > T_{if}$. This is the ice-albedo-temperature feedback.

The dependence of emissivity g on T has to express the greenhouse effect, and according to Sellers (1969) can be written :

$$g(T) = 1 - m \text{ tgh } (T/T_*)^6, \tag{35}$$

with m = 0.5, for 50 percent cloud cover, and $T_* = 1.9\ 10^{-15}\ K^{-6}$. This expression indicates that, as temperature increases, g decreases, so that the greenhouse effect becomes stronger. The shapes of $\alpha(T)$ and g(T) are reflected in the graphs of R_i and R_0 in Fig. 7.

3.4.1. Internal stability

Let us first consider the stability of the solutions of (34) to small perturbations in initial conditions.

The intersections of $R_i(T)$ and $R_0(T)$, or those of $R_i - R_0$ with the T-axis, determine the steady states, or equilibria, of model (34). We notice in Fig. 7 that there are three stationary solutions : T_1, T_2 and T_3. Hence, even as simple a model as the present one shows the possible existence of more than one climate, if we are willing to interpret the equilibria of the model as steady-state climates of the Earth.

In the real world, small deviations from a given temperature regime always appear, due to a multitude of mechanisms not included in the model. It is important, therefore, to investigate the stability of the model's solutions to small perturbations. Assume, for instance, that, at time t : $T = T_e + T'$. We are interested in knowing how the deviation $T'(t)$ evolves in time, for $\mu=1$.

If we assume $T'(0)$ suitably small, it suffices at first to consider a linearized equation for $T'(t)$, obtained by expanding (34) in T. If we use $dT_e/dt = 0$ and neglect the terms of higher order, such an expansion yields :

$$\frac{dT'}{dt} = \lambda_e T' \tag{36a}$$

where

$$\lambda_e = -(S'\alpha_e + \sigma g_e)/c \tag{36b}$$

$$\alpha_e = \left[\frac{d\alpha}{dT}\right]_{T=T_e} \qquad g_e = 4g(T_e)\,T_e^3 + \left[\frac{dg}{dT}\right]_{T=T_e} T_e^4 \tag{36c,d}$$

This is a linear ordinary differential equation for the deviation $T'(t)$ of $T(t)$ from the equilibrium value T_e.

The solution of Eq.(36) is :

$$T'(t) = T'_0 \exp[\lambda_1 t] \tag{37}$$

Hence T'will grow exponentially if $\lambda_1 > 0$ and decay to zero with time if $\lambda_1 < 0$. Thus T_1 is linearly stable if $\lambda_1 < 0$, and unstable if $\lambda_1 > 0$. A similar analysis holds for T_2 and T_3.

In Fig. 7, an equilibrium T_j, j = 1,2,3, can be seen to be stable if the corresponding slope of $R_i(T) - R_0(T)$ is negative at $T = T_j$, and unstable if it is positive. It follows that T_1 and T_3 are stable, while T_2 is unstable. Indeed, near T_1 and T_3, the energy balance (34) tries to restore T to its equilibrium value. Near T_2, quite the opposite happens.

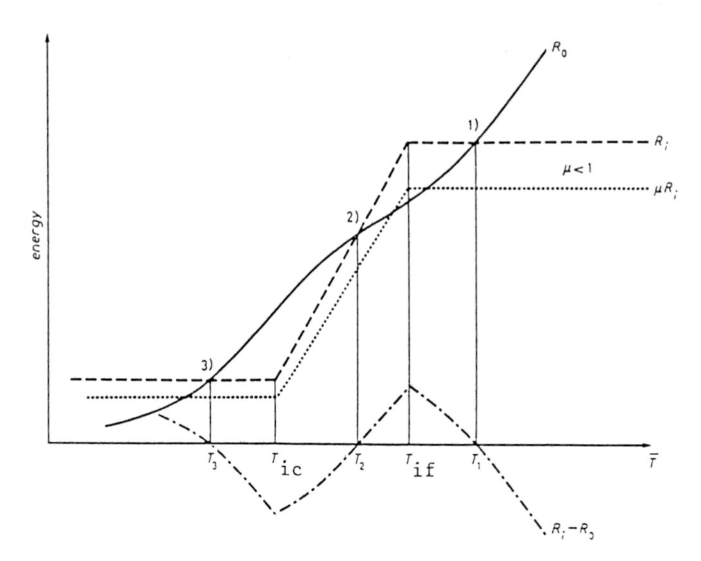

Figure 7. Energy components in a global energy balance model : R_0 is the outgoing infrared radiation, R_i is the absorbed incoming solar radiation R_i : $S'(1 - \alpha(T))$, S is the solar constant and the albedo α is temperature dependent. μ is a parameter describing the possible change in the incoming solar radiation. The three stationary solutions are given by T_1, T_2 and T_3; T_{ic} and T_{if} are extreme cases respectively for $T_2 = T_3$ ($\mu = \mu_d$) and $T_2 = T_1$ ($\mu = \mu_c$) (reproduced by permission of the author and the Societa Italiana di Fisica @ 1985, LXXXVIII Corso Bologna, Italy, from Ghil, 1985).

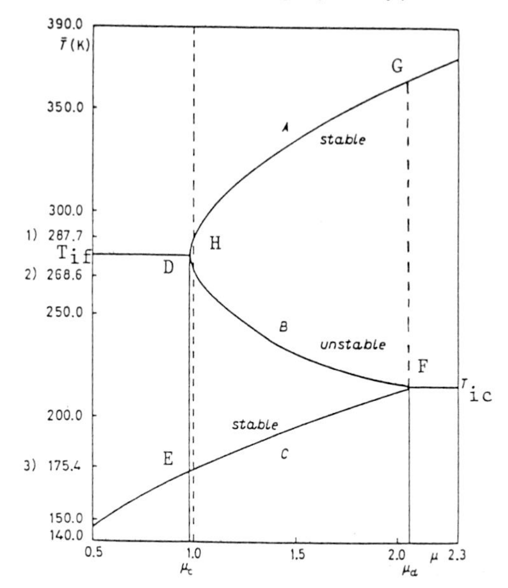

Figure 8. Dependence of the three stationary solutions of Fig. 7 and of their internal stability on the parameter μ (Budyko, 1982; Ghil, 1985).

This stability discussion suggests that it is reasonable to identify T_1 with the present climate; T_2, however, is not a good candidate for a colder climate, such as an ice age, since it is unstable and could never have persisted for any length of time. The equilibrium T_3 corresponds to a completely ice-covered Earth, or a "deep freeze". While the model shows it to be stable, it was never observed in the paleoclimatological record of the past.

3.4.2 Structural stability to changes in the parameters value

It remains to be seen how the existence of one or more solutions for Eq.(34), and their stability, can be affected by changes in the model's parameters. The most important among these is the value of the solar input, $\mu S'$. It could change as a result of variation in the Sun's energy output, in its distance from the Earth, or in the atmosphere's optical properties. Within our simple model, either one of these changes can be expressed by taking $\mu \neq 1$. Such a change will leave, in Fig. 7, the graph for μR_i be either above or below the curve representing R_i.

In Fig. 7, the situation for a certain value of μ less than 1 is represented by a dotted line. Clearly, T_1 and T_2 in this new situation lie closer to T_{if}, while T_3 lies more to the left of its value for $\mu=1$. If μ were to decrease further to a value μ_c, a critical situation would occur for which $T_2 = T_{if} = T_1$. For $\mu < \mu_c$, the solutions T_1 and T_2 disappear altogether, and T_3 is the only solution left.

If we recall now that T_1 represents the present climate within the model, it follows that, while the solar input μS decreases, the Earth's climate cools off slowly at first, as T_1 moves to the left. Then, as μ crosses the value μ_c, the temperature of the system would have to decrease dramatically to that of a completely ice-covered Earth, T_3. This corresponds to the passage from D to E in Fig. 8.

In the opposite situation, of μ increasing, T_1 moves to the right, while T_2 and T_3 approach T_{ic}. As μ increases through and beyond a critical value μ_d, T_2 and T_3 coalesce, then disappear, while T_1 keeps increasing.

It should be notice that, for any value of μ, the analysis of the stability for the model's equilibria can be carried out as it was for $\mu = 1$. Indeed, Eqs (36) become

$$\frac{dT'}{dt} = \lambda_j(\mu)\, T' \tag{38a}$$

$$\lambda_j(\mu) = -\,\{\mu\, S'\, \alpha_j(\mu) + \sigma\, g_j(\mu)\}\,/c \tag{38b}$$

$$\alpha_j(\mu) = [\frac{d}{dT}]_{T=T_j(\mu)} \tag{38c}$$

$$g_j(\mu) = 4g\,(T_j(\mu))\, T_j^3(\mu) + [\frac{dg}{dT}]_{T=T_j(\mu)}\, T_j^4(\mu) \tag{38d}$$

where $T_j(\mu)$, $j = 1,2,3$, is the position of solution T_j as μ increases or decreases. Clearly, each T_j is a continuous function of μ near $\mu = 1$, and, therefore, $\alpha_j(\mu)$ and $g_j(\mu)$ are continuous, so that λ_j depends continuously on μ near $\mu = 1$.

In particular, the sign of λ_j will not change near $\mu = 1$. The sign of λ_j is, in fact, the same as that of the slope of the curve $R_i(T;\mu) - R_0(T;\mu)$ in Fig. 7. This suggests that the sign of $\lambda_j(\mu)$ stays the same until the solution $T_j(\mu)$ coalesces with another solution, or until it disappears. Thus T_1 is stable to small perturbations in T for $\mu > \mu_c$, T_3 is stable for $\mu < \mu_d$, and T_2 is unstable for $\mu_c < \mu < \mu_d$. To summarize, T_1 and T_3 are stable, and T_2 is unstable, over the entire μ interval over which they exist.

Figure 8 shows the average temperature T of stationary solutions as a function of μ. Analysis of the T-slope of $\mu R_i - R_0$ infers that all solutions of (34) corresponding to branches A and C are stable, while branch B is unstable. This figure shows in fact the fundamental characteristic of non-linear systems. A slow decrease in the solar constant from initial conditions for the present day (μ=1, point H) means a gradual decrease in temperature until the point D is reached when a run-away feedback loop causes total glaciation and a rapid drop in temperature (jump from D to E). For $\mu < \mu_c$ only the "deep freeze" branch C exists. When the solar constant is then increased the process is not immediately reversed; the temperature rise slowly up to a point F where there is a jump to G; beyond $\mu = \mu_d$, only the "warm climate" branch A exists. The modelled climate exhibits therefore an hysteresis. Moreover, the line FD represents a branch of the solution which is physically unrealistic : while energy input increases, temperature decreases !

3.5. 1-D latitudinal EBM for the atmosphere-surface

In the case where we consider each latitude zone independently, the 0-D model can guide the study of the spatially 1-D.

We have seen that the radiation balance of the Earth-atmosphere system changes with latitude ϕ (Fig. 6, Berger, 1989, this volume). This latitudinal dependence of the net radiative balance gives rise to and is maintained by zonally averaged heat fluxes (Fig. 7, Berger, 1989, this volume). There are three kinds of heat transfer :

1. conduction heat flux is given by Fourier's law for solid media (Fick's law for the diffusion of trace constituents)

$$F_c = - k_c \nabla T \tag{39}$$

2. radiative heat transfer is negligible in horizontal heat transport but important in the vertical distribution of temperature.

3. advective heat flux in dynamical fluid system, Fa, is given by $\vec{V}.\nabla T$, where $\vec{V}$ is the velocity of the fluid which carries internal energy at the temperature T with it.

In the absence of internal sources of energy, temperature will thus change in time due to convergence or divergence of the heat flux at a given location.

It is clear that atmospheric and oceanic dynamics play an important role in the climatic system's local energy balance and temperature distribution via the velocity field $\vec{V}$. To determine $\vec{V}$ at the same time as T is a considerably more complicated task than computing T alone. Also, much more is known about T for climates different from the present one than about $\vec{V}$. When considering only the largest, planetary scales of the temperature field, $O(10^4$ km), and time scales longer than months and years, it is reasonable to attempt to eliminate the velocity field from our considerations, by using the so-called eddy diffusive approximation

$$- \nabla . F_a \sim \nabla . k_e \nabla T \tag{40}$$

In Eq.(40), k_e is an eddy diffusion coefficient being usually much larger than k_c. It often gives acceptable qualitative results when studying planetary-scale, long-term climate change, given the typical scales of atmospheric and oceanic eddies, $O(10^2$ versus $10^3)$ km and O(weeks versus months), respectively. The zonally averaged, meridional heat fluxes will therefore be expressed by combining (39) and (40) in :

$$- \nabla . F = \nabla . k \nabla T; \tag{41}$$

here $F = F_c + F_a$, $k = k_c + k_e$, and k is taken to be a function of latitude ϕ alone, $k = k(\phi)$, although it can also be a function of T and ∇T.

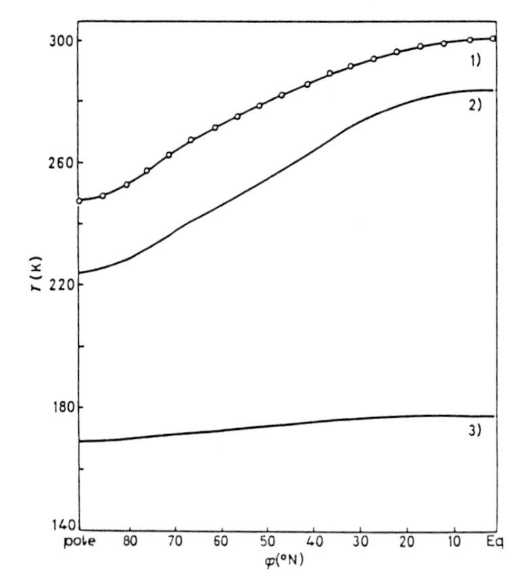

Figure 9. Temperature as a function of latitudes for the three equilibrium solutions of a 1-dimension zonally averaged energy balance model (reproduced by permission of the author and of the Societa Italiana di Fisica @ 1985, LXXXVIII Corso Bologna, Italy, from Ghil, 1985). This behaviour is typical of Sellers-Budyko 1-dimension models (1969), as it was also found by Berger (1977).

The energy balance of a zonal slice of the climatic system located at latitude ϕ and extending to the top of the atmosphere and down to a prescribed depth in the oceans and the continents, is thus governed by the following partial differential equation :

$$C(\phi)\ \frac{\partial T(\phi,t)}{\partial t} = \{\mu S'L(\phi)(1-\alpha(\alpha,T)) - \sigma g(\phi,T)\ T^4\}$$

$$+ \frac{1}{\cos\phi}\ \frac{\partial}{\partial\phi}\ [k(\phi)\ \cos\phi\ \frac{\partial T}{\partial\phi}] \qquad (42)$$

where $L(\phi)$ gives the normalised latitudinal distribution of solar radiation incident at the "top of the atmosphere" with latitudes (Berger, 1978); the second term to the right expresses $\nabla_h.F$ in the meridional direction according to (41). Computation of stationary solutions yields to 3 equilibria (Fig. 9). The model solution $T_1(\phi)$ closely matches the data for the present climate (open circles in the figure). The "deep freeze" $T_3(\phi)$ has a mean temperature T_3 about 100 K below that of the present and a much smaller pole-to-equator temperature difference.

We notice that both $T_1(\phi)$ and $T_2(\phi)$ straddle T_{ic}, the perennial snow line temperature, as well as T_{if}, the snow absence temperature. When (42) is still a model for annually averaged temperature, it is customary to take T_{ic} = -10°C and T_{if} = 10°C. Therefore, the relationship $T_3 < T_{ic} < T_2 < T_{if} < T_1$ for the mean equilibrium temperature obtained for the 0-D model does not hold in this model for all latitudes. However, the study of the internal stability of these equilibrium solutions to small perturbations shows, as expected from the 0-D model, that T_1 and T_3 are stable and T_2 is unstable. The discussion of the structural stability of the 1-D model is unfortunately not as simple as in the 0-D case (Ghil, 1985).

Qualitative agreement between the sensitivity of the present day climate simulated by both EBMs and GCMs of climate to insolation variations (1 K per percent change in solar input) offers striking confirmation to the simple story that EBMs are telling about climate.

3.6. Box-Energy balance Models

A more complex method for the heat transport is to consider each of the transporting mechanisms separately, with the flux divergence given by

$$\nabla.F = \frac{1}{\cos\phi}\frac{\partial}{\partial\phi}\ \{\cos\phi\ (F_0 + F_a + F_q)\} \qquad (43)$$

the three terms on the right representing transports of sensible heat due to ocean and atmosphere and of the latent heat :

$$F_0 = -K_0\ \frac{\partial T}{\partial y} \qquad (44a)$$

$$F_a = -K_a\frac{\partial T}{\partial y} + [\vec{V}]\ T \qquad (44b)$$

$$F_q = -K_q\ \frac{\partial q(T)}{\partial y} + [\vec{V}]\ q(T) \qquad (44c)$$

where the diffusion coefficient K are all functions of latitude and $[\vec{V}]$ is the zonally averaged wind speed. Obviously the more complicated the parameterization the more realistic one might expect the model to be : from the Newtonian cooling of Budyko where $\nabla.F$ is expressed in term of $(T(\phi) - \bar{T})$ to the eddy diffusive formula developed by Sellers.

Another way to complicate the system is to consider more than one subsystem. For example, an ice sheet model can be coupled to the response of the lithosphere to ice load and to an atmospheric EBM. Some models of this sort have been shown to exhibit internal variability : the components can interact to form a constantly varying climate even without such external forcing as the Milankovitch variations. However, other similar simulations with bedrock deformation and a calving mechanism have shown a satisfactory response to orbital variations. The concept of computing the energy budget of a few subsystems of the climate system can be extended to produce other forms of EBMs, the box models in particular which we are going to illustrate in this section.

To estimate the time-dependent response of the climate system to time-dependent trace-gas concentration changes, one particularly simple set of equations can be obtained by considering only two globally and annually averaged coupled atmosphere and ocean boxes (Tricot and Berger, 1987). The folowing equations can be written for the atmosphere (a) and the surface (s), respectively :

$$\rho_a C_{pa} \Delta Z_a \frac{\partial \Delta T_a}{\partial t} = \Delta LE + \Delta H + \Delta F_n - \Delta F\uparrow_a \quad (45)$$

$$\rho_s C_{ps} \Delta Z_s \frac{\partial \Delta T_s}{\partial t} = \Delta LE - \Delta H - \Delta F_n - r\Delta F_{sd} \quad (46)$$

where ΔX means the departure of X from its fixed equilibrium value; ρ_i, C_{pi}, ΔZ_i, T_i are the density, heat capacity, depth and temperature of the reservoir i (i=a or s), respectively; LE, H and F_n are the heat transfers from the surface to the atmosphere due to latent heat, sensible heat and net long-wave radiation; $F_n = F\uparrow_s - F\downarrow_s$ with $F\uparrow_s$ the upward long-wave radiation emitted by the surface and $F\downarrow_s$ the downward long-wave atmospheric radiation; $F\uparrow_a$ is the upward long-wave radiation at the top of the atmosphere; F_{sd} is the heat transfer between the mixed layer and the deeper ocean; r is the fraction of the Earth's surface covered by oceans (0.71).

In Eq.(46), ΔZ_s is an equivalent layer depth, obtained by taking a particular weighting of the actual mixed layer and land inertia. This equivalent mixed-layer depth allows the effect of the smaller land heat capacity to be included, assuming a finite wind mixing between the oceanic and continental atmospheres.

The following heat flux parameterizations can be used :

$$F\uparrow_s = \sigma T_s^4 \quad (47)$$

$$F\downarrow_s = \varepsilon \sigma T_a^4 \quad (48)$$

$$H = C_H (T_s - T_a) \quad (49)$$

$$LE = C_{LE} (e_s - e_a) \quad (50)$$

$$F\uparrow_a = A + BT_a \quad (51)$$

where σ is the Stefan-Boltzmann constant; ε is the equivalent emissivity of the sky as a function of cloudiness n (fixed to 0.55), T_a and relative humidity RH of the surface air; C_H and C_{LE} are the drag coefficients for the sensible and latent heat fluxes, respectively equal to 13 and 15.6 Wm^{-2} K^{-1}; e_s is the saturation water vapour pressure at surface temperature T_s; e_a is the atmospheric water vapour pressure using a fixed relative humidity RH (77%).

In all the simulations, the surface and atmospheric solar absorptions can be assumed fixed, giving a constant planetary albedo equal to 0.31 for a solar constant of 1368 W m^{-2}. On the other hand, in the parameterization of F_a we can simulate the radiative effect of trace gas changes (i.e. a change of atmospheric opacity) by modifying the value of the constant A in Eq.(51) which present day value was taken to be -168 Wm^{-2}.

From the definition of ΔQ_T (defined to be positive for any gain by the climatic system before any change in temperature), we have

$$\Delta Q_T = \text{perturbed net flux} - \text{initial net flux}$$

Assuming that $F\!\downarrow_{per} = F\!\downarrow_{ini}$

$$\Delta Q_T = F\!\uparrow_{ini} - F\!\uparrow_{per}$$

which, in terms of (51), leads to

$$\Delta Q_T = - \Delta A \qquad (52a)$$

and $$\lambda = B \qquad (52b)$$

Therefore, in the global model given by Eqs (45) and (46), the climatic feedback parameter λ is an adjustable parameter given by the chosen value of B in (51). For a particular choice of B, an equation similar to Eq.(19) gives directly the equilibrium surface air temperature response of the model, ΔT_{ae}, for a given radiative perturbation ΔQ_T which is assumed to be totally included in the atmosphere, neglecting the accompanying variation of $F_S^{\downarrow}$.

To solve Eqs (45) and (46), the heat exchange between the mixed layer and the deeper ocean must be explicitly formulated. Such heat transport results mainly from motion along constant-density (isopycnal) surfaces at low latitudes and convective overturning in the North Atlantic and Antarctic basins. A simplified formulation is obtained by using an advection-diffusion equation for the heat penetration in the deeper ocean. It is assumed that, on a global scale, the vertical mixing in the ocean interior can be parameterized as :

$$F_{sd} = \rho_s C_{ps} \left[K \frac{\partial T_d}{\partial z} + W (T_d - T_B)\right]_{Z=0} \qquad (53)$$

where T_d is the horizontally averaged deep ocean temperature below the mixed layer, T_B is the bottom water temperature ($\sim$1.2°C), Z = 0 is the level of the bottom of the mixed layer (Z>0 in the deeper ocean), K is a constant vertical eddy diffusion coefficient, W is a constant averaged upwelling velocity for the world ocean. This simplified formulation has been proposed (Hoffert et al., 1980) to hopefully incorporate the global effect of numerous heterogeneous regional mixing processes. Equation (53) is obtained by assuming that the polar water sinks with a constant temperature T_B and spreads instantaneously throughout the ocean bottom.

Subsequently, this cold flow is brought to the surface by upwelling over the wold ocean.

This equation (53) defines the so-called box advection-diffusion ocean model or BADOM. To obtain the evolution of the deep ocean temperature, the following relation is used :

$$\frac{\partial T_d}{\partial t} = K \frac{\partial^2 T_d}{\partial z^2} + W \frac{\partial T_d}{\partial z} \tag{54}$$

with the boundary conditions

$$T_d = T_s \qquad \text{at } Z = 0$$

$$T_d = T_B \ (K=0) \qquad \text{at } Z = D$$

where D is the depth of the ocean bottom.

The sea floor boundary condition equates the polar sea downwelling flux to the net (upwelling plus diffusion) upward flux into the world ocean

$$K \frac{\partial T_d}{\partial z} + W T_d = W T_B \rightarrow T_d = T_B \quad \text{at } Z=D \tag{55}$$

The steady-state solution of Eq.(54) is

$$T_d(Z) = T_B + (T_s - T_B) \exp\left(- \frac{W}{K} Z\right) \tag{56}$$

In equilibrium, the temperature profile of $T_d(Z)$ is adjusted such that the divergence of the upward advective heat flux balances the convergence of the downward diffusive heat flux at each point. Values of $W = 4$ m year^{-1} and $K = 0.6$ cm^2 s^{-1} are used to reproduce a vertical global temperature profile close to the GEOSECS (Geochemical Ocean Sections) world data (Hoffert et al., 1980).

In this section, results of a few transient simulations with the BADOM are presented for a period extending from the pre-industrial time (defined as the year 1850) up to the middle of the next century (the year 2050). The main goal is to illustrate the potential relative importance of some model uncertainties for the global climatic response. All simulations have been obtained using the global climate model described in (45 to 52) coupled with an advective-difffusive deep ocean model (53 to 55).

The time-dependent radiative forcing is estimated from the standard concentration scenarios given by Wuebbles et al. (1984) using the method given in Tricot and Berger (1987). The concentration of CO_2 and effective CO_2 are given as a function of time between 1850 and 2050 in Figure 10a (the effective CO_2 concentration is the CO_2 concentration which would have the same radiative effect as all greenhouse gases together). In the standard case, the pre-industrial CO_2 concentration is assumed to be 270 ppmv.

Equation (24) provides then the direct forcing ΔQ_T (Fig. 10a, right hand panel). The resulting equilibrium surface air temperature is subsequently provided by equation (19), for different values of the climatic feedback parameter λ (Fig. 10b). These values have been choosen to correspond to $\Delta T_{a,e}$ equal to 1, 3 and 5°C for a double CO_2 ($\Delta Q_T = 4.2$ Wm^{-2}); these λ values are thus respectively equal to 4.2, 1.4 and 0.84 Wm^{-2} K^{-1}.

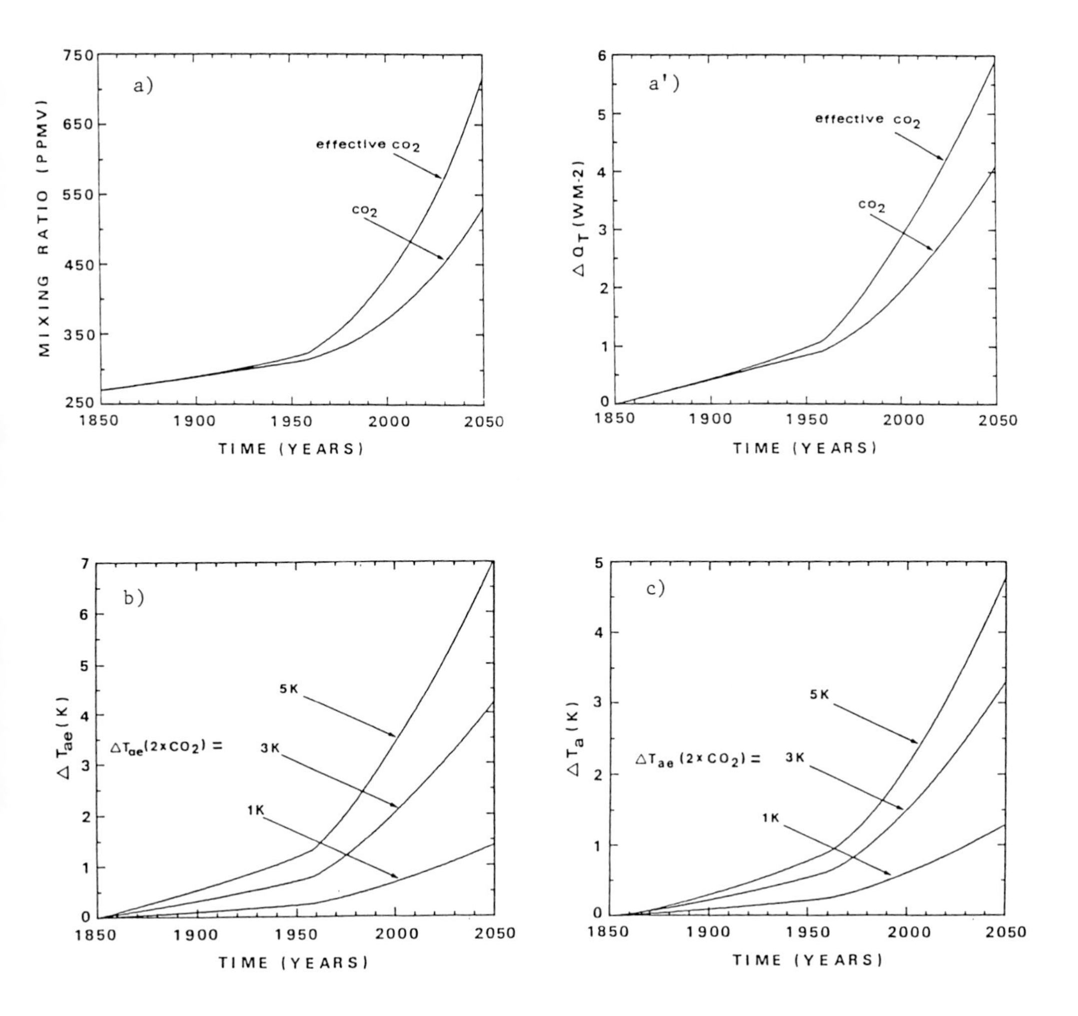

Figure 10. Transient response of the climate system to a time-dependent greenhouse gases forcing from 1850 to 2050 (Tricot and Berger, 1987). Panels (a) gives the change in the CO_2 and effective CO_2 atmospheric concentrations for a standard CO_2 scenario (the left hand panel is for the mixing ratio in ppmv and the right hand panel for the direct radiative forcing at the tropopause, ΔQ_T, in Wm^{-2}); (b) gives the equilibrium temperature change, ΔT_{ae}, for three values of the sensitivity parameters of the model corresponding to an increase in T_{ae} of respectively 1, 3 and 5 K for a doubling of CO_2 from pre-industrial level (270 to 540 ppmv); (c) gives the transient change in air temperature at the surface, ΔT_a, for the three cases of (b). Results in (b) and (c) are for the increase effective CO_2 concentration given in (a).

Equations (45 and 46) can now be integrated with ΔA and B of (51) being given by (52a and b).

Table 2 gives the equilibrium value of air and surface temperatures and of some heat fluxes computed by the model for the present climate and for an instantaneous CO_2 concentration doubling. The e-folding time of the model, τ_e, has also been computed by the model transient response to a step-function forcing for an instantaneous CO_2 doubling. Assuming an equilibrium increase of the surface air temperature $[\Delta T_{ae}(2xCO_2)]$ equal to 3K, this e-folding times is equal to 13 years and 14 years for the atmosphere and the equivalent mixed layer, respectively. From a set of simulations for a range of $\Delta T_{ae}(2xCO_2)$ between 1 K and 5 K, we found:

$$\tau_e \propto \Delta T^n_{ae}(2xCO_2) \text{ with } n \sim 1.5.$$

Table 2. Equilibrium global mean temperatures (in K) and heat fluxes (in Wm^{-2}) for the present climate and for a doubled CO_2 concentration. $2xCO_2$ values are given for a surface air temperature response equal to 3 K.

	Present	$2xCO_2$
Temperature (K)		
air	288.0	291.0
surface	289.3	291.8
Sensible flux (Wm^{-2})	16.7	11.0
Latent flux (Wm^{-2})	83.1	90.5
Long-wave flux (Wm^{-2})		
downward at the surface	335.9	352.0
upward at the top	236.0	236.0
$\Delta Q_T(Wm^{-2})$	—	4.2

To assess the model sensitivity to $\Delta T_{ae}(2xCO_2)$, Figure 10c shows T_a for three given values of $\Delta T_{ae}(2xCO_2)$: 1 K, 3 K and 5 K for the standard scenario. The transient response ranges between 0.4 and 1.4 K in 1980, between 0.7 and 2.6 K in 2010, and between 1.3 and 4.8 K in 2050. By comparison with the equilibrium temperature (Figure 10b), these transient responses indicate that the oceanic heat uptake reduces the model response by around 10 % for $\Delta T_{ae}(2xCO_2)$ = 1 K, 25 % for $\Delta T_{ae}(2xCO_2)$ = 3 K, and 35 % for $\Delta T_{ae}(2xCO_2)$ = 5 K. More precisely, this reduction factor is shown to decrease slightly with time, varying for example from 27 % in 1980 to 19 % in 2050 for ΔT_{ae} = 3 K, and shows a similar trend for other values of $\Delta T_{ae}(2xCO_2)$.

Moreover, these comparisons allow also to estimate the time delay for the global climate response from the instantaneous equilibrium response. By comparison with the curves of Fig. 10b giving the equilibrium responses, the transient responses of Fig. 10c indicate that the oceanic heat uptake delays the model response by 20-years roughly. For example, for ΔT_{ae} = 3K, the time lag is $\sim$15 years in 1980 and $\sim$20 years in 2030. It can be seen that this time lag increases with the model sensitivity for a CO_2 doubling. This behaviour can be expected from the relationship introduced above between the e-folding time of the model and its sensitivity for a CO_2 doubling. But it must be kept in mind that in the transient experiments, the radiative forcing is continuously varying and consequently the time lag during the transient experiment is not directly linked to the e-folding time defined for a given forcing.

These lags are mainly due to the deeper oceanic heat uptake while the mixed layer can be shown to introduce by itself only a small time delay on the equilibrium response, rather independently of the precise value of the mixed-layer depth (cf. Tricot and Berger, 1987, Fig. 8).

The transient experiment given in Fig. 11 defines our standard experiment. This figure gives the time-dependent responses of the surface air temperature (ΔT_a), the surface temperature (ΔT_s) and the deep-ocean mean temperature (ΔT_d). For this standard simulation, we have assumed an equilibrium surface air temperature increase $\Delta T_{ae}(2xCO_2)$ equal to 3 K. As illustrated in this figure, the surface air temperature warming is always larger than the surface temperature increase. Furthermore, the difference between the two warmings increases slightly with time. This is related to the non-linear increase of latent-heat exchange between the surface and the atmosphere when the temperature increases, despite the decreasing difference between the surface air and surface temperature (Table 2 shows indeed that the initial equilibrium value of T_S is larger than T_a and Figure 11 indicates that T_S increases less rapidly in time than T_a).

Our standard scenario for future greenhouse gas concentrations is only thought to give a reasonable extrapolation of present trace gas trends, without taking the range of future new energy policies into account. Clearly, the most significant climatic uncertainty for the near future is related to the future estimated CO_2 concentrations. As an illustation, Figure 12 gives the future ΔT_a for the three CO_2 scenarios proposed by Wuebbles et al. (1984), referred to as high, standard and low CO_2 scenarios. The range of calculated temperature warming is around 1.8 K in 2010 and between 2.9 and 4.2 K in 2050. Our ignorance of the exact future CO_2 concentration levels is alone sufficient to prevent us from making a precise quantitative prediction of the climatic change in the next century, even if the models used would be able to simulate in a reliable way the behaviour of the climate system at the human time scale.

These future scenarios for the trace gas concentrations are only indicative and even past trace gas concentrations are only known approximately. Keeping these uncertainties in mind, we have estimated a possible range for the global temperature change from the pre-industrial epoch to the middle of the next century using different assumptions. Our standard experiment (Fig. 11) gives a total surface air warming of 1.0 K due to the probable change in the concentration of the trace gases between the pre-industrial period and the present. The CO_2 alone induced warming amounts to 0.75 K, i.e. 75 % of the total warming. For a pre-industrial CO_2 concentration of 290 ppmv, the warming is 0.77 K and the CO_2-induced warming amounts to 0.50 K, i.e. 65 % of the total warming. Over the next 70 years, our standard experiment gives a possible global mean surface air warming of 2.3 K with the CO_2 contribution amounting to 1.6 K, i.e. 70 % of the total warming. Because of the simplified radiative scheme used in our computations, these relative contributions for CO_2 alone can be overestimated by roughly 15 %.

To take the present uncertainties in the climate sensitivity to trace gas changes into account, an assumed increase in the equilibrium temperature of between 1 K and 5 K seems reasonable for a CO_2 doubling. For such a range of equilibrium temperature increase, the global mean

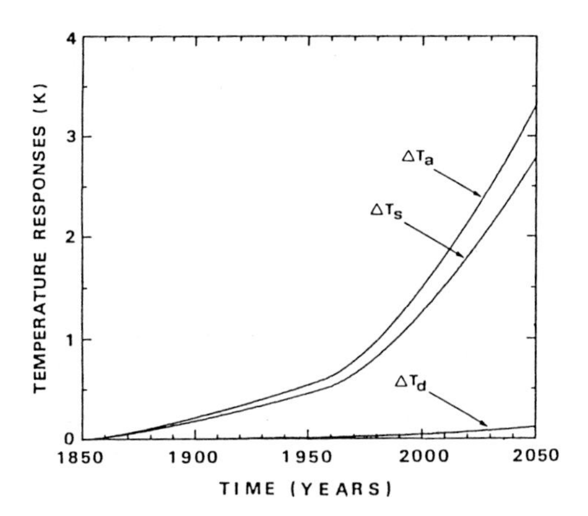

Figure 11. Transient response of surface air temperature, ΔT_a, surface temperature, ΔT_s, and mean deep ocean temperature, ΔT_d, for the experiment described in Fig. 10 corresponding to ΔT_{ae} $(2xCO_2)$ = 3 K (Tricot and Berger, 1987).

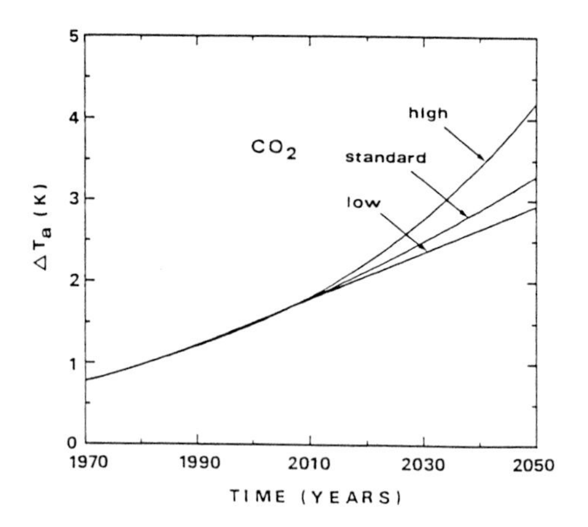

Figure 12. Transient response of the surface air temperature, ΔT_a, for the high, standard and low CO_2 scenarios of the climate system to the resulting effective CO_2 increase from 1970 to 2050. The changes ΔT_a refer to the 1850 level and are computed for $\Delta T_{ae}(2xCO_2)$ = 3K (Tricot and Berger, 1987).

surface air warming between 1850 and the present is estimated to range between 0.4 and 1.4 K for a pre-industrial CO_2 concentration equal to 270 ppmv. A pre-industrial CO_2 concentration of 290 ppmv lowers the range to between 0.3 and 1 K. For the same range of equilibrium temperature increase, the surface air warming is estimated to amount between 0.9 and 3.4 K over the next 70 years, for the standard trace gas scenarios of Wuebbles et al. (1984).

4. RADIATIVE-CONVECTIVE MODELS

Radiative-convective models emphasize the effects of variation of temperature with altitude z and allow analysis of stratospheric radiative feedbacks on tropospheric temperatures. As the name of this model type indicates, radiation and convection are treated explicitely. The radiation scheme is detailed and occupies the vast majority of the total computation time, while the convection is accomplished by a numerical adjustment of the temperature profile at the end of each time step.

4.1. Radiation computation

The principles involved in radiative computations in climate models can be illustrated by considering a very simple global model in which a single cloud or aerosol layer is spread homogeneously over the surface (Henderson-Sellers and McGuffie, 1987) (Fig. 13). Three main assumptions have been made in this simple model :

(i) there is no reflection of the upward-travelling short-wave radiation by the cloud; (ii) the surface emissivity has been set equal to unity; and (iii) the infrared absorptivity of the cloud is equal to ε. If we know the cloud albedo, α_c, its shortwave absorptivity, a_c, and its infrared emissivity, ε, the energy balances at the "top of the atmosphere", the cloud level and the surface respectively give the following set of equations :

$$S = \alpha_c S + \alpha_g (1-a_c)^2 (1-\alpha_c) S + \varepsilon\sigma T_c^4 + (1-\varepsilon)\sigma T_g^4 \qquad (57a)$$

$$a_c(1-\alpha_c)S + a_c\alpha_g(1-a_c)(1-\alpha_c)S + \varepsilon\sigma T_g^4 = 2\,\varepsilon\sigma T_c^4 \qquad (57b)$$

$$(1-\alpha_g)(1-a_c)(1-\alpha_c)S + \varepsilon\sigma T_c = \sigma T_g^4 \qquad (57c)$$

These equations can be solved with the elimination of α_g, the surface albedo, and T_c, the cloud temperature, to provide T_g, the temperature at the surface

$$\sigma\, T_g^4 = \frac{S(1-\alpha_c)}{2-\varepsilon}\,(2 - a_c)$$

For $S = 343\ W\ m^{-2}$ and a cloudless sky, a volcanic aerosol cloud and a water droplet cloud which characteristics are given in Table 3, one obtain respectively T_g = 283 K, 280 K and 288 K.

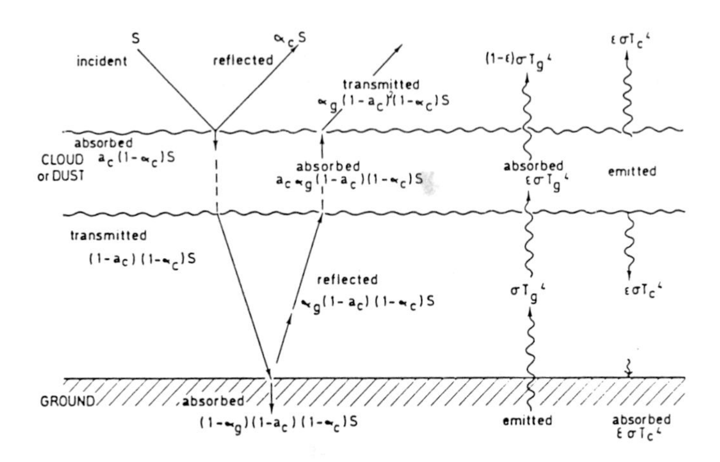

Figure 13. Box diagram of a radiative-convective model with the interactions between a cloud or dust layer and the surface for solar radiation (straight line) and infrared radiation (wavy lines) (reproduced by permission of the authors and of John Wiley and Sons from Henderson-Sellers and McGuffie, 1987).

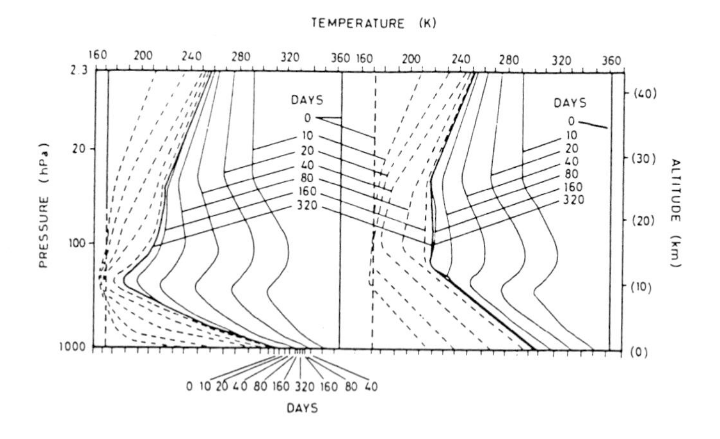

Figure 14. Approach to equilibrium state with a pure radiative (left-hand scale panel) and a radiative-convective (right-hand side panel) models. The solid and dashes lines show the approach from a warm and a cold isothermal initial atmosphere (reproduced by permission of the authors and of the American Meteorological Society from Manabe and Stricker, 1964).

Table 3. Characteristics of a cloudless sky, a volcanic aerosol cloud, a water droplet cloud and a high reflective cloud (Henderson-Sellers and McGuffie, 1987).

	α_c	a_c	ε	T_g
cloudless	0.08	0.15	0.4	283
volcanic	0.12	0.18	0.43	280
present cloud	0.30	0.20	0.90	288
high reflective cloud	0.40	0.20	0.90	277

It is interesting to point out that the introduction of a cloud cover approximately the same as that of the present has increased the calculated surface temperature because the greenhouse effect of the cloud is greater than the albedo effect. If, however, α_c were very slightly increased, the cloud albedo effect overcome the greenhouse effect leading to a cooling. The sensitivity of this simple model's climate illustrates the interconnected role of various parameters in radiative transfer calculations, in particular, the role of clouds in the climatic system. However, the treatment of atmospheric radiation in the different climate models is usually more complex, the major complicating features being detailed consideration of the dependence of scattering and absorption on wavelength and other atmospheric variables.

In the stratosphere, the globally averaged atmospheric temperature, T(z), is essentially in radiative equilibrium and hence determined by a local balance between solar heating and net longwave cooling. The solar heating depends mostly on ozone concentrations. Above the lowest layers in the stratosphere, the longwave cooling is largely controlled by carbon dioxide and is primarily dependent on local temperature. Radiative balance just above the tropopause is complicated by the importance of absorption (by CO_2, O_3, water vapor) of longwave radiation originating from warmer tropospheric layers.

4.2. Convective adjustment

Computation of globally averaged vertical temperature profiles determined solely from the vertical divergence of the net radiative fluxes yields very high surface temperatures and lapse rates in the lower troposphere considerably in excess of the mean value of -6.5 K km^{-1}. Radiative equilirium profiles are thus unstable to vertical convection and a rapid vertical transfer of heat will produce meaningful values of surface and vertical temperatures (Fig. 14). Thus in the troposphere, the vertical profile of T is determined primarily not by local radiative balance but rather by the vertical energy redistribution by moist convective processes which must be parameterized.

The classical study by Manabe and Strickler (1964) and Manabe and Wetherald (1967) introduced the assumption that, if the lapse rate exceeds the "critical rate" of 6.5°C km^{-1}, a convective adjustment would ensue and maintain the 6.5°C km^{-1}. Although most models use this value, other alternatives are the observed (closer to 5.5°C km^{-1}) or moist (saturated) adiabatic lapse rate (as in GCMs).

4.3. Sensitivity to lapse-rates

For a doubling of CO_2, a moist adiabatic convective adjustment yields surface temperature change smaller than does a constant 6.5 K/km lapse rate. Indeed, if a temperature change at the surface is extended upward along a moist adiabat, the magnitude of the change increases with altitude which amplifies the consequent changes of outgoing long wave flux. Thus a smaller surface temperature change is required to balance a given external change in atmospheric composition or heat input. It is a negative lapse rate feedback. For example, Hummel and Kuhn (1981) computed surface temperature changes 25 to 60% smaller when using the moist adiabatic lapse rate formulation. But, the exact warming depends on the cloud treatment; a low cloud induces a larger response.

Alternatively, if a model is heated by large amounts at high elevations, as in the nuclear war scenarios (Turco et al., 1983), then the lapse rate becomes less than the critical value, and moist adjustment no longer couples surface temperature to the troposphere. Under these conditions, large temperature changes can occur at the surface; the lapse rate feedback is positive.

4.4. Sensitivity to water vapour

Using a fixed absolute humidity in a $2xCO_2$ RC experiment, Hansen et al. (1981) found a 1.2°C temperature increase. In this basic model neither atmospheric water vapour nor clouds have changed in response to the temperature change and there is no feedback which means that λ corresponds to the feedback for blackbody radiation for a planet with no atmosphere and an albedo of 0.3 (λ_B = 3.75 Wm^{-2} K^{-1}). If relative humidity is now fixed, the vapour pressure will also increase following the temperature increase, which reflects surface evaporation. This introduces a positive feedback with a lower λ = 2.34 W m^{-2}°C and a higher temperature (ΔT = 1.94°C). The difference between the two models is the positive water vapour feedback as also ilustrated by Ramanathan (1981).

Indeed, Fig. 15 clearly shows that CO_2 increase acts on the surface temperature via three basic mechanisms

1. the direct radiative effect which enhances the downward infrared flux at the surface (CO_2 greenhouse effect stricto-sensu);

2. the increase of tropospheric temperatures and of the related downward infrared fux because of the absorption by tropospheric water vapour of some of the enhanced emission by CO_2 (overlapping between CO_2 and H_2O absorption bands);

3. the enhancement of the hydrological cycle as a consequence of the initial surface temperature perturbation. The greeenhouse effect of the tropospheric water vapour enhances further the downward infrared flux at the surface. This mechanism acts as an important positive feedback on the initial response to be associated to the initial radiative perturbations.

It is now well accepted that, provided these mechanisms are included, all models predict a global surface warming of about 2 for a doubling of the atmospheric CO_2 concentration (Schlesinger, 1984 and 1986).

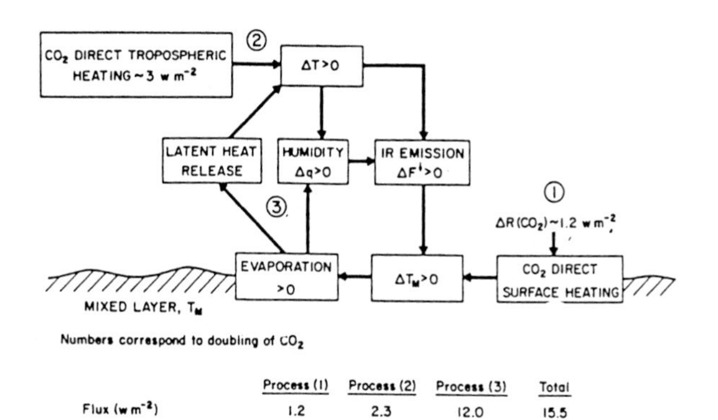

	Process (1)	Process (2)	Process (3)	Total
Flux ($w\,m^{-2}$)	1.2	2.3	12.0	15.5
Percent	8.0	15.0	77.0	
ΔT_s (Model dependent)	0.17	0.33	1.7	2.2

Figure 15. Schematic illustration of the ocean-atmosphere feedback processes by which CO_2 doubling warms the surface. The contribution by the various processes to surface radiative heating, the percentage contribution and to the computed surface warming are shown in the table. The results were obtained with a 1-D radiative-convective model. (reproduced by permission of the author and of the American Meteorological Society from Ramanathan, 1981).

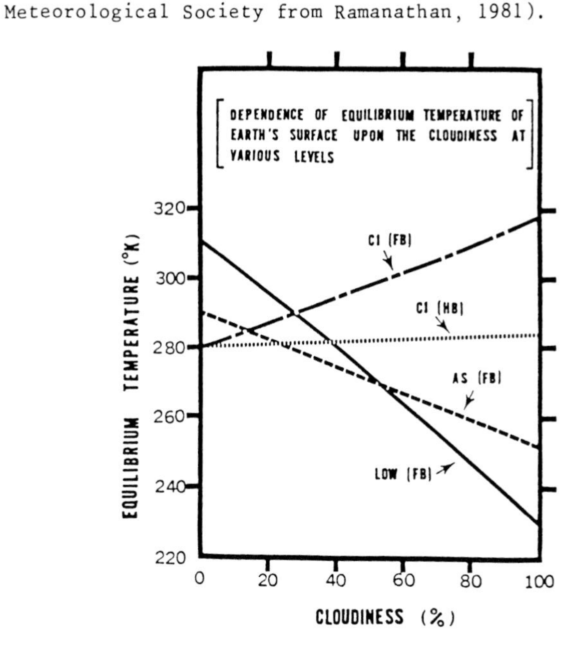

Figure 16. Equilibrium temperature at the Earth's surface from a radiative-convective model as a function of cloudiness (Ci : Cirrus; AS: altostratus; FB : full black; HB : half black) (reproduced by permission of the authors and of the American Meteorological Society from Manabe and Wetherald, 1967).

4.5. Sensitivity to clouds

Cloud can affect the radiation balance of a vertical column through the Earth-atmosphere system by reflecting a large amount of the incoming sunlight back to space, since, on a global average, the albedo of the cloudy part of the Earth is about 0.5, compared to the mean albedo of the cloudless fraction of the Earth of ~ 0.14.

Another well-known consequence of the presence of clouds in the atmosphere is a change in the upward flux of infrared radiation associated with atmospheric temperature. The temperature of the lower atmosphere is observed to decrease vertically at a nearly constant lapse rate. In the troposphere where a substantial amount of clouds exist, the lapse rate is about -6.5 K km^{-1}, if averaged over the whole globe. Since most clouds are very opaque to the planetary IR radiation, both an increase of the horizontal extent of the cloud coverage and increase of the effective height of the cloud tops reduce the upward flux of the IR radiation escaping from the Earth-atmosphere system to space.

In particular for an increase in the amount of global cloud cover, there are thus two competing opposite effects on the global radiation balance :

1) an increase in the planetary albedo causing a decrease in available solar energy (T_S decreases for new equilibrium)

2) a decrease in the IR radiative loss to space which implies a higher new equilibrium surface temperature.

A series of thermal equilibrium computations was performed with a 1-D Radiative-Convective model for various distributions of cloudiness (Manabe and Wetherald, 1967). Generally speaking, the larger the cloud amount, the colder is the equilibrium temperature of the Earth's surface, though this tendency decreases with increasing cloud height and does not always hold for cirrus (Fig. 16). Whether cirrus clouds heat or cool the equilibrium temperature depends upon both their height and their radiative properties (for full black cirrus, temperature increases with cloud amount).

Other factors, like the height of the clouds, also play a role. If a fixed cloud altitude is used, Manabe and Wetherald (1967) found a feedback $\lambda = 2.2$ W m^{-2} °C^{-1}. Alternatively, a fixed cloud-top temperature provides a positive feedback since the longwave flux from the cloud tops cannot change and the fluxes from the surface must change more. In this case, the cloud feedback on surface temperature is positive, λ being typically reduced by about 0.5 W m^{-2} C^{-1} from its value for fixed cloud-top altitude.

More recent results by Stephens and Webster (1981) confirm that the climatic impacts of clouds depend both in clouds altitude and in the liquid water path through the clouds. Especially, high thin clouds at low and middle latitudes in all seasons and all clouds at high latitudes in winter tend to warm the surface relative to the clear sky; in the same conditions, all other clouds tend to cool the surface. The summer high-latitude effects are similar to the low- and middle-latitude situation; i.e., a warming effect for the high thin clouds and a cooling one for a optically thicker mid- and low clouds.

Moreover the model sensitivity is compounded by surface albedo effects (Fig. 17). For a given cloud, a critical surface albedo may exist at which the cloud transits from cooling the surface relative to clear-sky conditions to warming the surface when the surface albedo is increased (for 35°N in winter, these critical albedos lie between 0.3 and 0.7 depending upon the type of clouds and their liquid content).

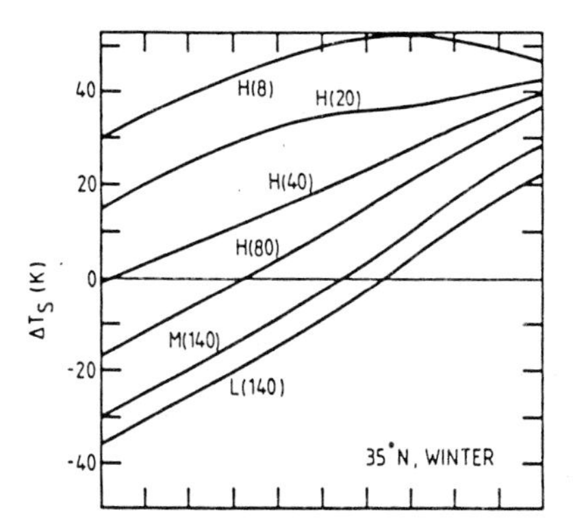

Figure 17. Surface temperature difference, ΔT_S, between various cloud cases with given water path (in parentheses, in gm^{-2}) and the clear atmosphere as a function of the surface albedo. The results are shown for 35°N in winter. The three cloud species are noted L for low clouds (913-854 mb), M for mid clouds (632-549 mb) and H for high clouds (381-301 mb). (reproduced by permission of the authors and of the American Meteorological Society from Stephens and Webster, 1981).

The sensitivity of the model indicates that great care is necessary in developing parameterizations for the radiative structure of clouds and for surface albedos for use in climate models. It is argued that simple EBM (e.g. Budyko, 1969) will be of limited use in climate research unless a cloud amount-surface temperature relationship is established. For example, the cloud-surface temperature effect will tend to buffer the Budyko surface temperature-surface albedo feedback if cloud amount increases with decreasing surface temperature, or is independent of surface temperature but will enhance the feedback if the cloud-surface temperature relationship were reversed.

Finally, the net effect of clouds on surface heating will also have to take into account the cloud liquid water content, the solar zenith angle, the local vertical distribution of temperature and radiatively active constituents, in particular.

5. CONCLUSIONS

Simple models of climate discussed here include horizontally averaged, one-dimensional radiative-convective and energy balance models. Others, more comprehensive include two-dimensional models (in which the horizontal momentum and heat transports by large scale atmosheric eddies are parameterized), and low resolution general circulation models (e.g., highly truncated spectral models).

Simple models have been useful for obtaining a preliminary estimate of climate sensitivity with relatively small expenditures of computer time. For example, they have been used for estimating the response of climate to an increase in atmospheric CO_2, other trace gases and aerosols. Their results are reasonably consistent with those from comprehensive general circulation simulations.

It is often difficult to interpret results from a three-dimensional climate model because non-linear interactions among many physical processes are incorporated in the computation. In such a situation, it is useful to use a simple model with fewer feedback mechanisms for the purpose of identifying the key processes which may be involved. On the other hand, the main problem associated with the use of simple models is the parameterization of the mass, momentum and energy transports by the atmosphere and oceans. A new result obtained from a simple model cannot thus be taken as definite proof unless it can be reproduced by more realistic detailed simulations of the climate system. However, we must also recognize that exact governing equations of the most comprehensive three-dimensional models are not known for all processes; as a consequence, individual numerical experiments can produce unexpectedly disappointing simulations because only part of a complex process has been captured. The spatial resolution is also subject to limitations set mainly by the available computing power. Various simplifications must thus also be introduced which depend upon the intended purpose of the model and the time scale of interest. An effective strategy for the study of the sensitivity and variability of climate therefore involves the use of both simple and comprehensive models.

ANNEX

A variation of CO_2 concentration perturbs not only the surface radiative balance but also the tropospheric and the upper atmospheric (stratospheric) radiative balances.

On a globally averaged basis, the surface and troposphere are strongly coupled by radiative and non radiative processes, including transfer of latent and sensible heat by convective and advective heat transports. Therefore, the surface warming is linked to the CO_2-induced radiative perturbation of the entire surface-troposphere system, ΔQ_T (sum of the surface and tropospheric radiative heatings). This heating is the change in the net radiative (longwave + solar) flux at the tropopause computed with surface and tropospheric parameters (including temperatures) prescribed at their unperturbed (before CO_2 variation) values. However, the stratospheric temperatures are allowed to come to equilibrium with the radiative perturbation due to the CO_2 increase. This temperature change is essentially instantaneous since the response time to heating changes in the stratosphere is a few weeks or less. During the stratospheric cooling (for $2xCO_2$), neither the surface (ocean) nor the troposphere, which is convectively coupled to the surface, have responded yet and T_s has not changed. Quickly (on a climatic time scale), the stratosphere reaches a radiative equilibrium and the net radiative budget at TOA equals the net radiative budget at the tropopause, ΔQ_T. The planet radiates less energy to space than in the case of present CO_2 concentration, because of the cooler stratosphere and greater mean altitude of emission from the troposphere (greenhouse effect). The energy gained by the earth at this time (ΔQ_T) will be used to warm the surface and the troposphere subsequently.

6. <u>REFERENCES</u>

Adem, J. (1964). On the physical basis for numerical prediction of monthly and seasonal temperature in the troposphere-ocean-continent system. Monthly Weather Review, 92, 91-104.

Adem, J. (1988). Possible causes and numerical simulation of the Northern Hemispheric climate during the last deglaciation. Atmosfera, 1, 17-38.

Arrhenius, S. (1896). On the influence of carbonic acid in the air upon the temperature of the ground. Philos. Mag., 41(5), 237-276.

Berger, A. (1977). Power and limitation of an energy balance climate model as applied to the astronomical theory of paleoclimates. Palaeogeography, -climatology, -ecology, 21, 227-235.

Berger, A. (1988). Milankovitch and Climate. Review of Geophysics, 26(4), 624-657.

Berger, A., and Andjelic, P. (1988). Milutin Milankovitch, père de la théorie astronomique des paléoclimats. Histoire et Mesure, III-3, 385-402.

Berger, A., Gallée, H., Tricot, Ch., Fichefet, Th., Marsiat, I. (1988). Transient response of the climate system to the astronomical forcing. In : Belgian Research on Global Change IGBP, A.H. Cottenie and A. Teller (Eds), 10-24, SCOPE Belgium, Académie Royale des Sciences, des Lettres et des Beaux-Arts de Belgique.

Berger, A. (1989). Basic concepts of climate modelling. This volume.

Budyko, M.I. (1969). The effect of solar radiation variations on the climate of the Earth. Tellus, 21(5), 611-619.

Budyko, M.I. (1982). The Earth's Climate : Past and Future. International geophysics Series, vol. 29 (W.L. Donn, Ed.), Academic Press, New York.

Callendar, G.S. (1938). The artificial production of carbon dioxide and its influence on temperature. Q.J.R. Meteorol. Soc., 64, 223-240.

Cess, R.D. (1976). Climate change : an appraisal of atmospheric feedback mechanisms employing zonal climatology. J. of Atmospheric Sciences, 33 n°10, 1831-1843.

Chervin, R.M. (1981). On the comparison of observed and GCM simulated climate ensembles. J. of Atmospheric Sciences, 38 n°5, 885-901.

Dickinson, R. (1986). How will climate change ? In : The Greenhouse Effect, Climatic Change and ecosystem. B. Bolin, B. Döös, J. Jaeger, R. Warrick (Eds), SCOPE 29, 207-270, John Wiley, New York.

Dickinson, R., and Cicerone, R.J. (1986). Future global warming from atmospheric trace gases. Nature, 319 (6049), 109-115.

Gates, W.L. (1981). Modeling of climatic changes. In : Climatic variations and Variability : Facts and Theories, A. Berger (Ed.), 435-459, D. Reidel Publ. Company, Dordrecht, Holland.

Ghil, M. (1981). Energy balance models : an introduction. In : Climatic Variations and Variability : Facts and Theories, A. Berger (Ed.), 461-480, D. Reidel Publ. Company, Dordrecht, Holland.

Ghil, M. (1985). Theoretical climate dynamics : an introduction. In : Turbulence and Predictability in Geophysical Fluid Dynamics and Climate Dynamics, M. Ghil, R. Benzi and G. Parisi (Eds), 347-402, North Holland, Amsterdam.

Gilchrist, A. (1979). Numerical modelling of the atmosphere. Reports on progress in Physics, 42(3), 503-545.

Hansen, J., Johnson, D., Lacis, A., Lebedeff, S., Lee, P., Rind, D., Russell, G. (1981). Climate impact of increasing atmosphere carbon dioxide. Science, 213 (4511), 957-966.

Hasselmann, K. (1976). Stochastic climate models, part 1, theory. Tellus, 28 n°6, 473-486.

Held, I.M., and Suarez, M.J. (1974). Simple albedo feedback models of the ice caps. Tellus, 6, 613-629.

Henderson-Sellers, A., and McGuffie, K. (1987). Introduction to Climate Modelling. J. Wiley and Sons, Chichester and New York.

Hoffert, M.I., Callegari, A.J., and Hsieh, C.T. (1980). The role of deep sea heat storage in the secular response to climatic forcing. J. Geophys. Res., 85, 6667-6679.

Hummel, J.R., and Kuhn, W.R. (1981). Comparison of radiative-convective models with constant and pressure-dependent lapse rates. Tellus, 33, 254-261.

IGBP Special Committee (1988). The International Geosphere-Biosphere Programme : A Study of Global Change. A Plan for Action. IGBP report n°4, IGBP Secretary, Stockholm, Sweden.

Lemke, P. (1977). Stochastic climate models, part 3. Application to zonally averaged energy models. Tellus, 29 n°5, 385-393.

Lorenz, E.N. (1955). Available potential energy and the maintenance of the general circulation. Telus, 7, 157-167.

Lorenz, E.N. (1967). The nature and theory of the General Circulation of the Atmosphere. WMO Publication n°218, TP 115, World Meteorological Organization, Geneva, Switzerland.

MacCracken, M.C., and Luther, F.M. (1985). Detecting the climatic effects of increasing carbon dioxide. DOE/ER-0235, US Department of Energy, Office of Energy Research, Washington D.C.

Manabe, S. (1983). Carbon dioxide and climatic change. Advances in Geophysics, 25, 39-82.

Manabe, S., and Strickler R.F. (1964). Thermal equilibrium of the atmosphere with a convective adjustment. J. Atmos. Sci., 21, 361-385.

Manabe, S., and Wetherald, R.T. (1967). Thermal equilibrium of the atmosphere with a given distribution of relative humidity. J. Atmos. Sci., 24, 241-259.

Manabe, S., and Bryan, K. (1969). Climate calculations with a combined ocean-atmosphere model. J. Atmos. Sci., 26, 786-789.

Manabe, S., and Wetherald, R.T. (1980). On the distribution of climate change resulting from an increase in CO_2 content of the atmosphere. J. Atmos. Sci., 37, 99-118.

Milankovitch, M.M. (1920). Théorie Mathématique des Phénomènes Thermiques produits par la Radiation Solaire. Académie Yougoslave des Sciences et des Arts de Zagreb. Gauthier-Villars.

Mitchell, J. (1989). Atmospheric General Circulation Models. This volume.

North, G.R., Cahalan, R.F., and Coakley, J.A.Jr. (1981). Energy balance climate models. Rev. Geophysics and Space Phys., 19(1), 91-121.

North, G.R., Mengel, J.G., and Short, D.A. (1983). Simple energy balance model resolving the seasons and the continents. Application to the astronomical theory of the Ice Ages. J. Geophysical Research, 88 C11, 6576-6586.

Ohring, G., and Adler, S. (1978). Some experiments with a zonally averaged climate model. J. Atmospheric Sciences, 35 n°2, 186-205.

Oort, A.H., and Peixoto, J.P. (1983). Global angular momentum and energy balance requirements from observations. Advances in Geophysics, 25, 355-490.

Peixoto, J.P., and Oort, A.H. (1984). Physics of Climate. Reviews of Modern Physics, 56(3), 365-429.

Peng, L., Chou, M.D., and Arking, A. (1982). Climate studies with a multi-layer energy balance model. Part I : Model description and sensitivity to the solar constant. J. Atmos. Sci., 39(12), 2639-2656.

Plass, G.N. (1956). Carbon dioxide theory of climatic change. Tellus, 8, 140-154.

Ramanathan, V. (1981). The role of ocean-atmosphere interactions in the CO_2 climate problem. J. Atmospheric Sciences, 38 n°5, 918-930.

Ramanathan, V., and Coakley, J.A. (1978). Climate modeling through radiative-convective models. Rev. of Geophys. and Space Phys., 16(4), 465-489.

Ramanathan, V., Cicerone, R.J., Singh, H.B., and Kiehl, J.T. (1985). Trace gas trends and their potential role in climate change. J. Geophys. Res., 90(D3), 5547-5566.

Ramanathan, V., Calis, L., Cess, R., Hansen, J., Isaksen, I., Kuhn, A., Lacis, A., Luther, F., Mahlmann, J., Reck, R., Schlesinger, M. (1987). Climate-chemical interactions and effects of changing atmospheric trace gases. Review of Geophysics, 25, 1441-1482.

Rind, D., Goldberg, R., and Ruedy, R. (1989). Change in climate variability in the 21st century. Climatic Change, 14, 5-37.

Robinson, A.R., and Baker, D.J. (Eds) (1979). Ocean models and climate models. Dynamics of Atmospheres and Oceans, 3, 81-521.

Saltzman, B. (1964). On theory of axially-symmetric time-average state of atmosphere. Pure Appl. Geophys., 57, 153-160.

Saltzman, B. (1978). A survey of statistical dynamical models of the terrestrial climate. Advances in Geophysics, 20, 183-304.

Schlesinger, M.E. (1984). Climate model simulations of CO_2-induced climatic change. In : Advances in Geophysics 26, B. Saltzman (Ed.), 141-237, Academic Press, New York.

Schlesinger, M.E. (1985). Analysis of results from energy balance and radiative-convective models. In : Projecting the Climatic Effects of Increasing Carbon Dioxide, M.C. MacCracken and F.M. Luther (Eds), 281-319, US Department of Energy, DOE/ER-0237.

Schlesinger, M.E. (1986). Equilibrium and transient climatic warming induced by increased atmospheric CO_2. Climate Dynamics, 1, 35-51.

Schlesinger, M.E. (1988). How to make models for behaviour of clouds. Nature, 336, 315-316.

Schlesinger, M.E., and Mitchell, J.F.B. (1987). Climate model simulations of the equilibrium climatic response to increased carbon dioxide. rev. of geophysics, 25 n°4, 760-798.

Schneider, S.H., and Dickinson, R.E. (1974). Climate modeling. Rev. Geophys. Space Phys., 12, 447-493.

Sellers, W.D. (1969). A global climatic model based on the energy balance of the Earth atmosphere system. J. Applied Meteorology, 8, 392-400.

Sellers, W.D. (1973). New global climatic model. J. Atmos. Meteor., 12(2), 241-254.

Semtner, A.J.Jr. (1984). Development of efficient, dynamical ocean-atmosphere models for climate studies. J. Clim. Appl. Meteor., 23, 353-374.

Semtner, A.J.Jr., and Chervin, R.M. (1988). A simulation of the global ocean circulation with resolved Eddies. J. of Geophys. Res., 93 C12, 15.502-15.522.

Shine, K.P., and Henderson-Sellers, A. (1983). Modelling climate and the nature of climates models : a review. J. of Clim., 3, 81-94.

Smagorinsky, J. (1963). General circulation experiments with the primitive equations. I. The basic experiment. Mon. Weather Rev., 91, 99-165.

Smagorinsky, J. (1983). The beginning of numerical weather prediction and general circulation modeling ; early recollections. Advances in Geophysics, 25, 3-37.

Starr, V.P. (1966). Physics of negative viscosity phenomena. McGraw Hill Book Company, New York.

Stephens, G.L., and Webster, P.J. (1981). Clouds and climate : sensitivity of simple systems. J. of Atmospheric Sciences, 38(2), 235-247.

Stone, P.H., and Mao-Sung Yao (1987). Development of two-dimensional zonally averaged statistical-dynamical model. Part II : the role of Eddy momentum fluxes in the general circulation and their parameterization. J. Atmos. Sci., 44 n°24, 3769-3786.

Tricot, Ch., Berger, A. (1987). Modelling the equilibrium and transient responses of temperature to past and future trace gases concentrations. Climate Dynamics, 2, 39-61.

Turco, R.P., Toon, O.B., Ackerman, T.P., Pollack, J.B., and Sagan, C. (1983). Nuclear winter : global consequences of multiple nuclear explosions. Science, 222 (4630), 1283-1292.

Turco, R.P., Toon, O.B., Ackerman, Th.P., Pollack, J.B., and Sagan, C. (1984). The climatic effects of a nuclear war. Scientific American, 251(2), 23-33.

Tyndall, J. (1863). On radiation through the Earth's atmosphere. Philos. Mag., 22(4), 160-194, 273-285.

Van Mieghem, J. (1973). Atmospheric Eneergetics. Oxford Univ. Press, Clarendon Press, Oxford.

Wang, W.C., Molnar, G., and Mitchell, T.P. (1984). Effects of dynamical heat fluxes on model climate sensitivity. J. Geophys. Res., 89 (D3), 4699-4711.

Washington, W.M., and Parkinson, C.L. (1986). An introduction to three-dimensional climate modeling. Univ. Sc. Books, Mill Valey, Calif. and Oxford Univ. Press, N.Y., 422pp.

Wigley, T.M.L. (1985). Carbon dioxide, trace gases and global warming. Climate Monitor, 13, 133-148.

Wuebbles, D.J., MacCracken, M.C., and Luther, F.M. (1984). A proposed reference set of scenarios for radiatively active atmospheric constituents. US Department of Energy, Carbon Dioxide Research, Division Tech. Report 015, Washington D.C.

RADIATIVE TRANSFER AND GREENHOUSE EFFECT

HANS-JÜRGEN BOLLE
Institut für Meteorologie, Freie Universität Berlin

Summary

The climate of the Earth as a planet is determined by the extraterrestrial solar irradiance and the averaged radiative properties of the Earth system, which consists of the atmosphere, the oceans, the land-surfaces with their biota, and the cryosphere. The geographical distribution of climates depends primarely on the inclination of the Earth's rotational axis with respect to the orbital plane, surface characteristics and the composition of the atmosphere. Radiatively active atmospheric constituents determine, how much of the solar flux reaches the ground. They also regulate the upwelling infrared radiative flux. At the surface a temperature is adjusted, which satisfies the energy balance requirements. To study climate processes and climate change it is therefore of eminent importance that the radiative transfer in the Earth's system, its change with atmospheric composition and structure as well as the interaction between radiation and the Earth's surfaces are correctly modelled. In this contribution some basic principles of these processes are discussed.

1. INTRODUCTION

If we consider time scales in the order of one decade the planet Earth can be assumend to be in at least approximate radiative equilibrium. This implies that the outgoing thermal infrared radiant energy at the top of the atmosphere equals in the average the solar radiant energy that is absorbed in the Earth's system.

The thermal infrared emission is governed by Planck's law, which describes the spectral distribution of the radiation emitted by a black body in thermal equilibrium. Only in relatively narrow spectral intervals and under cloudless skies radiation from the earth-surface can escape nearly directly to space. In other parts of the spectrum atmospheric trace gases like H_2O, CO_2, O_3, CH_4, N_2O and CFC's absorb the radiation from the surface and emit radiation according to their own temperature and the structure of their absorption bands. The radiation that finally reaches the top tmosphere is a mixture of radiation emitted and absorbed in all its levels according to the in situ values of the absorption coefficient and the temperature. In the atmosphere the temperature first decreases from the ground to the tropopause located at about 8 km altitude at high latitudes and 17 km in the tropics, and then increases again up to the stratopause at about 50 km height. In the presense of clouds the outgoing radiation would originate near the cloud-tops. Also aerosols affect the infrared radiation but to a much lesser extent than clouds. The transfer of infrared radiative energy from the surface to the top of the atmosphere is therefore a very complicated process, but nevertheless it is in principle well understood.

Solar radiation is partly absorbed in the atmosphere by ozone, water vapour, clouds and aerosols, and is scattered by molecules, aerosol particles and water droplets. Only a fraction of the solar radiation that reaches the ground is absorbed at the surface. The quantity that determines this fraction is the albedo, which is defined as the ratio of the outgoing hemispherical flux to the incomming flux. Only a small part of this energy is used for photosynthesis. Most of it is converted to heat and partitioned into three heat fluxes. One is directed downwards into the soil, and two are directed upwards into the atmosphere, the flux of sensible heat and the flux of latent heat. The upward directed heat fluxes cause convective processes that determine the average ("adiabatic") lapse rate of the troposphere. The tropopause, the height at which the lapse rate becomes zero, denotes the layer from where on radiative rather than convective processes govern the atmospheric temperature gradient. As a consequence the upper part of the atmosphere must be in the average in radiative equilibrium. This requires that the net total (solar and infrared) radiation flux at the tropopause is zero as it is at the top of the atmosphere. This fundamental fact has important consequences for the middle atmosphere in the case of climate change.

2. ENERGY BUDGET OF THE EARTH's SYSTEM

The irradiance S_o at the top of the atmosphere is 1367±1 W m^{-2}. This value was determined by measurements from satellites, rockets and balloons. It varies within less than 0.1% during one 11 years solar cycle. Maximum values close to 1369 W m^{-2} occurr at the solar maximum of which the next one is expected for 1991. Most of the solar energy is contained in the spectral range of 300 to 2500 nm. Only the fraction $(1 - \rho_p)$ of this energy is absorbed by the Earth system, the fraction ρ_p is reflected back to space and is called the mean planetary albedo. As long as the Earth remains in radiative equilibrium the total absorbed solar radiant energy $\pi R^2(1 - \rho_p)S_o$ (R = Earth radius) must be compensated by an equally large flux of outgoing longwave or thermal radiation $\Phi^+_{LW,\infty}$, which can be expressed as a Black Body radiation in terms of an equivalent planetary mean radiative temperature T_{eff} by

(1) $$\Phi^+_{LW} = 4\pi R^2 \sigma T^4_{eff} .$$

σ is the Stefan-Boltzmann constant which has the value 5.66961 10^{-8} W m^{-2} K^{-4}. Nota bene that the outgoing longwave infrared radiation results from the whole sphere and is not much different for day- and nighttime while the Earth's cross section for the solar radiation is that of a circle. The condition for radiative equilibrium then reads

(2) $$(1 - \rho_p)S_o/4 - \sigma T^4_{eff} = 0$$

While short term and seasonal deviations from this condition generally occur, equation (2) holds for averages over one or more years, if no internal changes take place in the Earth's system.

The climate relevant for the biosphere at the bottom of the atmosphere is primarely determined by the local radiation budget but may strongly be modified by advective transports of energy and by local hydrological conditions. For simplicity we consider here only the global average conditions at the bottom and the top of the atmosphere. The net radiation flux per unit area, $M_{R,o}^*$, at the surface is given by

$$M_{R,o}^{*} = M_{SW,o}^{-}(1 - \rho_o) + M_{LW,o}^{-} - M_{LW,o}^{+} \quad (3)$$

The $^+$ stands for upwelling, the $^-$ for downwelling radiation, SW for shortwave radiation which is equivalent to solar radiation, LW as before for longwave or thermal infrared radiation, ρ_o is the albedo of the surface. Most of the energy of the terrestrial longwave infrared flux is confined between the wavelengths of 2.5 and 50 µm. Spectral overlapping with the solar radiation is therefore small. Again the infrared fluxes can be approximated by an effective Black Body radiation:

$$M_{LW,o}^{+} = \varepsilon_o \sigma T_o^4 \quad (4a)$$

$$M_{LW,o}^{-} = \varepsilon_a^{-} \sigma T_a^4 . \quad (4b)$$

The emissivity ε has to be distinguished for upward and downward radiation. Because of the vertical temperature and water vapor gradients in the atmosphere the radiation emitted downwards by a finite layer is not equal to the radiation that emerges from the upper boundary of this layer. T_o is the temperature of the earth surface and T_a a weighted average atmospheric temperature.

The net radiation at the Earth's surface averaged over the whole globe and over some time is always positive in the sense that more solar energy is absorbed then infrared energy is lost. As indicated in Fig. 1 in our latitudes during summer the net thermal infrared flux may be in the order of 100 - 150 W m^{-2} while the daily average of the net solar flux is in the order of 300 W m^{-2} at an average surface albedo of 0.22 (the albedo changes under clear sky conditions slightly with the solar zenith angle, it is larger at low sun than at noon). Under undisturbed conditions (no heat advection, cloudfree sky) the temperatures at the surface, in the atmosphere and in the soil return after 24 hours to approximately the same values. This implies that (most of) the excess solar energy deposited at the surface must have been dissipated from here by a different mechanism than radiative transfer. In fact due to the absorption process radiation energy is transferred into heat and partitioned into different heat fluxes, the sum of which we can denote by $M_{H,o}^*$:

$$M_{R,o}^{*} - M_{H,o}^{*} = 0. \quad (5)$$

The storage of heat in the soil/ocean system or in the atmosphere is only intermediate and neglegible in the average over one year. All excess solar energy must therefore be transferred to the atmosphere as heat ($M_{H,o}^+$) and finally re-emitted to space at elevated levels as longwave thermal radiative energy. It is evident that the whole energy tranfer system depends on the distribution and concentration of constituents in the atmosphere, which either absorb solar energy or are involved in the radiative transfer in the infrared part of the spectrum.

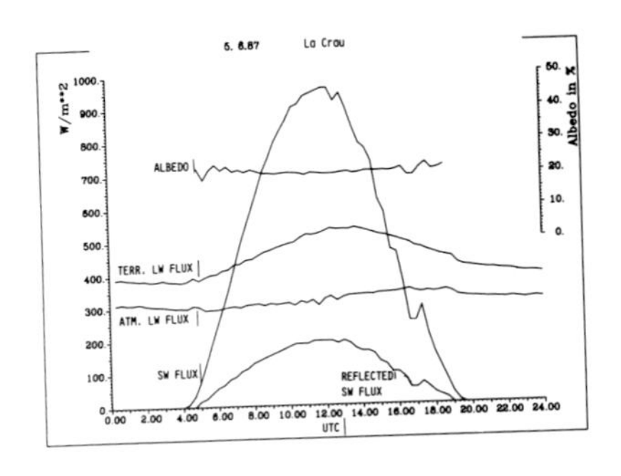

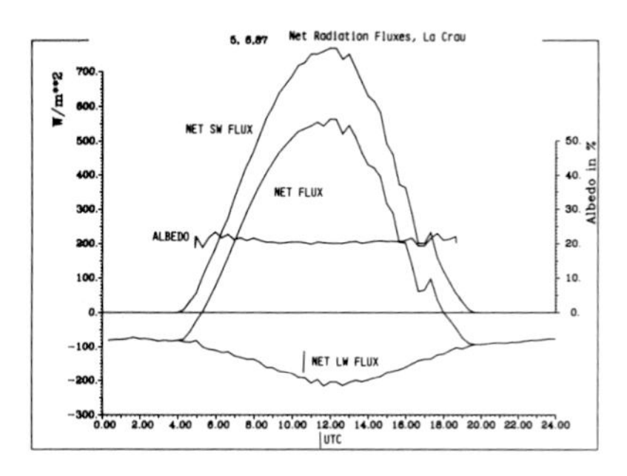

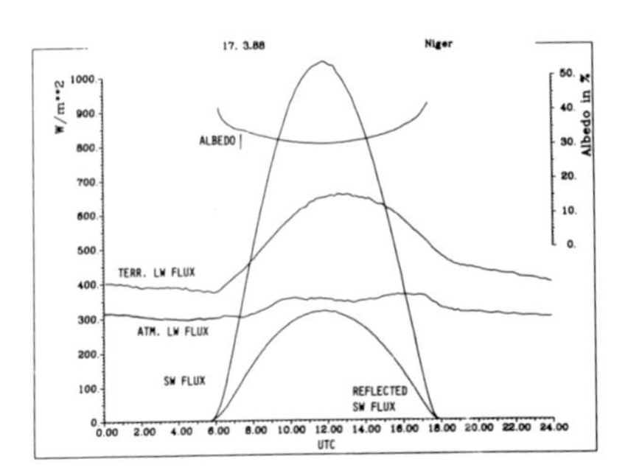

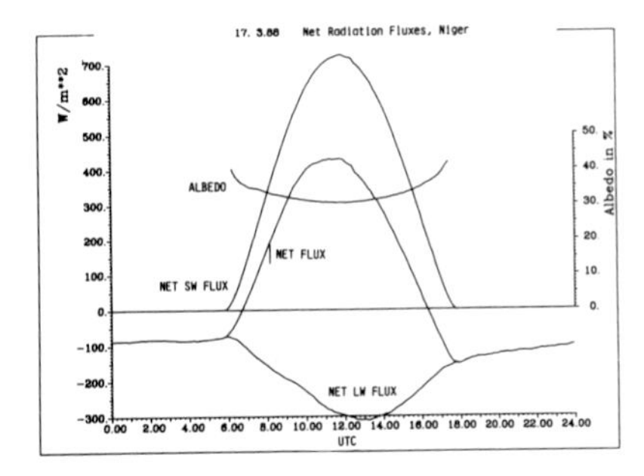

Fig. 1. The net radiation flux and its components measured at the ground in two locations: La Crau in South France (top), and Ibecetene in Niger (bottom).

3. DEPENDENCE OF SURFACE TEMPERATURE ON ATMOSPHERIC PROPERTIES.

In order to understand how a change in the composition of the atmosphere can affect the temperature at the surface, we have to solve equation (3) with respect to the surface temperature T_o. This will again be done in a very simple way in order to show how the principle works, without going into the details of the radiative transfer. First we rewrite equation (3) with the help of eqs. (3) and (5) where $M_{H,o}^*$ is replaced by $M_{H,o}^+$:

$$\varepsilon_o \sigma T_o^4 = M^+_{LW,o} = - M^+_{H,o} + M^-_{SW,o}(1 - \rho_o) + M^-_{LW,o}. \qquad (6)$$

The heat flux is still a function of T_o, which makes it difficult to solve eq. (6) for T_o. But since the longwave net flux is in the global average small compared with the net shortwave radiative flux, it may be sufficient to parametrize the heat flux into the atmosphere by the net shortwave flux:

(7) $$M^+_{H,o} = \chi M^-_{SW,o} (1 - \varrho_o).$$

The downward shortwave radiation can be related to the extraterrestrial solar flux in two ways. Either an effective transmission τ^* can be introduced to describe the losses of the shortwave radiation in the atmosphere, or the law of conservation of energy can be applied. In the first case it follows from equation (2)

(8) $$M^-_{SW,o} = \tau^* S_o/4 = \tau^* \sigma T^4_{eff}/(1 - \varrho_p)$$

In the second case the fraction of the solar flux that is absorbed in the atmosphere and at the ground must equal the net input to the system:

(9) $$M^-_{SW,o}(1 - \varrho_o) + \alpha S_o/4 = S_o(1 - \varrho_o).$$

α is the fraction of the extraterrestrial solar flux that is absorbed in the atmosphere by gases and aerosols.

Both expressions, eq. (8) and (9) are equivalent, if we consider that

(10) $$\tau^* = 1 - \alpha - \varrho_a$$

where $_a$ is the fraction backscattered to space by the atmosphere, and

(11) $$\varrho_p = {}_a + \tau^* \varrho_o.$$

By inserting the relations (4), (7) and (8) into eq. (6) it results:

(12) $$T^4_o = \tau^*(1 - \varrho_o)T^4_{eff}(1 - \chi)/[\varepsilon_o(1 - \varrho_p)] + T^4_a(\varepsilon^-_a/\varepsilon_o)$$

To determine the numerical value of T_o it is necessary to measure or to compute the quantities that enter eq. (12). ϱ_p and S_o have to be measured. Satellites are capable to provide this information. The presently valid value of ϱ_p is 0.29 ... 0.30. The measuring techniques will not be discussed here. Other quantities can be computed, if the atmospheric composition and the optical properties of the atmospheric constituents are known and if χ is determined [see Bolle II, this volume].

To derive the vertical structure of the atmosphere, the radiative energy absorbed in each atmospheric layer must be computed. This quantity is generally presented as a heating rate, which is the temperature change caused by the absorption (or emission) of radiant energy. Temperature changes due to exchange of radiant energy take place until a radiative equilibrium is achieved. In this case the absorbed radiative energy equals the radiation emitted according to its temperature by the atmospheric layer in question. An atmosphere in radiative equilibrium would have a very steep temperature gradient near the surface - mainly because of the strong decrease in water vapor density with height -, and a nearly isothermal cold middle atmosphere, if there would be no ozone. The presence of ozone in the Earth's atmosphere causes the warming of the middle atmosphere with peak temperatures at the stratopause at about 50 km altitude.

Because of the heat flux M_H from the surface into the atmosphere the lower part of the atmosphere is not in radiative equilibrium. The heating of the atmosphere from the bottom causes raising air - convection - which erases the steep temperature gradient of radiative equilibrium at the bottom and replaces it by a steeper adiabatic temperature gradient which is the result of thermodynamic equilibrium. At some altitude, however, the adiabatic lapse rate becomes larger than the radiative lapse rate. At this altitude convection is not effective anymore and a transition occurs from the adiabatic to the radiative lapse rate. This is the altitude where the temperature gradient goes through an isothermal layer to middle atmosphere conditions. This altitude is called the tropopause, which occurs at about 15 - 18 km in the tropics and 6 - 8 km in the arctic region.

4. RADIATIVE TRANSFER IN A CLOUDLESS ATMOSPHERE

4.1. General remarks.

The quantities by which the radiation properties of the atmosphere are described in equation (12) are τ^* respectively α and ρ_a for the shortwave part of the spectrum and $\varepsilon^-{}_a T_a^4$ or the equivalent $M^-_{LW,o}/\sigma$ for the infrared part. In this context it is only possible to outline some of the basic principles of radiative transfer, and it is also necessary to limit ourselves to the cloudless atmosphere.

4.2. Shortwave radiative transfer.

Most of the energy of the solar radiation is confined within the spectral range 0.2 to 2.5 µm. In Fig. 2 the extraterrestrial solar spectrum and its change by atmospheric absorption is demonstrated. The absorption is due to atmospheric ozone, water vapor and to a very small degree and quasi neglegible also by oxygen and carbon dioxide molecules. While the ozone absorption in the ultraviolett and visible range of the spectrum looks like a structured continuum, the water vapor absorption in the near infrared shows under high spectral resolution a distinct line structure.

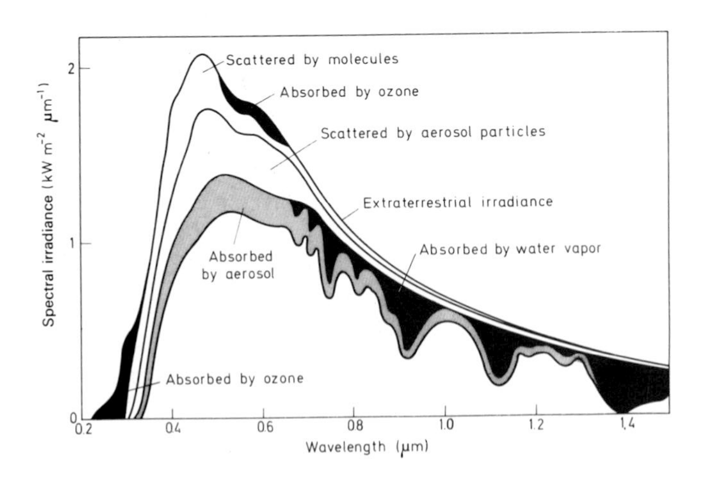

Fig. 2. The extraterrestrial solar radiation spectrum and its modification by the atmosphere.

The transitions that cause these absorption features are in the case of the ozone molecule electronic exitations or dissociative absorptions while in the near infrared range the vibration-rotation bands of the water molecule are responsible for the absorption. From known band strengths and the absorbing mass the amount of absorbed solar energy can be computed for each level of the atmosphere. About 19% of the extraterrestrial flux is absorbed by atmospheric gases.

Also aerosols absorb radiation (in the global average 4%) depending on their chemical composition (respectively their complex index of refraction) and their sizes. Clouds extinguish approximately the same fraction in relation to the extraterrestrial solar radiation.

Atmospheric gas molecules and aerosols have another property, which affect the absorbed radiative energy. This is the scattering of radiation, which forces photons to travel through the atmosphere for more than just one time. Molecules distribute the scattered radiation like dipoles emit electromagnetic energy (Rayleigh scattering). The scattering in forward and backward directions is equally large. At 0 and 180 degree scattering angle counted from the direction in which the primary beam proceeds, the radiation is unpolarized while at 90 and 270 degree it is fully polarized. Molecules always scatter at a single impact the same amount of energy in both, the upper and the lower hemispheres. This is quite different for aerosols. The larger the scattering particle is, the stronger is the scattering in foreward direction (Mie scattering). Aerosol scattering - as well as cloud droplet scattering - is highly asymmetric and therefore the partitioning of the scattered energy into the two hemispheres is also different. At a single impact a fraction of the incomming photons is absorbed by the aerosols. The ratio of scattered to impacting photons is called the single scattering albedo. If the number density, the size distribution and the single scattering albedo are known for an assemble of aerosols, the amount and the angular distribution of the scattered radiation in a volume of turbid air as well as the energy absorbed in this volume can be computed.

The transmission through the atmosphere in a spectral interval can be computed from Lambert-Beer's Law

$$\tau_D = (1/\Delta\nu)\int \exp(-\int \sigma ds)d\nu. \quad (13)$$

Here ν is the wavenumber, which is the inverse of the wavelength measured in cm^{-1}), σ the extinction coefficient (generally the sum of Rayleigh scattering, Mie scattering and absorption), and s the path length. The Rayleigh scattering coefficient σ_R is strongly wavelength dependent. It decreases with the forth power of the wavelength. The Mie scattering coefficient σ_M is much less wavelength dependent, it decreases with the 1.5th power.

The transfer of radiative energy in the atmosphere is complicated by the multiple traversal of the photons, some of which may even be reflected by the surface. Multiple pathways through the atmosphere are schematically indicated in Fig. 3. Here f and b are the fractions of radiation scattered into the hemispheres defined by the horizontal plane and the direction into which the primary beam proceeds respectively the opposite direction. $f = b = \frac{1}{2}$ for Rayleigh scattering but is different and dependent on the zenith angle of the primary beam for aerosol scattering. The letter r stands for the total scattered fraction of radiation and T for the flux of the direct and single scattered solar radiation reaching the surface. α and τ are the absorptance and the transmittance respectively and the indices D and F denote whether these symbols refer to the direct solar beam or to a diffuse flux of scattered and/or reflected shortwave radiation. This distinction is necessary because the direct solar beam enters the atmosphere at a specific zenith angle and passes a relative airmass approximately equal to the inverse of the cosine of the zenith angle while the diffuse flux travels at all possible angles, which for completely diffuse radiation is equivalent to an effective zenith angle of 53 degree or a relative air mass of 1.66. Fig. 3 also presents the formulae for the fractions of the solar energy that is transmitted to the surface, absorbed in the atmosphere or scattered back to space. Similar expressions would also hold if a cloud layer is introduced to the atmopshere, though this obviously complicates the problem to compute the fluxes.

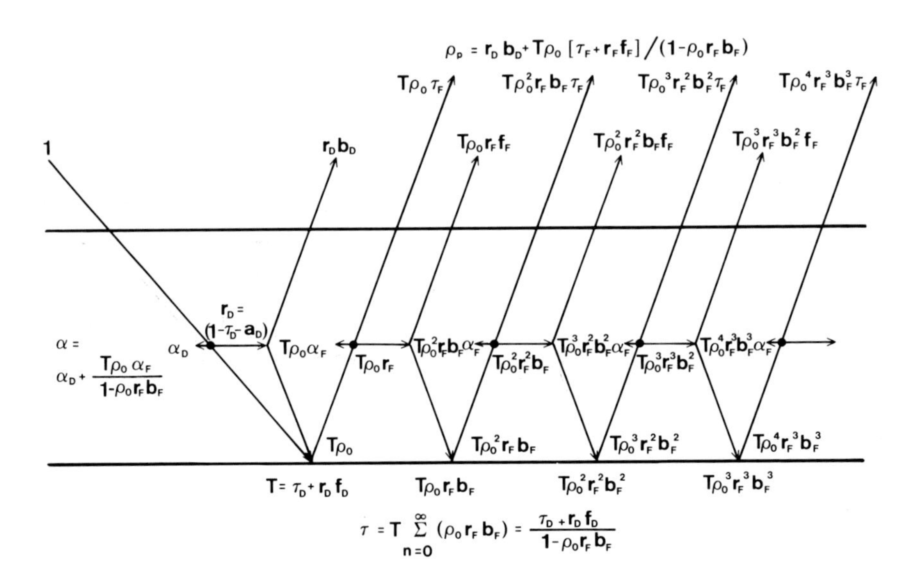

Fig. 3. Multiple scattering in the atmosphere. The incident extraterrestrial solar radiation flux is 1.

4.3. Longwave radiative transfer.

In the spectral region 3 to 30 µm, where most of the energy transfer in the thermal infrared occurs, the radiation fluxed are mainly determined by the vibration-rotation bands of gases which have a permanent dipole momemt like water vapor, carbon dioxide, ozone, methan, di-nitrogen-oxide and substances like the chloro-fluoro-carbons. These bands are centered at resonance frequencies for molecular vibrations (6.3 µm for H_2O, 4.3 and 15.0 µm for CO_2, 9.6 µm for O_3, 7.66 µm for CH_4 and 7.78 and 8.56 µm for N_2O) and combinations of these states. They spread from these frequencies to both sides because of simultanous changes in the many possible rotational states of the molecule. Water exhibits an additional feature. It is the only one of the atmospheric molecules with a pure rotation band that extends from its center near 200 µm into the region of 10 µm with individual lines but also with a quasi-continuum resulting from the superposition of the wings of the strong rotation lines. This quasi continuum is further complicated around 10 µm by the absorption of water vapor clusters $(H_2O)_n$ that form in the atmosphere.

The position of the many absorption lines as well as their strenths are fairly well known. Assuming a Lorentzian line shape, the absorption coefficient of one line can be expressed by

$$a(\nu) = (S[T]/\pi) * \alpha/([\nu - \nu_o]^2 + \alpha^2), \tag{14}$$

where ν is the wavenumber, ν_o the wavenumber of the line center, S the temperature dependent line strength and 2α the half widths of the line. $\alpha = \alpha(p,T)$ depends on pressure as well as - to a lesser degree - on temperature. This simple line shape formula has to be modified for wavenumbers more than a few cm^{-1} off the center and for higher layers of the atmosphere, where the Doppler effect becomes essential. To compute the absorption coefficient at any part of the spectrum one has in principle to sum over all lines that contribute to the absorption at this wavenumber.

In the atmosphere absorption and emission processes proceed from one layer to the next. The basic relations to compute the radiance of the atmosphere are summarized in Fig. 4. We consider first the spectral radiance dL originating in an infinitesimal layer dz. It is given by the Planck function B(T) for a black body (a cavity in thermal equilibrium) at temperature T - which is the maximum radiance that can be obtained from an infinite isothermal layer - multiplied by the absorption coefficient a and the emitting mass dm = $dz/\cos\xi$, where ξ is the zenith angle of the direction under observation.. Of this radiance the fraction $\tau(z,0)$ arrives at the surface. In order to compute the radiance from the whole atmosphere it has to be integrated from the surface to the top of the atmosphere. For the spectral radiance at the ground results the formula (with $\mu = \cos\xi$):

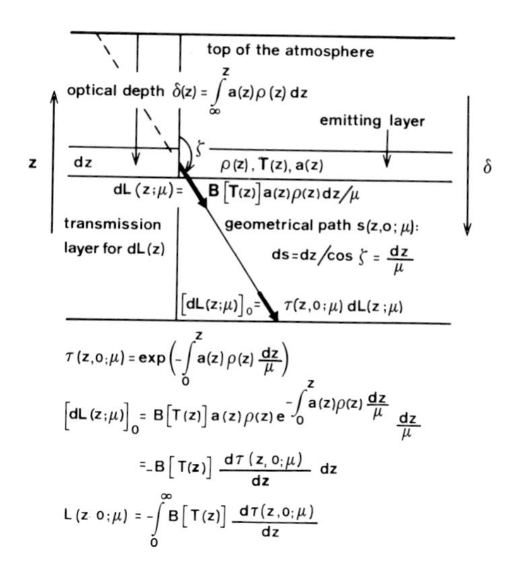

Fig. 4. Longwave radiative transfer through the atmosphere

$$L(\mu,z=0) = -\int_0^\infty B[T(z)][d\tau(\mu,z,0)/dz]dz \tag{15}$$

The spectral flux is obtained, if the radiance is integrated over the hemisphere. It can approximatively be computed, if the vertical optical depth δ (see Fig. 4) is replaced by 1.66δ. 1.66 is called the diffusity factor. Atmospheric thermal infrared spectra are shown in Fig. 5.

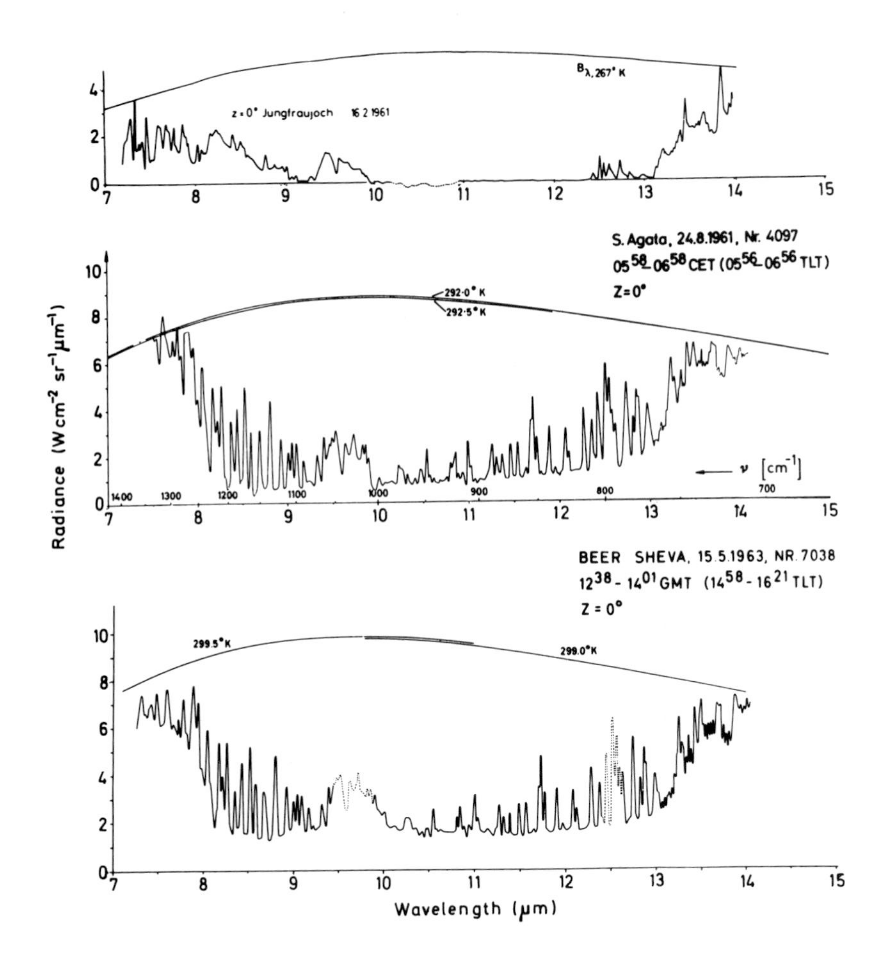

Fig. 5. Sky emission spectra measured at three locations: Jungfraujoch (3500 m), Switzerland, S. Agata, Italy and Beer Sheva, Israel.

The integration over the wavelengths is very time consuming if done line-by-line because the spectral resolution which has to be applied in the computation must be much smaller than the halfwidth of the lines. For the spectral integration therefore often averaged transmission functions or - with a slight modification of the derived equation - emissivities are used to compute the fluxes. These have carefully to be applied if the various emitters are taken into account: because spectral lines of different orign partly overlap, the result of a line-by-line computation differs in most cases from a simple multiplication of average emissivities.

4.4. The general equation of radiative transfer.

Differentiation of equatiom (15) yields with $\tau = \exp(-\delta/\mu)$, $\delta = a \cdot m$, m the mass in a vertical column,

$$\mu(dL/d\delta) = L(\delta) - B[T(\delta)], \tag{16}$$

which is known as the Equation of Radiative Transfer. This equation can generally be used if B is replaced by a general source Function J which includes also scattering and absorption processes.

5. EFFECT OF INCREASING CONCENTRATIONS OF RADIATIVELY ACTIVE SUBSTANCES ON THE ENERGY BUDGET AND THE TEMPERATURE AT THE SURFACE.

From eq. (12) it becomes immediately evident, that, with unchanged conditions at the top of the atmosphere, the surface temperature T_o must rise, if

- τ^* rises (less opaque atmosphere for solar radiation)
- ρ_o decreases (darker surface)
- the ratio χ of heat flux to net radiation decreases
- ε^-_o decreases (more deserts but this would be compensated by α_o, which would at the same time increase)
- ε^-_a increases (optical denser atmosphere in the thermal infrared respectively emission from lower and warmer layers of the atmosphere if T_a remains fixed)
- T_a increases.

The cross references between some of the quantities already show the complexity of the involved feedbacks. Furthermore α_p was treated as a constant, but if τ^* and/or ρ_o changes, also α_p would change, which would force T_{eff} to adjust to the new condition.

The radiative fluxes in the atmosphere as defined in sections 2 and 3 are schematically shown in Fig. 6. Here a cut was made at the tropopause to separatem the contributions of the troposphere, denoted by ' respectively index T, and of the middle atmosphere, denoted by " respectively index M. Following the discussion in section 1 we recall that the troposphere is in convective equilibrium and its temperature gradient is only changing if the water vapour concentration changes.

If the concentration of atmospheric gases with infrared absorption bands ("radiatively active gases") is increased as it is presently the case, the radiative transfer is changed. The absorbing matter in the atmosphere becomes larger, its emissivity increases and the effective emission height of the downward directed infrared radiation in the troposphere is lowered and shifted into levels with higher temperatures. The surface receives more radiation. Less radiation that originates at the surface can escape to space and is now used to warm the troposphere. The effect is a shift of the whole atmospheric lapse rate towards higher

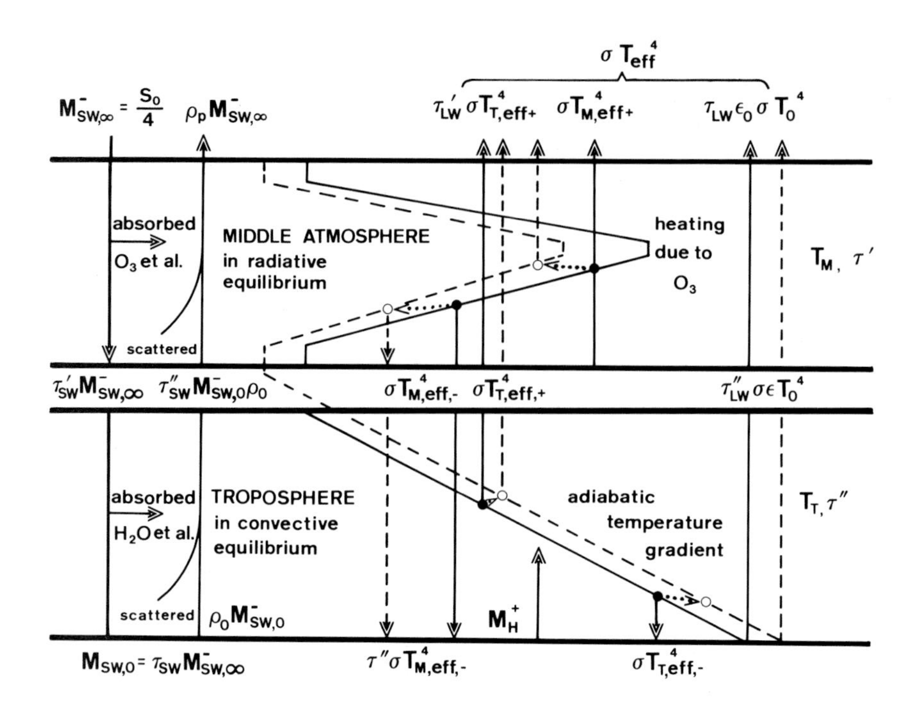

Fig. 6. Modification of radiative fluxes due to the increased greenhouse effect.

temperatures and a warming of the surface. Mainly due to the enhanced surface temperature the upwelling infrared flux at the tropopause will increase. This enhances momentary the outgoing flux at the top of the atmosphere. The Earth will then be in an transient stage and not in radiative equilibrium. In this stage it permanently looses more energy than it gaines from the sun. An equilibrium state can only be reached again, if the middle atmosphere compensates for the increased radiation from below. Since also in this region the emitting mass increases, the emitting layer for the downwelling radiation is shifted downward - but here into cooler regions - which consequently reduces the downward flux, while the effective emission layer for the outgoing radiation is shifted upwards into warmer layers near the stratopause. As a result the middle atmosphere's cooling against space is enhanced and its temperature decreases until its infrared losses are again in equilibrium with the gain of solar radiation.

But now the middle atmosphere is not in radiative equilibrium with respect to its lower boundary, the tropopause. Here the upward infrared flux is increased and the downwelling flux decreased which results in a net increase of the upwelling infrared flux. This could be corrected, if the tropopause shifts upward into cooler regions thus reducing the upward directed flux. This would be in accordance with the fact, that in a warmer atmosphere the convection may reach higher.

Much depends also on the water vapor in the atmosphere. If the relative humidity remains constant, the absolute humidity increases in the troposphere but decreases in the now cooler middle atmosphere. This would have four effects. The solar radiation is less absorbed in the stratosphere, its flux into the troposphere increases where the absorption increases and less radiation returns to space. The infrared fluxes from both the middle atmosphere downward and the troposphere upward would be reduced. The first because there is less emitting water vapor in the middle atmosphere and the second because the emission level is shifted upward into cooler regions.

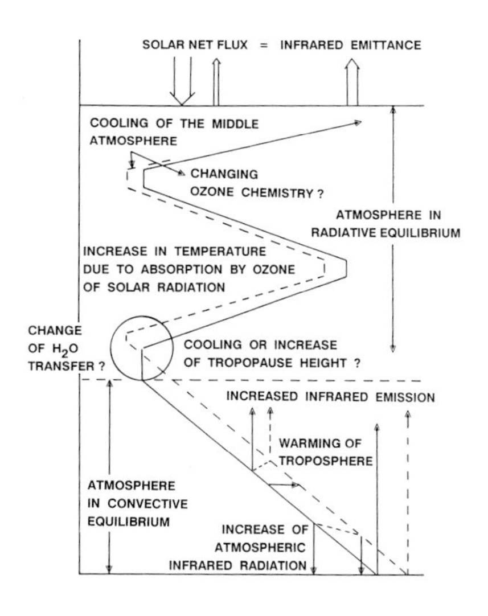

Fig. 7. Possible changes of the atmospheric structure due to the increasing greenhouse effect.

The downward radiation from the middle atmosphere may also be adjusted due to a vertical redistribution of radiatively active gases because of a temperature dependent photochemistry. A related aspect is the possible depletion of the ozone concentration. This would change the absorption of solar radiation in the middle atmosphere and would have impact on the temperature distribution in this layer. According to the above considerations this would introduce another feedback mechanism that has to be taken into account.

There are also other compensating effects possible. If cloudines increases due to an enhanced convection, there is less shortwave radiation penetrating to the surface which counteracts the heating by the increased amount of radiatively active gases in the thermal infrared. Thus the energy balance at the surface is completely or partly restored and by this also the outgoing infrared flux. Aerosols which are injected by volcanic eruptions into the stratosphere have the same effect. They scatter back to space additional shortwave radiation which is lost for the heating of the lower atmosphere and at the same time they warm the stratosphere by absorption of a small fraction of the radiant energy. This effect, which counteracts the warming by increasing concentrations of the radiatively active gases, lasts only for a few, up to about seven years after a volcanic eruption because of the sedimentation of the volcanic aerosols and can therefore only in times of strong volcanic activity reduce the incomming shortwave radiation over a longer period.

6. CONCLUSION

As a consequence of the changing radiative transfer the middle atmosphere must cool while the troposphere and the surface warm up with increasing concentration of radiatively active gases. Only this process was discussed in this paper. The resulting changes of the radiation fluxes affect the structure of the atmosphere and a number of possible feedback processes have to be studied in order to predict how the atmosphere umtimately adjusts to the changing radiation regime.

The accurate quantitative treatment of the various feedbacks is complicated by a number of facts. The radiatively active gases in question have different vertical and in some cases also horizontal distributions. Their absorption bands have different positions in the spectrum, which causes changes in the emission efficiencies because of the temperature dependence of the Planck function. It has also to be assumed that no other major changes occur in the radiative transfer, such as a change in cloudiness. It is also likely that with a change of the amount and distribution of atmospheric water vapour the slope of the atmospheric lapse rate changes. A possibly changing photochemistry has also to be accounted for. Difficult to predict is the situation at the tropopause. A temperature increase in the troposphere should enforce higher reaching convection with an elevated tropopause, and a smaller lapse rate in a wetter atmosphere would demand either a warmer or a higher tropopause. The latter would be in better agreement with the cooler middle atmosphere.

The radiation processes that are essential for climate studies could only be presented here in a very much simplified manner. Major problems involved, which could not be addressed, concern the radiative properties of clouds and the possible changes in cloudiness with increasing concentrations of radiatively active gases. For more detailed description of these processes a few references are given.

Literature:

H.-J. Bolle, Radiation and Energy Transport in the Earth Atmosphere System. In: O. Hutzinger, ed., Environmental Chemistry, Vol. 1, Part B, Springer-Verlag, Berlin, 1982.

Paltridge, G.W. and C.M.R. Platt, Radiative Processes in Meteorology and Climatology. Elsevier, 1976.

Ramanathan, V., R.J. Cicerone, H.B. Singh, and J.T. Kiehl, 1985: Trace Gas Trends and Their Potential Role in Climate Change, J. Geophys. Res. 90, D3, 5547-5566.

ATMOSPHERIC CHEMISTRY AND CLIMATE

Guy BRASSEUR
National Center for Atmospheric Research*
Boulder, CO 80307

Summary

The paper presents an overview of the chemical and photochemical processes in the atmosphere (troposphere and stratosphere) which are of importance for climate studies. Greenhouse gases, which contribute to the warming of the Earth's surface, are destroyed by oxidants (such as OH) which are produced through complex chemical processes. These may be altered by human activity. Models predict for the future substantial changes of the atmospheric composition and of the climatic conditions. Dramatic perturbations resulting from human activities (such as the ozone hole over Antarctica) are already observed.

1. INTRODUCTION

Chemical constituents in trace quantities play an important role for radiative transfer in the atmosphere. Hence, atmospheric dynamics and climate are sensitive to the delicate chemical processes occurring in the troposphere and even in the stratosphere.

Ozone, for example, which absorbs solar ultraviolet radiation, heats the atmosphere in the 20-60 km region and is therefore responsible for the existence of the stratosphere. Species such as carbon dioxide, water vapor, ozone, methane, nitrous oxide and chlorofluorocarbons absorb the thermal radiation emitted by the Earth. This so-called greenhouse effect produces a warming of the Earth's surface. The abundance of the greenhouse gases depends on a number of chemical and photochemical processes in the atmosphere as well as on the strength and the distribution of the eventual biological, oceanic and industrial sources of these gases.

The purpose of this paper is to review some of the connections between climate and atmospheric chemistry over time scales of decades to centuries. Because of the increasing emission of radiatively and chemically active gases, large and perhaps global changes are presently occurring in the atmosphere (e.g., the ozone hole over Antarctica) and their effects are predicted to even become more severe in the next few decades.

* The National Center for Atmospheric Research is funded by the National Science Foundation.

Human kind is conducting a giant chemical experiment in the atmosphere which should lead to substantial environmental and climatic perturbations on the global scale.

2. THE GREENHOUSE GASES

Because of the development in many parts of the world of industrial and agricultural activities, the release in the atmosphere of various chemical compounds is increasing rapidly. It is now well known for example that, primarily as a result of the burning of fossil fuels, the concentration of carbon dioxide, which was of the order of 285 ppmv in the beginning of the 19th century, is now (1989) approaching 350 ppmv. Beside a photodestruction mechanism, which takes place only above 80 km altitude, no chemical process destroys CO_2 in the atmosphere: this gas dissolves into the ocean, where it forms carbonate ions, or is incorporated in the biosphere through photosynthesis in the green plants.

The study of the global carbon cycle requires an accurate determination of the storage and exchanges of carbon compounds between the different parts of the geosphere. For example, the yearly transfer between the atmosphere and the ocean and between the atmosphere and the biosphere is roughly 13 percent and 10 percent of the atmospheric CO_2 content, respectively. The CO_2 input due to fossil fuel consumption and deforestation represents only 0.8 percent and 0.15 percent of the CO_2 content but, because the CO_2 system has not enough resilience to buffer these perturbations, about half of the carbon dioxide released in the atmosphere remains airborne. The anthropogenic perturbation, while small compared to the natural exchange fluxes, enhances by about 1.5 ppmv per year the global amount of carbon dioxide in the atmosphere.

Carbon dioxide, a linear molecule with a relatively simple absorption spectrum, has a strong absorption band at 15 μm. Since this band is located in a spectral region in which the emission of the terrestrial environment is very intense, it significantly affects the energy budget of the atmosphere. Another intense band at 4.3 μm has a less important influence because both solar and terrestrial emissions are weak at this wavelength.

Methane, whose average mixing ratio at the Earth's surface is presently of the order of 1.7 ppmv, is increasing by 0.8 to 1.0 percent per year. Methane concentrations were of the order of 1.15 ppmv in the middle of this century and as low as 0.35 ppmv during the ice ages. This gas is essentially of biological origin. As shown by Table 1, it is produced in rice paddies, swamps and lakes, in the tundra and the boreal marsh, in the guts of animals (cattle, termites and other insects), by biomass and solid waste burning, by natural gas losses and coal mining. About 85 percent of methane emitted at the surface is destroyed in the troposphere through oxidation by the hydroxyl radical (OH).

Methane has both direct and indirect effects on climate. It interacts directly with terrestrial infrared radiation, essentially through an absorption band at 7.66 μm and indirectly forming carbon dioxide, water vapor and ozone through chemical oxidation. These constituents further contribute to the greenhouse effect of the atmosphere.

Nitrous oxide is also produced at the Earth's surface. Analyses of ice cores show that the average concentration of this compound in the atmosphere was about 285 ppbv before the beginning of industrialization. The present value is nearly 310 ppbv with an increase of 0.8 ppbv per year. Natural sources of N_2O include microbiological

Table 1. Annual methane release rates for identified sources (Cicerone and Oremland, 1988)

Identity	Annual Release (Range) 10^{12}g CH_4
Enteric fermentation (animals)	65-100
Natural wetlands (forested and nonforested bogs, forested and nonforested swamps, tundra and alluvial formations)	100-200
Rice paddies	60-170
Biomass burning	50-100
Termites	10-100
Landfills	30-70
Oceans	5-20
Freshwaters	1-25
Methane hydrate destabilization	0-100
Coal mining	25-45
Gas drilling, venting, transmission	25-50
TOTAL	400-640

processes on the continent and in the ocean; anthropogenic productions are associated with burning of fossil fuels and the extensive use of nitrogen fertilizers. The major sink of N_2O is its photodissociation in the stratosphere. This gas also contributes to the greenhouse effect of the atmosphere.

The species with the largest relative increase rate in the atmospheric density are several halocarbons (e.g., the chlorofluorocarbons) as well as their decomposition products in the stratosphere. With the exception of methyl chloride (CH_3Cl) and, to a certain extend, of methyl bromide (CH_3Br), these compounds are almost exclusively of industrial origin. They are used for example as aerosol propellants, refrigerants, agents for blowing plastic foams, solvents, etc.

Fully halogenated halocarbons such as CCl_4 (CFC-10), $CFCl_3$ (CFC-11), CF_2Cl_2 (CFC-12), $C_2F_3Cl_3$ (CFC-113), $C_2F_4Cl_2$ (CFC-114), C_2F_5Cl (CFC-115) as well as the brominated compounds such as CF_2BrCl (halon-1211), CF_3Br (halon-1301), $C_2F_4Br_2$ (halon-2402) are photodecomposed by short wave solar ultraviolet in the stratosphere. The lifetime of these molecules is relatively long (about 70 years for CFC-11, 140 years for CFC-12, 100 years for CFC-113). Since the rate of release of these chemical compounds at the surface exceeds substantially their rate of destruction in the stratosphere, these gases accumulate in the atmosphere. Partially halogenated halocarbons such as CH_3CCl_3 (methylchloroform), $CHClF_2$ (HCFC-22), CF_3CHCl_2 (HCFC-123), CF_3CHClF (HCFC-124), CCl_2FCH_3 (HCFC-141b) and $CClF_2CH_3$ (HCFC-142b) are destroyed in large part in the troposphere by OH radicals. The lifetime of these compounds is generally of the order of 10 years.

The photochemical or chemical decomposition of these halocarbons leads to the production of chlorine atoms which contribute efficiently to the destruction of ozone in the stratosphere. The potential for ozone depletion is substantially larger in the case of the fully halogenated than in the case of partially halogenated halocarbons. Several molecules such as CF_3CHF_2 (HFC-125), CF_3CH_2F (HFC-134a), CF_3CH_3 (HFC-143a) or CHF_2CH_3 (HFC-152a) do not contain any chlorine or bromine atoms and are therefore not responsible for the destruction of ozone. All the fully or partially halogenated halocarbons contribute however to the greenhouse effect. Table 2 presents some important characteristics of these compounds including their ozone depletion potentials (ODP) relative to that of CFC-11.

3. THE GREENHOUSE EFFECT

On a global and annual basis, the energy gain resulting from the absorption of solar radiation in the climatic system is balanced by the emission to space of terrestrial radiation produced by the Earth's surface, by several radiative trace gases in the atmosphere as well as by clouds and aerosols. This radiation is emitted in the thermal infrared at wavelengths larger than 4 μm. Longwave radiation emitted by the surface is partly intercepted by the atmosphere and re-emitted at its own temperature. To describe this effect from a purely conceptual point of view, let's consider (Fig. 1) the exchange of energy in a single isothermal atmospheric layer which includes a radiatively active gas (e.g., carbon dioxide). The temperature of this gas is noted T, the

Table 2. Average mixing ratios, global trends, global emissions, lifetimes and ozone depletion potentials of atmospheric halocarbons (approximate values).[1] Data correspond to the early 1980's.

Compound	Average mixing ratio (pptv)	Global trend (%/yr)	Global emission (10^9g/yr)	Lifetime (yrs)	Ozone depletion potential (relative to CFC-11)
CFC-11	200	5	330	70	1.0
CFC-12	320	5	440	100	0.9
CFC-113	20	12	140	100	0.9
CFC-114	11	6	20	300	0.7
CFC-115	4	5	5	500	0.4
$CC\ell_4$	140	1	95	60	1.2
$CH_3CC\ell_3$	130	6	520	7	0.15
HCFC-22	60	11	200	20	0.05
HCFC-123				2	0.02
HCFC-124				7	0.02
HCFC-141b				10	0.1
HCFC-142b				25	0.05
Halon 1211	1	20	5	15	2.2[2]
Halon 1301	1		8	110	13.2[2]
Halon 2402					6.2[2]

[1] Based on information by Wuebbles (1986), Fabian (1986), Wuebbles and Edmonds (1988), Prinn (1988) and CMA (personnal communication).

[2] Approximate values variable with the amount of chlorine present in the atmosphere.

absorptance of the layer at wavelength λ is $\Delta\tau(\lambda)$ and the temperature of the surface is T_s. The emission function at wavelength λ and temperature T is then given by

$$B(\lambda, T) \sim \lambda^{-5} exp(-960/T).$$

The outgoing flux F above the layer is the sum of the emission by the surface, after absorption in the layer, and of the emission by the layer itself:

$$F(\lambda, T_s, T, \Delta\tau) = \pi B(\lambda, T_s)\,[1 - \Delta\tau(\lambda)] + \pi B(\lambda, T)\Delta\tau(\lambda)$$

or

$$F(\lambda, T_s, T, \Delta\tau) = \pi B(\lambda, T_s) + \Delta F(\lambda, T_s, T, \Delta\tau)$$

where

$$\Delta F(\lambda, T_s, T, \Delta\tau) = \Delta\tau(\lambda)\,[B(\lambda, T) - B(\lambda, T_s)] \cdot \pi$$

Thus, if the surface temperature (T_s) is larger than the temperature of the atmospheric layer (T), the outgoing flux is reduced ($\Delta F < 0$) by the presence of the radiatively active gas and hence radiation is trapped in the atmosphere. Figure 2 shows an observed spectrum of the outgoing longwave emission over the tropics. The signature of the surface temperature (about 290 K) is visible in the spectral regions where atmospheric absorption is negligible. The two strong absorption bands due to carbon dioxide near 15 μm and to ozone near 9.6 μm are clearly visible and are a direct manifestation of the greenhouse effect. The signature of water vapor is also present. Table 3a shows the radiative energy being trapped by the atmosphere, as calculated by Dickinson and Cicerone (1986) for 1985 conditions, while Tables 3b and c present an estimation by

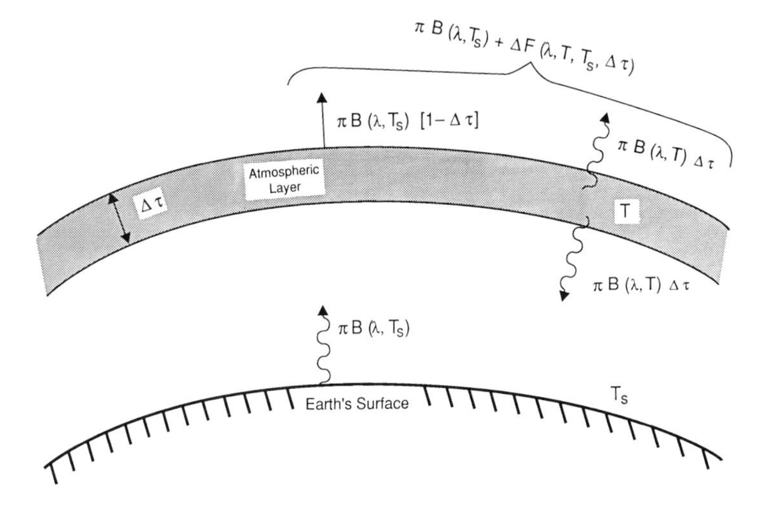

Figure 1. Schematic representation of the greenhouse effect by a single atmospheric layer.

Table 3a. The current 1985 trapping of thermal infrared radiation (ΔQ) by current tropospheric trace constituents

Gas	Current Concentration	ΔQ_{total} ($W\ m^{-2}$)
Carbon dioxide	345 p.p.m.	~ 20
Methane	1.7 p.p.m.	1.7
Ozone	10-100 p.p.b.	1.3
Nitrous oxide	304 p.p.b.	1.3
CFC-11	0.22 p.p.b.	0.06
CFC-12	0.38 p.p.b.	0.12

The term ΔQ_{total} is the change of net radiation at the tropopause if the given constituent is removed, but the atmosphere is otherwise held fixed.

Table 3b. Estimated pre-industrial trace gas concentrations and implied change in thermal trapping to 1985

Gas	Pre-industrial Concentration	$\Delta Q_{pre-industrial}$ ($W\ m^{-2}$)
Carbon dioxide	275 p.p.m.	1.3
Methane	0.7 p.p.m.	0.6
Tropospheric ozone (below 9 km)	0-25% less	0.0-0.2
Nitrous oxide	285 p.p.b.	0.05
CFC-11	0.00 p.p.b.	0.06
CFC-12	0.00 p.p.b.	0.12
TOTAL		~ 2.2

Table 3c. Scenario for trace gas concentrations in the year 2050 and implied increased trapping of thermal radiation from 1985

Gas	Year 2050 scenario	ΔQ_{2050} ($W\ m^{-2}$)
Carbon dioxide	400-600 p.p.m.	0.9-3.2
Methane	2.1-4.0 p.p.m.	0.2-0.9
Tropospheric ozone (0-12 km)	15-50% more	0.2-0.6
Nitrous oxide	350-450 p.p.b.	0.1-0.3
CFC-11	0.7-3.0 p.p.b.	0.23-0.7
CFC-12	2.0-4.8 p.p.b.	0.6-1.4
TOTAL		2.2-7.2

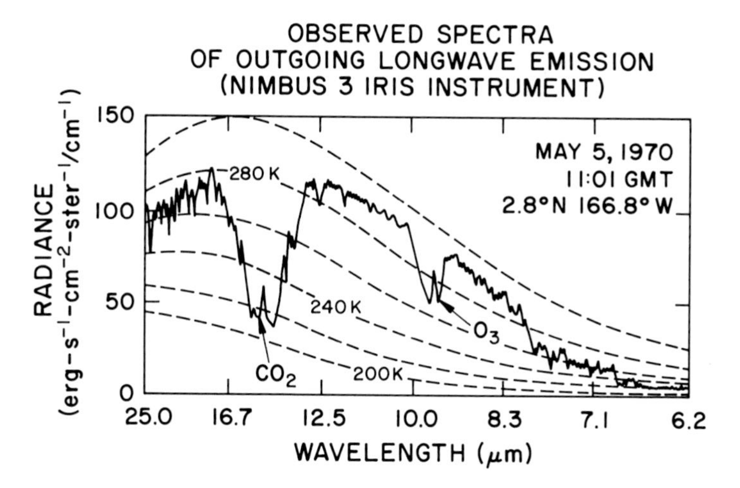

Figure 2. Spectrum of the infrared emission taken by the IRIS instrument on board the Nimbus 3 satellite (Hanel et al., 1972). The dashed lines indicate the effective radiation temperature as a function of wavelength.

the same authors of the change in thermal trapping between the preindustrial era and 1985, and between 1985 and 2050 respectively, based on specified scenarios for the concentration increases of several greenhouse gases. The global mean surface temperature is determined by a balance between the outgoing long-wave radiative energy and the solar energy absorbed by the Earth-atmosphere system (about 236 Wm^{-2}). This surface temperature on the average is about 35 K higher in the presence of the greenhouse gases than would be observed in the absence of these gases. Thus, increasing emissions in the atmosphere of gases with absorption bands in the infrared contribute to increase the average surface temperature and hence to induce climatic changes.

4. NUMERICAL MODELS

Numerical models are now commonly used to describe the radiative, dynamical and chemical processes in the atmosphere and, more specifically to predict climatic changes resulting from the increase of trace gases in the atmosphere. Different model formulations are possible to make these types of studies: Energy balance models, for example, determine the variation of the surface temperature with latitude by calculating the different terms contributing to the energy balance in each latitude zone. One-dimensional radiative-convective models derive the vertical distribution of the temperature by modeling explicitly the radiative processes in the atmosphere and accounting for rapid vertical heat exchanges in regions which are convectively unstable. Two-dimensional models account explicitly for dynamical processes in a zonally average framework and represent the surface processes as a function of latitude. Global general circulation models include in a three-dimensional framework most physical processes believed to be important for the atmosphere. The fundamental dynamical equations are solved numerically with a spatial resolution which is compatible with available computer resources. The most complex models include some parameterization of important coupling mechanisms such as ocean-atmosphere interactions, cloud radiation feedback or chemistry radiation interactions.

Predictions made by these models show that a doubling of the amount of carbon dioxide in the atmosphere (which is expected to be appear before year 2050) should raise the global mean surface temperature by 2 to 5 K (depending on the model and the type of climatic feedbacks included). The transient response of the climate system depends crucially on the response of the ocean, which is not yet fully understood. The effect of plausible increases in the density of other gases (including tropospheric ozone) is also substantial as can be seen in Figure 3. Donner and Ramanathan (1980) have for example calculated that the presence of 1.5 ppmv of methane in the atmosphere causes the globally averaged temperature to be 1.3 K higher than in the absence of this gas.

Finally, a prediction of the temperature change with the three-dimensional climate model of Hansen et al. (1988) (Fig. 4) suggests that the greenhouse warming should be clearly identifiable in the 1990's and that the predicted temperature changes are sufficiently large to have major impacts on the biosphere and on the populations. The curves shown in Figure 4 correspond to a case where the annual increase rate in CO_2 and other greenhouse gases remains approximately constant at the present level. The particular effect of non-CO_2 trace gases is visible. The prediction of regional climate changes requires the development of high resolution models based on an multidisciplinary approach including atmospheric and oceanic dynamics, biology, chemistry, ecology, hydrology, etc.

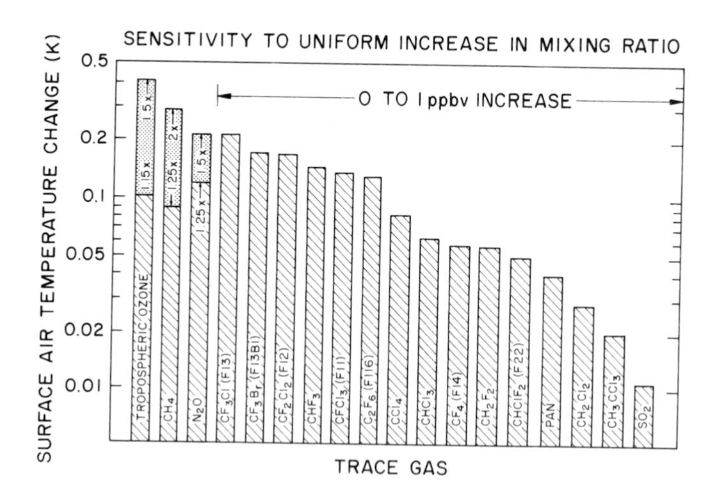

Figure 3. Change in surface air temperature, completed by Ramanathan et al. (1975), as a result of increase in trace gas concentrations. Tropospheric O_3, CH_4 and N_2O increases are indicated in the figure. The mixing ratio of the other gases is increased uniformly from 0 to 1 ppbv.

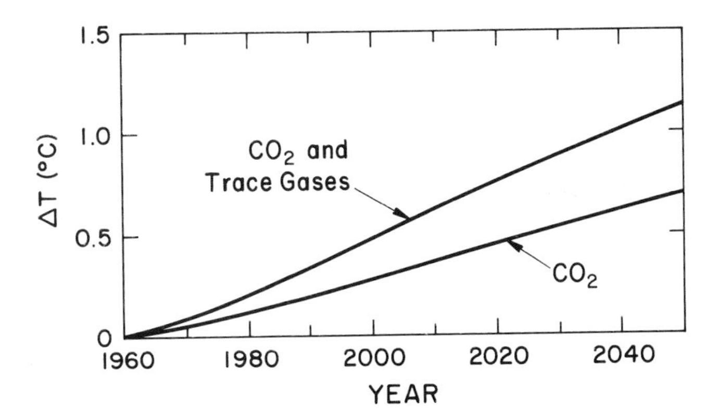

Figure 4. Equilibrium greenhouse warming calculated as a function of time by Hansen et al. (1988) with no climate feedback. The scenario used in this model calculation is such that the annual increase of the greenhouse climate forcing remains approximately constant at the present level. The effect of trace gases other than CO_2 is visible.

5. ATMOSPHERE-BIOSPHERE INTERACTIONS

The atmosphere and the terrestrial surface are linked through fluxes of energy and the emission or uptake of many trace gases. A large fraction of the species released in the atmosphere and contributing to the greenhouse effect or affecting the atmospheric composition are produced in whole or in part by microbiological processes in soils and water. The biosphere contributes to the formation of methane (CH_4), nitrous oxide (N_2O), carbon monoxide (CO), methylchloride ($CH_3C\ell$), ammonia (NH_3), hydrogen cyanide (HCN), dimethylsulfide (CH_3SCH_3), etc. The relation between the type of vegetation and the emission rate of species is poorly understood and an inventory of the surface and ocean sources of different gases, including their variations in space and time, needs to be established. For example, it is important to determine how these sources are expected to change as a result of potential variations in climatic conditions. The existence of possible positive or negative climatic feedbacks involving changes in the biosphere should be identified. Global ecological models need to be coupled to atmospheric and climate models to account not only for important factors such as modifications in the surface albedo or in the evapotranspiration of the vegetation but also for possible regulations by the biosphere of the emission of chemically and radiatively active trace gases.

6. CHEMICAL PERTURBATIONS IN THE ATMOSPHERE

Greenhouse gases released in the atmosphere do not only affect the radiation balance in the atmosphere; they also alter the global composition of the atmosphere, in particular the distribution of ozone. For example, the production of chemically active radicals such as those belonging to the nitrogen, hydrogen or chlorine families, which are a product of the decomposition of chemical compounds produced at the Earth's surface (source-gases), contribute significantly to the destruction of ozone in the stratosphere and in the mesosphere. A thinning of the ozone layer would lead to increased penetration in the atmosphere of harmful solar ultraviolet radiation with substantial environmental consequences at the Earth's surface. Furthermore, enhanced levels of ultraviolet radiation will increase the photochemical activity in the troposphere and particularly the processes affecting the production and destruction rates of species such as ozone and OH. More generally, a modification of the oxidizing capacity of the troposphere will affect the destruction rate and consequently the lifetime and the density of a variety of species, including the greenhouse gases and chemical pollutants. Chemical perturbations are therefore expected to have consequences not only on the entire climatic system but also on the global environment. For example, the observed increase in the atmospheric density of carbon monoxide is expected to lead to a reduction in the concentration of the hydroxyl radicals. Since the lifetime of species such as methane is determined by the amount of OH in the troposphere, the observed increase in the methane density could be due in part to a decrease in the OH concentration. Furthermore, ozone itself is a greenhouse gas, a change in its concentration, especially in the lower stratosphere, could produce a change in the surface temperature (Fig. 5a). Other important problems for climate studies are the influence of increasing densities of methane on the water budget near the tropopause and hence on the formation of radiatively active cirrus clouds, and the transformation of gaseous sulfur compounds into aerosol particles, which affect radiation in the atmosphere. Finally, since the mixing ratio of water vapor in the troposphere is highly variable with temperature, climatic changes (e.g., due to CO_2 perturbation) will affect the abundance of H_2O and consequently of hydroxyl radicals (Fig. 5b). We will now describe briefly the most important chemical and photochemical processes which control the composition of the troposphere and of the stratosphere.

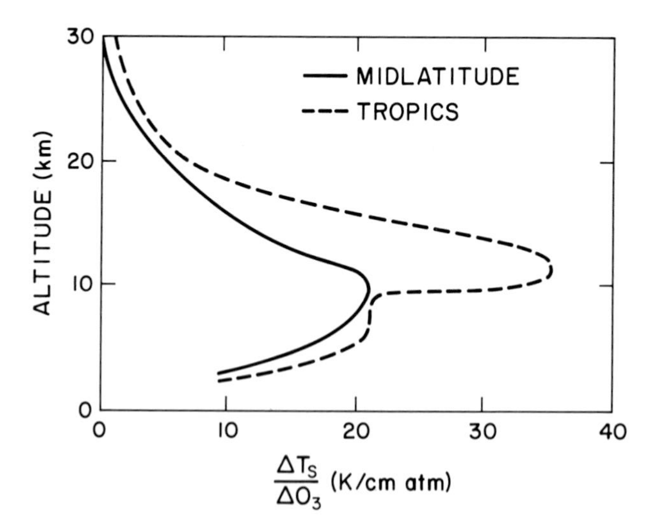

Figure 5a. Change in surface temperature calculated by Wang et al. (1980) as a function of altitude at which ozone is perturbed (mid-latitude and tropics).

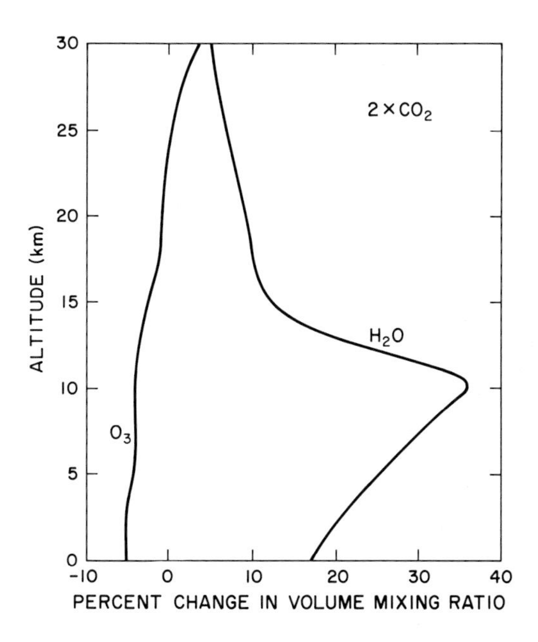

Figure 5b. Indirect effect of CO_2 doubling on H_2O and O_3 below 30 km altitude (Callis et al., 1983).

7. TROPOSPHERIC CHEMISTRY

Perhaps the most important role of the troposphere from the chemical point of view is its ability to oxidize gases which are released at the surface. The most important oxidant is the OH radical but other oxidants are the peroxy radicals (HO_2) – or higher forms RO_2, where R is an alkyl such as CH_3, nitrogen trioxide, ozone, chlorine or oxygen atoms. Oxidation in the aqueous phase is also of significance for species which hydrolyze in droplets, the most likely compound to play a key role in this process being hydrogen peroxide (H_2O_2).

The hydroxyl radical, the most active oxidation agent, is formed in the troposphere by the reaction between water vapor and the electronically excited oxygen atom

$$H_2O + O(^1D) \rightarrow 2\,OH. \tag{1}$$

The existence of the $O(^1D)$ atom arises from the photodissociation of ozone by solar radiation at wavelength less than 310 nm

$$O_3 + h\nu \rightarrow O(^1D) + O_2. \tag{2}$$

Hydroxyl radicals are destroyed by a reaction with carbon monoxide

$$OH + CO \rightarrow CO_2 + H \tag{3}$$

followed by a recombination of atomic hydrogen with molecular oxygen, which produces the peroxy radical HO_2:

(4) $$H + O_2 + M \rightarrow HO_2 + M,$$

where M (N_2 or O_2) is a third body that allows conservation of momentum. In the presence of a sufficiently large amount of nitric oxide (NO), HO_2 is converted back to OH by

(5) $$HO_2 + NO \rightarrow OH + NO_2.$$

Subsequently, in the presence of sunlight, nitrogen dioxide (NO_2) is photodissociated

(6) $$NO_2 + h\nu \rightarrow NO + O$$

and an ozone molecule is produced by

(7) $$O + O_2 + M \rightarrow O_3 + M.$$

A reaction similar to (5)

(8) $$NO + RO_2 \rightarrow NO_2 + RO,$$

where the organic peroxy radical RO_2 is produced by the oxidation of methane and other hydrocarbons, also converts NO into NO_2 and contributes to the formation of ozone in the troposphere.

Table 4. Global and tropical budget of tropospheric carbon monoxide (10^{12} gm CO/yr) from Seiler and Conrad (1987)

	Global	Tropics (30°S-30°N)
Sources		
Technological sources	640 ± 200	—
Biomass burning	1000 ± 600	800 ± 500
Vegetation	75 ± 25	60 ± 20
Ocean	100 ± 90	50 ± 45
CH_4 oxidation	600 ± 300	400 ± 200
Oxidation of nonmethane hydrocarbons	900 ± 500	600 ± 300
Production by soils	17 ± 15	10 ± 9
Total Production	3300 ± 1700	1900 ± 1100
Sinks		
Oxidation by OH	2000 ± 600	1200 ± 400
Uptake by soils	390 ± 140	105 ± 35
Flux into stratosphere	110 ± 30	80 ± 20
Total Decomposition	2500 ± 750	1400 ± 430

In a poor–NO_x environment, the reconversion of HO_2 into OH occurs through

$$HO_2 + O_3 \rightarrow OH + 2O_2 \quad (9)$$

and, in this case, ozone is consumed rather than produced.

The largest production of carbon monoxide (see Table 4) results from the oxidation of methane and other natural and anthropogenic hydrocarbons. Since CO is also produced by incomplete combustion, fossil fuel burning, forest clearing, savanna burning, usage of wood as fuel represent substantial sources. CO is also emitted by vegetation and oceans. This gas which is responsible for the conversion of OH into HO_2 is also destroyed by OH. Other tropospheric losses are due to uptake by soils and transport into the stratosphere. Since as much as 60 percent of the CO production could be related to anthropogenic processes, and since the lifetime of these molecules is only of the order of 2 months, the observed mixing ratios are significantly different in both hemispheres (typically 150-200 ppbv in the Northern hemisphere and 50 ppbv in the Southern hemisphere). Seasonal cycles in surface concentrations are observed and probably associated — at least in part — with variations in the vegetation and biomass burning.

An accurate budget of nitrogen oxides in the troposphere is still difficult to establish because of uncertainties in the different sources (Fig. 6) which have been identified: fossil fuel burning at the surface, emission by aircraft in the troposphere, lightning activity primarily in the tropics, deforestation, intrusion of odd nitrogen from the stratosphere. Because the average lifetime of NO_x is of the order of a day, the NO–rich environments should be found in the temperate regions of the Northern hemisphere and in the boundary layer in the tropics. Long-range transport of odd nitrogen can however happen, if NO_2 is converted into organic nitrates (such as peroxyacetylnitrate or PAN). The lifetime of PAN with respect to its decomposition is approximately 1 hr. at 300 K, 2 days at 273 K and 150 days at 250 K. Thus, at temperatures below 285 K or so, PAN becomes a stable product by which odd nitrogen can be transported over long distances in the troposphere and subsequently released as NO_2 in regions with temperatures above 285 K.

To summarize, the concentration of OH is controlled, to a large extent, by the presence of ozone, carbon monoxide and nitrogen oxides. This radical, which is a major oxidant in the troposphere, determines the lifetime of several gases emitted at the surface and, for example, prevents the build up in the atmosphere of many hydrocarbons. Ozone is produced photochemically in the troposphere in the presence of methane and other hydrocarbons, carbon monoxide and nitrogen oxides. The sources of these gases are unevenly distributed over the globe and strongly influenced by human activity. The emission of these species is increasing with time, essentially in industrialized regions. A direct consequence of these human activities is a substantial increase in the density of tropospheric ozone in the Northern hemisphere as well as an enhanced oxidation capacity of the troposphere. Table 5 gives a summary of parameters related to selected trace gases in the troposphere (Ehhalt, 1986).

8. STRATOSPHERIC CHEMISTRY

Perhaps the easiest way to summarize the chemical and photochemical processes occurring in the stratosphere is to discuss the formation and destruction mechanisms of ozone in this region of the atmosphere.

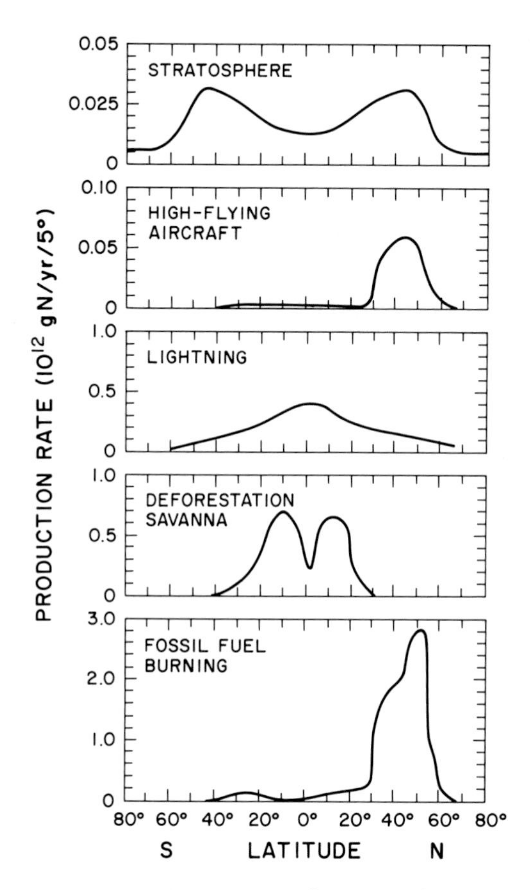

Figure 6. Esimate of the production rate of NO as a function of latitude for five sources, according to Ehhalt and Drummond (1982). Mean values.

Table 5. Current tropospheric mixing ratio, lifetime, global burden, increase rate and man's contribution to global sources for selected trace gases (Ehhalt, 1986)

Gas	Mixing Ratio (ppbv)	Lifetime (years)	Burden (10^6 tons)	Increase Rate (%/year)	Anthropogenic Sources (%)
CH_4	1700	10	3900	1	60
N_2O	305	100	1900	0.2	30
CO	100	0.2	400	1 ?	60
O_3	40	0.1	270	1 ?	50 ?
NO_2	variable (0.5)	0.002	(0.3)	1	60
CF_2Cl_2	0.35	90	6	5	100
$CFCl_3$	0.20	70	3.9	5	100

Ozone is produced by a three-body reaction

$$O + O_2 + M \rightarrow O_3 + M, \tag{10}$$

following the photolysis of molecular oxygen (O_2), which produces 2 oxygen atoms (O)

$$O_2 + h\nu(\lambda < 242nm) \rightarrow O + O. \tag{11}$$

Ozone is also photolyzed by UV and visible light

$$O_3 + h\nu \rightarrow O + O_2 \tag{12}$$

but the oxygen atom recombines almost immediately with O_2 to reform ozone (see Eq. 10). The net loss of odd oxygen (O_x is either O or O_3) occurs through a recombination of ozone and atomic oxygen

$$O + O_3 \rightarrow 2O_2. \tag{13}$$

Because of the strong temperature dependence of the rate of this reaction, the direct destruction mechanism (13) introduces a negative feedback between ozone and temperature, particularly in the upper stratosphere. Recombination process (13) can be catalyzed by different radicals present in the atmosphere. For example, above 55 km or so, the most efficient destruction mechanisms of O and O_3 involves the action of hydrogen compounds through the following cycles (Bates and Nicolet, 1950)

$$\begin{array}{rcl} H + O_3 & \longrightarrow & OH + O_2 \\ OH + O & \longrightarrow & H + O_2 \\ \hline \text{Net}: \; O + O_3 & \longrightarrow & 2O_2 \end{array}$$

and

$$\begin{array}{rcl} OH + O & \longrightarrow & H + O_2 \\ H + O_2 + M & \longrightarrow & HO_2 + M \\ HO_2 + O & \longrightarrow & OH + O_2 \\ \hline \text{Net}: \; O + O & \longrightarrow & O_2 \end{array}$$

Hydroxyl radicals and hydrogen atoms are produced by dissociation or oxidation of water vapor, methane and molecular hydrogen. These gases are originating at the Earth's surface and transported into the middle atmosphere.

In the stratosphere, the destruction of odd oxygen is catalyzed by the presence of nitrogen oxides (Crutzen, 1970)

$$\begin{array}{rcl} NO + O_3 & \longrightarrow & NO_2 + O_2 \\ NO_2 + O & \longrightarrow & NO + O_2 \\ \hline \text{Net}: \; O_3 + O & \longrightarrow & 2O_2 \end{array}$$

Nitric oxide is produced *in situ* in the stratosphere by oxidation of nitrous oxide, a gas produced at the surface and transported into the stratosphere. Another catalytic destruction of ozone (Stolarski and Cicerone, 1974)

$$Cl + O_3 \longrightarrow ClO + O_2$$
$$ClO + O \longrightarrow Cl + O_2$$

$$\text{Net}: \; O_3 + O \longrightarrow 2O_2$$

results from the presence of chlorine produced essentially by UV photodecomposition of industrially produced chlorofluorocarbons (Molina and Rowland, 1974). In the lower stratosphere and in the troposphere, the most efficient ozone destruction mechanism is due to a cycle involving the presence of hydrogen radicals (Hampson, 1974)

$$OH + O_3 \longrightarrow HO_2 + O_2$$
$$HO_2 + O_3 \longrightarrow OH + 2O_2$$

$$\text{Net}: \; 2O_3 \longrightarrow 3O_2$$

Figure 7 presents a schematic diagram of the chemical transformations of trace gases occurring in the stratosphere and indicates how these mechanisms affect ozone. This diagram shows that the source-gases produced at the Earth's surface and penetrat-

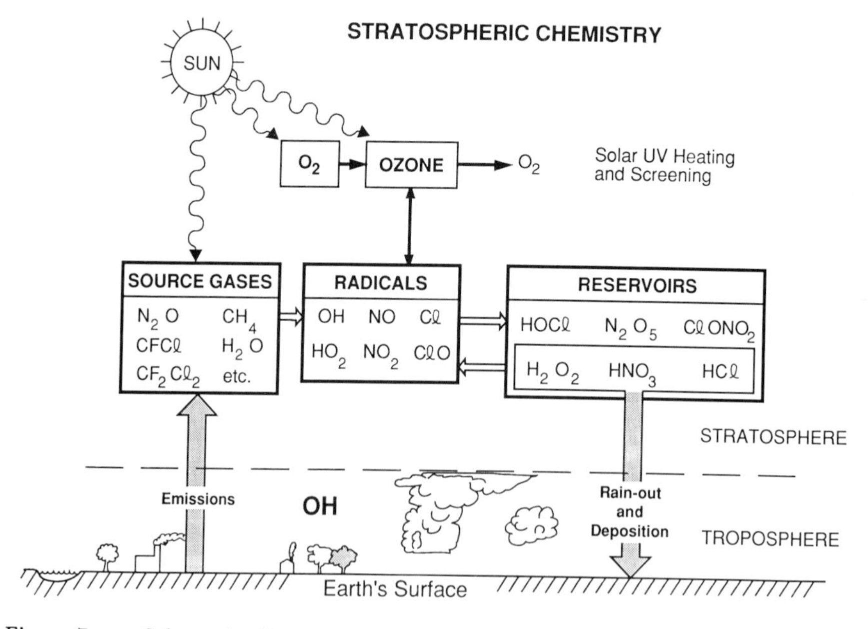

Figure 7. Schematic diagram of chemical and photochemical conversions in the stratosphere in relation with ozone chemistry.

ing in the stratosphere, when photolyzed or oxidized, lead to the formation of active radicals (OH, NO, Cℓ, etc.), which contribute to the destruction of ozone. In the presence of sunlight, these radicals are nearly in equilibrium with more stable constituents (named reservoirs) such as HNO_3, $C\ell ONO_2$, $HOC\ell$, N_2O_5, $HC\ell$, H_2O_2. These chemical reservoirs are inactive towards ozone and, since their chemical lifetimes are relatively long in the lower stratosphere, their concentrations are significantly higher than those of the fast-reacting radicals and their distributions are affected by transport in this region of the atmosphere.

The density of ozone in the stratosphere and consequently the ozone column amount as well as the depth of penetration of solar ultraviolet light and the heating balance in the middle atmosphere are highly dependent on the chemical reactions affecting all these minor constituents.

9. PERTURBATIONS OF THE ATMOSPHERIC COMPOSITION

Estimation of possible changes in the chemical composition of the atmosphere can be made by numerical simulations using sophisticated coupled chemical – dynamical – radiative models of the atmosphere. Different scenarios of the growth in the emission of source gases, changes in solar activity, and plausible increase in the aerosol content of the atmosphere should be considered to assess the climatic impact of these natural and anthropogenic forcings. In this overview, only particular cases will be briefly discussed.

Figure 8 shows, for example, the change in the meridional distribution of temperature for a doubling of the carbon dioxide amount calculated by the two-dimensional model of Brasseur et al. (1989). Besides the warming of the troposphere resulting from the greenhouse effect, a cooling of the stratosphere due to increased infrared emission to space is clearly observable. The largest temperature decrease is found near the stratopause, particularly at high latitude (-8 K at the equator, -10 K at the summer pole and -17 K at the winter pole).

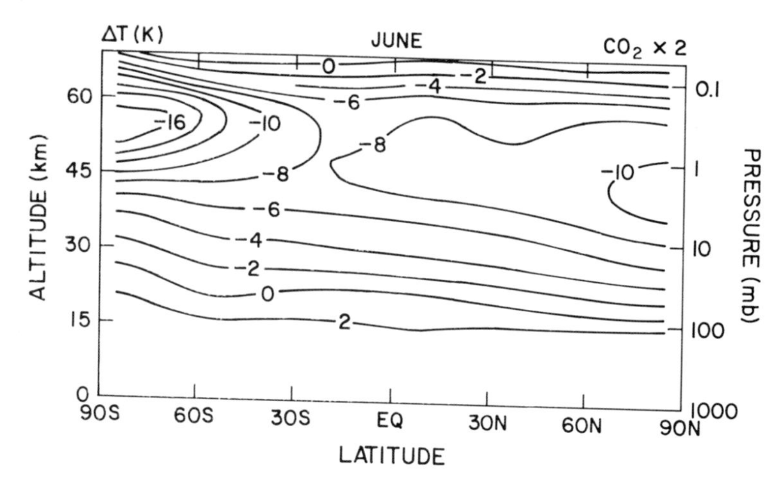

Figure 8. Calculated change in the temperature of the middle atmosphere for a doubling of CO_2 (June conditions).

Because of the temperature dependencies of the rate constants involved in the destruction of ozone, the cooling of the stratosphere resulting from a CO_2 doubling tends to increase the concentration of ozone and hence its column abundance. Figure 9 shows the percentage change in the vertically integrated ozone concentration as a function of latitude and season. According to the model, total ozone should increase on the average by about 1 percent; increases larger than 2 percent are however predicted at high latitude in spring. This seasonal modulation results from several factors including atmospheric transport of ozone-enriched air from the upper to the lower stratosphere.

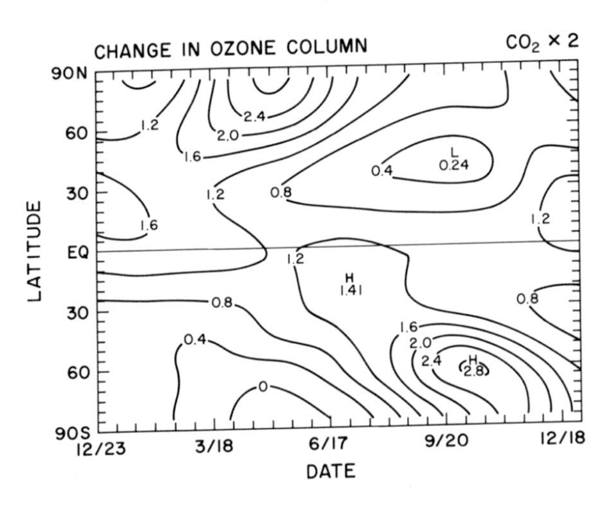

Figure 9. Percent change in the ozone column abundance as a function of latitude and season for a doubling of CO_2.

Another model simulation attempts to describe the effect of the chlorofluorocarbons on ozone. Figure 10 shows the meridional distribution of the percentage change in the ozone density for an increase of odd chlorine from 2.8 to 7.5 ppbv. Such an increase in $C\ell_x$ corresponds to a change from present-day conditions to a situation reached in the middle of the 21st century if the emission of the chlorofluorocarbons do not increase more than authorized by the Montreal Protocol on the protection of the ozone layer. Reductions larger than 50 percent are predicted in the upper stratosphere at high latitudes. A substantial decrease in temperature near 50 km altitude, due to the reduced absorption of ultraviolet radiation, is also found. Because the largest ozone depletion takes place above the layer where the density of this gas is highest, the calculated reduction in the ozone column for this particular case is limited to about 2 percent at the equator but reaches more than 6 percent at the poles in Spring. It should be indicated however that the effect of heterogeneous chemistry (on the surface of liquid or solid particles present in the atmosphere) is not taken into account in this model simulation. These processes however are known to be important over Antarctica, where, in the presence of polar stratospheric clouds, they initiate a complex chemistry leading to a springtime "ozone hole," and perhaps also at other latitudes, especially after large volcanic eruptions.

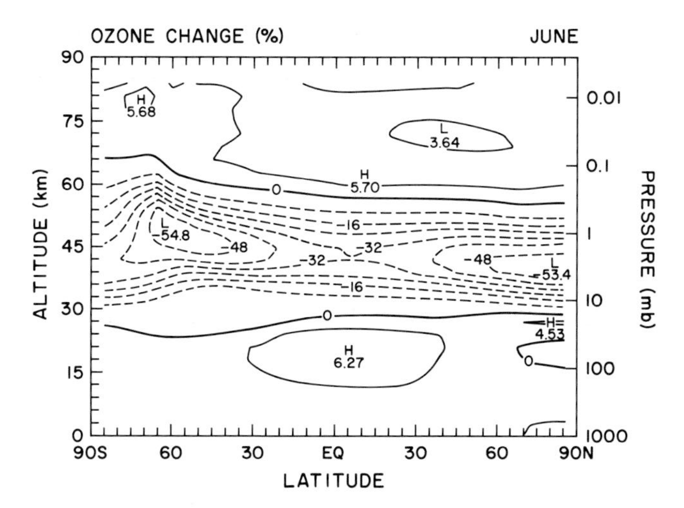

Figure 10. Meridional distribution of the relative change in the ozone density for an increase in the surface concentration of chlorofluorocarbons (from 220 to 800 pptv for CFC-11, from 418 to 2200 pptv for CFC-12). June conditions.

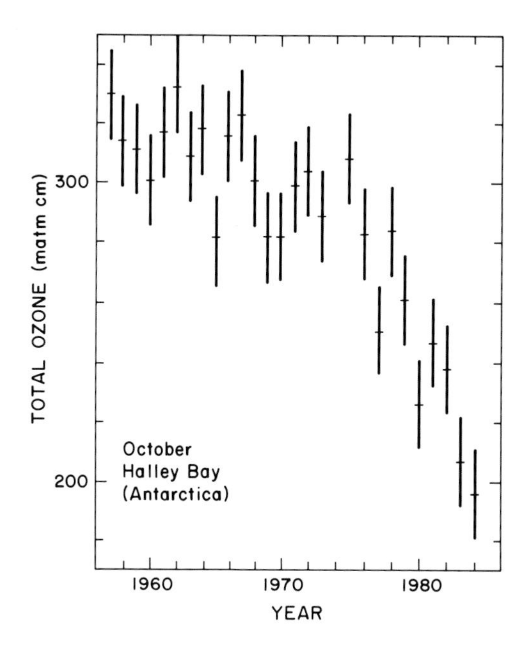

Figure 11. Monthly averaged total column abundance over Halley Bay in October from 1958 to 1984 (Farman et al., 1985).

10. ANTARCTIC OZONE

In 1985, scientists of the British Antarctic Survey (Farman et al., 1985) reported that the ozone column measured in October over the station of Halley Bay in Antarctica had gradually decreased by as much as 40 percent between 1979 and 1984 (Fig. 11). Similar decreases were reported at other scientific stations, for example Syowa (Chubachi and Kajiwara, 1986) and the South Pole (Komhyr et al., 1986). Satellite measurements (Stolarski et al., 1986) showed that this dramatic ozone depletion takes place from September to November in a region extending over the entire Antarctic continent. It was also observed that the ozone reduction is confined in a region ranging from 12 to about 22 km altitude.

Different explanations have been suggested to account for this unexpected and probably unprecedented change in the ozone climatology of the Southern Hemisphere. It appears today, as shown for example by recent observations made from a high-altitude aircraft, that active chlorine (e.g., $C\ell O$) in the lower stratosphere is about 100 times more abundant inside the polar vortex of the Southern hemisphere than at mid-latitudes. At the same time, the polar vortex appears to be denitrified and dehydrated. Under these conditions, ozone can be destroyed by the following cycle (Molina et al., 1987)

$$\begin{array}{rcl} 2(Cl + O_3 & \longrightarrow & ClO + O_2) \\ ClO + ClO + M & \longrightarrow & Cl_2O_2 \\ Cl_2O_2 + h\nu & \longrightarrow & ClOO + Cl \\ ClOO + M & \longrightarrow & Cl + O_2 + M \\ \hline \text{Net}: \; 2O_3 & \longrightarrow & 3O_2 \; , \end{array}$$

which does not require the presence of atomic oxygen and can therefore become efficient at low altitudes, if the concentrations of $C\ell O$ are sufficiently high (about 1 ppbv).

The existence of high $C\ell O$ densities in the lower stratosphere over Antarctica is believed to be associated with the presence of polar stratospheric clouds, which are produced in these regions because of the extremely low temperatures (less than 195 K) occurring in the polar vortex of the Southern hemisphere during winter. Molina et al. (1987) as well as Wofsy et al. (1988) have suggested that, if $HC\ell$ molecules dissolve in the ice crystals of the polar stratospheric clouds, a reaction of gas phase $C\ell ONO_2$ with dissolved $HC\ell$

$$C\ell ONO_2(g) + HC\ell(s) \rightarrow C\ell_2(g) + HNO_3(s)$$

would produce nitric acid in the condensed phase, while $C\ell_2$ would be released in the gas phase. As the Sun returns over Antarctica, $C\ell_2$ would be photolyzed and active chlorine ($C\ell$, $C\ell O$) produced in sufficiently large quantity to destroy ozone. The details of the clouds formation are not yet fully understood. Winter dynamics creates the conditions for the temperature to become sufficiently low for these clouds to be produced. The polar vortex is probably nearly isolated from the rest of the stratosphere. The possible intrusion of air in the vortex on small scales should however be investigated. Although the ozone hole seems to be particular to conditions prevailing

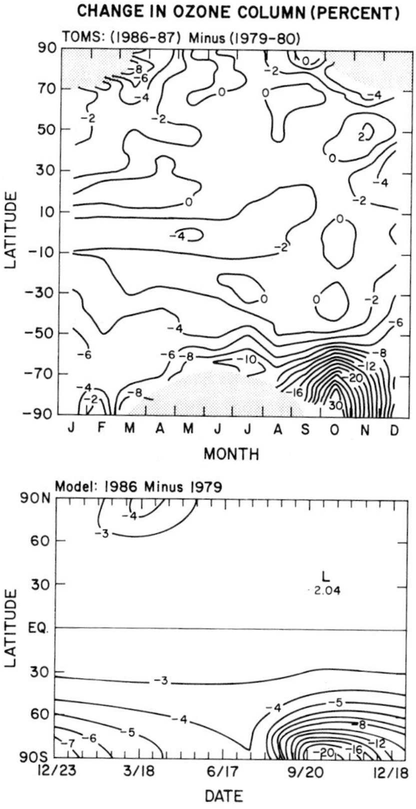

CHANGE IN OZONE COLUMN (PERCENT)

Figure 12a. Relative change in the ozone column abundance measured by the TOMS instrument (after correction based on information provided by ground-based Dobson instruments) between 1979–1980 and 1986–1987. The observed values have been averaged over 2 years to remove the effect of the quasi-biennial oscillation.

b. Relative change in the ozone column abundance derived by a chemical-radiative-dynamical two-dimensional model in which a crude parameterization of heterogeneous processes in the polar region has been introduced.

near the South Pole, it is important to determine if the processes involved could play a role in other regions of the world and if large ozone reductions could be produced, for example, in the Arctic region. It is also necessary to determine if the region of large ozone reduction could become wider or deeper and to what extent the air with low ozone could be diluted towards lower latitudes and affect the entire Southern hemisphere. As the ozone hole is linked to increasing abundances of chlorine, it can be stated that the springtime hole over Antarctica should be persistent over several decades, even if the release in the atmosphere of chlorofluorocarbons was dramatically reduced. This is a direct consequence of the long atmospheric lifetime (about 100 years) of the CFCs. On the other hand, the ozone column abundance in the hole is not expected to further decrease in a substantial way because, in the region where depletion mechanisms are important, almost all of the ozone is destroyed each spring. It is nevertheless crucial to determine if the extent in altitude and latitude of the polar stratospheric clouds remains unchanged from year to year.

11. GLOBAL OZONE TREND

Monitoring of total column ozone abundances has been performed for more than 30 years. The determination from these observations of a possible long-term trend is difficult to achieve due to existing dynamical oscillations and variations in solar activity (which affect the photochemical production of O_3), and because of the poor geographical coverage of the ground-based instruments measuring ozone and possible instrument drift. A careful analysis of the data however shows that, at mid- and high latitudes in the Northern hemisphere, the total ozone amount has decreased by about 2 to 3 percent over the 1970-1986 period. Satellite data, corrected for instrument drift through a normalization with the well calibrated ground-based instruments, suggests a trend of about -2.5 percent in the region between 65°N and 65°S for the period 1979 to 1986. As shown by Fig. 12a, the ozone reduction is of the order of 1-2 percent at the equator but generally increases with latitude, especially in winter and spring. A clear hemispheric asymmetry is visible. The appearance of the ozone hole over Antarctica during the 80's is clearly shown. These observations are broadly consistent with a model estimation of the ozone change between 1979 and 1986 (Fig. 12b), if this model includes a parameterization of heterogeneous processes in the regions where the temperature is sufficiently cold for polar stratospheric clouds to be produced. The observed magnitude of the ozone reduction in the hole, however, is larger than predicted by the model. However, because of interannual variations in the temperature over the poles, the "depth" of the ozone hole fluctuates from year to year, while the model provides an average value that does not account for these interannual variations of dynamical origin. The observed depletion in global ozone during the 1979-1986 period is believed to be due, in part, to the increasing concentrations of chlorine in the atmosphere (of industrial origin) and, in part, to the decrease over this period of solar activity (Brasseur et al., 1988), except for high latitude winter and spring when the ozone decrease is attributed only to anthropogenic activity.

12. CLOSING REMARKS

By releasing in the atmosphere large amounts of gases of industrial origin, by burning increasing quantities of fossil fuel, humankind is progressively changing on a global scale the conditions prevailing in the atmosphere as well as the Earth's climate. Atmospheric chemistry plays a major role in the determination of the present and future climatic conditions because of the interactions between trace gases and radiation. Furthermore, important couplings between the different components of the geosphere occur through mass transfer of chemical compounds. The International Geosphere-Biosphere Program will represent an important milestone in the investigation of the Earth's system as a whole and of the complex processes – including chemical processes occurring in the atmosphere – that drive climatic changes.

REFERENCES

Bates, D. and M. Nicolet, The photochemistry of water vapor, *J. Geophys. Res., 55*, 301-327, 1950.

Brasseur, G., M. H. Hitchman, P. C. Simon and A. De Rudder, Ozone reduction in the 1980's: A model simulation of anthropogenic and solar perturbation, *Geophys. Res. Lett., 12*, 1361-1364, 1988.

Brasseur, G., M. H. Hitchman, M. Dymek, M. Pirre and E. Falise, An interactive chemical, dynamical, radiative two-dimensional model of the middle atmosphere, in preparation, 1988.

Callis, L. B., M. Natarajan and R. E. Boughner, On the relationship between the greenhouse effect, atmospheric photochemistry and species distribution, *J. Geophys. Res., 88*, 1401-1426, 1983.

Chubachi, S. and R. Kajiwara, Total ozone variations at Syowa, Antarctica, *Geophys. Res. Lett., 13*, 1197-1198, 1986.

Cicerone, R. J. and R. S. Oremland, Biogeochemical aspects of atmospheric methane, *Global Biogeoch. Cycl.*, in press, 1988.

Crutzen, P. J., The influence of nitrogen oxides on the atmospheric ozone content, *Quart. J. Roy. Met. Soc., 96*, 320-325, 1970.

Dickinson, R. E. and R. J. Cicerone, Future global warming from atmospheric trace gases, Nature, 319, 109-115, 1986.

Donner, L. J. and V. Ramanathan, Methane and nitrous oxide: Their effects on terrestrial climate, *J. Atmos. Sci., 37*, 119-124, 1980.

Ehhalt, D. H., Atmospheric trace gases: Interactions with the biosphere, in Climate-Vegetation Interactions (C. Rosenzweig and R. Dickinson, eds.), UCAR, Office for Interdisciplinary Earth Studies, Report OIES-2, 1986.

Ehhalt, D. H. and J. W. Drummond, NO_x sources and the tropospheric distribution of NO_x during STRATOZ III, in Tropospheric Ozone (I.S.A. Isaksen, ed.), Reidel Pub. Co., Dordrecht, NATO Asi Series, 1988.

Fabian, P., Halogenated hydrocarbons in the atmosphere, in The Handbook of Environmental Chemistry, *Vol. 4* (O. Hutzinger, ed.), Springer Verlag, Heidelberg, 1986.

Farman, J. C., B. G. Gardiner and J. D. Shanklin, Large losses of total ozone in Antarctica reveal seasonal $C\ell O_x/NO_x$ interaction, *Nature, 315*, 207-210, 1985.

Hanel, R. A., B. J. Conrath, V. G. Kunde, C. Prabhakara, I. Revah, V. V. Salomonson and G. Wolford, The Nimbus 4 infrared spectroscopy experiment. 1. Calibrated thermal emission spectra, *J. Geophys. Res., 77*, 2629-2641, 1972.

Hansen, J., I. Fung, A. Lacis, D. Rind, S. Lebedeff, R. Ruedy, G. Russell and P. Stone, Global climate changes as forecast by Goddard Institute for Space Studies three-dimensional model, *J. Geophys. Res., 93*, 9341-9364, 1988.

Komhyr, W. D., R. D. Grass and R. K. Leonard, Total ozone decrease at South Pole, Antarctica, 1964-1985, *Geophys. Res. Lett., 13*, 59-62, 1987.

Molina, M. J. and F. S. Rowland, Stratospheric sink for chlorofluoromethanes: Chlorine atom catalyzed destruction of ozone, *Nature, 249*, 810-814, 1974.

Molina, M. J., T. L. Tso, L. T. Molina and F.C.-Y. Wang, Antarctic stratospheric chemistry and chlorine nitrate, hydrogen chloride: Release of active chlorine in the Antarctic stratosphere, *Science, 238*, 1253-1257, 1987.

Prinn, R. G., How have the atmospheric concentrations of the halocarbons changed?, in The Changing Atmosphere (F. S. Rowland and I.S.A. Isaksen, eds.), J. Wiley and Sons, New York, 1988.

Ramanathan, V. R., R. J. Cicerone, H. B. Singh and J. T. Kiehl, Trace gas trends and their potential role in climate change, *J. Geophys. Res., 90*, 5547-5566, 1985.

Seiler, W. and R. Conrad, Contribution of tropical ecosystems to the global budget of trace gases, especially CH_4, H_2, CO and N_2O, in The Geophysiology of Amazonia (R. E. Dickinson, ed.), J. Wiley and Sons, New York, 1987.

Stolarski, R. S. and R. J. Cicerone, Stratospheric chlorine: A possible sink for ozone, *Can. J. Chem., 52*, 1610-1615, 1974.

Stolarski, R. S., A. J. Krueger, M. R. Schoeberl, R. D. McPeeters, P. A. Newman and J. C. Alpert, Nimbus 7 satellite measurements of the springtime Antarctic ozone decrease, *Nature, 322*, 808-811, 1986.

Wang, W. C., J. P. Pinto and Y. L. Yung, Climatic effects due to halogenated compounds in the Earth's atmosphere, *J. Atmos. Sci., 37*, 333-338, 1980.

Wofsy, S. C., M. J. Molina, R. J. Salawitch, L. E. Fox and M. B. McElroy, Interactions between $HC\ell$, NO_x and H_2O ice in the Antarctic stratosphere: Implications for ozone, *J. Geophys. Res., 93*, 2442-2450, 1988.

Wuebbles, D. J., Chlorocarbon emission scenarios: Potential impact on stratospheric ozone, *J. Geophys. Res., 88*, 1433-1443, 1983.

Wuebbles, D. J. and J. Edmonds, A primer on greenhouse gases, U.S. Dept. of Energy DOE/NBB-0083, TR040, Washington, D.C., 1988.

THE PHYSICS AND DYNAMICS OF THE CLIMATE SYSTEM:

SIMULATION OF CLIMATE

John F B Mitchell
Dynamical Climatology Branch
Meteorological Office, Bracknell

Summary

The physical basis of numerical modelling of climate is outlined, using atmospheric models as an example. The use and verification of climate models is described, illustrated with data from simulations made with Meteorological Office models.

1. INTRODUCTION

It is well known that the climate of the earth has been very different from present, as for example during the major Ice Ages. Awareness of the possibility that man can change the atmospheric environment has been heightened in the last few years by publicity concerning "Acid rain", "The ozone hole" and the "Greenhouse effect". The advent of electronic computers has made it possible to develop 3-dimensional models of the atmosphere and oceans which have been used to investigate these phenomena. In these two lectures, we will consider the simulation of global climate, past, present and future.

Climate may be regarded as the statistical distribution of temperature, rainfall and other such parameters over a period of time at a given location. Thus, for example, we are not only concerned with the long-term mean monthly temperatures, but also the variability from day to day, and from year to year. In assessing the impacts of changes in climate, the changes in the frequency of extreme events such as frosts, gales, droughts and heatwaves are as important as the changes in mean parameters. As the climate varies over all timescales (Figure 1), the length of the averaging period will vary from 30 years or so for contemporary climate (for example, Schutz and Gates, 1971), through millennia when variations in orbital parameters are being considered, to geological timescales.

In order to simulate climate it is evident from the definition just given that we must be able to model temporal and spatial variability. Thus we are compelled to use 3-dimensional general circulation models (GCMs) of the atmosphere and oceans. Such models are expensive to run and it is often difficult to analyse their behaviour. Hence models of lower dimension have been used to investigate a wider range of parameter changes than is possible with a full model, or identify the dominant mechanisms in a GCM experiment.

Climate models, unlike numerical weather prediction models, cannot be validated against a large number of test cases. Although climate has varied in the past, the causes of these variations are not generally well known and our knowledge of past climates themselves is imprecise because of uncertainties in the interpretation of palaeoclimatic data. Thus, the usual scientific cycle of observation, making an hypothesis, validating the hypothesis against further independent observations and so on is difficult to achieve. It is therefore particularly important that climate models should be firmly based on physical principles and validated against whatever data there is available. These two topics are discussed further in the remainder of this paper.

2. ATMOSPHERIC MODELS AND THEIR PHYSICAL BASIS

In this section, the physical basis of climate models is discussed using atmospheric models as an example. A similar approach is taken in modelling the oceans, sea-ice and the land surface, but these components will not be discussed further. The purpose of this section is to outline the principles used in constructing a climate model as opposed to providing a comprehensive review of current climate models. Examples will be taken predominantly from the Meteorological Office climate model (Wilson and Mitchell, 1987a, henceforth referred to as WM; Slingo et al, 1988).

2.1 Dynamics and physical parametrizations

In a numerical model, the atmosphere is represented by values of wind, temperature and humidity which are held at various levels of grid points. For example, the model used by WM has 11 levels in the vertical and horizontal spacing of 5° latitude by 7.5° longitude (Figure 2) typical of current climate models. At any time t the state of the atmosphere is described by about 10^5 variables, and the values at some future time $t+\Delta t$ can be determined using the prognostic equations described below. Δt is of order 20 minutes. The Met. Office climate model (WM) requires 4.5 hours of elapsed time on Cyber 205 for each simulated year.

The prognostic equations are based on the laws of classical physics, and the local rates of change $\frac{\Delta Y_i}{\Delta t}$ at each location and level $\underline{X}$ are given by

$$\frac{\Delta Y_i}{\Delta t} = F_i(\underline{Y},\underline{X},t) \qquad Y_i = u,v,T,q \qquad (1)$$

where u and v are the east-west and north-south components of wind, T is temperature, q is humidity, F_i is a function of all the variables $\underline{Y}$, and their spatial gradients at that location. Equation (9.2) may be expanded.

$$\frac{\Delta Y_i}{\Delta t} = \underbrace{\underline{\nabla}\cdot \underline{V}\, Y_i}_{(a)} + \underbrace{Z}_{(b)} + \underbrace{D}_{(c)} + \underbrace{N}_{(d)} \qquad (2)$$

Where the term

(a) is the change in Y_i due to transport by the flow
(b) represents the effects of pressure gradients and the earth's rotation in the wind equation, and of adiabatic vertical motion in the temperature equation

(c) are smoothing (diffusive) terms
(d) are the sources and sinks of momentum ($i=u,v$), heat ($i=T$) and moisture ($i=q$).

The left-hand side and the terms (a) and (b) are often referred to as the large scale dynamics. The equations for u and v may be transformed into 2 equations for divergence and the vertical component of vorticity, and so may be solved using spectral techniques (see, for example, Schlesinger 1988): otherwise they are solved using finite difference techniques. In solving the equations, care must be taken that there are no spurious sinks of heat, moisture or momentum. This is particularly important for models used in climate change experiments, since the sizes of the perturbations typically considered are small (4 Wm^{-2} in the case of doubling carbon dioxide concentrations (Lal and Ramanathan, 1984)). The diffusive terms (c) are required to prevent energy accumulating in the smallest scales resolved by the model grid and producing unmeteorological solutions. These terms can be regarded as representing small-scale motions not resolved by the grid.

The final term in 2 covers all the small-scale processes not incorporated in (c), including boundary and surface processes, radiation and convection. For some of these phenomena, notably radiation, there exists an accepted theoretical basis, and accuracy is limited by the availability of computer resources and the necessity of using grid-box values of parameters which in reality may vary substantially within a grid box. For other processes, there may be no accepted underlying theory, or the theory may be valid only for limited idealized conditions. The subgrid scale parametrizations may be derived on the basis of one or more of the following:

a. Simplified forms of the exact equation (if known)
b. Numerical experiments on a much finer mesh than feasible in the GCM.
c. Observations from the atmosphere
d. Laboratory experiments
e. Sensitivity studies made with general circulation models.

To illustrate the diversity of these approaches, the derivation of the radiative and convective parametrizations are considered in more detail.

In principle, the equations of radiative transfer are well known (see for example Paltridge and Platt, 1976) and detailed measurements of the absorption of radiation as a function of wavelength have been made for the principal atmospheric absorbers. Thus the accuracy of radiative fluxes under clear skies is limited only by computational effort. Since the explicit integration of the full equations for the vertical flux of long-wave radiation F(z)

$$F(z) = \int_0^\infty d\nu \int^\infty \frac{d\mu}{\mu^3} \int_z^\infty dz' \Pi B(\nu,z') \frac{d}{dz'} \exp \{-\mu \int \sum_i S_i(z'') f_i(\nu,z'') dz''\}$$

(from Rodgers, 1977) includes a double integration over the vertical co-ordinate z, integrations over all angles represented by μ and all frequencies ν and a summation over all spectral lines i, much effort has been devoted to simplifying the equations and devising efficient mathematical methods of solution. The Met. Office model currently uses an emissivity approximation which involves a single integration over the vertical co-ordinate and a summation over 7 broad bands (Slingo and Wilderspin, 1986). The radiative properties of water clouds can be derived theoretically provided the droplet-size distribution is known. Ice clouds are comprised of non-spherical particles which may vary in shape and orientation, making it difficult to estimate the radiative properties theoretically. The major uncertainties in radiative effects of ice and water cloud arise in determining cloud properties (height, thickness, extent, water content and droplet (particle) characteristics) from grid scale model variables.

Much of the vertical transport of heat, moisture and momentum is the result of vertical motions on scales much smaller than a model grid box such as cumulus convection. Unlike radiative transfer, there is no generally accepted underlying theoretical basis for parametrizing convective motions. However, some general principles can be applied. If the atmosphere becomes statically unstable, the vertical profile will adjust towards the dry adiabat (or moist adiabat if saturation occurs), conserving moist static energy, $C_pT + Lq + gz$. This is the basis of adjustment schemes, such as the moist convective adjustment scheme of Manabe and Strickler (1964) which instantaneously removes convective instability, and the Betts-Miller scheme (Betts and Miller, 1984) which relaxes the vertical profile towards reference profiles defined from observational data. Other schemes (Arakawa and Schubert, 1974; Lyne and Rowntree, 1976, used in current Met. Office models) distinguish explicitly between small-scale ascent in convective towers and large-scale subsidence and warming of environmental air. A third approach taken by Kuo (1974) links convection directly to the large-scale convergence of moisture in a vertical column, including that due to surface evaporation, repartitioning the moisture supply between precipitation and moistening the column. In all but the moist convective adjustment scheme, there are explicit tunable parameters which may be chosen by testing the scheme against observed data such as those from the GARP Atlantic Tropical Experiment (GATE, Lyne et al, 1976) or by carrying out sensitivity experiments using the full model. A shortcoming of the former approach is that the large-scale convergence of heat and moisture is not well known, whereas in sensitivity experiments, errors in other parametrizations may degrade the performance of the convection scheme. Further guidance may be obtained from observations and numerical simulations of individual convective cells.

The factors guiding the development of various parametrizations used in a recent version of the Met. Office climate model (WM) are shown in Table 1. Note that apart from the snow free land reflectivities which are derived from observational data, adjustable parameters are prescribed with the same value over the whole globe, although separate values may be allocated to land and sea points. Hence there is a limit to the tuning possible in order to optimize the simulation of regional climate.

2.2 Using a climate model

Certain features, referred to here as boundary conditions, must be specified before a GCM can be integrated. For an atmosphere-only model, these include carbon dioxide and ozone concentrations, sea surface temperatures and sea-ice extents, orography and time of year. The normal procedure is then to integrate the model over the period of interest and look at the long-term statistics of the simulation. In the case of a climate change experiment, a second simulation is made changing only the

relevant boundary conditions (for example carbon dioxide concentrations) and the long-term characteristics of the two simulations are compared. Note that even with fixed boundary conditions, the simulation will vary in time due to the inherent variability of climate. Hence it is necessary to perform statistical tests on the differences between the control and perturbed simulations in order to establish that the probability that such differences could have arisen by chance is small.
The length of simulation will depend on a variety of factors. First, the time taken for the model to adjust to the prescribed boundary conditions will vary from a week or so for prescribed sea surface temperature experiments to decades or even centuries in the case of coupled ocean-atmosphere models. Second, the length of time for which prescribed boundary conditions are valid will also vary, from a month or so for sea surface temperature anomalies to thousands of years in the case of orography. Third, the integrations must be sufficiently long (or numerous) to allow the statistics of the simulated climate to be defined. In practice, the length of a simulation is often limited by the computational resources available.

In the above approach, there are certain underlying assumptions whose validity may affect the relevance of the results to observed climate.

(a) A "mean" or quasi-equilibrium state exists in nature and in the model. As discussed in the introduction, the climate of the earth has varied considerably over a wide range of timescales. Thus in the real world at least, we do not know if statistically stationary climatic states exist. The problem of the stability of climatic states has been addressed using "simple" models (see for example Fraedrich 1978, Ghil and Childress, 1987) and will not be considered further here.

(b) The "mean" state is unique, and independent of the initial state. Lorenz (1975) suggested that the earth's climate may have more than one stable state, and that transitions between these states may be rare (climate is almost "intransitive"). It follows that the choice of initial conditions may determine which of the stable states evolve. For example, Wetherald and Manabe (1975) found that a simulation made with an idealized climate model which commenced with an ice covered surface remained ice-covered, whereas when the initial conditions were free of ice, only polar latitudes became ice-covered.

(c) The model converges, and converges to the correct solution. Although mathematical models of climate may be based on sound physical principles, this is not sufficient to guarantee that the model will produce the correct solution. We cannot check the convergence of such complex models using mathematical stability analyses, so the only way to check for convergence is to integrate the model and look at the results.

2.3 Verification

The validity of the model can be assessed on a variety of timescales against observational data. As climate models can be regarded as a development of numerical weather prediction models, one can specify initial conditions from meteorological analyses, perform short-range forecasts and compare the evolution of synoptic features in the model with those observed (for example, Carson and Cullen, 1976). A similar approach can be taken using long-range forecasts (Mansfield, 1986), though after a period of a week or so, the correspondence between individual simulated and observed features will be lost, and the forecasts of necessity become statistical rather than deterministic.

The traditional method of calculating atmospheric GCMs has been to compare the simulated means with climatological data (Manabe and Holloway, 1975; Mitchell, 1983). For example, by comparing a 4-year mean of simulated mean sea-level pressure from the version of the Met. Office GCM described by Slingo et al (1988) with observational data (Figure 3) one can see that the model reproduces all the main features of the observed circulation including the equatorial and mid-latitude pressure troughs, and the subtropical anticyclones which form over the ocean in summer and extend over land in winter. Note however that the simulated pressure ridge over the Rockies is weaker than observed, and the weak trough over Central Europe is much too strong in the model.

The variability in the model may also be assessed: Figure 4 shows the simulated and observed high pass filtered 500 mb heights which give an indication of the strength and position of the storm tracks. Again the agreement with observation is good, though the Atlantic storm track is weaker than observed, and penetrates too far south at the European end. One can assess variability on longer timescales including year to year in a similar manner.

The detailed evolution of the seasonal cycle can be assessed by comparing mean monthly grid point data with observations from suitable climatological stations (Reed, 1986). In Figure 5, the mean monthly midnight surface temperatures at a model grid point over eastern England are compared with observed 2 m minimum temperatures from Cambridge (Wilson and Mitchell, 1987b). Note the colder than observed spring temperatures. This is associated with more frequent than observed cold spells in the model (Figure 6) which are a result of too many or too strong outbreaks of easterly flow off the cold Eurasian continent (Reid, 1986).

The above approach may be used to establish that the mean climate and level of variability in the model are correct. In order to test that the structure of the model's response to specific perturbations is correct, one can examine the response of the model to observed changes in boundary conditions such as sea surface temperature anomalies, and compare the response with that observed. Palmer and Mansfield (1986) have examined the simulated and observed response to different sea surface anomalies in the tropical Pacific, and Folland et al (1986) have examined the relationship between sea surface temperatures and Sahel rainfall, using observations and numerical experiments. A weakness in this approach to model verifications is that in reality factors other than the changes in boundary conditions specified in the model may contribute to observed changes in the atmosphere, including inherent atmospheric variability.

The ability to simulate contemporary climate accurately is a necessary but not a sufficient condition for a model to produce reliable estimates of climate change. It is possible that in choosing values of disposable parameters which give an accurate simulation of present-day climate, the response of the model to changes in climate may be distorted. Thus, attempts have been made to simulate climates in the past, including the last glacial maximum 18,000 years before present (18 K bp) (Manabe and Broccoli, 1985) and the last northern hemisphere summer insolation maximum

(9 K bp), considered in the following chapter. Manabe and Broccoli compared the changes in surface temperature between 18 K bp and present from two different versions of the model with a palaeoclimatic estimate of the temperature change (Figure 7). However, the investigation was inconclusive because of both the small differences between the results from the two versions of the models and the uncertainties in the palaeoclimatic data.

3. OTHER COMPONENTS OF THE CLIMATE SYSTEM

The approaches used in developing atmospheric models may be applied to other components of the climate system including sea-ice, the oceans and continental ice-sheets. There is considerably less data from the ocean than from the atmosphere. In addition, synoptic eddies in the ocean have a typical length scale of 30 km as opposed to 1000 km typical of the atmosphere and so are not resolved by the current generation of global ocean GCMs. Thus ocean GCMs are even more difficult to validate than atmospheric models. Since the atmosphere, ocean and continental ice-sheets respond on timescales of days, centuries and thousands of years respectively, special care must be taken when representing the interaction between them. Schlesinger (1979) has listed several ways in which an atmospheric model may be coupled to an ocean model, and Bryan (1984) has tested methods for accelerating the convergence to equilibrium of ocean climate models. Alternatively, Hasselmann (1979) has suggested parametrizing the effect of quickly changing elements of a climate model (for example, the atmospheric component of a model which includes the deep ocean).

4. SUMMARY

There is a growing interest in the simulation of climate, both in order to predict future changes in climate due to man's activities, and to understand past climatic change. Climate models are based on the laws of classical physics, though processes which occur on scales smaller than the model grid (or timestep) must be approximated (parametrized). This may be done on the basis of theory, observations and laboratory and numerical experiments. Validation of models includes detailed verification of individual components of the model, and the comparison of results from the full model over a variety of timescales with observational data. Certain assumptions concerning the stability of climate and its dependence on specified boundary conditions are implied in the current use of GCMs in climate simulations.

5. ACKNOWLEDGEMENTS

This paper is an enlarged and updated version of a paper "The physical basis of climate modelling" which appeared in "Palaeoclimatic Analysis and Modelling" (Ed Ghazi) published by D Reidel, 1983.

6. REFERENCES

Arakawa, A. and Schubert, W. H. 1974 Interaction of a cumulus cloud ensemble with large-scale environment. Part I. J. Atmos. Sci., 31, 674-701.

Betts, A. K. and Miller, M. J. 1984 A new convective adjustment scheme. ECMWF Technical Report No 43, 65 pp.

Bryan, K. 1984 Accelerating the convergence to equilibrium of ocean-climate models. J. Phys. Oceanogr. 14 (4) 666-673.

Carson, D. J. and Cullen M. J. P. 1976 Intercomparison of short-range numerical forecsts using finite difference and finite element models from the UK Meteorological Office (Paper presented at the joint DMG/AMS International Conference on the Simulation of large-scale Atmospheric Processes, Hamburg, 30 August-4 September 1976). Met O 20 Tech. Note II/81, Meteorological Office, Bracknell.

Clarke, W. C. 1982 Carbon Dioxide Review: 1982 (Ed. W. C. Clarke) Clarendon Press, Oxford; Oxford University Press, New York. 469 pp.

Folland, C. K., Palmer, T. N. and Parker, D. E. 1986 Sahel Rainfall and Worldwide sea temperatures, 1901-85. Nature, 320, 602-607.

Fraedrich, K. 1978 Catastrophes and resilience of a zezro-dimensional climate system with ice-albedo and greenhouse feedback. Quart. J. R. Met. Soc. 105, 147-167.

Ghil, M. and Childress, S. 1987 Topics in Geophysical Fluid Dynamics; Atmospheric Dynamics, Dynamo Theory, and Climate Dynamics. Applied Mathematical Sciences 60, pp 485, Springer Verlag, New York.

Hasselmann, K. 1979 On the problem of multiple time-scales in climate modelling. In "Man's Impact on Climate" (Eds. Back, Pankrath and Kellogg). Developments in Atmospheric Science, 10, 43-55. Elsevier (Amsterdam, Oxford, New York).

Joles, P. D., Wigley, T. M. L., Folland, C. K., Parker, D. E., Angell, J. K, Lebedeff, S. and Hansen, J. E. 1988 Evidence for global warming in the past decade. Nature, 338, 790.

Kuo, H. L. 1974 Further studies of the parametrization of the influence of cumulus convection on large scale flow. J. Atmos. Sci., 31, 1232-1240.

Lal, M. and Ramanathan, V. 1984 Effects of moist convection and water vapour radiative processes on climate sensitivity. J. Atmos. Sci. 41, 2238-2249.

Lorenz, E. 1975 Climatic Predictability, WMO — GARP Publication Series No. 16, Geneva, 132-136.

Lyne, W. H. and Rowntree, P. R. 1976 Development of a convective parametrization using GATE data. Met O 20 Technical Note II/70, Meteorological Office, Bracknell.

Lyne, W. H., Rowntree, P. R., Temperton, C. and Walker, J. 1976 Numerical modelling using GATE data. Met. Mag. 105, 261-271.

Manabe, S. and Broccoli, A. O. 1985 A Comparison of Climate Model Sensitivity with data from the last glacial maximum. J. Atmos. Sci. 42, 2643-2651.

Manabe, S. and Holloway, J. L. Jnr. 1975 The Seasonal variation of the hydrologic cycle as simulated by a global model of the atmosphere. J. Geophys. Res., 80, 1617-1649.

Manabe, S. and Strickler, R. F. 1964 Thermal equilibrium of the atmosphere with a convective adjustment. J. Atmos. Sci. 21, 361-385.

Mansfield, D. A. 1986 The skill of dynamical long-range forecasts, including the effect of sea-surface temperature anomalies. Quart. J. R. Met. Soc., 112, 1145-1176.

Mitchell, J. F. B. 1983 The seasonal response of a general circulation model to changes in CO_2 and sea surface temperature. Q.J.R. Met. Soc., 109, 113-152.

Palmer, T. N. and Mansfield, D. A. 1986 Wintertime Circulation Anomalies During Past El Nino Events, using a High Resolution General Circulation Model II. Variability of the seasonal mean response. Quart J.R. Met. Soc., 112, 639-660.

Paltridge, G. W. and Platt, C. M. R. 1976 Radiative proceses in Meteorology and Climatology. Developments in Atmospheric Science, 5. Elsevier, New York, 318 pp.

Reed, D. N. 1986 "Simulation of temperature and precipitation over eastern England by an atmospheric general circulation model". Journal of Climatology 6, 233-253.

Roeckner, E., Schlese, U., Biercamp, J. and Loewe, P. 1987 Cloud optical depth feedbacks and climate modelling. Nature, 329, 138-139.

Rodgers, C. D. 1977 "Radiative Processes in the Atmosphere". In proceedings of ECMWF Seminars. "Parametrization of Physical Processes in the free atmosphere", 5-66.

Schlesinger, M. E. 1979 Comments on ocean-atmosphere coupling and discussion of the paper "A global ocean-atmosphere model with seasonal variation: Possible application to a study of climate sensitivity". Climate Research Institute, Report No. 39, Oregon State University.

Schlesinger, M. E. 1988 (Ed) Physically-based modelling and simulation of climate and climatic change. Vol. 1. NATO Advanced Study Institute Series. (Reidel, in Press).

Schutz, C. and Gates, W.L. 1971 Global Climatic data for surface, 800 mb, 400 mb (January), R-915-ARPA, RAND, Santa Monica.

Slingo, A. and Wilderspin, R. C. 1986 Development of a revised longwave radiation scheme for an atmospheric general circulation model. Q.J.R. Meteorol. Soc., 112, 371-386.

Slingo, A., Wilderspin, R. C. and Smith, R. N. B. 1988 The effect of improved physical parametrizations on simulations of cloudiness and the earth's radiation budget in the tropics. Submitted to Q.J.R. Meteorol. Soc. (Also Met O 20 DCTN 65).

Wetherald, R. T. and Manabe, S. 1975 The effects of changing the solar constant on the climate of a general circulation model. J. Atmos. Sci. 32, 2044-2059.

Wilson, C. A. and Mitchell, J. F. B. 1987a A Doubled CO_2 Climate Sensitivity experiment with a GCM including a simple ocean. J. Geophys. Res., 92, 13315-13343.

Wilson, C. A. and Mitchell, J. F. B. 1987b Simulated Climate and CO_2 — induced climate change over western Europe. Climate change, 10, 11-42.

7. Suggested Bibliography

Physically-based Modelling and Simulation of Climate and Climate Change Vol.1 (1988) Schlesinger, M. E. (Ed), NATO Advanced Study Institute Series, (Reidel, in Press). Excellent chapters by Bourke on spectral methods, and Arakawa on finite differences, as well as other components of climate models.

Numercial Weather Prediction and Dynamical Meteorology (1980) Haltiner, G. J. and Williams, R. T. (John Wiley). Generally good on numerical methods, but out-dated on physical parametrizations.

An Introduction to Three-Dimensional Climate Modelling (1986) Washington, W. M. and Parkinson, C. L. There are not many books on climate modelling, and this is the best to date.

The Global Climate (1984) Houghton, J. T. (Ed). (Cambridge University Press) Contains a series of articles by different authors on various aspects of the climate system.

Table 1 Choice of parameters in the atmospheric component of the model used by Wilson and Mitchell (1987)

	Parameter valid over globe	Tuning Experiments cf climatology	Theory	Detailed Models	Field Observations	Laboratory Experiments
Dynamics						
Diffusion Coefficient	X	X	(X)			
Time Smoothing	X	X				
Radiation						
Gaseous absorption	X		X	X	X	X
Cloud amounts	X	X			X	
Cloud properties	X	X			X	
Surface Reflectivity					X	
Convection						
Parcel size	X	X			X	
Entrainment and Detrainment	X	X	X		X	X
Evaporation	X	X	X		X	
Boundary Layer						
Drag coefficient	X		X		X	X
Surface roughness	X				X	
Other						
Soil moisture	X				X	
Heat flux through ice	X		X	X	X	X

FIGURE CAPTIONS

Figure: 1 (a) Estimated global mean surface temperature over the last 850,000 years (from Clarke, 1982)

(b) Observed anomalies in global mean surface temperature 1901-1987 (from Jones et al, 1988).

2 The horizontal grid used by WM. This is typical of current climate models. The horizontal spacing is 5° latitude by 7.5° longitude.

3 Mean sea level pressure (contours every 4 mbs)

(a) January, observed. (Schutz and Gates, 1971)

(b) December to February, simulated using a high resolution version of the Met. Office GCM (see Slingo et al, 1988)

4 High pass filtered variance of 500 mb heights for the northern hemisphere for December to February. Contours every 1 mb.

(a) From a high resolution version of the Met. Office GCM.

(b) Observed, from Met. Office operational analyses, 1983-1986.

5 Monthly mean surface temperature over Eastern England. Solid line, climatological 2 m screen minimum temperatures at Cambridge; dashed line, simulated ground temperatures 00Z from a 5-layer atmospheric model. (Wilson and Mitchell, 1987b). The crosses are the observed mean monthly minimum.

6 Frequency distribution of surface temperature over central/eastern England (mean, annual and semi-annual cycle removed) at 0.5 K intervals.
Light line — From 3 years of Central England Temperature Series
Heavy line — From a 3-year simulation with a 5-layer atmospheric model (Reed, 1986).

7 Latitudinally averaged changes in surface temperature between 18 K bp and present. Crosses, sea surface temperature changes deduced from palaeoclimatic data. Solid and dashed lines, as simulated using a climate model with a mixed-layer ocean using prescribed and model generated cloud amounts respectively. (From Manabe and Broccoli, 1985).

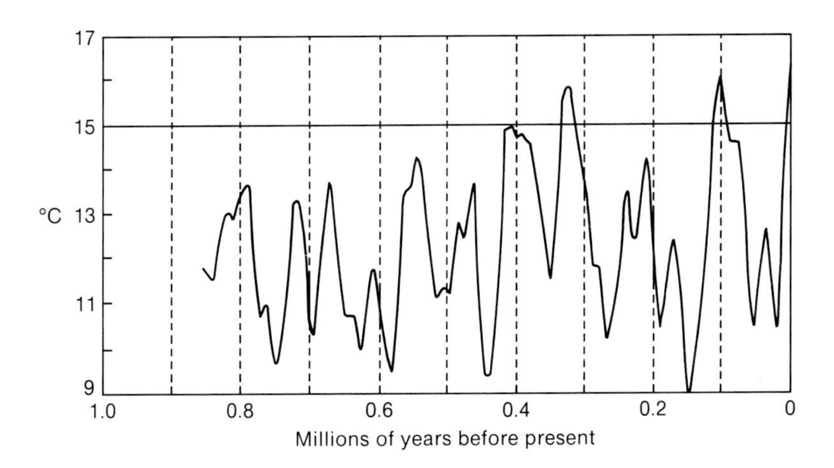

Figure 1 (a)

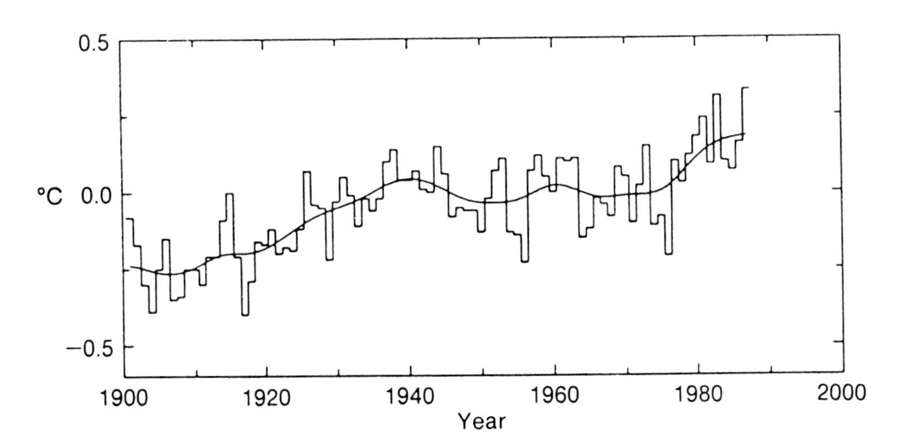

Figure 1 (b)

(a) Estimated global mean surface temperature over the last 850,000 years (from Clarke, 1982)

(b) Observed anomalies in global mean surface temperature 1901-1987 (from Jones et al, 1988).

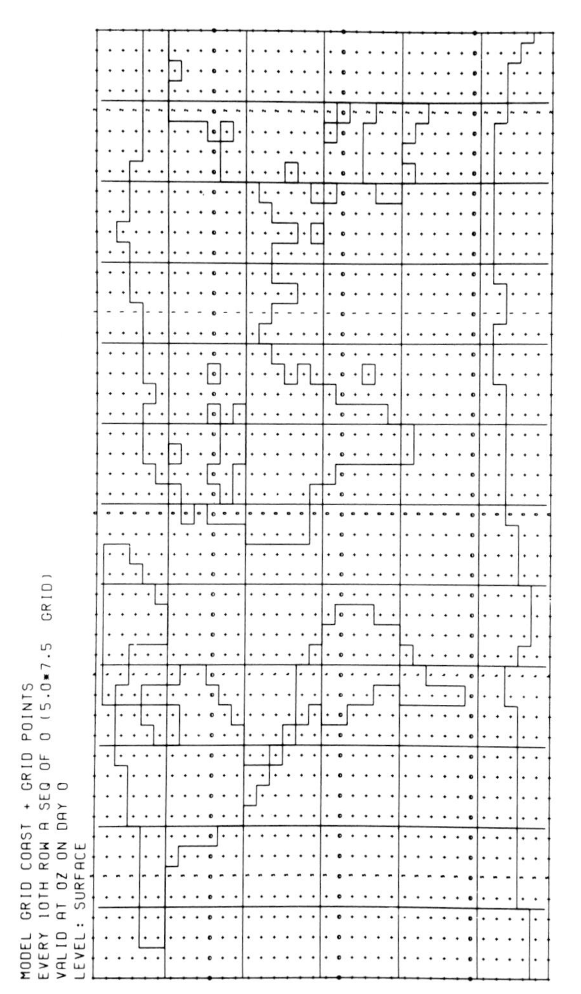

Horizontal resolution of the 11-Layer Model when integrated on the 5 x 7.5 degree latitude-longitude grid.

Figure 2

The horizontal grid used by WM. This is typical of current climate models. The horizontal spacing is 5° latitude by 7.5° longitude.

Mean-sea-level pressure (mb)

Observed (January)

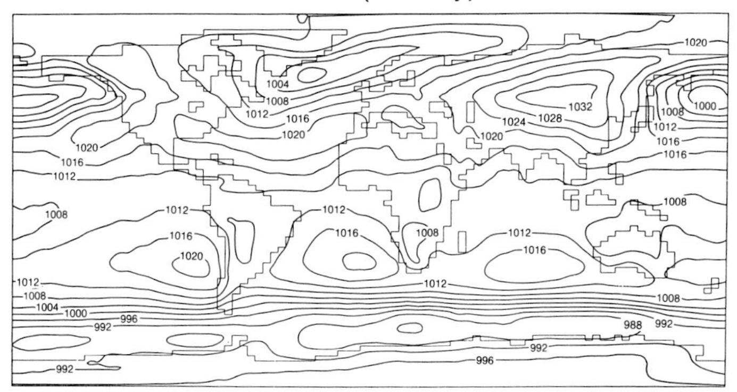

Model (December to February)

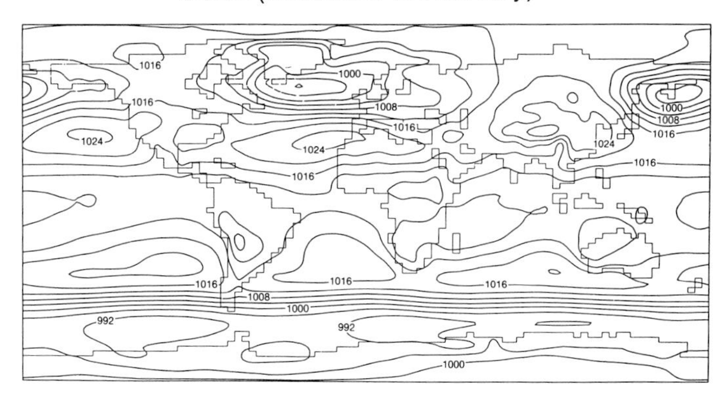

Figure 3 (a) and (b)

Mean sea level pressure (contours every 4 mbs)

(a) January, observed. (Schutz and Gates, 1971)

(b) December to February, simulated using a high resolution version of the Met. Office GCM (see Slingo et al, 1988)

High pass filtered variance of 500 mb heights
(December to February)

Simulated

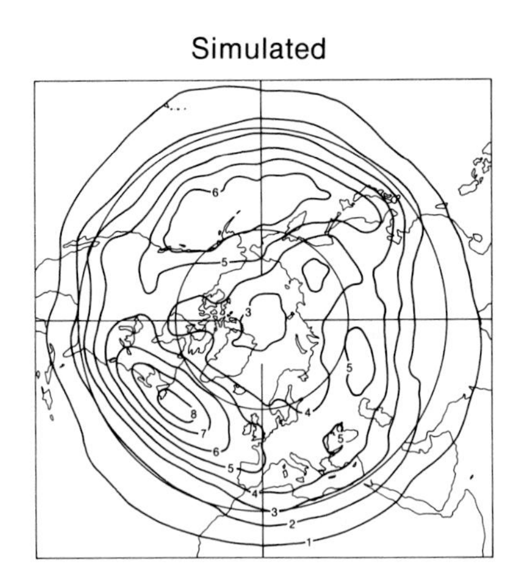

Observed

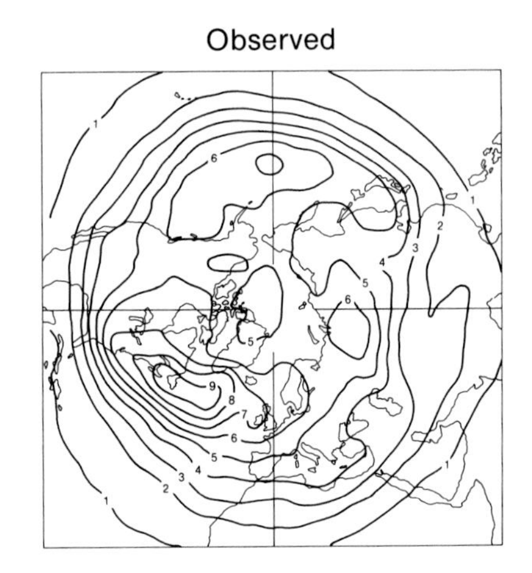

Figure 4: (a) (b)

High pass filtered variance of 500 mb heights for the northern hemisphere for December to February. Contours every 1 mb.

(a) From a high resolution version of the Met. Office GCM.

(b) Observed, from Met. Office operational analyses, 1983-1986.

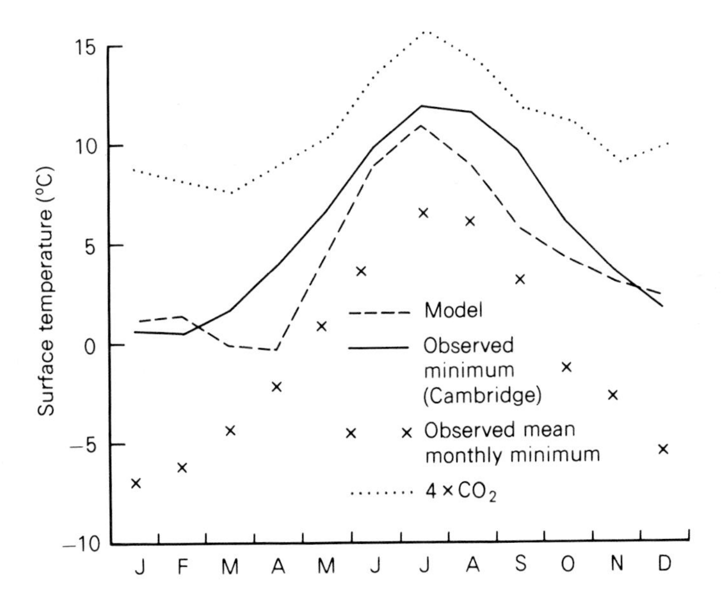

Figure 5

Monthly mean surface temperature over Eastern England. Solid line, climatological 2 m screen minimum temperatures at Cambridge; dashed line, simulated ground temperatures 00Z from a 5-layer atmospheric model. (Wilson and Mitchell, 1987b). The crosses are the observed mean monthly minimum.

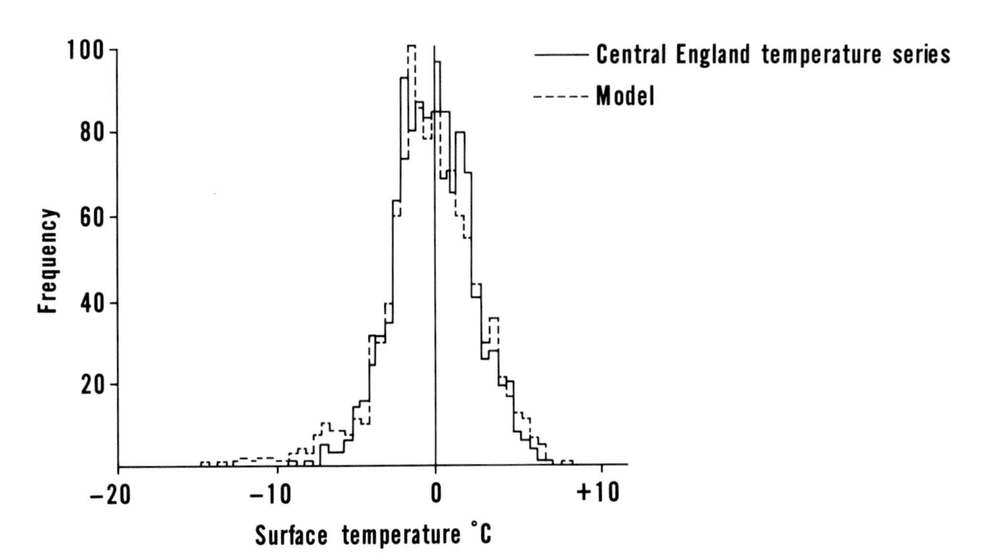

Figure 6

Frequency distribution of surface temperature over central/eastern England (mean, annual and semi-annual cycle removed) at 0.5 K intervals.
Light line — From 3 years of Central England Temperature Series
Heavy line — From a 3-year simulation with a 5-layer atmospheric model (Reed, 1986).

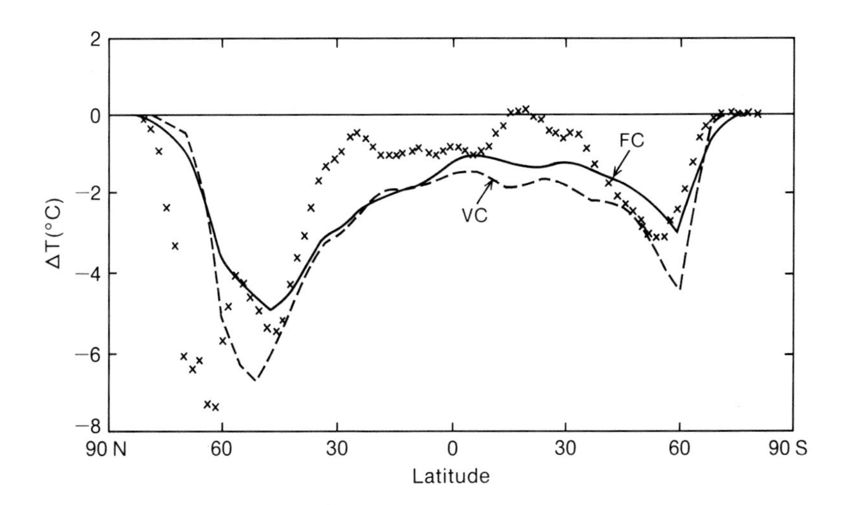

Figure 7

Latitudinally averaged changes in surface temperature between 18 K bp and present. Crosses, sea surface temperature changes deduced from palaeoclimatic data. Solid and dashed lines, as simulated using a climate model with a mixed-layer ocean using prescribed and model generated cloud amounts respectively. (From Manabe and Broccoli, 1985).

OCEAN MODELS AND OCEAN-ATMOSPHERE COUPLING

H. CATTLE
Dynamical Climatology Branch
Meteorological Office
United Kingdom

Summary

The nature of ocean (and sea ice) models used for general circulation model studies of climate and climate change is outlined, with particular emphasis on the Bryan/Semtner/Cox ocean model, which is that most extensively used to date in coupled atmosphere-ocean model simulations which include a deep ocean. The problem of the spin up of ocean and coupled atmosphere-ocean models is discussed and available techniques surveyed. Recent results from published studies using coupled atmosphere-deep ocean models of the impact of increased levels of atmospheric carbon dioxide are summarised.

1. INTRODUCTION

The oceans are important for studies of climate and climate change for a variety of reasons. For example:

- they carry a significant amount of the poleward transfer of heat within the ocean-atmosphere system

- they possess a large thermal inertia

- they determine and affect the atmospheric climate through the sea surface temperature and its variations and the extent and variability of the sea ice cover. These are themselves affected by the atmosphere via the air-sea exchanges.

In other words, the atmosphere, oceans and sea ice form interacting constituent parts of the climate system. As a consequence, if models of climate are to be capable of predicting climate change, not only the atmosphere, but also the oceans and sea ice must be modelled. The problem of determining the change in climate to a given perturbation (for example, the impact of increases in greenhouse gas concentrations) can be naturally divided into two parts: study of the *equilibrium climate response* and determination of the *transient response*. The equilibrium response is the eventual change in climate to the given perturbation (for example a doubling of effective carbon dioxide (CO_2) concentrations) after a period of time long enough for the climate to have settled down to an equilibrium state. The transient response is the evolution of the changes with time as the perturbation is applied (for example, immediately following the instantaneous CO_2 doubling, or, as is happening in reality, a via continuous change in the levels of greenhouse gas concentrations. In the case of greenhouse gas warming, the transient response for a given increase in the trace gases is always less, at any given time, than the final equilibrium response because of large thermal inertia of the oceans which slows down the rate of warming. For this reason, modelling the transient response to climate change requires modelling of the global ocean throughout its depth. We shall here consider the nature of the ocean and, briefly, the sea ice models which have been coupled to atmospheric general circulation models and used for studies of climate and climate change.

There are a number of issues to be borne in mind when considering the nature and uses of coupled ocean-atmosphere models.

Firstly, correct representation of the surface fluxes at the air-sea (or ice) interface is of crucial importance to the coupling. However, this is a complex matter. The ocean is driven by the effects of the wind stress on its surface and by the net surface heat flux (the sum of the surface net solar and infra-red radiative fluxes and the surface sensible and latent heat fluxes) and precipitation less evaporation difference (fresh water flux). At the same time these fluxes are directly or indirectly influenced by the ocean or ice surface temperature and ice extent which are themselves determined by the impact of the fluxes on the surface layers of the ocean and on the ice surface. Feedbacks in the coupled system may lead to amplification of errors in the simulations by the component models of either the surface temperatures (see e.g. Roberts and Cattle, 1990 (1)), the fluxes, or, most likely, in both of these.

Secondly, not only must the surface exchanges across the air-sea interface be correctly represented, but so also must their redistribution through the ocean. Just as for atmospheric models (see the lecture on 'Atmospheric GCMs' by Mitchell, this volume), this requires parametrisation of appropriate physical processes, including the turbulent transfer of heat, momentum and salt through the oceanic boundary layer (the ocean "mixed layer") and convective overturning to deeper layers. In high latitudes, the presence of sea ice on the ocean surface also needs to be represented. As discussed below, these processes are as yet fairly poorly represented in the ocean models used for climate studies.

In considering the limitations of the present generation of ocean models it is important to recognise that the characteristic space and time scales for the ocean are very different to those for the atmosphere. Thus, for example, the "synoptic scale" eddies of the ocean (the equivalent of atmospheric cyclones and anticyclones) have horizontal scales of only tens of kilometers or so, compared to the thousand kilometre scale of their atmospheric counterparts. Further, whilst the upper ocean responds rapidly to imposed atmospheric forcing, the deep ocean response time is of the order of centuries to thousands of years. Limitations of computing power mean that the current generation of coupled climate models are unable to have anything like the resolution necessary to explicitly represent the oceanic eddy fields, whilst special techniques must be used to spin up the coupled system if its equilibrium climate is to be established. Both of these aspects will be discussed in more detail below. However, we note here that global eddy resolving ocean-only models are beginning to be a reality, as demonstrated by Semtner and Chervin (1988) (2).

2. OCEAN MODELS USED FOR STUDIES OF CLIMATE AND CLIMATE CHANGE

As already implied, because the thermal relaxation time of the deep ocean is of the order of centuries or more, equilibrium experiments with models of the deep ocean are potentially very expensive computationally. As a consequence, many of the experiments exploring the equilibrium response of climate have been made with atmospheric models coupled to a representation of the ocean as a slab of water of constant depth (h), usually of order 50 metres (see, e.g, Weatherald and Manabe, 1986 (3); Hansen et al, 1984 (4); Washington and Meehl, 1984 (5); Schlesinger and Zhao, 1988 (6); Wilson and Mitchell, 1987 (7)). In early experiments, advection of heat by the ocean was ignored, but in later experiments, this was allowed for by prescribing a heat convergence (C), assumed constant for both present day and enhanced CO_2 climates (see e.g Wilson and Mitchell, 1987 (7)). Surface temperature (T_S) in the model then changes with time (t) according to an equation of the form:

$$\partial T_S/\partial t = (Q_N + C)/\rho_w c_p h \qquad (1)$$

where Q_N is the net heating across the ocean surface, ρ_w is the density of seawater (assumed constant) and c_p is the specific heat of seawater.

As already noted, to study the transience of climate change, it is necessary to use dynamical models which represent the ocean to its full depth. Hansen et al (1988) (8) have used a low resolution atmospheric model run with a mixed layer model coupled diffusively to a deep ocean to simulate the time-dependent response to a gradual increase in trace gases. A number of models, however, have been run, in particular in studies of the impact of increased levels of CO_2, using a full dynamical ocean model. To date, only one basic model has been widely employed for such GCM studies of climate change, that due to Bryan and Cox, based on the work of Bryan (1969) (9) (see also Semtner, 1974 (10); Cox, 1984) (11). The ocean model is based on the equations for conservation of horizontal momentum, mass, heat and salt, together with an the hydrostatic equation and an appropriate equation of state for seawater, of the form:

$$\partial \mathbf{V}/\partial t = -L\mathbf{V} - f\mathbf{k} \times \mathbf{V} + (1/\rho)\boldsymbol{\nabla} p + \varkappa\partial^2\mathbf{V}/\partial z^2 + A_M\nabla^2\mathbf{V} \qquad (2)$$

$$\nabla.\mathbf{V} + \partial w/\partial z = 0 \qquad (3)$$

$$\partial T/\partial t + LT = \varkappa\partial^2 T/\partial z^2 + A_H\nabla^2 T \qquad (4)$$

$$\partial S/\partial t + LS = \varkappa\partial^2 S/\partial z^2 + A_H\nabla^2 S \qquad (5)$$

$$\partial p/\partial z = -\rho g \qquad (6)$$

$$\rho = \rho(T, S, p) \qquad (7)$$

Here, $\mathbf{V}$ is the horizontal current vector with components (u, v); $L = (\mathbf{V}.\boldsymbol{\nabla})\mathbf{V}$; $\boldsymbol{\nabla} = \mathbf{i}u\partial/\partial x + \mathbf{j}\partial/\partial y$; $\nabla^2 = \partial^2/\partial x^2 + \partial^2/\partial y^2$; $\mathbf{i}$, $\mathbf{j}$, $\mathbf{k}$ are unit vectors in the x, y, and z directions respectively; f is the Coriolis parameter; $\varkappa$ is the vertical coefficient of diffusion and A_M and A_H are horizontal coefficients for diffusion of momentum and heat respectively.

Equation (7) represents the equation of state for sea water by which the temperature is non-linearly related to the potential temperature, T, salinity, S and pressure, p (equivalent to depth). In principle, and given suitable boundary conditions, equations (2) to (7), written in suitable finite difference form, enable the horizontal components of the currents, the potential temperature and the salinity to be stepped forward in time, and for the density, ρ, the pressure and the vertical component of velocity, w, to be diagnosed for each timestep.

The model is formulated on a latitude-longitude grid and any distribution of levels can be chosen in the vertical. The appropriate geography is set by a suitably specified dataset of ocean depths (with land points taken as zero) specified in terms of the number of model levels to which the ocean extends at any given grid point. Thus a variable bottom topography (of depth H) is allowed for, with respect to which the flow is assumed parallel to the slope, so that, on the ocean bottom:

$$w = -(u/a \cos\varphi)\partial H/\partial\lambda - (v/a)\partial H/\partial\varphi \qquad (8)$$

where a is the radius of the earth, λ is longitude and φ is latitude. No flow of heat or salt is allowed through the sides or bottom of the model ocean and a 'no slip' condition for momentum $[(u,v) = 0]$ is applied at horizontal boundaries. The solution of the numerical equations for momentum is complicated by the imposition of a 'rigid lid' condition, $w = 0$, at the upper boundary, which has the advantage that, at least for the relatively coarse resolutions that have been used to date for climate studies, fast moving gravity waves are removed from the solution, thus enabling longer timesteps to be taken than would otherwise be the case. This requires reformulation of the model equations, by eliminating the surface pressure at the upper boundary of the model and splitting the velocity field into its barotropic (vertically integrated) and baroclinic parts. These are then calculated separately, the barotropic part via the solution of a Poisson equation for the vertically integrated flow.

Referring back to equations (2), (4), and (5), it will be noted that these include terms for horizontal diffusion of momentum, heat and salt. It has sometimes been said that these allow for representation of oceanic eddies unresolved by the model grid. However, for the grids used to date in coupled model simulations (typically of order 5° x 5°, similar to that of the atmospheric models to which they are coupled), the horizontal diffusion coefficient for momentum (A_M) must be chosen solely on grounds of computational stability (see Bryan et al, 1975) (12). This results in values for this parameter several orders of magnitude higher than can be inferred, for example, from regional scale eddy resolving models. As a consequence the models are highly viscous, with the result that simulated ocean currents are markedly weaker than those observed. It is not until the grid length is brought down to a size of the order a degree of latitude and longitude that values of A_M chosen for computational reasons correspond to those found in reality. The coarse horizontal resolution also means that features such as the western boundary currents (Gulf Stream, Kuroshio) are not properly resolved.

The vertical resolution for the oceanic component used in published coupled model studies to date varies from 4 to around 10 unequally-spaced levels in the vertical. As yet, specific representation of mixed layer physics has not been included in the oceanic component of these coupled models (it is currently being included into the Meteorological Office model): the surface momentum, and net heat and water fluxes pass directly into the topmost model layer, from which they transfer into the deeper ocean by diffusion or vertical convection. As yet, oceanic convective processes are very crudely represented by an adjustment scheme which acts to vertically mix temperature and salinity uniformly on the grid scale when the grid-square averaged density stucture becomes convectively unstable.

3. SEA ICE

Ice-albedo feedback effects make it important the coupled models also include an interactive model of sea ice. Such climate models as have been run to date have all used relatively simple thermodynamic representations of sea ice based on that of Semtner (1976) (13). Some (e.g. Bryan, et al., 1988 (14)) also include allowance for variable ice concentrations over a grid square and a simple representation of the effects of ice convergence on ice thickness. A common feature of sea ice simulations in these models is that they tend to produce ice extents which are too extensive in the northern hemisphere, with too little ice cover in the southern hemisphere. Synchronous runs of such models (in which atmosphere, ocean and sea ice models are run on the same timescales, as opposed to the asynchronous approach used by Bryan et al. (14) - see below) show a gradual disappearence of the southern hemisphere sea ice from the climatologically initialised extents on a timescale of several years (see Gates et al, 1985 (15)). The reasons for this are not yet clear, but may be that the omission of a proper representation of ice dynamics plays a role.

4. SPIN UP OF OCEAN AND COUPLED OCEAN-ATMOSPHERE MODELS

Simulations of present day climate using coupled models have been carried out, for example, by Manabe et al (15) and Bryan et al. (12) (both 1975); Manabe et al (1979) (17); Washington et al. (1980) (18) and Gates et al (1985) (15) (see also Cattle and Roberts (1990) (1). A feature of the coupled ocean-atmosphere system already noted is the disparity in the timescales for adjustment between the atmosphere, the upper ocean layers and the deep ocean. Present day computing power enables runs of atmospheric models to be carried out for simulations of perhaps decades in length, depending on the resolution used, and more if a dedicated supercomputer is available. The same is true for ocean models of similar resolution (assuming the same timestep is used). Thus runs of ocean and atmosphere-deep ocean general circulation models to equilibrium (over integration periods of perhaps thousands of years) are only computationally feasible if steps are taken to drastically reduce the amount of computer time needed for such runs. This requires the use of so-called asynchronous spin up techniques. Within the ocean model itself, this can be done using a longer timestep for temperature and salinity than for velocities. For a given temperature and salinity timestep, Δt, this is equivalent to modifying the finite difference version of Equation (2) by a factor α (>1) to yield:

$$\alpha \Delta \mathbf{V}/\Delta t = -L\mathbf{V} - f\mathbf{k} \times \mathbf{V} + \ldots. \qquad (8)$$

The thermal and freshwater adjustment times within the deep ocean can also be decreased by modifying the thermal and salinity equations by a depth dependent factor, $\gamma(z)$ (≤ 1) to give (Bryan and Lewis, 1979 (19)):

$$\gamma \Delta T/\Delta t = -LT + \ldots \qquad (9)$$

$$\gamma \Delta S/\Delta t = -LS + \ldots \qquad (10)$$

In the upper ocean, γ takes values of unity and subsequently decreases with depth.

Such techniques have the effect of distorting the dynamics of the waves in the model, which is acceptable provided that the detailed character of the spin up is unimportant and an equilibrium solution is sought (when $\partial/\partial t \rightarrow 0$). Strictly, therefore, the technique is only valid for simulations (or those parts of the model ocean) which exclude consideration of the seasonal variation. For further details, the reader is referred to Bryan (1984) (20) (see also Killworth et al., 1984 (21)).

A similar technique can be applied to enable tha atmospheric component to be included when running in coupled mode. Thus the problem of the long timescale for adjustment of the deep ocean was overcome by Bryan et al (1982, 1988) ((22), (14)) by running the simulation for annual mean conditions and allowing the atmosphere, the upper ocean and the deep ocean to be integrated with respect to time using appropriate timescales for each component. In this way, the atmospheric component of the model was numerically integrated over the equivalent of 8.2 years, the upper ocean for 1250 years and the deep ocean for 34000 years.

A number of other techniques for accelerating the approach to equilibrium of coupled models have been either tried or proposed. Washington et al (1980) (18) fitted annual and semiannual harmonics to 10 day sample surface forcing data taken from runs of their atmospheric model for the months of January, April, July and October (a 30 day sampling period was used towards the end of their integration, as equilibrium was approached). The fitted data were then used to run the ocean and ice models for 5 years. Sea surface temperature data and ice extents from the final year of the ocean model run were used to provide the bottom boundary conditions for the next series of atmospheric sample runs (climatological values were used at the start of the integration). The ocean was forced using monthly mean atmospheric model surface stresses and appropriate values for the net heat input into the ocean, In deriving the latter, the solar flux absorbed at the surface, the downward infrared flux, the boundary layer temperature, T_A, and specific humidity, q_A (and presumably the boundary layer wind speed, V_A), taken from the preceeding atmospheric model run, were prescribed as a function of time during the oceanic part of the integration. The upward longwave and surface sensible and latent heat fluxes (F_U, H_S and LE respectively (where L is here the latent heat of vaporisation and E the evaporative flux), were then recomputed using the latest ocean model surface temperature (T_S) as the integration of the ocean model proceeded, via formulae of the form:

$$F_U = \varepsilon\sigma T_S^4 \quad (11)$$

$$H_S = \rho_A C_H V_A (T_S - T_A) \quad (12)$$

$$LE = \rho_A C_E V_A L (q_S - q_A) \quad (13)$$

Here, ε is the longwave emissivity of the ocean surface, σ the Stefan-Boltzmann constant, ρ_A the air density and C_H and C_E appropriate exchange coefficients for sensible heat and water vapour respectively. q_S is the saturation mixing ratio at the sea surface temperature, T_S. Computation of these fluxes as the ocean model integration proceeds, rather than use of the atmospheric model fluxes themselves directly is necessary to prevent the buildup of large and unrealistic errors in the sea surface temperature, since the negative feedbacks so induced tie the surface temperature to that of the atmospheric model air temperature, so preventing the two models from drifting too far apart.

Another technique was used by Manabe et al. (1979) (17) in a coupled model run with seasonally varying isolation in which surface boundary conditions for the two models came from libraries of, for running the ocean model, monthly mean atmospheric forcing data and, for running the atmospheric model, sea surface temperatures and sea ice thickness. These libraries were then updated as 1 day runs of the atmospheric model were alternated with annual cycle runs of the ocean model. The dataset of sea surface temperatures and ice thicknesses were updated at the end of each ocean model run, and the dataset of atmospheric model parameters each time the atmospheric model had completed one month. A further method has been proposed by Gates (Schlesinger, 1979) (23) whereby the coupled and ocean models are run in alternate phases. The coupled model run is carried out synchronously (i.e. the ocean and atmosphere models are run coupled together at the same rate) which then provides the data for a longer ocean-only run over several seasonal cycles. For a critique of these techniques, see the review by Schlesinger (1979) (23). More recently, energy balance models of varying complexity have been used by Dickinson (1981) (24), Schneider and Harvey (1986) (25), Harvey (1986) (26) and Roberts (1988) (27) to investigate the computational accuracy of methods of asynchronously coupling atmosphere and ocean models. Harvey (26) investigated asynchronous coupling strategies for runs of coupled models with inclusion of seasonal forcing and concluded that periods of synchronous running were highly desirable, whilst Roberts (27), in his investigation of the Gates' strategy, concluded that the synchronous phase should last for more than a year, to give the models time to adjust to the new ocean state.

5. SOME RESULTS FROM COUPLED MODEL SIMULATIONS OF CLIMATE CHANGE

As yet, coupled atmosphere-deep ocean models show substantial errors in their simulations of present day climate. As already indicated, currents in the low resolution ocean models used in the current generation of coupled models are too weak. In addition, the modelled sea surface temperature distribution, for example, shows errors of up to several degrees C, over much of the world oceans, of the same order as the simulated equilibrium sea surface temperature changes due to a doubling of CO_2 (see the lecture by Mitchell on 'Climate Model Results', this volume). The strength of the seasonal thermocline (the region of vertical temperature decrease at the base of the seasonal mixed layer) is also often weaker than observed. Whilst these factors need to be born in mind in interpreting the results of climate experiments with such models, nevertheless, a number of interesting results are beginning to emerge from their use. As already noted, these have primarily concentrated of assessing the impacts of increasing levels of CO_2 concentrations see, for example, Bryan et al. (1982) (22) (see also Spelman and Manabe, 1984 (28); Bryan and Spelman, 1985 (29)), Schlesinger et al, (1985) (30) and Bryan et al. (1988) (14). To conclude, we briefly discuss some of the results of these investigations below.

The integrations of Bryan et al. (1982) (22) and Bryan et al (1988) (14) both used an idealised land-sea distribution, in the former case with north-south symmetry. The Bryan et al. (1988) experiment, in which CO_2 levels were instantaneously doubled included north-south asymmetry and a circumpolar channel in the southern hemisphere. At *equilibrium*, the atmospheric response was consistent with equilibrium results of atmosphere-slab ocean integrations (see the lecture by Mitchell on 'Climate model results', this volume) . The ocean model showed the enhanced surface warming in the polar regions to be mixed downward and advected equatorward to give a pronounced reduction in vertical temperature gradient at mid-latitudes.

The *transient response* in Bryan et al's (1988) experiment, following the instantaneous doubling of CO_2 was, however, found to be markedly different for northern and southern hemispheres. In the southern hemisphere, the response of sea surface temperature lagged substantially behind that of the northern hemisphere. This resulted from the presence of a deep overturning cell in the southern hemisphere (also present in the control, present day climate run) at around 60°S in the circumpolar channel, the downwards flow to the northern side providing an important downward pathway for the heat and the upward flow continuing to supply deep cold deep water to the surface, which was untouched by the switch on event.

The results of such experiments using idealised geometry must be viewed with some caution. Thus, for example, Bryan et al. (1988) note that their model contained no counterpart of North Atlantic deep water formation. However, they also mention that later experiments which include the actual geometry of the world ocean and realistic deep water production in the northern hemisphere exhibit the same asymmetry in the transient response between hemispheres.

Coupled atmosphere-deep ocean models have the potential to be used to explore the the question as to just what pathways and through what processes the ocean takes up the CO_2-induced warming. This question has been explored, for example by Bryan and Spelman (1985) (29) and Schlesinger and Jiang (1988) (31). Schlesinger and Jiang found the CO_2 induced heat input to be everywhere into the ocean, but not uniformly so with latitude. Penetration of surface heating was found to be a minimum in the tropics and a maximum in the subpolar latitudes of both hemispheres. In the tropics the heat penetration was minimised by upwelling of cold water; in the subtropics, the CO_2-induced heating was transported downward by the existing downwelling there. In the high latitude regions convection although suppressed by the CO_2 heating of the ocean surface, was seen to nevertheless play a very important role its penetration. Schlesinger and Jiang comment that the warming of the ocean surface at high latitudes, by stabilising the oceanic stratification and reducing the intensity of the convective overturning, reduces the ability of the ocean to loose heat to the atmosphere in high latitudes, and results in a net uptake of heat by the global ocean. They note, however, that this conclusion could be modified by inclusion in the models of the effects of brine rejection (which occurs as sea water freezes) on the intensity of convection. Clearly high latitude processes and the parametrization of convective mixing in ocean models require more detailed parametrizations than have been used to date.

6. CONCLUDING REMARKS

It is evident that marked improvements are needed in the simulation of the ocean component in coupled models. This must come both through improvements to the models themselves, in particular by improved representation of the processes of ocean mixing, by use of higher horizontal and vertical resolution and by improvements to the simulated fluxes across the air-sea interface. In respect of the latter, further development of atmospheric models is needed too. Improvements to the quality of the embedded sea ice models are also necessary. The use of deep ocean models coupled to atmospheric models is essential, however, if we are to model the transient response of climate to continuously increasing levels of greenhouse gases, for example. Results from coupled models are already beginning to indicate interesting differences in the response to increased CO_2 warming between hemispheres not evident in the equilibrium experiments. There is still a considerable way to go, however, before the more detailed regional responses can be assessed using such models.

REFERENCES

(1) CATTLE, H. and ROBERTS, D.L. (1990). Simulation of sea ice in a coupled ocean-atmosphere model. Report of the first session of the Sea Ice Numerical Experimentation Group, Washington, 23-25 May 1989. WCRP/WMO. To appear.

(2) SEMTNER, A.J and CHERVIN, R.M. (1988). A simulation of the global ocean with resolved eddies. J. Geophys. Res., 93, C12, 15502-15222.

(3) WETHERALD, R.T. and MANABE, S. (1986). An investigation of cloud cover change in response to thermal forcing. Climatic Change, 8, 5-23.

(4) HANSEN, J., LACIS, A., RIND, D., RUSSELL, G., STONE, P., FUNG, I., RUEDYAND, R., and LERNER, J. (1984). Climate sensitivity and analysis of feedback mechanisms. In: 'Climate sensitivity'. Ed. J. Hansen and T. Takahashi, Geophysical Monograph 29, 130-163. American Geopysical Union, Washington D.C.

(5) WASHINGTON, W.M. and MEEHL, G.A. (1984). Seasonal cycle experiment on the climate sensitivity due to a doubling of CO_2 with an atmospheric general circulation model coupled to a simple mixed layer ocean model. J Geopys. Res., 89, 9475-9503.

(6) SCHLESINGER, M.E. and ZHAO, Z. (1988). Seasonal climatic changes induced by doubled CO_2 as simulated by the OSU atmospheric GCM/mixed - layer ocean model. J. Climate, 2, 459-495.

(7) WILSON, C.A. and MITCHELL, J.F.B. (1987). A doubled CO_2 sensitivity experiment with a GCM including a simple ocean. J Geophys. Res., 92, 13315-13343.

(8) HANSEN, J., FUNG, I., LACIS, A., RIND, D., LEBEDEFF, S., RUEDY, R. and RUSSELL, G. (1988). Global climate changes as forecast by Goddard Institute for Space Studies three-dimensional model. J. Geophys. Res., 93, D8, 9341-9364.

(9) BRYAN, K. (1969). A numerical method for the study of the circulation of the world ocean. J Comp. Phys., 4, 347-376.

(10) SEMTNER, A.J., JR. (1974). An oceanic general circulation model with bottom topography. Dept. of Met. University of California, Technical Report No.9. (11) COX, M.D. (1984). A primitive equation, 3-dimensional model of the ocean. Geophysical Fluid Dynamics Laboratory Ocean Group Technical Report No. 1.

(12) BRYAN, K., MANABE, S. and PACANOWSKI, R.C. (1975). A global ocean-atmosphere climate model. Part II. The oceanic circulation. J. Phys. Oceanogr., 5, 30-46.

(13) SEMTNER, A.J., JR. (1976). A model for the thermodynamic growth of sea ice in numerical investigations of climate. J. Phys. Oceanogr., 6, 379-389.

(14) BRYAN, K., MANABE, S. and SPELMAN, M. J. (1988). Interhemispheric asymmetry in the transient response of a coupled ocean-atmosphere model to a CO_2 forcing. J. Phys. Oceanogr., 18, 851-867.

(15) GATES, W. L., HAN Y-J., and SCHLESINGER, M. E. (1985). The global climate simulated by the OSU coupled atmosphere-ocean GCM. In: 'Coupled ocean-atmosphere models'. Ed. J. Nihoul, Elsevier, Amsterdam, pp 131-151.

(16) MANABE, S., BRYAN, K. and SPELMAN, M. (1975). A global ocean-atmosphere climate model. Part I. The atmospheric circulation. J. Phys. Oceanogr., 5, 3-29.

(17) MANABE, S., BRYAN, K. and SPELMAN, M. (1979). A global ocean-atmosphere model with seasonal variation: possible application to a study of climate sensitivity. Dyn. Atmos. Oceans, 3, 393-426.

(18) WASHINGTON, W.M., SEMTNER, A.J., JR., MEEHL, G.A., KNIGHT, D.J. and MAYER T.A. (1980) A general circulation experiment with a coupled atmosphere, ocean and sea ice model. J. Phys. Oceanogr., 10, 1887-1908.
(19) BRYAN, K and LEWIS, L.J. (1979). A water mass model of the world ocean. J Geophys. Res. C5, 2503-2517.
(20) BRYAN, K. (1984). Accelerating the convergence to equilibrium of ocean-climate models. J. Phys. Oceanogr., 14, 666-673.
(21)KILLWORTH, P.D., SMITH, J.M. and GILL, A.E. (1984). Speeding up ocean circulation models. Ocean Modelling, No. 56.
(22) BRYAN, K., KOMRO, F.G., MANABE, S. and SPELMAN, M.J. (1982). Transient response to increasing carbon dioxide. Science, 215, 56-58.
(23) SCHLESINGER, M.E., (1979). Comments on ocean-atmosphere coupling and discussion of the paper "A global ocean-atmosphere model with seasonal variation: possible application to a study of climate sensitivity" Dyn. Atmos. Oceans, 3, 427-432.
(24) DICKINSON, R.E. (1981). Convergence rate and stability of ocean-atmosphere coupling schemes with a zero-dimensional climate model. J. Atmos. Sci., 38, 2112-2120.
(25) SCHNEIDER, S.H and HARVEY, L.D.D. (1986). Computational efficiency and accuracy of methods for asynchronously coupling atmosphere-ocean climate models. Part I. Testing with a mean annual model. J. Phys. Oceanogr., 16, 3-10.
(26) HARVEY, L.D.D. (1986). Computational efficiency and accuracy of methods for asynchronously coupling atmosphere-ocean climate models. Part II. Testing with a seasonal cycle. J. Phys. Oceanogr., 16, 11-24.
(27) ROBERTS, D.L. (1988). An investigation of the Gates asynchronous coupling strategy using a simple energy-balance model. Meteorological Office Dynamical Climatology Technical Note, DCTN 64.
(28) SPELMAN, M.J. and MANABE, S. (1984). Influence of oceanic heat transport upon the sensitivity of a model climate. J. Geophys. Res., 89, 571-586.
(29) BRYAN, K. and SPELMAN, M.J. (1985). The ocean's response to a CO_2-induced warming. J. Geophys. Res., 90, C6, 11679-11688.
(30) SCHLESINGER, M. E., GATES, W.L. and HAN, Y-J. (1985). The role of the ocean in CO_2-induced climate warming: preliminary results from the OSU coupled atmosphere-ocean GCM. In: 'Coupled ocean-atmosphere models'. Ed. J. Nihoul, Elsevier, Amsterdam, PP 447-478.
(31) SCHLESINGER, M.E. and JIANG, X. (1988). The transport of CO_2-induced warming into the ocean: an analysis of simulations by the OSU coupled atmosphere-ocean general circulation model. Climate Dynamics, 3, 1-17.

LAND ICE AND CLIMATIC CHANGE - AN INTRODUCTION

JOHANNES OERLEMANS
Institute of Meteorology and Oceanography
University of Utrecht, Princetonplein 5
UTRECHT, The Netherlands

Summary

Land ice plays an important role in the earth's climate system. Climatic variations on longer time scales are well documented by the extent of the large northern hemisphere ice sheets. End moraines, erratic boulders, etc. have learned us a lot about the timing and severeness of ice ages. On a smaller scale, mountain glaciers also serve as very sensitive indicators of climatic changes.

In this lecture an outline is given on the aspects of glaciers and ice sheets that are particularly important with regard to climatic change. After a general introduction a few very simple mathematical models describing ice sheets and glaciers will be discussed to give a feeling for orders of magnitude.

In the last section the potential contribution of glaciers and ice sheets to sea-level rise is discussed in connection with the greenhouse warming.

1. DISTRIBUTION OF LAND ICE

Land ice in the form of glaciers and ice sheets occurs in many parts of the world. At higher latitudes glaciers are found at lower altitudes, but much also depends on local conditions (in particular, *continentality*). About 90% of all land ice is stored on the Antarctic continent, Greenland takes care of about 9.5% and the many glaciers and small ice caps do not contribute more than 0.5%, in fact. This does not necessarily mean that small ice bodies are unimportant. They generally have much shorter response times, and on a time scale of decades or centuries they may even dominate the scene.

Melt of all ice sheets and glaciers would lead to a rise in sea level of about 90 m, that is, if the areal extent of the world ocean would remain the same. Taking into account the increase in ocean surface that would take place *and* the fact that part of the present ice is found below sea level, this figure reduces to 80 m. The potential importance of a small shift in the land ice balance is thus clear: a 1% change in ice volume would lead to a 0.8 m change in sea level!

Ice sheets can be very thick. The maximum ice thickness measured so far is 4700 m, on East Antarctica. The mean thickness of the Antarctic and Greenland ice sheets is about 2490 and 1790 m, respectively. Mountain glaciers are much thinner, ice depth typically is a few hundred meters. This is due to the larger slope of the bed on which a mountain glacier flows. The downward pulling force acting on the ice is larger, resulting in faster flow and smaller ice thickness. A situation generally seen on plateaus is the existence of a rather flat ice sheet, with ice discharge through a number of so-called outlet glaciers. These descend from the plateau to lower land and are similar to mountain glaciers. Examples are the Vatnajökull Ice Cap on Iceland, the Patagonian Ice Cap (South America), and also large parts of the margin of the Greenland Ice Sheet.

When climate is sufficiently cold and the nourishment with snow sufficiently large, glaciers may flow all the way to the coast to form *calving glaciers*. At the coast icebergs break off. The larger ones may last for years and may travel far, thus forming a threat to shipping in some areas. There are also places where the ice does not break off immediately when it starts floating, but extends to form an *ice shelf*. The largest present ice shelves are the Ross and Ronne-Filchner ice shelves in Antarctica. They have linear dimensions of hundreds of kilometers, and are up to 1000 m thick. So at the grounding line (see Figure 1), the bedrock is far below sea level. Sometimes ice shelves may run aground to form *ice rises*. Here basal drag becomes important again, and ice rises thus can obstruct the smooth spreading of an ice shelf. Ice rises may exert a *back stress* on the main ice sheet, resulting in smaller ice velocities and a thicker inland ice sheet.

Ice sheets and glaciers have a marked influence on the landscape. As witnessed by the deep fjords in for instance Scandinavia, Greenland and Antarctica, glaciers have a large eroding capacity. On the other hand, they may bring large amounts of sediment to their marginal regions. Push and side moraines often form spectacular elements in the mountain landscape, and can tell a lot about the former extent of glaciers.

Readers interested in a detailed morphological description of glaciers and ice sheets may consult Wilhelm (1957). Geomorphological aspects are treated for instance in Sugden and John (1976).

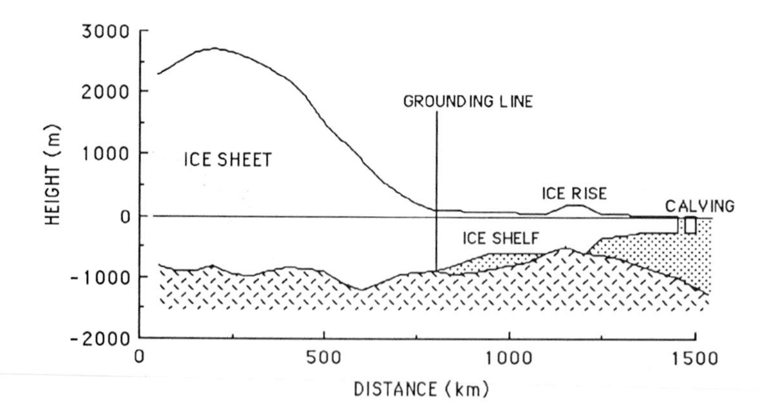

FIGURE 1. Cross section through a *marine* ice sheet. The bed of the main ice sheet is far below sea level, and an ice shelf is formed. Ice rises may occur when the shelf runs aground. At the edge of the ice shelf large tabular icebergs are produced (calving).

2. MASS BALANCE

In all glaciers and ice sheets the movement is from higher to lower parts. Ice flows from the *accumulation* region, where the annual mass balance at the surface is positive, to the *ablation* region, where the balance is negative. Those regions are separated by the *equilibrium line.* The equilibrium-line altitude E is often used as a first basic parameter to describe the climatic conditions.

Various processes contributing to accumulation can be identified. Snow fall is most common, and a pack of snow gradually densifies to form *firn* (old snow that survived a summer) and, after several years, glacier ice. In the accumulation region melting may occur in summer, of course. However, in most cases the melt water does not run-off, but penetrates into the snow. Refreezing then warms the snow/firn layer through release of latent heat. Sometimes *superimposed ice* is formed. This occurs at places with a large seasonal cycle (i.e. at higher latitudes). Melt water refreezes on bare ice, and the vertical structure of the ice layers may become very complex (useless for some purposes like obtaining specific records from an ice core). In contrast, in the very cold climate of central Antarctica accumulation is mainly by the steady sinking of tiny ice crystals, and there is never any melting.

In the ablation zone a glacier looses its mass. It is general practice to compare the contributions from radiation and from the turbulent energy flux in the melting process. On most glaciers radiation is dominating, but the turbulent heat flux may play an important role in more maritime climates, in particular at the end of the summer, when solar radiation is less intense and warm air masses are advected from the sea.

A plot of mass balance versus elevation is very helpful in understanding the mass budget of a glacier. Figure 2 shows an example. Normally, the balance gradient is more or less constant below the equilibrium line, and drops off at higher elevations. In many cases even the balance itself starts to decrease beyond a certain critical elevation. This is due lower moisture

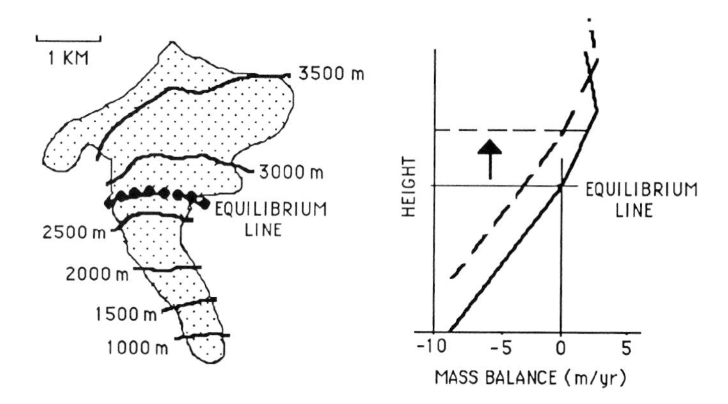

FIGURE 2. A typical mountain glacier. At the right hand side the mass balance is shown as a function of height; it is expressed in m water equivalent per year. To *first* order, a climatic change can be represented by a shift of the equilibrium line, while the mass balance profile remains unchanged.

content of the atmosphere (lower air temperature) and sometimes also to snow drift, carrying mass away from higher more exposed terrain.

For a glacier to be in equilibrium, mass continuity requires that

$$(1) \qquad \int_a M \, da + C = 0 \, .$$

Here M is the mass balance and a the area of the glacier, C denotes the calving rate in the case that calving in a lake or sea occurs. Since the rate of ablation can be much faster than the rate of accumulation, it follows from eq. (1) that the accumulation zone is generally much larger.

On the Antarctic Ice Sheet ablation is negligibly small and the balance is essentially between accumulation and calving. On the Greenland Ice Sheet, ablation and calving are roughly of equal importance. The difference in the mass balance characteristics of these ice sheets are due to differences in temperature. The Antarctic continent is 20 to 30 K colder than Greenland.

A number of mass-balance measurements from glaciers are plotted in Figure 3. Here Nigardsbreen and Engabreen (both located in Norway) span the widest altitude range. Like Peyto Glacier (west side of Rocky Mountains, Canada), these are maritime glaciers. The balance gradient is large and the maximum accumulation too. So these glaciers are active - ice velocities must be large to maintain the balance. Tuyuksu glacier is a continental glacier (southern USSR) with lower accumulation and somewhat lower balance gradient. Hintereisferner is a glacier in the central Alps. Its balance gradient over the lower part is comparable to that of the Scandinavian glaciers. Like for Engabreen, accumulation tends to decrease with altitude in the uppermost part. Some additional profiles are shown in Figure 4, now for more polar regions. The accumulation rates are much smaller. The profile for Antarctica is along a transect in East Antarctica. Here the accumulation rate reaches a maximum value close to the coast and then decreses steadily with altitude, down to values of 0.02 m/yr only ! The differences in balance

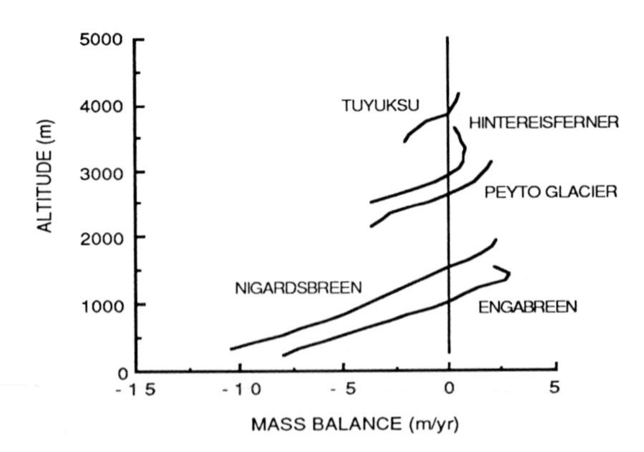

FIGURE 3. Examples of long term mass-balance profiles from various glaciers.

gradient (in the ablation zone) between Nordbo Glacier (southern Greenland), White glacier and Devon Ice Cap (both northern Canda) are large. No general theory is available that explains such differences, although in many cases specific factors can be seen to be important. Exposure, gradients in surface albedo associated with morainic material or atmospheric dust, advective heat fluxes from bare ground, etc., are such factors.

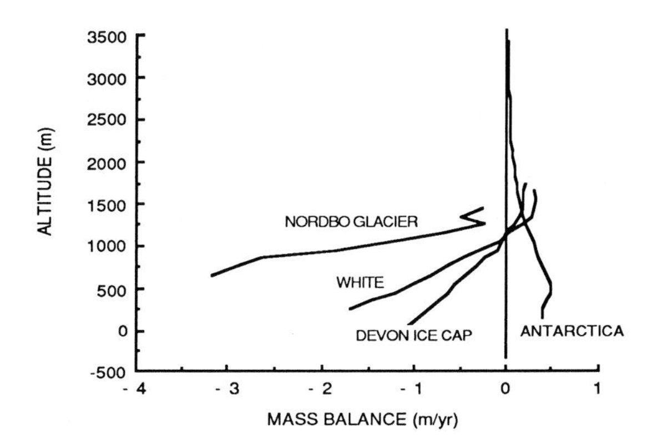

FIGURE 4. Mass balance profiles in polar environments.

3. FLOW OF GLACIERS AND ICE SHEETS

Ice bodies spread out due to the action of gravity. There are basically two modes of flow: *deformation* and *sliding*. To some extent ice behaves as a viscous fluid, and internal layers can shift relative to each other. This leads to a velocity profile in which the velocity decreases with depth. At the glacier bed the velocity then reduces to zero (frozen bed), or to the sliding velocity (basal ice at melting point), see Figure 5. The mean ice velocity parallel to the bed (U) is related to the *driving stress* according to

$$(2) \qquad U = U_d + U_s = k_1 H \tau^3 + \frac{k_2 \tau^3}{N}$$

The subscripts d and s refer to deformation and sliding, respectively. N is the normal load on the bedrock (weight of ice column minus pressure of subglacial water). The parameter k_1 expresses how viscous the ice is. This depends mainly on ice temperature, crystal fabric, and impurity content.. Attempts have been made to obtain a theoretical expression for the sliding parameter k_2, but a concensus has not yet been reached. Both k_1 and k_2 should be considered as empirical parameters, varying from glacier to glacier

From eq. (2) it is obvious that ice velocity increases very strongly with driving stress (to third power). This reflects the strongly nonlinear character of ice deformation: little deformation when forces are small, very large deformation when forces are large. As a result, ice flows in such a way that a tendency towards constant stress exists, i.e. sH = constant. This critical stress is termed the *yield stress*, and it is of the order of 1 bar (100 hPa). So when the slope is steep ice thickness will be smaller. In fact, the assumption of constant stress (*perfect plasticity*) can be used to make a first estimate of glacier thickness based on the surface slope. We will also use the assumption in a later section.

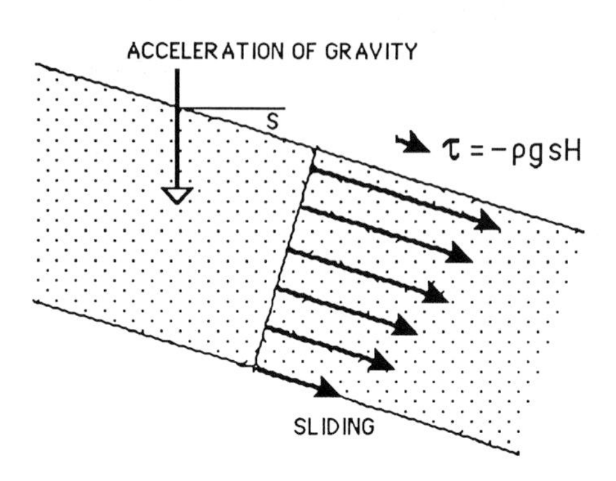

FIGURE 5. Both sliding and deformation contribute to the mass flux. The driving stress τ increases linearly with surface slope s and ice thickness H. The gravitational constant is denoted by g, ice density by ρ.

The volume flux of ice is given by H**U**, where **U** now is surface slope). Ignoring differences in ice density for the moment, the local rate of change of ice thickness is thus given by

$$\frac{\partial H}{\partial t} = -\nabla \cdot \mathbf{U} H + M \tag{3}$$

This equation forms the basis for many numerical studies of the the dynamics of glaciers and ice sheets. It is frequently used in one-dimensional form in *flow-line models*, where the dynamics of a glacier are calculated along a central line down the surface slope (but still taking into account varying lateral geometry).

The contribution of sliding to the vertical mean velocity differs enormously from glacier to glacier, and also through the seasons. In summer, the motion of a temperate (ice at melting point everywhere) valley glacier can be almost entirely by sliding. In the large ice sheets, deformation is generally far more important, except in the fast outlet glaciers and the ice streams. The process of sliding is still poorly understood, and the theories that have been developed are of an empirical nature.

For a more extensive introduction to glacier motion, see Paterson (1981) and, for the joggers, Hutter (1984).

4. SENSIVITY OF GLACIERS AND ICE SHEET TO CLIMATIC CHANGE

In this section we study, with the aid of some extremely simple models, how glaciers and ice sheets may react to small changes in their climatic environment. To obtain a qualitative understanding, numerical models are not needed.

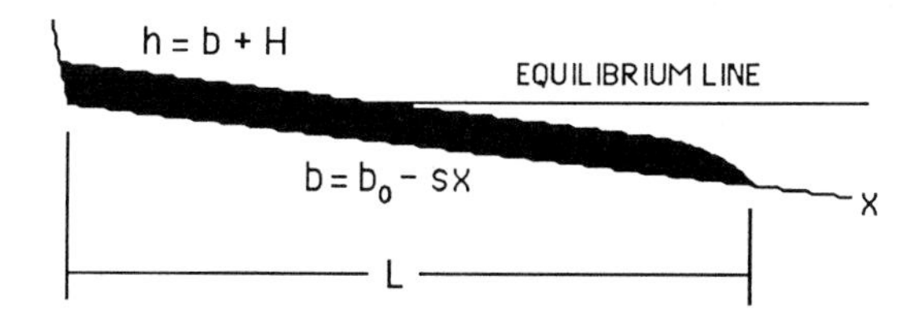

FIGURE 6. A glacier on a bed of constant slope s. Elevation of bed is b, of the ice surface h. L is the length of the glacier.

As a start, we consider a valley glacier on a bed with constant slope, see Figure 6. Next it is assumed that the mass balance is a linear function of height relative to the equilibrium-line altitude E, i.e. M = a(h-E). So a is the balance gradient. According to eq. (1), the glacier is in equilibrium when

$$\int_L M\,dx = a\int_L (H + b_0 - sx - E)\,dx = 0 \tag{4}$$

This equation is easily integrated and solved for L:

$$L = 2(H^* + b_0 - E)/s \tag{5}$$

Here H* is the mean ice thickness. It is noteworthy that in this case the solution does not depend on the balance gradient. Next the perfect-plasticity approximation is used, implying that $H(dh/dx) = \rho g/\tau^* = \Lambda$, where τ^* is the yield stress. For a yield stress of 100 hPa, which is a typical value, Λ is about 11 m.

The expression H(dh/dx) is constant along the glacier, so it can directly be replaced by sH*. Substituting this in eq. (5) then yields:

$$L = 2(\Lambda/s + b_0 - E)/s \tag{6}$$

Several conclusion can be drawn from this. First of all, for given b_0 and E, glaciers will be much longer when the slope of the bed is small. Secondly, the ice thickness - mass balance feedback, reflected by the first term in the brackets, may be significant, in particular when the bed slope is small. This is further illustrated in Table I, where glacier length is given for various slopes.

TABLE I. Equilibrium length of a valley glacier of constant width on a uniform slope. L' represents the glacier length when the effect of glacier thickness on the mass balance is left out. Parameter values used are: E = 2800 m, b_0 = 3300 m, Λ = 11 m.

slope s:	0.05	0.1	0.2	0.3	0.4
Λ/s (m):	220	110	55	36.7	27.5
L (m):	28800	12200	5550	3578	2638
L' (m):	20000	10000	5000	3333	2500

Formally, the sensitivity of a glacier to climatic change can be defined as dL/dE, which, according to eq. (6), equals 2/s. So the sensitivity is also larger when the slope is smaller.

So far we only discussed equilibrium states. However, the climatic environment changes in a continuous fashion, and the response time of a glacier is important. Small glacier may react within one or two decades, large glaciers (> 10 km, say) may need a century to approach equilibrium. Also, the response time will be shorter when the mass turnover is large. To calculate the response of a particular glacier to time-dependent forcing, numerical models have to be used.

Next we turn to *ice sheets*. The simplest possible case is an ice sheet on a horizontal bed. The driving stress then equals ρgH(dH/dx). Using the perfect plasticity approximation again, we have H(dH/dx) = Λ. This can be integrated to give:

$$H\,dH/dx = \Lambda \longrightarrow H = \sqrt{2\Lambda x} \tag{7}$$

The solution is for 0<x<L/2. The other half (L/2<x<L) is the mirror image. The ice-sheet profile thus obtained is shown in Figure 7.

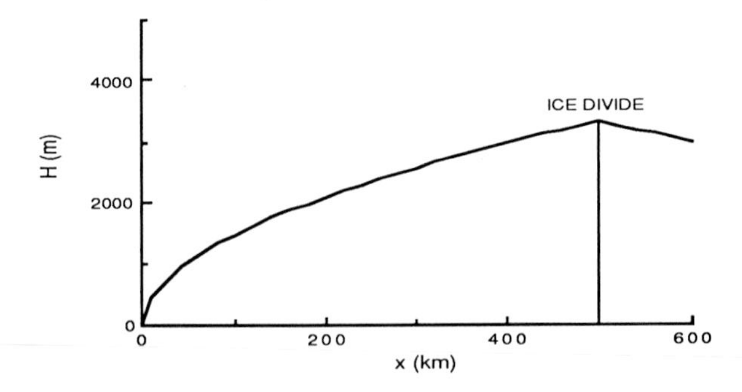

FIGURE 7. Profile of a perfectly plastic ice sheet (Λ = 11 m, L = 1000 km). The profile only depends on L, not on the mass balance! The axis of symmetry is at x = 500 km (ice divide).

It should be realized that the parabolic profile found above really is a first-order approximation. Since the effect of longitudinal stress gradients is not taken into account, the solution cannot be realistic close to the ice divide and at the edge of the ice sheet. A direct consequence of the assumption of perfect plasticity is the fact that the ice sheet profile does not directly depend on the mass balance. For given size L, the profile is fixed.

When the mass balance is known as a function of x, mean ice velocities can be calculated. Equilibrium requires that

$$(8) \qquad UH = \int_{L/2}^{0} M(x)\,dx \quad \longrightarrow \quad U = \frac{1}{\sqrt{2\Lambda x}} \int_{L/2}^{0} M(x)\,dx$$

The velocities thus obtained are frequently referred to as *balance* velocities. A comparison with actual velocity measurements (which have to be converted to vertical mean velocities, a major problem) may then show whether the ice sheet is in balance or not.

The perfectly plastic model can be used to demonstrate the character of the ice thickness - mass balance feedback for large ice sheets. We consider a flat bounded surface (an island in the polar ocean, say), of size L. Again, the mass balance is a linear function of height relative to the equilibrium-line altitude [$M = a(h-E)$]. When an ice sheet is present, its total net mass balance will be

$$(9) \qquad M_{tot} = a \int_{0}^{L} [H(x)-E]\,dx = a\left[\tfrac{2}{3}\sqrt{\Lambda L} - E\right]$$

When the expression in square brackets becomes negative, i.e. when the equilibrium-line altitude rises above the mean surface elevation H_m, the ice sheet is doomed to disappear. Interestingly, for $0<E<H_m$ two stable states are possible (an ice sheet with fixed profile, or no ice at all). This can be illustrated in a solution diagram, see Figure 8. The heavy lines represent stable situations. When the equilibrium line is 'below sea level', there must be an ice sheet. When the equilibrium line is above H_m, an ice sheet cannot exists. In between there are two possibilities.

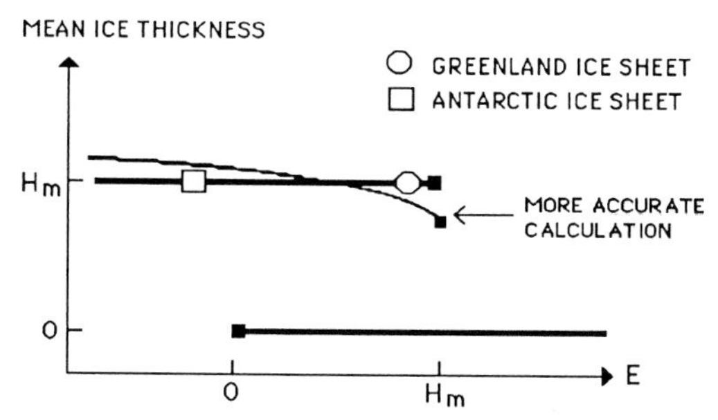

FIGURE 8. Solution diagram for the perfectly plastic cie-sheet model. Critical points are indicated by black squares.

In the diagram the basic difference between the Greenland and Antarctic ice sheets can be illustrated nicely. For present conditions, E = 1400 m for Greenland, placing it close to the critical point. When all the ice would be removed, no large ice sheet would form again ! Conditions in the Antarctic are much colder, and the ice sheet would certainly form again under the present climatic conditions.

The model discussed above is extremely simple, and only useful to obtain a qualitative understanding. Calculating the solution diagram with a more sophisticated numerical ice-sheet model leads to essentially the same picture. The result is shown in the figure. Now the mean ice thickness increases when the equilibrium-line altitude decreases. Although not shown here, the profile is not parabolic anymore.

In reality continents are not perfectly flat, E varies from place to place, and the physical properties of ice are not the same everywhere. So it is impossible to derive the response of an ice sheet to climatic change with such simple models as discused above. Numerical models are needed that can handle detailed geometry and time-dependent forcing, but it is outside the scope of this lecture to discuss these in any detail. Some references: Budd and Jenssen (1975), Mahaffy (1976), Oerlemans and Van der Veen (1984), Ritz et al. (1982), Reeh et al. (1985), Birchfield and Grumbine (1985), Oerlemans (1986), MacAyeal and Thomas (1986), Dahl-Jensen and Johnsen (1986), Van der Veen (1986).

4. RECENT CHANGES

Except for the last decades, very few direct observations of changes in the extent of glaciers are available. However, for some glaciers in the Alps and Scandinavia it has been possible to construct records of glacier length back to the 18th or even 17th century. Drawings, etches and paintings, written documents, as well as terminal moraines to infer exact positions of a glacier snout, have been used to achieve this. In Figure 9 some well-known records are shown.

On the longer time scale, the behaviour of the glaciers is remarkably uniform. The steady retreat over the last hundred years, in particular, shows up in all records. It is now well-accepted that this retreat has been of a world-wide nature (Meier, 1984). Most glaciers reached their neoglacial maximum between 1600 and 1900 AD, and at many places the maximum stand is seen from trimlines. Although the amplitude of the curves in Figure 9 differs from glacier to glacier, it should be remarked that this is mainly a geometric effect. A glacier with a very wide accumulation basin and a narrow tongue (Nigardsbreen, for instance), will exhibit large fluctutations in front position for small changes in climatic conditions. When the records are adjusted for such effects, the retreat becomes very similar (Oerlemans, 1988).

Many outlet glaciers of the Greenland Ice Sheet also retreated (Weidick, 1967). This is known from descriptions and pictures, and trimlines far above present ice levels further illustrate this. It seems likely that the retreat of those outlet glaciers is a consequence of increased melting rates on the outlet glaciers. Although less accumulation on the main ice sheet could cause a similar retreat, this would probably be a much slower process. In any case, nothing is really known about changes in the inland ice during the last few centuries [this applies to Greenland *and* Antarctica].

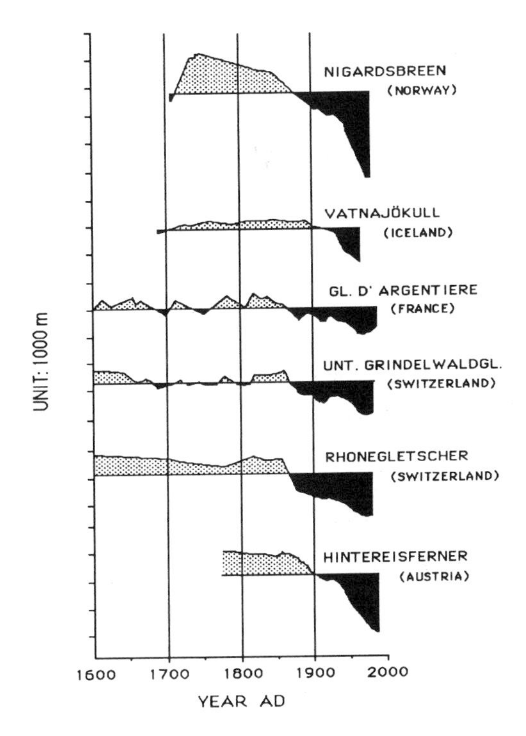

FIGURE 9. Fluctuations in the length of some glaciers over the last centuries.

Two question appear essential at this point, namely:

(1) Why did glaciers retreat so much during the last hundred years ?

(2) To what extent is the distribution of land ice in balance with the *present* climatic conditions ?

The gradual increase in global temperature over the last 100 years (about 0.5 K, e.g. Ellsaesser et al., 1986) has been put forward frequently as the cause of glacier retreat. However, when looking more carefully the matter is not so simple. We could use a statistical relation between equilibrium line altitude and temperature, based on interannual variations in a mass balance series, to estimate the change in E over the last 100 or 150 years. According to an analysis by Kuhn (1988), a 0.5 K temperature increase would correspond to 40 m higher altitude of the equilibrium line (this does not include the effect of associated increase in atmospheric counterradiation, for reasons to be discussed later). Experiments with numerical models for several glaciers have shown that a +40 m change in E only explains half of the observed retreat only.

Recently, a number of workers have tried to simulate historic glacier variations over the last 3 or 4 centuries (Nigardsbreen: Oerlemans, 1986; Glacier d'Argentiere: Huybrechts et al., 1988; Rhonegletscher: Stroeven et al., 1988), by calculating the evolution of a glacier along a flow line, while forcing the equilibrium line to move up and down in proportionality to a *local* climatic indicator like tree-ring width, (proxy) temperature and/or (proxy) precipitation. The results of these studies were not very good. Models can be tuned to some extent, but not in such a way that both the neoglacial maximum and the retreat over the last 100 years show up.

Better results are obtained with global forcing function (Oerlemans, 1988), derived with simple energy balance model. The upper two curves in Figure 10 show forcing functions imposed to the model. The forcing is in terms of perturbed radiation balance at the *surface*. It consists of a contribution from greenhouse gases (following Wigley, 1987) and from volcanic activity as inferred from a Greenland ice core (Hammer et al., 1980). The model calculates perturbations of mixed-layer temperature (shown), land temperature, atmospheric temperature, and radiation budget over a melting ice surface. The mixed-layer provides inertia, and changes in temperature and radiation balance are not in phase. The associated changes in equilibrium-line altitude are then obtained from an analysis of records of mass balance and climatolocigal data from Hintereisferner (Kuhn, 1988), as shown in the lower curve. The mixed-layer temperature appears as a smoothed curve, while the ELA signal has more short-term variability because it depends on mixed-layer temperature *and* changes in the radiation balance. See Oerlemans (1988) for more detail.

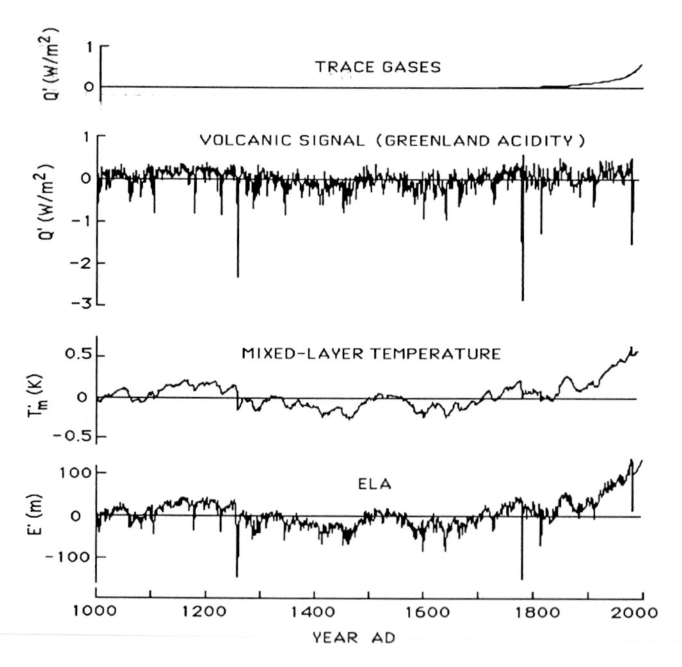

FIGURE 10. Radiative forcing from greenhouse gases (upper curve) and volcanic activity (second curve) as used in a simple energy balance model. The resulting perturbations of mixed-layer temperature and equilibrium-line altitude are shown in the lower curves.

The next step is to impose the ELA-curve to a numerical glacier model. In the case discussed here this was done for a schematic valley glacier, with a fairly wide accumulation basin and a more narrow tongue, having an equilibrium length of about 13 km. Since a glacier model needs time to equilibrate, the integrations were done from 1000 AD onwards. A typical result is shown in Figure 11. Concerning the reord from 1600 AD, the model glacier shows maximum stands in the 17th century, around 1850 and in the beginning of this century. A comparison with Figure 9 reveals that this is quite a nice simulation of the observed 'average' record. Experiments with an alternative volcanic forcing function, namely Lamb's dust veil index (Lamb, 1970) gave a remarkably similar result.

Running an experiment *without* the greenhouse forcing (not shown, see Oerlemans, 1988) leads to a substantially different result: the 1850-1980 retreat as shown in Figure 11 is *halved* ! So, *if* the forcing functions are correct, one could conclude that a major part of the glacial retreat over the last century is due to greenhouse warming. The fact that, for reasons explained earlier, the mass balance is so sensitive to changes in the global radiation balance seems to make glaciers good detectors of the greenhouse effect. It is important to realize, however, that this result is to a large extent directly related to the relative amplitudes of the forcing functions (Figure 10).

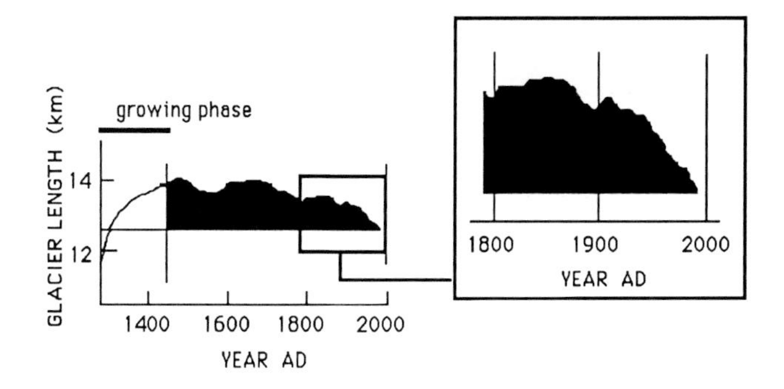

FIGURE 11. Result from a numerical glacier model forced by the ELA-curve from Figure 10. Note the reproduction of maximum stands around 1850 and in the beginning of this century, and the steady retreat.

Concerning recent changes on the large ice sheets we know very little. According to Budd and Smith (1985), a direct estimate of the current balance of Antarctica (based on accumulation data and velocitiy measurements) is very inaccurate. A 20% imbalance would not be detected by the data available at present. This imbalance corresponds to a rate of sea-level change of 1.2 mm/yr ! In fact, it would be amazing if the Antarctic Ice Sheet was very close to equilibrium with the present climate. A number of long time scales are involved (coupling of temperature and deformation in the ice, isostatic response of bedrock), so adjustment to the last glacial-interglacial transition (15000 BP) still takes place.

The scarce data from the Greenland Ice Sheet indicate that it is not far out of balance (Kostecka and Whillans, 1988), although temperature in the deepest layers is still 2 to 5 K below the equilibrium value (Dahl-Jensen and Johnsen, 1986). However, the recent retreat of outlet glaciers suggests that the ablation zone as a whole probably suffers from increased melting and is not in balance.

5. LAND ICE AND FUTURE SEA LEVEL

Much debate is going on concerning the future effect of greenhouse warming on land ice and the consequent changes in global sea level. Sea level has risen by 10 to 15 cm during the last hundred years (Barnett, 1983; Gornitz et al., 1982), and according to Meier (1984) the retreat of mountain glaciers and small ice caps could have contributed 2 to 4 cm. Thermal expansion of ocean water, associated with the 0.5 K warming, is thought to be responsible for another 5 cm (Wigley and Raper, 1987). A substantial part of the observed sea-level rise thus seems to be 'explained'.

The diagram in Figure 12 shows the most important factors concerning greenhouse warming and sea level.

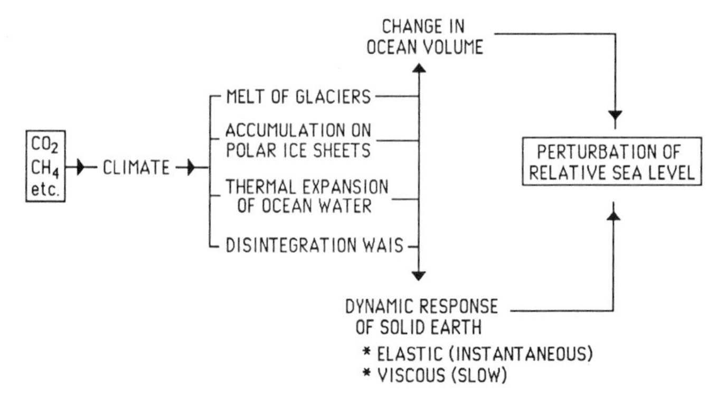

FIGURE 12. Most important factors contributing to sea-level change on a time scale of decades to centuries (WAIS = West Antarctic Ice Sheet).

It is not an easy task to quantify the various contributions, but several attempts have been made (e.g. EPA, 1983; Robin, 1986; Van der Veen, 1986, Oerlemans, 1989). Here a detailed account of the merits and uncertainties of such estimates will not be given. The following is just a brief discussion of the results of Oerlemans (1989), as summarized in Figure 13.

The temperature scenario from the Villach II meeting (fall 1987) has been taken as starting point. It is shown in Figure 13. Note that it shows global mean temperature relative to the 1850 value, which is considered 'undisturbed'. The uncertainty in the scenario has been estimated and is indicated by vertical bars. The probability of being outside the bars is about 1/3.

Figure 14 shows the contributions to sea-level change for the most likely temperature scenario. Some background and arguments are given in the following.

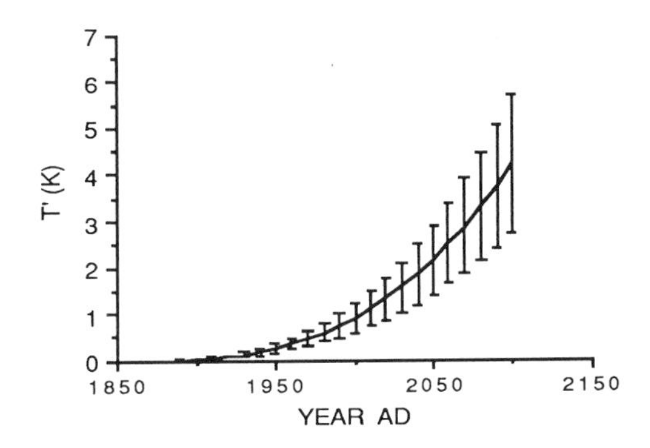

FIGURE 13. Temperature scenario from the Villach-II meeting. Vertical bars span a 2/3 probability.

a. Melt of mountain glaciers. It will be obvious that, in view of the large 'observed' sensitivity of mountain glaciers and small ice caps, a substantial retreat has to be expected even in case of a moderate warming. The very large number of glaciers, most of which are unmeasured, makes it impossible to do a detailed calculation. The estimate shown in Figure 14 was derived by integrating the equation

$$\frac{dV}{dt} = \alpha (T - T_R) \left[V - V_R \exp\{-(T - T_R)/\beta\} \right] \tag{10}$$

Here V is ice volume, V_R and T_R initial (1850 AD) ice volume and global surface temperature. The constants α [taken as 0.06 (yr K)$^{-1}$] and β [3.5 K] involves a characteristic response time and sensitivity, respectively. There is ambiguity in chosing values for these parameters, of course, and much work is needed to improve on this. However, eq. (10) is at least consistent from a dynamics point of view: it has a time scale, and melt is proportional to temperature perturbation and remaining ice volume. According to this 'model', glacier melt would give a +30 cm contribution to sea-level change in the year 2100.

b. Accumulation on polar ice sheets. Increasing temperature will lead to different responses of the Antarctic and Greenland ice sheets, at least as far as surface mass balance effects are concerned. This is due to the large difference in climatic conditions as discussed earlier. Accumulation on the Antarctic continent will probably increase, whereas melting will still involve a negligible fraction of the mass budget as long as the warming is restricted to a few degrees K. On the Greenland ice sheet melting rates will certainly go up and lead to a negative mass budget, although there may be some compensation by increased accumulation in the interior. When interest is in the next hundred or two hundred years only, the problem can be considered as static: the change in mass balance is proportional to the temperature perturbation, so the total effect can be obtained from the time integral of T'. In terms of sea level changes, the following values have been used here:

Greenland:	+0.5 ± 0.5 mm / (yr K)	based on Ambach (1980), Bindschadler (1985), Ambach and Kuhn (1988)
Antarctica:	-0.5 ± 0.5 mm / (yr K)	based on Oerlemans (1982), Lorius et al. (1985)

Note that the effects of the Greenland and Antarctic ice sheets tend to cancel, and that the estimated error is very large.

c. Thermal expansion of ocean water. The most extensive study of this has been carried out by Wigley and Raper (1987), and was referred to earlier. Here a simple diffusion model has been used to produce the curve labelled 'expansion' in Figure 14. This diffusion model can easily be tuned to give results that are very similar to those of Wigley and Raper. The relevant equation reads (z is depth, θ' temperature perturbation, K diffusivity):

$$\frac{\partial \theta'}{\partial t} = \frac{\partial}{\partial z}\left[K\frac{\partial \theta'}{\partial z}\right]; \quad K = K_0 \exp(-z/L) \qquad (11)$$

The diffusivity is assumed to decrease exponentially with depth with an e-folding length scale L. Values used are: K_o = 0.00025 m^2/s and L = 500 m. with the temperature scenario from Figure 13 as a time-dependent boundary condition at the upper surface, eq. (11) can easily be integrated. The resulting expansion can then be calculated from standard formula's for the physical properties of sea water (e.g. Neumann and Pierson, 1966).

d. Instability of the West Antarctic Ice Sheet (WAIS). Parts of ice sheets grounded far below sea level may be very sensitive to small changes in sea-level or melting rates at the base of adjacent ice shelves (Mercer, 1978;

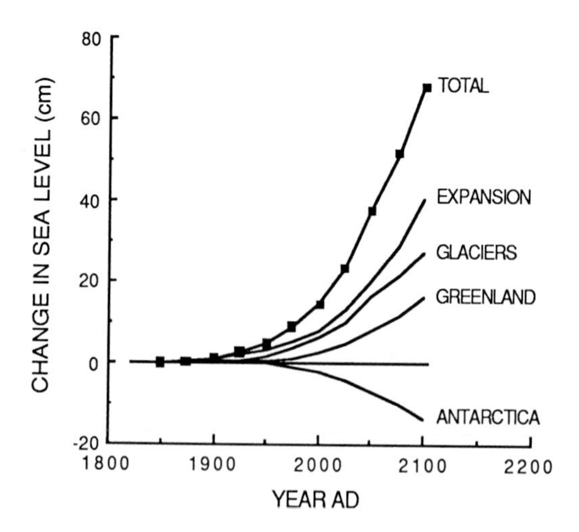

FIGURE 14. Projected change in sea level for the temperature scenario shown in Figure 13.

Thomas et al., 1979; Hughes, 1981; Lingle, 1985; Van der Veen, 1986). At present, the WAIS has a configuration that does not appear as 'very stable'. A schematic picture has already been presented in Figure 1. In case of a climatic warming, enhanced melting rates (at the base of the ice shelves) could lead to disappearance of ice rises. Reduced back stress and larger ice velocities may result, which subsequent thinning of the grounded ice and grounding-line retreat. It is hard to make quantitative statements about this mechanism. Attempts have been made to model the ice sheet-shelf system and to study its sensitivity (Thomas et al., 1979; Lingle, 1985, Van der Veen, 1987; Budd et al., 1987). A major problem in modelling of the ice sheet-ice shelf system is to deal in a proper way with the stress field. In the vicinity of the grounding line all components of the stress field can become of comparable magnitude, making it impossible to use rigorous simplifications. Even a flow-line model is already quite complicated. Here we mention an extensive study of Van der Veen (1986). He arrives at the conclusion that earlier estimates of the sensitivity of West Antarctica were too large. Budd et al. (1987) give a fairly detailed discussion on the response of the West Antarctic Ice Sheet to a climatic warming. Their considerations are based on a large number of numerical experiments with flow-band models. According to these experiments, very large ice-shelf thinning rates (10 to 100 times present values) would be required to cause rapid disintegration of the West Antarctic Ice Sheet. For a probably more realistic situation of a 50% increase in ice-shelf thinning rate for a one-degree warming (order-of-magnitude), the associated sea-level rise would be about 0.1 mm/yr for the coming decades. This is a small number and has not been included in Figure 14, also since the uncertainties are very large.

At the present state of knowledge, the uncertainty in the projection of future sea level is of the same order of magnitude as the 'signal' itself. The implications become clear when the results are presented in terms of threshold probabilities. Table II shows an example (from Oerlemans, 1989).

year:	2025	2050	2075	2100	**trend = 10 cm/100 yr**
-50 cm	>>99	>>99	>99	99	
-20 cm	>99	>99	98	98	
-10 cm	99	98	97	96	
0 cm	96	96	95	94	
+10 cm	87	89	92	91	
+20 cm	63	79	87	88	
+30 cm	34	65	80	83	
+40 cm	12	49	71	78	
+50 cm	3	24	61	71	

TABLE II. Threshold probabilities for a sea-level stand *above* the value indicated in the left column (relative 1985 AD). The table shown here is for a +10 cm/century background trend.

Here the calculation has been done for an additional trend in sea-level rise of 10 cm per century. Such a (local) trend can be associated with crustal movements of isostatic or tectonic origin, or with compaction of the soil.
Table II makes clear that the uncertainty is large, and that we need a much more accurate picture of the polar ice caps to improve projections of future sea level.

Acknowledgement
I thank Roderik van de Wal for useful comments on the manuscript.

References

Ambach W (1980): Anstieg der CO_2-Konzentration in der Atmosphäre und Klimaänderung: Mögliche Auswirkungen auf den Grönlandischer Eisschild. *Wetter und Leben* **32**, 135-142.

Ambach W and M Kuhn (1988): Altitudinal shift of the equilibrium line in Greenland calculated from heat balance characteristics. In: *Glacier Fluctuations and Climatic Change.* (ed.: J Oerlemans), Reidel, xx pp. [In press]

Barnett T P (1983): Recent changes in sea level and their possible causes. *Climatic Change* **5**, 15-38.

Bindschadler R A (1985): Contribution of the Greenland Ice Cap to changing sea level: present and future. In: *Glaciers, Ice Sheets and Sea Level: Effects of a CO_2-induced Climatic Change.* National Academy Press. (Washington), 258-266.

Birchfield G E and R W Grumbine (1985): "Slow" physics of large continental ice sheets and underlying bedrock and its relation to pleistocene ice ages. *J. Geophys. Res.* **90** (B13), 11294-11302.

Budd W F and D Jenssen (1975): Numerical modelling of glacier systems. *IAHS-AISH Publ. No. 104*, 257-291.

Budd W F, B J McInnes, D Jenssen and I N Smith (1987): Modelling the response of the West Antarctic Ice Sheet to a climatic warming. In: *Dynamics of the West Antarctic Ice Sheet* (eds: C J van der Veen and J Oerlemans), Reidel, 321-358 .

Budd W F and I N Smith (1985): The state of balance of the Antarctic Ice Sheet, an updated assessment 1984. In: *Glaciers, Ice Sheets and Sea Level: Effects of a CO_2-induced Climatic Change.* National Academy Press. (Washington), 172-177.

Dahl-Jensen D and S J Johnsen (1986): Palaeotemperatures still exist in the Greenland ice sheet. *Nature* **320**, 250-252.

Ellsaesser H W, M C MacCracken, J J Walton and S L Grotch (1986): Global climatic trends as revealed by the recorded data. *Reviews of Geophysics* **24**, 745-792.

EPA-report (1983): *Projecting future sea level rise methodology, estimates to the year 2100 and research needs.* US Environmental Protecton Agency, 126 pp.

Gornitz V, L Lebedeff and J Hansen (1982): Global sea level trend in the past century. *Science* **215**, 1611-1614.

Hammer C U, H B Clausen and W Dansgaard (1980): Greenland ice sheet evidence of post-glacial volcanism and its climatic impact. *Nature* **288**, 230-235.

Hughes T (1981): The weak underbelly of the West Antarctic Ice Sheet. *J.*

Glaciology **27**, 518-525.
Huybrechts P, P de Nooze and H Decleir (1988): Numerical modelling of Glacier d'Argentiere and its historic front variations. In: *Glacier Fluctuations and Climatic Change* (ed.: J Oerlemans), Reidel, in press.
Hutter K. (1984): *Theoretical Glaciology*. Reidel, Dordrecht.
Huybrechts Ph and J Oerlemans (1988): Thermal regime of the East Antarctic Ice Sheet: a numerical study on the role of the dissipation-strain rate feedback with changing climate. *Annals of Glaciology* **11**, in press.
Kostecka J M and I M Whillans (1988): Mass balance along two transects of the west side of the Greenland ice sheet. *J. Glaciology* **34**, 31-39.
Kuhn M (1988): The response of the equilibrium line altitude to climatic fluctuations. In: *Glacier Fluctuations and Climatic Change* (ed.: J Oerlemans), Reidel, in press.
Lamb H H (1970): Volcanic dust in the atmosphere; with a chronology and assessment of its meteorological significance. *Philos. Trans. R. Met. Soc.* **A266**, 425-533.
Lingle C S (1985): A model of a polar ice stream, and future sea-level rise due to possible drastic retreat of the West Antarctic Ice Sheet. In: *Glaciers, Ice Sheets and Sea Level: Effects of a CO_2-induced Climatic Change.* National Academy Press. (Washington), 317-330.
Lorius C, D Raynaud, J R Petit, J Jouzel and L Merlivat (1984): Late glacial maximum Holocene atmospheric and ice thickness changes from ice core studies. Annals of Glaciology 5, 88-94.
Mahaffy M W (1976): A three-dimensional numerical model of ice sheets: tests on the Barnes Ice Cap, Northwest Territories. *J. Geophys. Res.* **81**, 1059-1066.
MacAyeal D R and R H Thomas (1986): Numerical modelling of ice-shelf motion. *Annals of Glaciology* **3**, 189-194.
Meier M F (1984): Contribution of small glaciers to global sea level. *Science* **226**, 1418-1421.
Mercer J H (1978): West Antarctic ice sheet and CO_2 greenhouse effect: a threat of disaster. *Nature* **271**, 321-325.
Neumann G and W J Pierson (1966): *Principles of Physical Oceanography.* Prentice Hall, 545 pp.
Oerlemans J (1982): Response of the Antarctic Ice Sheet to a climatic warming: a model study. *J. Climatology* **2**, 1-11.
Oerlemans J (1986a): An attempt to simulate historic front variations of Nigardsbreen, Norway. *Theor. Appl. Climatol.* **37**, 126-135.
Oerlemans J (1988): Simulation of historic glacier variations with a simple climate-glacier model. *J. Glaciology*, in press.
Oerlemans J (1989): A projection of future sea level. *Climatic Change*, in press.
Oerlemans J and C J van der Veen (1984): *Ice Sheets and Climate.* Reidel, 217 pp.
Paterson W.S.B. (1981): The Physics of Glaciers. Pergamon Press, Oxford.
Reeh N, S J Johnsen and D Dahl-Jensen (1985): Dating the Dye 3 deep ice core by flow model calculations. In: *Greenland Ice Core: Geophysics, Geochemistry, and the Environment.* Geophysical Monograph 33 (AGU), 57-65.
Ritz C, L Lliboutry and C Rado (1982): Analysis of a 870 m deep temperature profile at Dome C. *Annals of Glaciology* **3**, 284-289.
Robin G deQ (1986): Changing the sea level. In: The Greenhouse Effect,

Climatic Change and Ecosystems (Eds.: B Bolin, B R Döös, J Jäger and R A Warrick), John Wiley & Sons (New York), 323-359.
Stroeven A, R van de Wal and J Oerlemans (1988): Historic front variations of the Rhone glacier: simulation with an ice flow model. In: *Glacier Fluctuations and Climatic Change* (ed.: J Oerlemans), Reidel, in press.
Sugden D.E. and B.S. John (1976): *Glaciers and Landscape*. Edward Arnold, London, 376 pp.
Thomas R H, T J D Sanderson and K E Rose (1979): Effects of a climatic warming on the West Antarctic Ice Sheet. *Nature* **227**, 355-358.
Weidick A (1967): Observations on some holocene glacier fluctuations in West Greenland. *Meddr Gronland* **165**, 202 pp.
Wigley T M L (1987): Relative contributions of different trace gases to the greenhouse effect. *Climate Monitor* **16** (1), 14-28.
Wigley T M L and S C B Raper (1987): Thermal expansion of sea water associated with global warming. *Nature* **330**, 127-131.
Wilhelm F. (1975): *Schnee- und Gletscherkunde.* Walter de Gruyter (Berlin, New York), 434 pp.
Veen C J van der (1986): Numerical modelling of ice shelves and ice tongues. *Ann. Geophysicae* **4B**, 45-54.
Van der Veen C J (1987): Longitudinal stresses and basal sliding: a comparative study. In: *Dynamics of the West Antarctic Ice Sheet* (eds: C J van der Veen and J Oerlemans), Reidel, 223-248.

LAND-SURFACE - VEGETATION - CLIMATE FEEDBACKS

HANS-JÜRGEN BOLLE
Institut für Meteorologie, Freie Universität Berlin

Summary

The interaction between the soils, the vegetation and the atmosphere is determined by the exchange of energy, momentum, water and other substances. This paper deals primarely with the energy and water fluxes. Their nature and computation procedures are described. Finally it is discussed how these interactions may vary under changing climatic conditions and what kind of feedback to the atmosphere can be expected from a change of land-surface characteristics.

1. INTRODUCTION

To assess the feedbacks between the land-surfaces, vegetation and climate one has to consider at the one hand the sensitivity of climate for changes at the land-surfaces induced by the action of man and on the other hand the effect that climate change will have in addition to the action of man on vegetation as well as soils and how these changes may then feed back to the climate system. Since vegetation is primarely dependent on water and the temperature regime, it is evident that the hydrological cycle plays an important rôle in these processes. It is a most complicated climate sub-system of which only some basic features can be discussed in this context.

Climate models are presently not capable to take full account of the processes that occur at the land surfaces. Especially the approximations of the hydrological cycle and the geographical resolution are very crude. Nevertheless a number of predictions have been made how the temperature and soil moisture distribution might look like if the present carbon dioxide concentration in the atmosphere is doubled. For the European-African sector (Fig. 1) most of the models agree with warming to be maximum in winter and stronger at high latitudes than at lower ones and that there will be higher precipitation than today at high latitudes mainly in winter. The strong enhancement of the warming effect during winter may result partly from a positive snow feedback: the warmer the northern climates the less the snow coverage and the lower the albedo, which allows more shortwave radiation to be absorbed. In the Mediterranean it is very likely that the precipitation slightly decreases in winter as well as in summer. The maximum increase of temperature as well as soil moisture must be expected in north-eastern Europe. This would strengthen the humidity gradient between a dryer south-western Europe and a wetter north-eastern Europe and reduce the temperature gradients especially in winter.

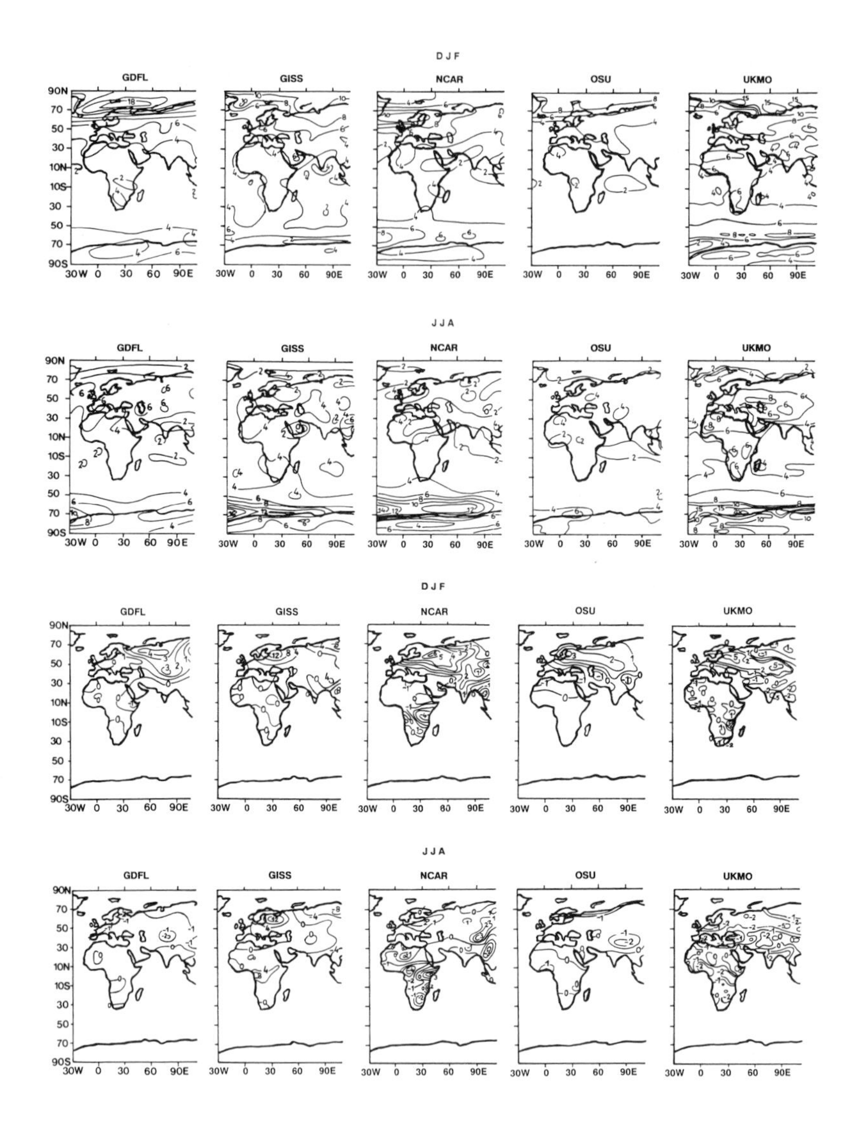

Fig. 1. An intercomparison of the predictions of five climate models for the European-African sector for twice the present CO_2 concentration in the atmosphere after Schlesinger [1]. Upper part: Temperature change in degree. Lower part: Soil water change in cm. The models are: GDFL - Geophysical Fluid Dynamics Laboratory, GISS-Goddard Institute for Space Studies, NCAR - National Center of Atmospheric Research, OSU - Oregon State University, UKMO-United Kingdom Meteorological Office.

On the basis of such predictions one can estimate the change in vegetation by shifting the for many species well known boarders of optimum existence according to the new climate scenarios. This gives a first order approximation how climate change may affect the future distribution of vegetation but it takes into account neither the possible feedback of such shifts, nor a possible adaptation of the plants to a changing environment. In a finer analysis the south-to-north gradient of the radiation regime, seasonal differences at different latitudes and the different soils have to be taken into account in determining the favourable environmental conditions for plants in a new climate scenario.

A major question is still to be solved: why do models predict for a warmer climate a geographical scenario that differs e.g. for northern Africa from the conditions of about 7,000 years ago at the time of the climate optimum? This climate period compares favourably with the temperature regime predicted for doubling of CO_2 but not with the predicted moisture regime that reminds more of the dry glacial then the wet interglacial conditions. Around 7,000 B.P. there was vegetation in the now and for the future predicted deserted areas of e.g. the eastern Sahara [2]. Of course, the starting conditions were different. 10,000 B.P. the Earth approached the warm period from the glacial era. Now we are according to the constellation of the Earth's orbital parameters close to the end of a natural warm peak, and additional warming is caused by a change in the longwave infrared radiative transfer due to the interaction of mankind with the climate system. One reason for the increased humidity in the Sahara during the last climate optimum was pointed out by Kutzbach [3]. Increased heating of the land surfaces relative to the surrounding oceans due to the about 6% increased solar irradiance during June to August could have resulted in a decrease of the geopotential height at the 900 hPa level over the continents with respect to the oceans and an increase of monsoons that transport water vapor from the oceans to the land enhancing there the rainfall. Increasing concentrations of water vapor over the continents would in addition reduce the radiative losses due to infrared radiation and favour the warming. But so far no clear and physically convincing explanation was given for the differences between 7,000 BP and the model predictions for 2030 AD. Maybe the reasons have to be sought in the imperfect modelling of the water cycle, because the oceans and the ice shields are not interactively included, or because the insight into processes occurring at the land surfaces and sub-surface water flows is not well enough advanced.

Already in a very simple analysis of the energy budget and the greenhouse effect (Bolle, 1989, in this volume, referred to as Bolle I) appear two quantities, which are directly connected with the land-surface and the vegetation: the surface albedo and the ratio of the heat flux to the net solar radiation at the surface. We shall therefore start our analysis with these quantities. The first step therefore will be to estimate the importance for climate development of albedo changes that can be induced by vegetation changes and then we shall deal with the heat fluxes.

2. GLOBAL CLIMATE SENSITIVITY TO SURFACE ALBEDO CHANGES

The sensitivity of climate with respect to albedo changes can be estimated by a very simple model. Starting with equation (12) of Bolle I and replacing the surface albedo ρ_0 by the area weighted sum of the land, the ocean and the cloud albedo, one can directly derive the sensitivity parameter of the surface temperature with respect to the land albedo.

(1) $$\rho_o = \Sigma \rho_{o.i} A_i$$

$i = 1$ (land), o (ocean) and c (clouds),

where A_i is the area fraction.

The land surface covers about 30% of the earth surface. If we assume that the 50% global cloudiness is equally distributed over land and sea, the effective land surface is 0.15. Differentiation of equation (I,12) with respect to $\rho_{o.1}$ leads to

(2) $$dT/d\rho_{o.1} = - \tau^*/(1 - \bar{\rho}_p)(\bar{T}_{eff}/T)^4(1 - \rho)(A_e/4\pi R^2)(T/4)$$

If we multiply eq. (2) with $\rho_{o.1}$ we obtain the sensitivity parameter for a relative change of the albedo. Inserting the values for the present climate in eq. (2) yields approxinately

(3) $$dT_o = - 0.8 \, d\rho_{o.1}/\rho_{o.1}$$

If one assumes an average land-surface albedo of 0.25 the global temperature effect of an albedo change of +1% would be - 0.032 K. This is a very small number, but a 15% albedo increase would have an effect on the temperature of the same order of magnitude with opposite sign as the 70 ppm-v increase of CO_2 within the last 100 years. We have not regarded here the reduction of snow and ice cover that will accompagny the warming of the planet and will have a positive feedback to the greenhouse effect.

A more accurate estimate of the effect of albedo changes on climate would have to take into account the geographical distribution and the relative insolation of the areas suffering from albedo changes.

3. THE NATURE OF THE ENERGY FLUXES AT THE LAND SURFACES.

In eq. (I.5) the conversion of radiative energy into heat was introduced without distinguishing between the different heat fluxes that have their orign at the surface. This distinction becomes very important if we consider regional climates. Sensible heat fluxes into the atmosphere transport heat by turbulent convection from the surface to adjacent atmospheric layers and thus heats the atmosphere starting from the surface more or less directly on top of the area where the radiative energy was converted to heat. The latent heat or water vapor flux penetrates through most of the planetary boundary layer without any heating effect and releases the energy at the condensation level where clouds are formed with their high albedo. The water vapor may also be transported into higher levels of the troposphere and may be released as rain in lee of the evaporating surfaces. It is therefore important to know, how the absorbed radiative energy is partitioned into sensible and latent heat. One fraction of the heat is furthermore directed into the soil where it heats the ground, is stored at daytime and released to the atmosphere during nighttime, when it has an effect on the atmosphere different then without this intermediate storage. In spring a small fraction of this heat remains each day in the ground and is released again in autumn.

The part of the instantanous net radiation flux that is converted into heat (I.5) equals the sum of the heat flux into the ground (M_G), the small amount of energy used for photosynthesis in the plants (M_{PH}), and the sensible (M_{SH}) as well as the latent (M_{LH}) heat fluxes into the atmosphere:

$$(4) \qquad M_R^* + M_{SH} + M_{LH} + M_G + M_{PH} = 0$$

Of the total amount of radiant energy incident on a plant, only 1 to 7% is used to build up organic matter [4]. In a global average this leads to a number in the order of 0.1% of the solar radiation that is used for photosynthesis.

To obtain a sensible heat flux it is necessary that a thermal gradient is build up between the surface and the medium to which the heat should flow. The magnitude of a sensible heat flux into the atmosphere depends also on the wind speed which is responsible for the transportation of the heat from the surface due to the generation of turbulence. Evaporation from a bare soil can only occur if (i) water is available, (ii) energy is reaching the water to evaporate it, and (iii) a moisture gradient is present to transport the vapor away from the surface. Again the wind speed regulates the rate at which this occurs.

In a soil which is dry at the surface the conversion of water into vapor can also occur at somewhat lower layers to which the diurnal heat wave reaches down to provide the necessary energy. The picture becomes more complicated if vegetation is present. Biologically active vegetation extracts water from deeper layers, the root zone, and transports it to the cells in the leaves. Near the lower surface of the leaves, in the mesophyll, the partial pressure of water vapor normally is close to saturation but the actual flux into the atmosphere is regulated by the stomata that act like resistors with respect to the water vapor flux. Evaporation in the presence of plants can be described by the water potential that exists at any level of the micro-routes the water takes on its way from the soil to the leaves and the resistances in these routes that limit the flux (Fig. 2). The water potential (= the chemical potential of water) is expressed by

$$(5) \qquad \Phi = P - \pi + \rho_w gh$$

$\rho_w = m_w/V_w$, π = osmotic pressure, P = pressure in excess of atmospheric pressure, which becomes for the atmosphere:

$$(6) \qquad P = RT \ln e/e^* + \rho_w gh$$

e^* = saturation water vapor pressure and h the height to which the water is elevated with respect to a reference level in the soil.

The resistors that are indicated in Fig. 2 cover a wide range. Minimum stomatal resistances at low wind speeds for small and moderate sized leaves are typically 100-200 s m^{-1}, for some species down to 30 and up to 500 s m^{-1} (mesophytes) or even 6,000 (xerophytes). The actual resistance from the inner part of a leaf to the atmosphere consist of four components: the intercellular air-space resistance, the stomatal resistance, the cuticular resistance and the leaf-boundary-layer resistance. The

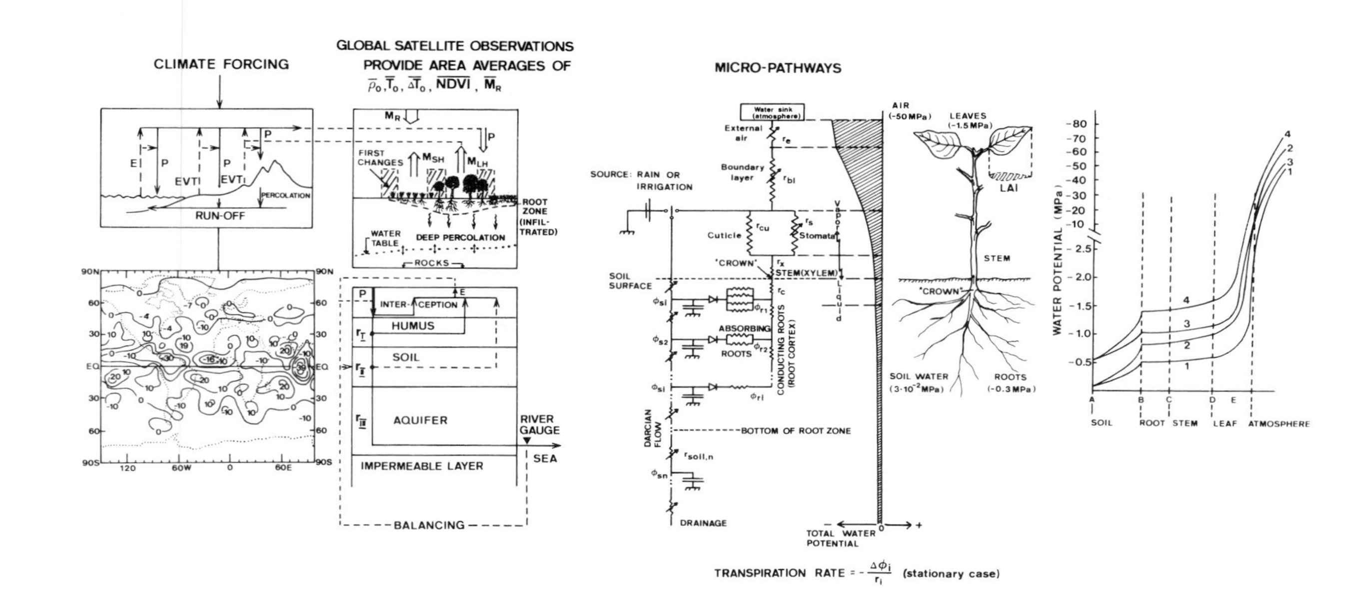

Fig. 2. Routes of water vapor and water in the atmosphere, the soil and plants. At left the large scale water vapor transport from the oceans to the continents is indicated and at the right the micro routes in the soil and plants (after Hillel [4]).

effective leaf resistances are generally in the order of 100 - 400 s m^{-1} but for some species of pines 2-3,000 s m^{-1}. For the leaf-boundary-layer resistance values range between 150 - 300 s m^{-1} in quiet air and smaller than 50 s m^{-1} already for wind speeds in the order of 0.25 m s^{-1}.

The effective leaf resistance depends on the water potential, the solar irradiance, the temperature, the carbon dioxide partial pressure and the humidity of ambient air. They are therefore clearly dependent on environmental conditions and climate.

5. MATHEMATICAL FORMULATION OF THE HEAT FLUXES

5.1. Heat Flux into the Ground

The heat flux into the soil is a conductive flux that depends on the value of its thermal conductivity,

(7) $$M_G = \lambda dT/dz$$

The divergence of this heat flux equals the heating rate of the soil

(8) $$dM/dz = d^2T/dz^2 = C\ dT/dt.$$

It depends on the heat capacity $C = c_s\rho_s$ which is a function of the volume fractions of minerals (VF_m), organic material (VF_o) and water content (VF_w): $C \approx 0.48VF_m + 0.60VF_o + VF_w$. The density varies between 1.3 kgm^{-3} for organic material and 2.6 for minerals. Average values for C are 2.0 - 2.5.10^6 $Wm^{-3}K^{-1}$. If the temperature at the surface would be sinusoidal with an amplitude of ΔT_o (which is strictly speaking not the case), the heat flux at the surface is given by

(9) $$M_g(0,t) = \Delta T_o\sqrt{\omega c\rho\lambda}\quad \sin(\omega t + \pi/4 + \varphi_o).$$

With a knowledge of the thermal inertia $\sqrt{\omega c\rho\lambda}$ the flux into the ground can be calculated for given amplitudes ΔT_o for the diurnal temperature wave.

5.2 Heat fluxes into the Atmosphere

For the heat fluxes into the atmosphere different formulations are in use. In the lowest layer of the atmosphere, the Planetary Boundary Layer (PBL), no stationary large scale vertical flow can develop because of the solid lower boundary. Thus an upward transport of mass has to be replaced by a downward transport of an equal mass. The turbulent transports into the atmosphere exceed by far the conductive fluxes which are only important for the thin interfacial sublayers that surround each obstacle at the surface. Due to the nature of this essentially turbulent process heat fluxes can be described by up- and downward transports of energy.

The turbulent fluxes are defined by the time average over the fluctuations of the vertical wind component w' multiplied with the quantity of the transported energy, ρc_pT respectively ρLq, with L being the latent heat of evaporation, q the specific humidity. If the vertical velocity is splitted into an average term $\overline{w}$ and a fluctuating term w' a quantity χ

is transported vertically together with the mass m at the rate

(10) $$M^+ = (\overline{w} + w')\chi^+$$

With the downward movement a different amount of the quantity χ is transported downward. If we also split the quantity χ into an average value and a fluctuating term

(11) $$\chi = \bar{\chi} + \chi'$$

the difference between the upward and the downward transport is

(12) $$M = M^+ - M^- = (\overline{w} + w')(\bar{\chi} + \chi').$$

For sensible heat the quantity χ has the value c_pT per unit of mass, for latent heat Lq, where q is the specific humidity in the air parcel. For the two heat fluxes we obtain by averaging eq. (12) over time and neglecting small terms:

(13a) $$M_{SH} = \rho c_p \overline{w'\chi'}$$

(13b) $$M_{LH} = \rho L \overline{w'q'}.$$

Because there exists no theory to relate w' directly to other meteorological parameters that can more easily be measured, one usually substitutes the physical definition of the fluxes by bulk formulas. These are derived from similarity principles and dimensional analysis: if the variables that describe a process can be combined into a dimensionless quantity the existence of a functional relationship is possible though the function must usually be determined experimentally. The variables of turbulent fluxes combine into a constant k, the von Karman constant, which has been determined to be 0.4 or at least close to this number. From these considerations follows for the dynamical sublayer of the Planetary Boundary Layer - that is the layer disconnected from the surface roughness elements but close enough to the surface so that Coriolis forces do not yet play a rôle and where the profiles are logarithmically -:

(14a) $$M_{SH} = -\rho c_p K_{SH}(d\theta/dz)$$

(14b) $$M_{LH} = -\rho K_{LH} L(dq/dz)$$

with

(15) $$K_{LH} \approx K_{SH} = k^2 u(z) \quad (\ln\{(z - d)/z_{o.i}\})^{-1}$$

(i = v for water vapor, h for sensible heat and m for momentum). The z_o values kann be slightly different for the three fluxes. d_o is the effective height of the roughness elements (displacement height). The potential temperature θ is introduced here because only if the lapse rate deviates from the adiabatic one defined by θ a heat transport can occur. In the lower parts of the planetary boundary layer $T \approx \theta$.

In the whole surface sublayer extending from the basis of the dynamical sublayer up to the height where the Coriolis parameter becomes important the buoyancy resulting from the effective vertical density gradients has to be accounted for and the flux parameters do not usually combine to the von Karman constant but to a flux profile function that depends on the height $z - d_o$ and the Monin-Obukhov length defined by

$$L = -u_*^3 \rho / gk[(M_{SH}/T_a c_p) + 0.61\ M_{LH} \tilde{L}] \quad (16)$$

u_* is the friction velocity defined by the square root of shear stress divided by the density. It can be determined from

$$u_* = k(z - d_o)(du/dz). \quad (17)$$

The heat fluxes then modify to

$$M_{SH} = -\rho c_p K_{SH}(d\theta/dz)\Phi_h(z/L)^{-1} \quad (18a)$$

$$M_{LH} = -\rho K_{LH} L(dq/dz)\Phi_v(z/L)^{-1} \quad (18b)$$

with

$$\Phi_h(z/L)^{-1} \approx \Phi_v(z/L)^{-1} \approx (1 - 16z/L)^{1/2} \quad (19)$$

Below the dynamic sublayer, in the interfacial sublayer, the logarithmic profiles are not any more valid. Here the turbulent transfer becomes dependent on the type of the obstacles and a similar procedure has to be applied as for the surface sublayer but with different parametrization. dT/dz and dq/dz can be measured with large effort only. In practise therefore the following approximations are used:

$$M_{SH} = - c_p \rho D_{SH}(T_o - T_r) = c_p \rho u_r C_{h.r}(\theta_s - \theta_r) \quad (20a)$$

$$M_{LH} = - c_p \rho D_{LH} L(q_o - q_r) = \tilde{L} \rho u_r C_{e.r}(q_s - q_r) \quad (20b)$$

The r refers to a reference level, either 2 or 10 m above ground and the D's are now transfer coefficients that are integrated over the height interval o ... r taking also into account the diffusion that takes place in the thin layer directly at the surface, a few mm thick, with the diffusion coefficient a_a:

$$D = 1/\int_0^r [dz/(a_a+K)] = C_{i.r} u_r \quad (21)$$

The D's depend on the wind speed. If this is separately accounted for, these transfer coefficients can be expressed by wind independent numbers, $C_{h.r}$ (Stanton number) and $C_{e.r}$ (Dalton number), respectively.

An alternative representation of the fluxes uses the concept of resistances (Fig. 2) that can in principle more easily be extended to systems of increasing complexity. It is especially useful for the computation of evaporation from non-wetted surfaces for which q_s cannot be defined. In some cases it is possible to replace q_o by deeper layer, which is saturated, and represent the intermediate layer by a resistance. Any flux can be computed by dividing the related potential-difference by a resistance like in Ohm's law. The specific humidity at the surface of a dry leaf is unknown, but in the sub-stomatal cavities saturation is to be assumed. The transfer through the stomata can be characterized by a stomatal resistance r_{st}. From the leaf to the upper boundary of the plant canopy another resistance can be introduced, r_{lc}, and from the canopy to the reference level in the surface sublayer of the atmosphere a resistance r_{ca} has to be applied. r_{1c} and r_{ca} can be combined to r_{av}, which characterizes the transfer from the stomata opening to the reference level.

In correspondence to eq. (21) the resistance $r_{a.v}$ is e.g. defined by

(22) $$r_{av} = 1/u_r C_{e.r}^{1/2}.$$

The latent heat flux becomes with this notation

(23) $$M_{LH} = \rho \tilde{L}(q_s^* - q_r)/(r_{st} + r_{av})$$

or with the notation r_c = canopy resistance and $r_a = u_r/u_*^2$ = aerodynamic resistance:

(24) $$M_{LH} = \rho \tilde{L}(q_s^* - q_r)/(r_c + r_a).$$

The discussion in this section shows that all heat fluxes can individually be computed provided the atmospheric structure and the properties of the surface are known. It is, however, a necessary condition that eq. (4) is fulfilled. This energy budget equation can be used as an additional information if not all structural parameters of the atmosphere or of the surface are explicitely known.

The latent and the sensible heat fluxes can furthermore be related to each other by the Bowen Ratio which is under the assumption that $K_{SH} = K_{LH}$ given by:

(25) $$\beta = M_{SH}/M_{LH} = (c_p/L)(\theta_o - \theta_r)/(q_o - q_r).$$

With $q = \rho_v/\rho_a = (0.622e/R_dT)/(p/R_wT) = (0.622e/p)(R_w/R_d)$ and $\theta = T$ (which is valid in the dynamical sublayer and below) eq. (25) can be rewritten as

(26) $$\beta = \gamma(R_d/R_w)(T_1 - T_2)/(e_1 - e_2).$$

Here $\gamma = c_p p/0.622L = 0.67$ hPa for $p = 1013.25$ hPa. Furthermore the ratio of the differences of saturation pressures $e_a^* = e^*(T_a)$ at two levels and the temperatures measured at these levels is a known quantity, e.g.:

(27) $$(e_s^* - e_a^*)/(T_s - T_a) = \Delta = de^*/dT = 0.622Le^*/R_dT^2$$

(Clausius-Clayperon equation). Introducing eq. (27) into eq. (25) results in

(28) $$\beta = (\gamma/\Delta)[1 - (e_a^* - e_a)/(e_s - e_a)]$$

for a wet surface when $e_s^* = e_s$. From eq. (4) follows with $M_{SH} = M_{LH}/\beta$:

(29) $$(M^* - M_G) - M_{LH}(1 + \beta) = 0$$

and further

(30) $$(M_R^* - M_G) = M_{LH}\{1 + (\gamma/\Delta)[1 - (e_a^* - e_a)/(e_s - e_a)]\}.$$

This can also be written:

(31)

$$M_{LH} = (M_R^* - M_G)/\{\Delta/(\Delta+\gamma)\} - \{\gamma/(\Delta+\gamma)\}(e_a^* - e_a)/(e_s - e_a)M_{LH}.$$

This is the Penman equation, which is valid for wet surfaces.

For the ratio γ/Δ the following numbers are valid:

T(°C):	-20	-10	0	+10	+20	+30	+40
γ/Δ :	5.864	2.829	1.456	0.7934	0.4549	0.2749	0.1707

The term $M_{LH,A} = (\gamma/\Delta)(e_a^* - e_a)/(e_s - e_a)M_{LH}$ is referred to as the "drying power of the air". It can be expressed by

(32) $$M_{LH,A} = f(u_r)L(e_a^* - e_a) = 0.622L\ \rho p^{-1}C_e u_r(e_a^* - e_a),$$

and for neutral conditions

(33) $$f(u_r) = 0.622k^2u_r/\{R_dT_a\ln[(z_q - d_o)/z_{ov}]\ln[z_u - d_o)/z_{om}]\}$$

$$= ku_*\rho(q_a^* - q_a)/\{(e_a^* - e_a)\ln[(z_a - d_o)/z_{ov}]\}$$

For a dry leaf eq. (31) has to be modified by the factor $(r_{st} + r_{av})/r_{av}$ since the resistance to the wet surface is enhanced by r_{st}. For a dry soil a similar procedure is possible if the resistance to a lower lying layer of water can be estimated:

(34) $$(e_s - e_a) = (e_a^* - e_a)\{r_{av}/(r_{st} + r_{av})\}.$$

With this modification and the approximation

(35) $$M_{LH,A} = \rho L(q_s^* - q_r)/(r_{st} + r_{av})$$

it follows with $q = \rho_v/\rho$, $\rho_v = 0.622e/R_dT$, $\rho = (p/R_dT)(1 - 0.378e/p)$, $\gamma = c_pp/0.622L$:

(36) $$M_{LH} = [\Delta(M_R^* - M_G) + \rho c_p(e_a^* - e_a)/r_{av}]/[\Delta + \gamma(1 + r_{st}/r_{av})].$$

This is the modification of the Penman - Monteith equation given by Brutsaert [6] (Monteith originally used the canopy resistance r_c and $r_a = (u_*^2/u_r)^{-1}$ instead of r_{st} and r_{av}).

While these equations for the latent heat flux can readily be applied for climatological daily averages or neutral conditions the "drying power of the air" has to be adjusted for instantaneous values with respect to the stability of the atmosphere:

(37) $$M_{LH,A} = ku_*\rho(q_a^* - q_a)/\{\ln[(z_a - d_o)/z_{ov}] - \Psi_{sv}(z_a - d_o)/L\}$$

(38) $$u_* = ku/[\ln\{(z - d_o)/z_{om}\} - \Psi_{sm}(z/L)].$$

L is again the Monin - Obukhov length. The modified Penman equation can be solved by iteration. First neutral conditions ($\Psi_{sv} = 0$) are assumed and u_* and M_{LH} is computed. Then M_{SH} can be estimated. From u_*, M_{LH} and M_{SH} the Monin - Obukhov length is determined and the profile functions

$$\psi_{sv}(z/L) = 2 \ln[(1 + x^2)/2] \tag{39}$$

$$\psi_{sm}(z/L) = 2 \ln[(1 + x)/2] + \ln[(1 + x^2)/2] - 2 \operatorname{arctg}(x) + \pi/2 \tag{40}$$

can be computed with $x \approx 16$. Now the computation can be repeated with the new profile functions until the solution is stable.

6. INTERACTION WITH CLIMATE

In Fig. 3 some exemplary conditions are schematically illustrated for the size and ratio of the heat fluxes. If a bare dry soil is present the albedo is determined by the the soil characteristics. In dry regions the soil is often highly reflective but also dark surfaces occur like the laterit in some areas of Africa. Some examples of spectral albedos are shown in Fig. 4. The radiation energy absorbed by the dry surface is partitioned into two sensible heat fluxes, one directed into the atmosphere and one into the soil. The surface is heated up by the absorbed solar and longwave radiation until the emitted longwave flux M_{LW}, the heat flux into the soil (stipelt in Fig. 3) and the sensible heat flux into the atmosphere (dashed in Fig. 3) are in equilibrium. This requires that first a temperature gradient is build up between the surface and the lowest layers of the atmosphere respectively the uppermost layers of the soil, which allows the sensible heat fluxes to develop. Only if the heat flux into the soil hits moist soil, the "front" of the soil moisture, water is evaporated and the vapor can reach the surface through the micro-pathways of the soil. In that case the sensible heat flux as well as the surface

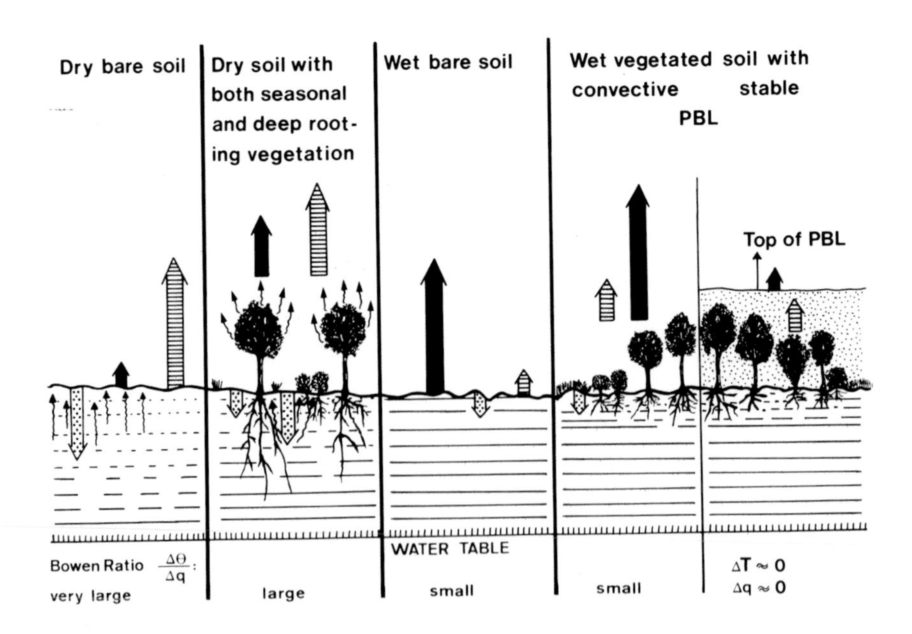

Fig. 3. Heat fluxes over different terrain.

Fig. 4. Examples of spectral ground albedo.

Top:
Open circles - dusty bush,
dots - dry Cram-Cram grass,
triangles - sand dune,
squares - dark brown laterite.

Middle:
Measurements made at about 100 m distance at an
irrigated (1, 4, and 5) and a
dry (2 and 3)
grassland area of La Crau, France.

Bottom:
Temporal development of the spectral albedo of winter wheat during three weeks in May/June. Hildesheimner Börde, Federal Republic of Germany.

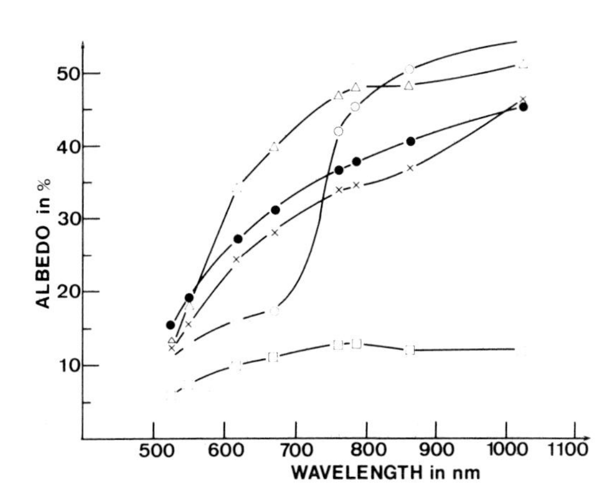

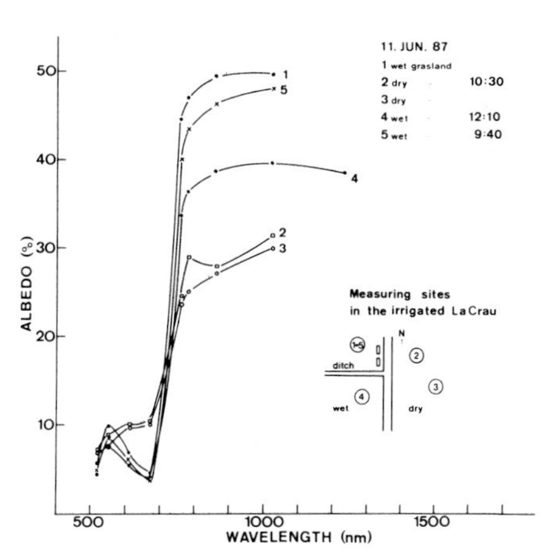

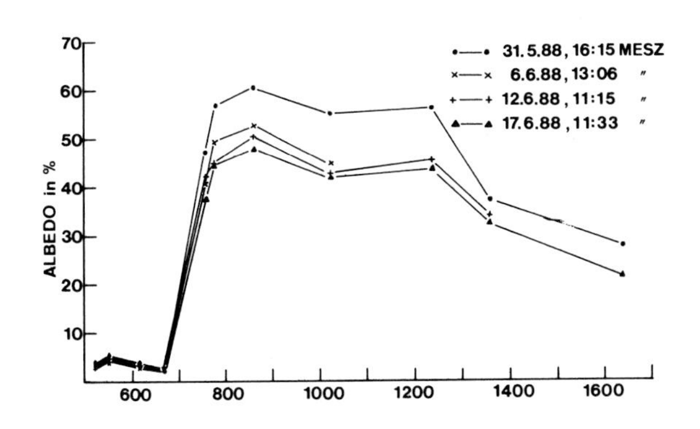

temperature are reduced on account of the latent heat flux. In dry climates the latent heat flux (full arrow in Fig. 3) is generally very small compared with the sensible heat flux.

This picture changes if sparse vegetation is present in the dry area, vegetation which may extract water from deeper layers that either store precipitated water from the rainy period or are moistured by sub-surface aquifer systems. This vegetation acts as a water pump and even if the soil surface is completely dry, some water is transpired through the leafs of the vegetation.

In the case of a moist bare surface first the water is evaporated before a significant raise of the surface temperature can take place. The latent heat flux is the largest of the energy fluxes. Wet surfaces generally have lower albedo as in dry state. Consequently more shortwave radiative energy is absorbed and converted into heat. Because most of this heat is latent heat, the surface temparature is not raising as high as it would in the dry case. Therefore also the outgoing longwave flux is reduced with respect to the dry case and even more energy must leave the surface as heat than under dry conditions.

This albedo effect may even be enhanced if vegetation with its low shortwave reflectance is present. However, also the opposite may be true, especially for dark soil, because of the high near infrared reflectance of green and very productive plants (Fig. 4). If wet soil is covered by (dense) vegetation, this vegetation due to its leaves also increases the area for evaporation and transpiration, which may enhance the latent heat flux. At the same time the canopy temperature raises above the air temperature due to the biological activity, which also increases the sensible heat flux.

As stated earlier the development of heat fluxes require gradients to be build up first, gradients of temperature respectively moisture. If the moist area is superimposed by wet and hot air in a stable planetary boundary layer (PBL), the air may be incapable to accept heat, the "atmospheric demand" is very small. In that case more moisture can be accomodated only if the top of the boundary layer raises - and this is what occurs, if it is heated at its lower boundary, - and/or if water vapour is blown off the top of the PBL into the free troposphere respectively advected horizontally into dryer regions.

With a changing climatic regime the processes at the surface can be influenced in one or the other way. The important climatic parameters are cloudiness - that affects the radiation regime - and precipitation - that affects the hydrological regime as well as the albedo. The absolute value of the temperature is of minor importance for the processes at the land surfaces because only temperature gradients determine the magnitude of the fluxes. The absolute temperature may, however, affect the vegetation and this would indirectly have impact on the partitioning of the solar energy into the heat fluxes.

Experiments with General Circulation Models confim these qualitative considerations. For changes in the water regime and albedo in the Saharean region Sud and Molod [6] found the following. The reduction of the solar input due to an increase in surface albedo and as a consequence the reduction of the sensible heat flux from a dry surface causes a relative cooling inside the PBL where the sensible heat is deposited, and as a consequence sinking aloft and moisture divergence at the surface which suppress moist convection and reduces local rainfall.

In Fig. 2 the movement of water was schematically shown. Often it invades the continents from the sea with humid air that is forced upward over the continents due to heating or by mountains providing precipitation in coastal near zones. As it enters central parts of the continents the air would dry out if not new water vapor is provided from the surface with its vegetation, which is partly intercepted rainwater but partly also taken from underground aquifer systems. If this vegetation is removed or changed also the supply of water to the atmosphere is reduced and in lee of this zone the precipitation will be changed. This may lead to a change in the local climate, a drying out of such areas or in a final stage to desertification. If at the same time the little precipitation falls within a short time in large rates, the sudden impact of rain to an area of reduced vegetation leads to erosion which can manifest itself in a lack of minerals or even structural changes of the land surfaces. The dryer parts of the land are then in addition less protected against wind erosion.

Reduced precipitation may also in a first phase increase the demand for irrigation, which for some time provides for additional evaporation but draws on the water reserves and finaly causes a drop of the groundwater level.

Dryer soil also has a lower heat capacity with the consequence that during the summer less heat will be accumulated, which is missing in autumn to compensate for the reduced incomming radiation. In a dryer area the seasonal amplitude is therefore less damped by the storage of heat in the early summer and its release in fall.

The change of vegetation at the continents does also have an effect on the roughness of the surface. The roughness enters the equations of the heat fluxes into the atmosphere by the z_0 value. This value determines the height of the boundary layer and the transfer of momentum between the surface and the atmosphere. This momentum transfer is important for the tropospheric wind system and thus for the general circulation. A decrease of the bulk aerodynamic transport resistance of the PBL for a smoother surface would reduce all heat fluxes. In order to maintain the rate of heat transport that is required for an equal amount of net radiative flux it would tend to increase the temperature and moisture gradients above the surface. At the same time the wind stress is reduced and windspeed increased. This produces the moisture convergence in the PBL into a surface low which, in turn, reduces moist convection and rainfall [6]. Models show that a large scale change of the roughness such as occurred in Africa between the climate optimum about 7,000 years ago and now could be one reason for the change of the wind system over Africa.

Vegetation is furthermore actively involved in both the chemical cycles of the climate system and the hydrological cycle. The water and the carbon dioxide cycle are interconnected by the photosynthetically active radiation (PAR) in the biomass production in the plants. There exist three different mechanisms how the plant can build up biomass. The most common process is that of C_3 plants. They combine water pumped up through the roots and carbon dioxide inhaled through the stomata openings into C_3 compounds. Highly efficient C_4 plants apply a different chemical reaction, which leads to molecules with four C atoms. In very hot climates, when water is available at a minimum, plants developed that take in the carbon dioxide during nighttime, fix it until solar radiation becomes available and combine it with water under the influence of PAR during daytime (Crassulacean Acid Metabolism, CAM) forming C_4 compounds in a very similar way than C_4 plants do it. In this case the transpiration losses are minimized and these plants do not contribute substantially to the water budget.

6. CONCLUSION

The processes that occur at the land-surfaces are in principle well understood and can locally quantitatively be described in detail if the boundary conditions are known. This becomes more difficult if area averages of a structured landscapes are required. The feedback between a changing climate and changing land-surface characteristics covers a range of scales and can presently only marginally be described by models. First approaches to model this interactive system will be presented in another contribution of this volume.

If temperature rises it can be concluded from Fig. 1 that e. g. in the Mediterranean will be less precipitation. Vegetation will then be reduced, albedo will increase, less shortwave radiation will be converted into mainly sensible heat. These already now persistent summer conditions will be enhanced and last longer during the year. These inland processes may cause at the coasts an increased entrainment of wet air with the sea breeze and thus increase the regional climatological gradients.

Literature:

[1] Schlesinger, M. E.: Model Projections of the Climatic Changes Induced by Increased Atmospheric CO_2. Presentation at the NATO Symposium Climate and Geosciences, Louvain-la-Neuve, Belgium, 22-27 May 1988.

[2] Pachur, H.-J. and S. Kröpelin, 1987: Wadi Howar: Paleoclimatic Evidence from an Extinct River System in the Southeastern Sahara. Science **237**, 298-300.

[3] Kutzbach, J. E., 1981, Science **214**, 59

[4] Gates, Biophysical Ecology, Springer Verlag, New York, 1980.

[5] Brutsaert, W.: Evaporation into the Atmosphere, D. Reidel, Dordrecht, 1984.

[6] Sud, Y.C. and A. Molod, 1988: A GCM Simulation Stuidy of the Influence of Saharan Evapotranspiration and Surface-Albedo Anomalies on July Circulation and Rainfall. Monthly Weather Review **116**, 2388-2400.

SATELLITE DATA FOR CLIMATE STUDIES

G. Dugdale
Department of Meteorology, University of Reading

1. INTRODUCTION

Satellites offer a global view of the earth and atmosphere unachievable by more conventional observing platforms. So, in principle satellites give us a data source for the monitoring of the climate and its changes, for inputting to global models to give us a better understanding of climate and for ensuring the proper representation of sub-grid scale processes in these models.

Prior to this course I had intended to review the various data bases derived from satellite data which are available and applicable to studies of different aspects of climate and its impact. However, after mixing with the participants in the first few days of the course it became clear that few had any grounding in satellite remote sensing. There is no doubt that satellite derived data is proving a useful tool in climate studies and that its application in this field will increase. However, satellite data has its own limitations and biases. In these circumstances it may be more important to the participants in this course, who are mainly young scientists starting their research careers, to have an understanding of how satellite data is acquired and treated so that they can judge its suitability for any particular application. This paper will therefore give a brief review of the principles and practice of current satellite remote sensing and discuss how the meteorologically significant quantities are derived from the outputs of the satellite borne instruments. The implications of these considerations for the different uses of satellite data will be discussed. Clearly, in a paper corresponding to a single lecture, this topic can only be introduced.

In a paper of this type it is inappropriate to try to give a comprehensive set of references; instead, at the start of each section, one or two basic texts or reviews are recommended which will allow the reader to continue to the next stage of detail in the topic of his interest.

2. THE PRINCIPLES OF SATELLITE REMOTE SENSING

Texts: Introduction to Environmental Remote Sensing, Barratt & Curtis (1983). Chapman & Hall.
Remote Sensing in Meteorology, Oceanography & Hydrology, Cracknell (Ed) (1981). Ellis Horwood.

2.1. Background

Satellite borne sensors react to the electro-magnetic radiation received from the earth-atmosphere system. The whole science of remote sensing is concerned with relating that radiation arriving at the satellite to the physical property which it is desired to investigate.

2.2 Radiation sources

Radiation eminating from the earth-atmosphere may have originated from it or may be backscattered radiation from an outside source; these sources are either the sun (solar radiation) or a transmitter on board the satellite itself. The last case is known as "active" remote sensing and utilises RADAR principles. Figure 2.1 below represents the electro-magnetic spectrum and indicates which regions are commonly used for satellite remote sensing and which frequency bands are commonly used for the remote sensing of climatologically significant quantities.

Figure 2.1

The E - M spectrum

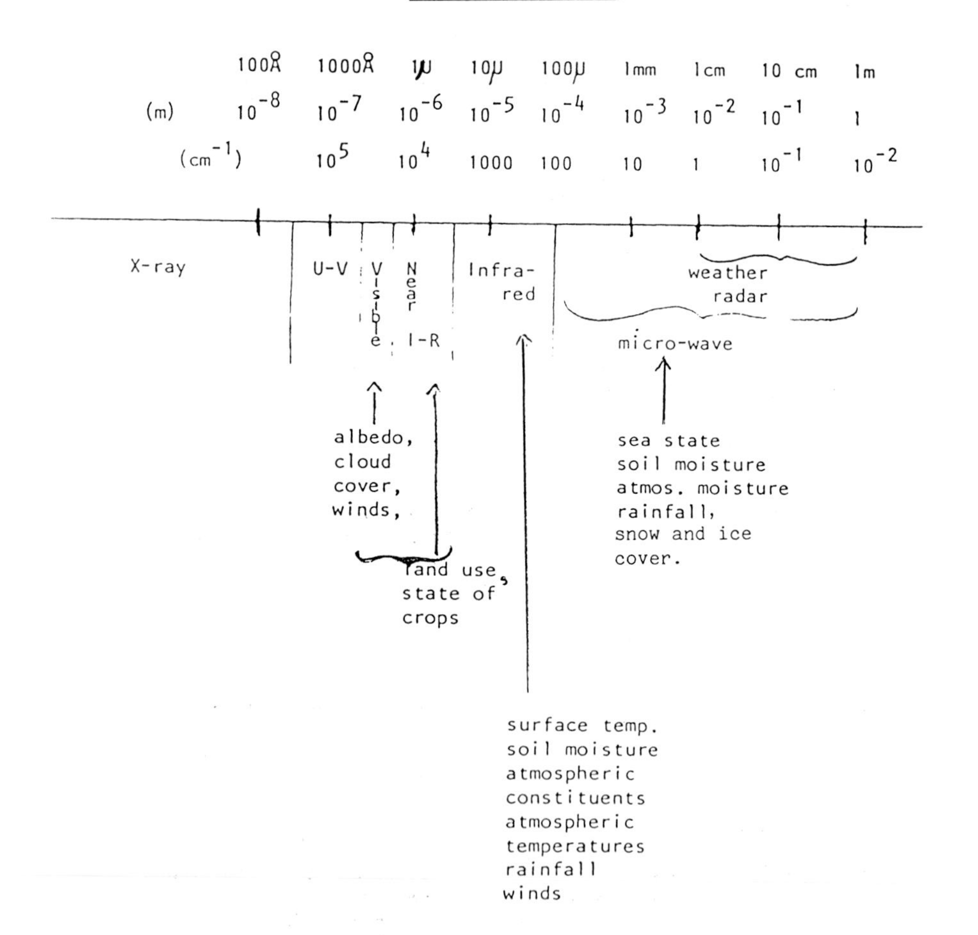

It is not the intention of this paper to discuss the physics of remote sensing but a few notes on the terminology and important aspects of the interactions between matter and e-m radiation are necessary for an understanding of the techniques and possibilities and limitations of satellite remote sensing.

When e-m radiation impinges on a surface it may be absorbed, transmitted or scattered. Materials which absorb all impinging radiation at all wavelengths are known as black bodies. Generally bodies are only black over part of the spectrum at other regions the fraction of radiation absorbed is known as the absorption coefficient. Black bodies also emit radiation at all wavelengths according to physical laws (the Planck function). Figure 2.2(a) shows the normalised black body radiation from bodies at about 245 and 6000K. Increasing the temperature of a black body causes it to emit more energy at all wavelengths, also the wavelength of maximum emission decreases with increasing body temperature. The total emission varies as the fourth power of the body temperature. So in figure 2.2(a) the unnormalised area under the curves would have a ratio of about 3.6×10^5. Bodies emit the same fraction of the black body radiation as they absorb of incident radiation at the same frequency. Scattering is the name given to interactions between radiation and matter which result in a change in direction of the radiation but no change in the frequency or energy. Scattering is strongly wavelength dependant, increasing as the size of the scatterer reaches the wavelength of the incident radiation. Hence we see preferential scattering in the atmosphere of the blue (short wavelengths) of solar radiation. In the absence of clouds the scattering of the thermal wavebands (5 to 50 μm) in the atmosphere is negligable. Similarly clouds droplets cause only slight scattering in the microwave bands.

The absorption (and emission) bands of the major atmospheric gases are illustrated in figure 2.2(b)&(c). Note that electronic bands occur in the ultra-violet region while rotation or rotation/vibration bands dominate in the thermal infra-red.

In the microwave region there are both water vapour and oxygen bands. Emission from solid surfaces in the microwave region is strongly influenced by the texture and moisture content of the surfaces.

2.3 Radiation & meteorological data.

We are interested in a wide range of physical properties of the earth's surface and of the atmosphere. Generally the radiation emitted or backscattered to a satellite radiometer is not directly related to the property in which we are interested. Hence, we have to use models (algorithms) to deduce the property from the radiometric data. These techniques allow us to measure most of the features of interest. Surface temperatures can be measured in wavebands in which the atmosphere is transparent if the surface emissivity is known. Atmospheric temperature profiles can be calculated from a series of measurements in narrow spectral intervals in the emission bands of uniformly mixed atmospheric gases. Global albedo and outgoing long wave radiation can be measured. Winds can be deduced from cloud motion. Land surface properties can be inferred from the differences in reflectivity of natural surfaces over the visible (and perhaps micro-wave) bands. Etc.

Figure 2.2

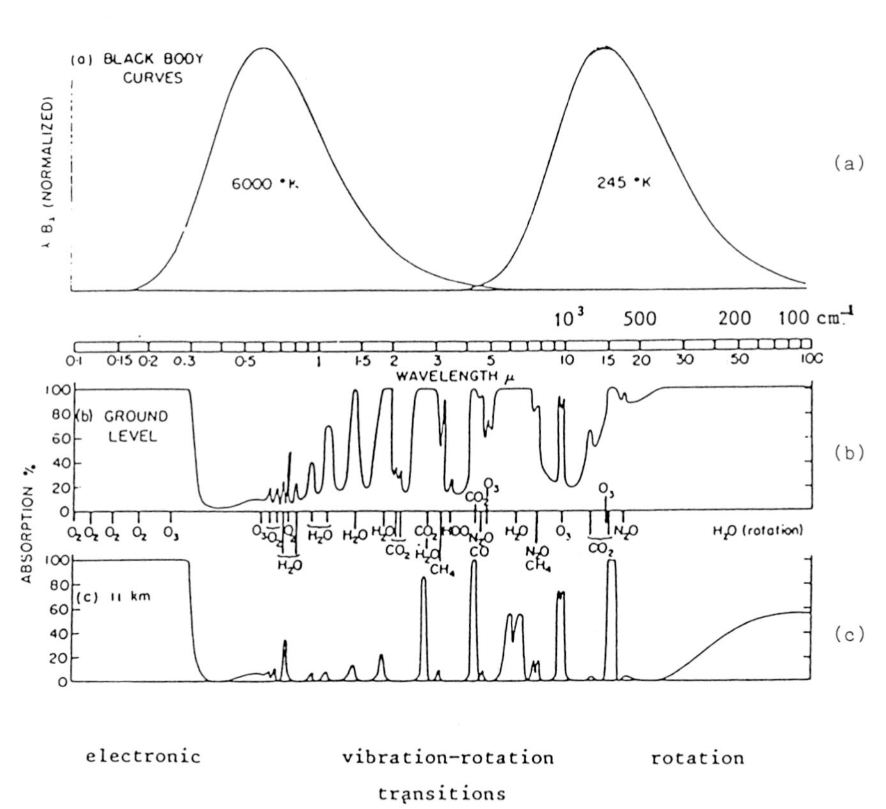

Atomospheric absorptions **(and emissions)** after Goody, R.M., Atmospheric Radiation, p.4.

(*a*) Black-body curves for 6000° K and 245° K. (*b*) Atmospheric gaseous absorption spectrum for a solar beam reaching ground level. (*c*) The same for a beam reaching the temperate tropopause. The axes are chosen so that areas in (*a*) are proportional to radiant energy. Integrated over the earth's surface and over all solid angles the solar and terrestrial fluxes are equal; consequently, the two black-body curves are drawn with equal areas beneath them. An absorption continuum has been drawn beneath bands in (*b*). This is partly hypothetical because it is difficult to distinguish from the scattering continuum, particularly in the visible and near infra-red spectrum. Conditions are typical of mid-latitudes and for a solar elevation of 40°.

Almost invariably the conversion from radiometric data to meteorological information will include approximations, assumptions or data from other sources which may themselves contain errors. Even the most direct measurements such as, for instance, the solar radiation reflected by the earth-atmosphere is not usually measured directly but over only part of the solar spectrum, corrections have to be made for the missing "tails" of the waveband making assumptions about the reflectivity at these wavelengths. Similarly, although the sea is effectively a black body, satellite measurements of sea surface temperature have to be corrected for the absorption of the outgoing radiation by the atmosphere and for its emission because there are no wavebands at which the atmosphere is completely transparent. More details of these techniques and their implications are given in section 4.

3. SATELLITES AND THEIR SENSORS

Text: Climatology from Satellites. Barrett (1974). Methuen.

3.1 The main types of environmental sensing satellites.

a) Polar orbiting meteorological satellites. These satellites orbit the earth at about 1000km altitude which gives a period of about 100 minutes. The plane of orbit is tilted slightly from the earth's axis and is fixed relative to the sun-earth line. This ensures that on each orbit the satellite crosses any line of latitude at a fixed solar time and views the daylight and night side of the earth once each orbit.

The width of the swath scanned by the radiometers on each orbit must be at least 25° at the earth's surface in order that consecutive swaths should overlap and give global coverage twice each day; once in daylight and once at night.

b) Geostationary meteorological satellites. These satellites orbit the earth at about 36,000km altitude which gives a twenty-four hour period of rotation. The orbit is in the plane of the earth's equator and the direction of rotation is the same as that of the earth. Hence the satellite remains stationary relative to the earth and can be positioned over any preselected point on the equator.

The high altitude of such satellites limits the spatial resolution with which the earth can be viewed. However, because they are geostationary frequent images of the same area can be obtained. Typically forty-eight images per day are taken.

The earth is scanned by the rotation of the satellite about its axis which is parallel to the earth's axis. The width of the scan line on the earth's surface is from one to five kilometres at the satellite sub point. A stepped motor changes the latitude of the scan line each rotation of the satellite to build up the complete image of the earth.

c) Earth resources satellites - a series of experimental and operational satellites of which LANDSAT is the best known. The altitude and orbits are similar to the polar orbiting meteorological satellites but the radiometers carried have a much narrower field of view. This gives a ground resolution of a few tens of metres but restricts the area covered to a narrow strip beneath the orbit so that it takes about eighteen days to build up a complete global image.

So we see that there has always a compromise to be made between spatial and temporal resolutions. The type of satellite to be used to study any particular phenomenon depends on the size and time scale of that phenomenon.

3.2 Some properties of satellite instrumentation.

a) The detectors. Early satellites often used television type "cameras" to view the earth. These have been replaced almost entirely by radiometers which scan a narrow strip of the earth, the output from which is sampled to give the required resolution along the strip. The scan may be achieved by the rotation of the satellite or by swinging the radiometer or its optical system. The former method is used for geostationary satellites and was used in the early polar orbiting satellites. Later orbiters are controlled so that the instrument console always points towards the earth and the radiometers usually scan perpendicular to the satellites path across the earth.

Radiometers may observe the electromagnetic radiation reflected or emitted from the surface below over a wide range of frequencies. These may include several bands in the visible and near infra-red, one of more bands in the thermal infra-red and micro-wave bands. The uses of the different bands will be discussed in later sections of this paper.

b) Spatial resolution. The spatial resolution of a radiometer is limited by the diffraction limit and by the signal to noise ratio.

The diffraction limit is imposed by the size of the aperture or objective of the radiometer and may be expressed as $\alpha = 1.22\lambda/D$ when λ is the wavelength of the radiation, D the aperture diameter and α the maximum angular resolution.

The spatial resolution is also limited by the energy available from the source area within the waveband being observed. If this energy is too low the signal will be masked by the noise generated within the radiometer and its associated electronic circuits.

Figure 3.1

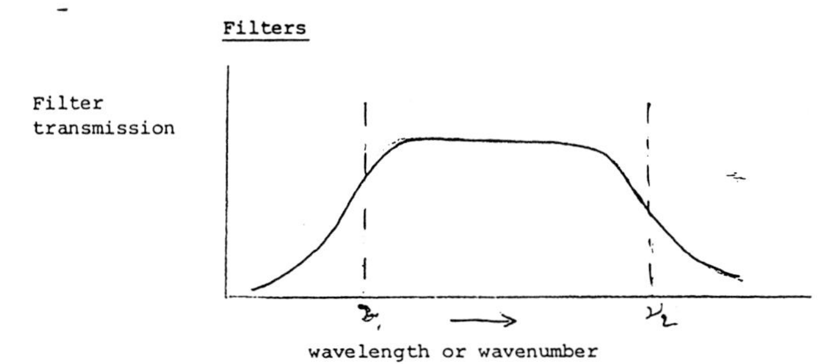

Detector output for a signal of constant energy per unit spectral interval

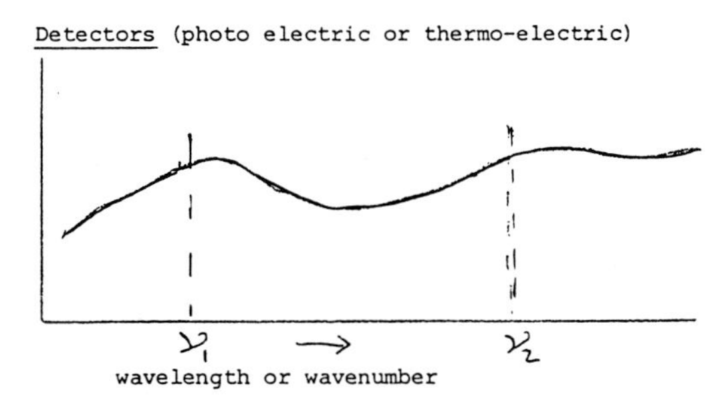

Amplifiers and electronics

Amplification

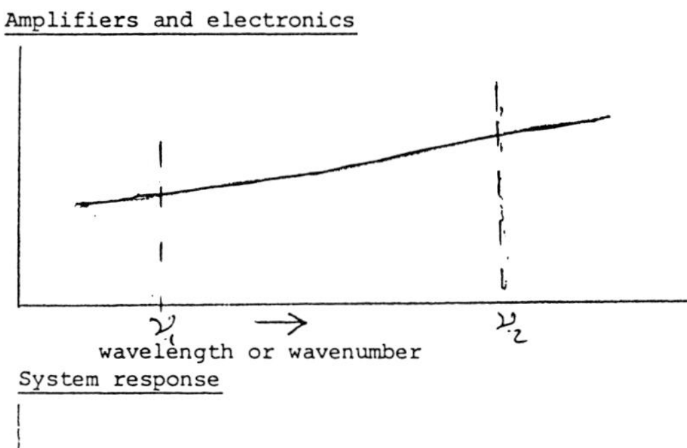

System response

System response (S_ν)

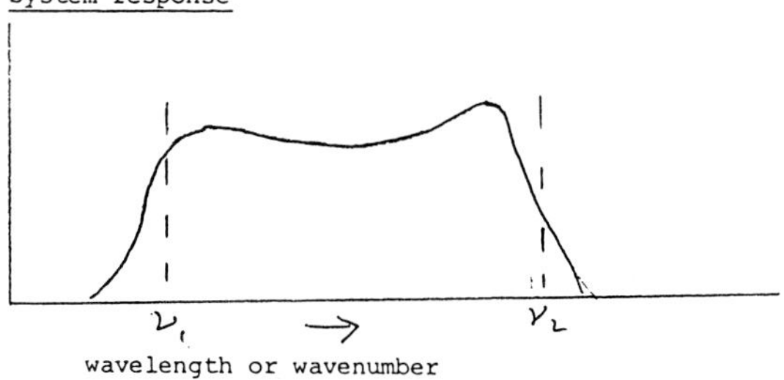

The output of the radiometer is given by $\int_\nu N_{I\nu} S_\nu \, d\nu$
where N_I is the input radiance

c) Wavelength resolution. For may purposes the information which can be deduced from radiometric data depends on how narrow a waveband can be observed. Again, the narrower the waveband the lower the signal to noise ratio so this eventually limits the resolution. Also the characteristics of the filters used to define the waveband introduce limits on how narrow the band can be.

d) Sensitivity of a radiometer system. The sensitivity of the radiometer depends on several factors including the characteristics of the optics and filters, the detectors and the amplifiers. This is illustrated in figure 3.1.

4. INFERRING METEOROLOGICAL QUANTITIES FROM SATELLITE DATA

Texts: Remote Sounding of Atmospheres. Houghton (1984). Camb. Univ. Press.
Remote Sensing of Cloudy Atmosphere. Ed. Henderson-Sellars (1984). Taylor & Francis.
Remote Sensing Applications in Meteorology and Climatology, Ed. Vaughan, NATO ASI Series.

4.1 Introduction

In this section some of the methods used for the extraction of the most important meteorological quantities from satellite data are reviewed. Any one of them could be the subject of a text book. What is attempted here is to highlight the principles and where in the extraction processes uncertainties may be introduced into the final estimate. It is hoped that these examples may serve to illustrate most of the possible sources of error as well as the potential of the technique.

4.2 Ground surface temperature

The principle here is to measure the outgoing radiation from the earth's surface in wavebands in which the surface is "black" and the atmosphere is transparent. Subject to radiometric errors one could then invert the radiance received at the satellite to give the precise surface temperature. Unfortunately the atmosphere is not transparent in any waveband. The clearest areas within the zone of strong emission from the earth's surface are around 4 μm and 11 μm (see figure 2.2(b)) but even here corrections must be made for the absorption and emission by the intervening atmosphere. These corrections can be precise if the profiles of concentration and temperature of the absorbing gases are known. There is usually uncertainty in the vertical distribution of water vapour so corrections may be applied using climatological data or climatological data adjusted by satellite derived water vapour estimates. However, the latter are most prone to error in the boundary layer which is very important when atmospheric corrections are calculated.

Another technique for deriving atmospheric corrections is the "split window" ie. observing the radiance from the surface in two wavebands in which the atmospheric absorption is different. This allows the correction to be calculated to a fraction of a degree K provided the emissivity of the surface is constant across both wavebands. This is the case with water but not generally over dry land surfaces. The problem of surface temperature sounding over land is compounded by the atmospheric correction increasing as the discontinuity between surface and air temperature increases. In summmary, in clear skies sea surface temperatures can generally be measured, after correction, to about 0.3K but daily maximum temperatures over dry land in strong sunshine may be in error by 3K in addition to any error due to the surface being a non-black emitter.

The presence of clouds prevents the use of the thermal I-R bands for surface measurements, though micro-wave measurements are still possible but carry the penalty of poor spatial and temperature resolution. The presence of sub-pixel clouds or thin cirrus which affect the measured temperature can be difficult to identify in thermal I-R imagery. Multi-spectral methods have to be employed to remove such contaminated data.

4.3 Atmospheric temperature profiles

If we know the vertical distribution, the radiative properties of the atmospheric gases and the vertical temperature profile we can calculate the energy emerging from the top of the atmosphere from the atmosphere below. By selection a portion of the spectrum in which only one uniformly mixed gas is emitting the problem resolves to a single variable, the temperature profile. The sounding of the atmosphere is a question of inverting the equation and deducing the temperature profile from measurements of the outgoing radiation. From a single measurement a profile cannot be reconstituted because the measured radiance is the integral of the emission from all atmospheric layers and by the earth's surface, attenuated by the absorption in the layers above. If the atmospheric absorption (emission) is high then most of the radiation leaving the atmosphere will have originated near the top of the atmosphere. Conversely, if the absorption is low much of the emergent energy will have originated from the earth's surface. So, by taking a series of measurements at intervals on the edge of an absorption band we obtain measurements which have components originating at different heights in the atmosphere. The calculated fraction of the outgoing radiation from different levels in an isothermal atmosphere at a particular wavelength is the "weighting function" associated with that wavelength. Figure 4.1 illustrates three sets of weighting functions for radiometers flown on the Nimbus 6 Satellite.

Figure 4.1

Weighting functions (gradient of transmission with respect to log pressure) for instruments sounding the temperature of the lower atmosphere on the Nimbus 6 satellite: (*a*) 15 μm channels of HIRS. (*b*) 4.3 μm channels of HIRS, (*c*) channels of SCAMS (c.f. § 6.9; from Smith and Woolf 1976 and Staelin *et al.* 1975).

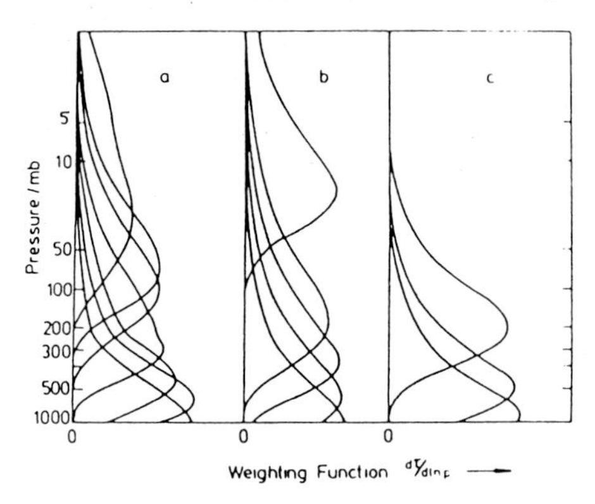

It is seen that up to seven channels are used. However, the data from each channel is not independant in that there is significant overlap in the weighting functions. The sharper the peak in the weighting function the better the vertical resolution of the temperature sounding. Sharp peaks are associated with narrow wavebands, hence low energy and a decrease in the signal to noise ratio of the radiometer. Much effort has gone into the design of filters to allow good resolution with adequate energy. The subject of retrieval of the temperature profile is complex and several techniques are used. However, it is clear that without the use of additional data temperature profiles cannot do more than represent the mean temperature of a similar number of atmospheric layers that there are weighting functions and that the overlap of the weighting functions further reduces the vertical resolution of the deduced temperature profile. It is common practice to use the satellite atmospheric data in conjunction with sea surface temperature data and with climatological or adjacent radio-sonde data to derive the temperature profiles. Recent tests have shown that the result of including satellite temperature profiles over the oceans in the ECMWF forecast model make little difference to the northern hemisphere forecasting capability but do show positive results in the data-sparse southern hemisphere.

4.4 Surface vegetation cover

In principle the earth resources satellites with their high spatial resolution offer the best tool for measuring vegetation cover, type, height, vigour and seasonal variations: all quantities which influence the fluxes of heat & water from the surface to the atmosphere. However, because of the low return frequency of resource satellites to the same zone they cannot reliably monitor changes on time scales less than several months. Instead the visible and near infra-red bands of radiometers on the operational meteorological satellites are used. The methods rely on the change in reflectance of green vegetation from below 0.2 to 0.5 or higher as one moves from the red to the near infra-red wavelengths. In the absence of clouds daily coverage of the globe is possible. However, in many parts of the world cloudy days prevail. Also, to obtain daily coverage, the radiometers scan the swath of the earth's surface beneath their path viewing the surface at angles up to 50° from the vertical. Hence, the atmospheric effects on the data change along each scan line. To minimise these effects the radiometric data is normalised and the highest normalised value over several days taken as the index of vegetative cover. The normalised vegetation index is defined as

$$NVI = \frac{\text{Near IR} - \text{red}}{\text{Near IR} + \text{red}}$$

The typical reflectance characteristics of grass and soils are indicated in figure 4.2 below.

Various problems with the NVI include poor sensitivity at low surface vegetation cover, lack of calibration, partial cloud contamination of pixels, and lack of stable calibration against uniform unchanging surfaces. Despite this the NDVI does offer an available data set giving global coverage which allows seasonal and interannual vegetation changes to be monitored.

Figure 4.2

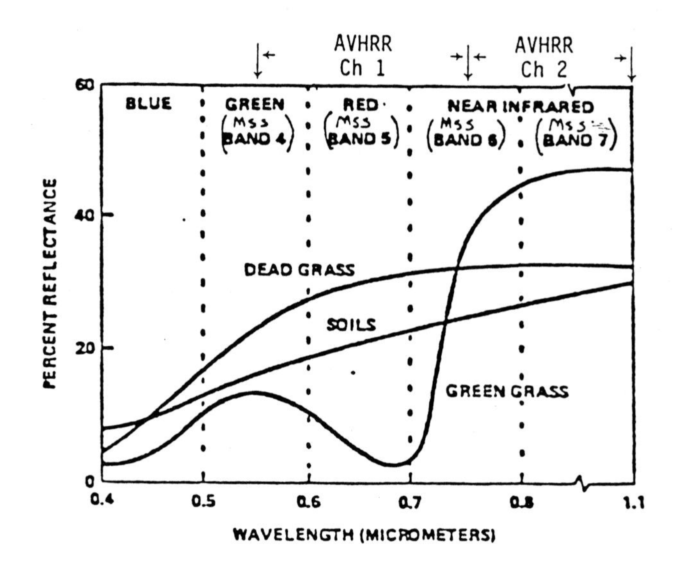

Typical reflectance of herbaceous vegetation and soil from 0.4 to 1.1 micrometers. Bands 4, 5, 6, and 7 are for Landsat MSS and Channels 1 and 2 are for NOAA 6 and 7 AVHRR. From: Lautenschlager and Perry (1981).

5. USING SATELLITE DATA IN CLIMATE STUDIES

Texts: Remote sensing applications in meteorology & climatology: Vaughan (Ed); (1987) NATO ASI Series.

Satellite sensing of a cloudy atmosphere: Henderson-Sellars (Ed) 1984. Taylor & Francis.

In the long term it may be expected that satellite data will be used for monitoring climate change but global data sets that are currently available do not span an adequate period for this. Meanwhile the available data sets and those expected to be available in the next few years have an invaluable role in the studies of inter-annual climate changes and in developing and validating climate models. This is particularly true of studies into feedback effects and tele-connections. It is recognised that several characteristics of the earth/atmosphere which must be properly represented in climate models cannot be monitored by traditional meteorological instrumentation. Important among these are: the cloud field, small changes in which make large differences in the energy balance of the surface; the land surface processes which control the flow of heat and water from the land to the atmosphere; and the extent of ice and snow cover over land and sea which again reflect much incoming solar radiation and also act as long and medium term water stores.

International efforts to provide techniques for monitoring land surface parameters and data on global cloud fields are underway. Meanwhile, there are relevant global data sets which can be used. The Earth-Radiation Budget Data Set makes available data since 1978 on regional and global albedo, outgoing longwave infra-red radiation flux and net radiation. Vegetation index (see section 4.4) data sets are available from 1982 through NOAA. The European Space agency is archiving a cloud data set for the Meteosat viewed disc, and the U.S. Air Force have been archiving global cloud data for several years. An analysis of the latter is described in text edited by Vaughan above. There are several data sets of winds derived from satellite observed cloud vectors.

In the next few years the advent of operational microwave sensors will increase the range of routinely available satellite sensed global data to include sea surface winds, soil moisture (in the upper few cm.), raifall over oceans, ice cover and snow cover and thickness. The prospects are exciting but the researcher must always be aware of the limitations of the data acquisition system and the assumptions made in processing the radiometric data to produce the data sets. Any data from polar orbiting satellites may contain a diurnal time bias because of the overpass time of the satellite and changes in data over the years may be due to changes in satellite or to radiometer drift. Visible and thermal infra-red information on ground surface conditions is biased to clear sky conditions. The poor spatial resolution of most micro-wave data may fail to detect important small scale phenomena. Satellite cloud vector winds give no information on clear sky winds. Unless these and other possible shortcomings of the data taken are into account we may produce some strange and erroneous climatologies.

EUROPEAN SCHOOL OF CLIMATOLOGY AND NATURAL HAZARDS

Course on

"Climatic Change and Impacts: A General Introduction"

4. THE GREENHOUSE GAS INDUCED CLIMATE CHANGE

CARBON DIOXIDE: ITS NATURAL CYCLE AND ANTHROPOGENIC PERTURBATION

Ulrich Siegenthaler
Physics Institute, University of Bern
Sidlerstrasse 5
3012 Bern, Switzerland

1. INTRODUCTION

Carbon dioxide is, besides water, the main nutrient for plants and therefore for life on earth. In consequence of fossil fuel burning and human impact on the land biota, the atmospheric concentration of carbon dioxide is steadily increasing, which may lead to long-lasting changes of the global climate. These two facts explain the strong interest of scientists from many disciplines in this gas and its natural cycle.

This chapter deals with the cycling of CO_2 between the atmosphere and other global reservoirs. Its atmospheric concentration is largely determined by the ocean, so that emphasis is naturally given to air-sea exchange and the oceanic carbon cycle. However, the discussion would not be complete without also shortly considering the exchange with the terrestrial biosphere. Other carbon-containing compounds, such as CH_4 or CO, are not discussed here, since CO_2 behaves essentially as an inert gas in the atmosphere, decoupled from the cycles of other trace gases.

After a general description of natural reservoirs and fluxes, the anthropogenic CO_2 increase and its modelling are discussed, as well as CO_2-induced global warming. Finally, natural variations of atmospheric CO_2 are considered. Obviously, this chapter cannot be a comprehensive treatment of all important aspects of carbon dioxide in nature. Rather, the aim is to provide some insight into the processes that determine the natural level of atmospheric CO_2 and its variability as well as the fate of man-made CO_2, and to report on some recent developments in the field. The interested reader can find more detailed information in a number of reviews. "Tracers in the Sea" (13) is an excellent textbook and review of the oceanic cycles of carbon and other elements. SCOPE 16 (8) deals with modelling aspects of the anthropogenic perturbations and contains compilations of oceanic and atmospheric data. Sundquist (68) critically reviewed carbon cycle data. "Carbon Dioxide Review 1982" (16), "Changing Climate" (43), and the book by Liss and Crane (39) are all excellent reviews, the former two with emphasis on climatic effects of CO_2 and trace gases.

2. THE NATURAL CYCLE OF CARBON DIOXIDE

2.1. Reservoirs, Fluxes, Residence Times

Data on the distribution and fluxes of carbon are given in Table 1. The numbers are affected with uncertainty, some more than others; more information on this can be found in the review-papers cited above.

The amounts of carbon in the various reservoirs and the fluxes between these must have fluctuated somewhat already before the perturbation by human activities, but here are reasons (for instance sedimentary records) for assuming that they were, if averaged over appropriate time periods, more or less constant. Therefore, steady state is often assumed in geochemical studies. This permits to apply the useful concept of mean residence time. The mean residence time of molecules entering a reservoir is defined in the usual statistical way, as the average timespan between entrance and exit. It can be shown (46) that the mean residence time is given by

$$\tau = M/F$$

where M is the mass in the reservoir and F the (total) influx (= outflux) into the reservoir. Sometimes, residence times are calculated with respect to only one specific influx or outflux; in this way the importance of different fluxes can easily be compared. In Table 1, some residence times are also indicated.

More than 99 percent of the carbon present in the atmosphere is in the form of CO_2. The pre-industrial concentration must have been near 280 ppm by volume (see paragraph 3.1). The present atmospheric concentration is about 20 percent above that value. Of the rapidly exchanging reservoirs, the ocean is the largest one, containing about 60 times as much carbon as the atmosphere, most of which is dissolved inorganic carbon or "total CO_2" (ΣCO_2), including HCO_3^-, $CO_3^=$ and CO_2. The sediments are by far the largest carbon pool, but the exchange with the other reservoirs is so slow that it can be neglected on time scales of 10^3 years or less.

The residence time of CO_2 in the atmosphere with respect to exchange with the oceans is 8 years and that with respect to exchange with the biosphere is similar, so that the overall atmospheric residence time is about 4 years. The figures indicated for photosynthesis on land correspond to net primary production (NPP), excluding the short-time (hours to days) fixation and respiration: these are included in the gross primary production (GPP) rate which is about twice as large as NPP that only NPP is efficient for biological uptake of a perturbation like bomb-produced ^{14}C, but not fixed carbon that is returned to the atmosphere by plant respiration after a very short period. Compared to the land biosphere, the mass of the marine biosphere is negligible. Marine organisms have a very short lifetime but a large productivity, so that the rate of photosynthesis in the sea is comparable to that on land. The material from dead marine organisms is recycled with high efficiency, and only 0.1 to 0.2 percent of the organic carbon produced totally settles down and is buried in the sediments.

The isotopes of carbon provide valuable information on processes and time rates. Isotope fractionation connected to physico-chemical and biological processes is responsible for differences in natural $^{13}C/^{12}C$ ratios. These are usually given in delta-notation

$$\delta^{13}C = \frac{^{13}R\ (\text{sample}) - {}^{13}R\ (\text{standard})}{^{13}R\ (\text{standard})} \cdot 1000\ ‰ \qquad (1)$$

with $^{13}R = [^{13}C]/[^{12}C]$. The standard used internationally is PDB (originally for "Pee-Dee belemnite" (17).

The radioactive carbon isotope ^{14}C has a half-life of 5730 yr (mean

Table 1. Major global carbon reservoirs and fluxes. Main sources: Bolin et al. (10); Bolin (8); Clark (16).

Reservoirs			10^{15} g C
Atmosphere CO_2:	Before 1850 ca. 280 ppm		594
	1980 338 ppm		717
Other gases:	CH_4 1.5 ppm		4
	CO 0.1 ppm		
(Troposphere: 80 %, stratosphere: 20 % of atmospheric mass)			
Oceans:	Inorganic C (ΣCO_2)		37,400
	Dissolved organic matter	ca.	1,000
	Biomass		3
Land biosphere:	Living		560
	Soil, humus (pre-historic: $200\text{-}500.10^{15}$ g C more)		1,500
Groundwater:			450
Sediments:	Inorganic C	ca.	60,000,000
	Organic C	ca.	12,000,000
Fossil fuels:		ca.	5,500

Fluxes (gross)		10^{15} g C/yr
Atmosphere-ocean, CO_2 exchange		78
Atmosphere-land biota, photosynthesis/respiration (NPP)		65
Marine photsynthesis		45
Sedimentation in oceans		0.2
Volcanism	ca.	0.07
Fossil fuel combustion, 1980		5.3

Residence times: τ = mass/flux	
Atmosphere (pre-industrial): total exchange	4 yr
exchange with ocean only	8 yr
exchange with biosphere only	9 yr
Living land/biosphere: photosynthesis/respiration	11 yr
Marine biosphere: photosynthesis/respiration	0.07 yr
Oceans: exchange with atmosphere, total flux	490 yr
sedimentation only	180,000 yr
Atmosphere + biosphere + oceans: sedimentation	210,000 yr

life 8267 yr, decay constant $\lambda = 1/8267$ yr). ^{14}C is, as ^{13}C, subject to isotope fractionation processes which affect the isotope ratio $^{14}C/^{12}C$ (= $^{14}/R$) twice as much as the ratio $^{13}C/^{12}C$. A ^{13}C corrected $^{14}C/C$ ratio is often considered

$$^{14}R_{corr} = {}^{14}R\ (1 - 2\ (\delta^{13}C + 25\ ‰)) \qquad (1)$$

i.e. the actual $^{14}C/C$ ratio is corrected to a $\delta^{13}C$ value of -25 ‰, as typical for wood. In geophysical work, ^{14}C concentrations are often given as

$$\Delta^{14}C = (^{14}R\ corr\ /^{14}R\ (standard) - 1)\ .\ 1000\ ‰ \qquad (2)$$

where the standard activity approximately corresponds to the natural atmospheric ^{14}C level (i.e. $\Delta^{14}C \approx 0$).

Some typical isotopic values are given in Table 2. The ^{14}C age of mean deep sea water with respect to the warm surface is about 1000 yr.

Table 2. ^{13}C and ^{14}C in natural reservoirs (pre-industrial situation).

	$\delta^{13}C$ (‰)	$\Delta^{14}C$ (‰)
Atmospheric CO_2	- 6.4	0
Land vegetation	- 25	- 0
Ocean, ΣCO_2: warm surface water	2	- 50
deep sea (average)	0.5	- 160

2.2. Air-sea Exchange of CO_2

There is a continuous exchange of CO_2 between atmosphere and ocean, and under natural conditions, the global invasion and evasion fluxes must very nearly balance each other. The magnitude of the exchange flux can be estimated in three ways: using natural or bomb-produced ^{14}C, and using radon-derived air-sea transfer velocities.

1. In steady state, the decay of ^{14}C in the ocean must be balanced by the net inflow from the atmosphere. The latter is equal to the difference between gross inflow, $R_a R_o A_s$, and gross outflow, $R_s F_o A_s$, where R_a, R_s are the mean (^{13}C-normalized) ^{14}C concentrations in atmosphere and in surface ocean water, F_o is the mean (gross) CO_2 exchange flux per m^2 of ocean surface, and A_s the global ocean surface. The balance can be written as:

$$\lambda R_{oc} N_{oc} = F_o (R_a - R_s) A_s$$

whence:

$$F_o = \frac{\lambda R_{oc} N_{oc}}{(R_a - R_s) A_s} \tag{3}$$

λ = is the ^{14}C decay constant, R_{oc} and N_{oc} are the average ^{14}C concentration and the total mass of carbon dissolved in the ocean. The atmospheric ^{14}C concentration, R_a, is set 100%; then $R_s \approx 95$ % and $R_{oc} \approx 84$ %. With $N_{oc} = 3.84 \cdot 10^{19}$ g C and $A_s = 3.62 \cdot 10^{14}$ m^2, we obtain $F_o = 216$ g C m^{-2} yr^{-1} = 17.9 mol m^{-2} yr^{-1}, or a global air-sea exchange rate of $78 \cdot 10^{15}$ g C yr^{-1}. The error of this number is about $\pm$ 25 % and mainly stems from the uncertainty of about 1 % in R_s and therefore in R_a-R_s (= 5 $\pm$ 1 %).

2. In an analogous way, the mean air-sea exchange flux can be estimated from the inventory of bomb-produced ^{14}C in the ocean, as obtained based on the GEOSECS tracer data. Again the driving force is proportional to the difference between the ^{14}C concentrations of

atmosphere and surface waters, which now are functions of time and are reasonably well known from observation (Figure 1).

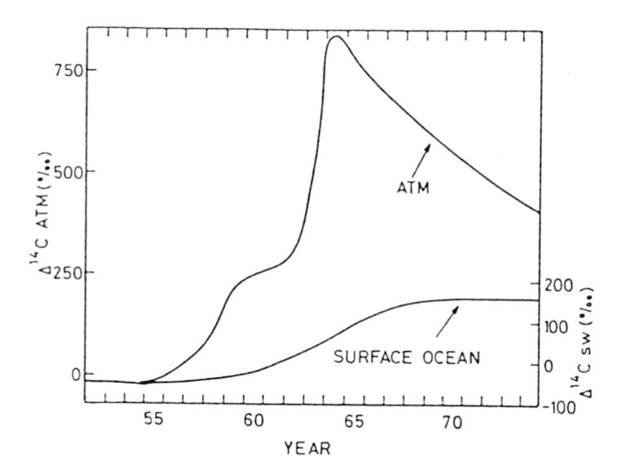

Figure 1. ^{14}C of atmospheric CO_2 (left-hand scale) and in ocean surface water (right-hand scale) in response to the input of bomb-produced ^{14}C. After Broecker and Peng (13).

Different estimates of the mean air-sea CO_2 exchange flux, based on bomb ^{-14}C, converge in the range 20-23 mol m^{-2} yr^{-1} (e.g. (65)). These values are valid for the period 1962 - 1972, when atmospheric bomb-^{14}C was high above oceanic values; then the CO_2 concentration was about 320 ppm. The corresponding pre-industrial value is obtained by noting that the flux is proportional to the partial pressure of CO_2, the pre-industrial value of which was about 280 ppm. Correspondingly the above range corresponds to a pre-industrial exchange flux of 17.5 to 20 mol m^{-2} yr^{-1}. This is in good agreement with the result from the distribution of natural ^{14}C. Note that oceanic bomb-^{14}C represents an integral over about 10 years, and natural ^{14}C over about 500 years (oceanic residence time of CO_2). Thus we conclude that the mean pre-industrial CO_2 exchange flux was 18 mol m^{-2} yr $\pm$ 25 %. For a pre-industrial atmospheric concentration of 280 ppm (in dry air) and calculated with a global mean sea surface temperature of 20°C, eq. (4) (below) yields an approximate mean transfer velocity of $4.9 \cdot 10^{-5}$ m s^{-1} (= 17.5 cm h^{-1}).

3. The radon-deficit method yields local and instantaneous gas transfer velocities. Based on the results that Peng et al. (49) obtained during GEOSECS, a mean pre-industrial CO_2 exchange flux of about 12 mol m^{-2} yr^{-1} can be derived (see next paragraph). This is somewhat lower than the ^{14}C-derived values; a probable reason is that the radon-derived transfer velocities are not representative for mean global conditions, especially not for high-wind conditions occuring in winter in temperate and high latitudes.

2.3. Regional Variability of Air-sea Flux

In a natural steady-state situation the global net CO_2 flux from air to sea must have been near zero, except for a very small unbalance connected to river input minus sedimentation of carbon. Regionally, however, surface water can be supersaturated or undersaturated in CO_2 with respect to the atmospheric concentration. Then there is a non-zero net flux between air (index a) and sea (index s) which can be represented as difference between invasion flux $F_{a,s}$ and evasion flux $F_{s,a}$:

$$F_{net} = F_{a,s} - F_{s,a} = k(s\,C_a - C_w) = k\,s(C_a - pCO_2) \qquad (4)$$

where k is the gas transfer velocity related to liquid-phase concentrations, s the solubility of CO_2, C_a and C_w are CO_2 concentrations in air and in bulk surface water. (38, for a review of the physics of gas exchange).

Instead of C_w, usually the equilibrium partial pressure $pCO_2 = s^{-1} C_w$ is considered; correspondingly the gas-phase related transfer velocity, k s, instead of the liquid-phase related transfer velocity, k, should be considered.

C_a and pCO_2, as used in (4), are partial pressures in moist air of 100 % relative humidity at the temperature of surface water, since the air is saturated with water vapour directly at the interface. Therefore, atmospheric concentrations referring to dry air must be converted to C_a according to

$$C_a = C_{a,dry} (1-e_s/p_a) \tag{5}$$

where e_s is the saturation water vapour pressure and p_a the barometric pressure at sea-level. The correction may be important for the net fluxes calculated using eq. (4).

2.3.1 Dependence of transfer velocity on temperature and wind

The evasion flux (for given C_w) is proportional to k (transfer velocity with respect to liquid-phase concentrations), the invasion flux (for given C_a) proportional to k s (transfer velocity w.r.t. gas-phase concentrations). The transfer velocity k can approximately be represented as $k = (D/\nu)^n$ f(turbulence), where D = diffusion coefficient for CO_2 in water, ν = kinematic viscosity of water (ν/D = Sc, Schmidt number), $\nu \simeq 0.5$ in presence of capillary waves (26) and probably for average oceanic conditions, and f(turbulence) summarizes the dependence of k on the near-interfacial turbulence caused for instance by wind. $(D/\nu)^{0.5}$ increases, solubility decreases with rising temperature. Between 0°C and 30°C, s varies by a factor 0.40, $(D/\nu)^{0.5}$ by a factor 2.2, and their product by a factor 0.87. Thus, the direct effect of temperature is small for k s and for the CO_2 flux for given pCO_2 and atmospheric concentration. Laboratory experiments and field work (76) show that transfer velocity increases with wind more strongly than proportional to wind speed. Annual mean wind speed above the ocean is nearly a factor 2 larger in high latitudes (> 50°) than in tropical regions. Thus, we may expect that k s is higher in high than in equatorial latitudes by roughly a factor of two due to the effect of wind, while the temperature difference should cause a slight reduction. The expected overall effect is an increase towards high latitudes.

2.3.2. Estimate of regional variability

Some of the quantities determining the CO_2 flux exhibit considerable regional variability. In order to see the influence of that variability, I estimated mean values for the Atlantic and Pacific Oceans and for the latitude zones > 50°N, 50°N - 10°N, 10°N - 10°S, 10°S - 50°S and > 50°S (excluding the Pacific region > 50°N for which there are essentially no data on pCO_2 and k); the subdivision was taken over from SCOPE 16 (8) . The available field data are not complete enough to allow the calculation of reliable regional mean values, and the exercise should therefore only be considered as a sort of sensitivity test.

Basis data and results are given in Figure 3. CO_2 transfer velocities were obtained by averaging the radon-based results of Peng et al. (49) (values for actual water temperatures, not normalized to 20°C) and multiplying by $(D(CO_2)/D(Rn))^{0.5} = 1:36^{0.5}$. Sources for pCO_2 data are a map of GOESECS data of Takahashi et al. (73) (a meridional profile for the pacific Ocean is shown in Figure 2), results obtained by R. Weiss during the FGGE experiment in 1979 (30) and the papers of Roos and Gravenhorst (53) and Smethie et al. (63). pCO_2 data were converted to ppm in moist air where necessary and corrected to 1973. Areas and temperatures were calculated from the atlas of Levitus (36).

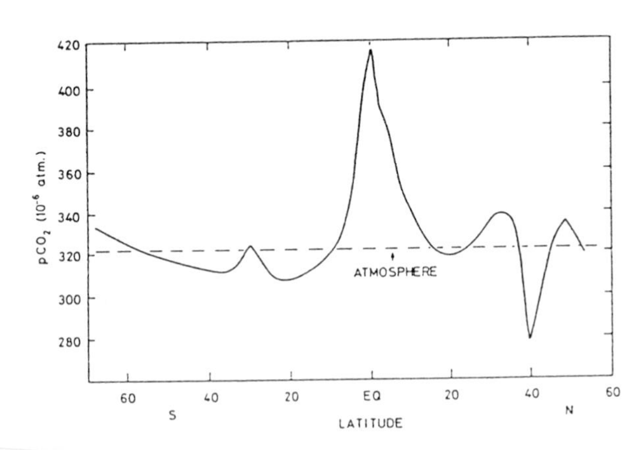

Figure 2. North-south section of pCO_2 in surface water of the Pacific Ocean. The equatorial maximum is due to upwelling of CO_2-rich subsurface water. After Broecker et al. (14).

Exchange fluxes and transfer velocities

The average k values derived from radon data vary between about $2.5 \cdot 10^{-5}$ and $5 \cdot 10^{-5}$ m s^{-1} (9 and 18 cm h^{-1}), the invasion fluxes between 9 and 29 mol m^{-2} yr^{-1}. Mean values (areally weighted) are $3.9 \cdot 10^{-5}$ m s^{-1} (14 cm h^{-1}) and 13.9 mol m^{-2} yr^{-1}. For a pre-industrial concentration of 280 ppm (instead of 329 ppm in 1973), the invasion flux would be 11.8 mol m^{-2} yr^{-1}, considerably less than the value obtained from ^{14}C, as mentioned above. The present discussion is based on the (unproved) assumption that the radon results still approximately reproduce the geographical variation of mean annual transfer velocities.

The transfer velocity k does not show a pronounced latitude dependence; high values are observed in the southern temperate and the Antarctic part of the Pacific Ocean, but not in the Atlantic Ocean (Figure 3). Thus, the direct temperature dependence of k may be partly compensated by higher wind speeds in cooler regions. In contrast, the gas-phase related transfer coefficient, k s, and the gross CO_2 fluxes are significantly larger in high southern latitudes than elsewhere. The calculated mean invasion flux at high northern and southern latitudes (>50°) is 22.1 mol m^{-2} yr^{-1}, a factor of 1.77 higher than for the rest of the ocean. The different behaviour of k and k s (or gross flux) is a consequence of the temperature dependence of the solubility.

Net air-sea fluxes

Large positive fluxes, i.e. net evasion from the ocean, occur in tropical latitudes because of high pCO_2 in surface water (Figure 3). In temperate latitudes the ocean is generally a natural CO_2 sink. The high latitude regions are, according to Figure 3, strong sinks. However, pCO_2 data are rather sparse there, so that the calculated fluxes are affected with large uncertainty. The calculated global net flux is very near to zero; it corresponds to a net oceanic source of $0.3 \cdot 10^{15}$ g C yr^{-1}, while a sink of about $2 \cdot 10^{15}$ g C yr would be expected from carbon cycle models (see below). The reason for the discrepancy is probably that the assumed regional values of pCO_2 and transfer velocity are affected with error; furthermore, the Indian Ocean is not included in this analysis.

It should be emphasized that the calculated net fluxes do not indicate in any way whether the considered regions are efficient or not as sinks of anthropogenic CO_2. It can be expected that the presently observed map of pCO_2 differences to the atmospheric concentration roughly corresponds to the natural pCO_2 distribution.

Takahashi (70) also estimated net air-to-sea fluxes and ended up with a mean global value of 0.49 ± 0.49 mol m^{-2} yr^{-1}. His calculations are based on slightly different pCO_2 values and on a constant gas-phase related transfer velocity k s.

2.4. Marine Carbonate Chemistry

The chemical equilibria between the dissolved carbonate species can be written as follows (for a detailed discussion see e.g. Broecker et al. (13):

$$CO_{2,aq} + H_2O \overset{K_1}{=} H^+ + H\,CO_3^- \tag{6}$$

$$H\,CO_3^- \overset{K_2}{=} H^+ + CO_3^= \tag{7}$$

where $CO_{2,aq}$ stands for dissolved CO_2 including H_2CO_3. K_1 and K_2 are equilibrium constants. In addition, the solubility s links pCO_2 to concentration of $CO_{2,aq}$. Instead of the concentrations of $CO_{2,aq}$, HCO_3^-, $CO_3^=$ and H^+, a set of four other parameters can conveniently be observed: pCO_2, $\sum CO_2$, alkalinity A and pH = $-10 \log\left[H^+\right]$. $\sum CO_2$ and alkalinity (negative charge contributed by weakly dissociated acids) are defined by

$$\sum CO_2 = \left[CO_{2,aq}\right] + \left[HCO_3^-\right] + \left[CO_3^=\right] \tag{8}$$

$$A = \left[HCO_3^-\right] + 2\left[CO_3^=\right] + B(OH)_4^- + \ldots \approx \left[HCO_3^-\right] + 2\left[CO_3^=\right] \tag{9}$$

There exist two equilibrium constraints, equations (6) and (7). Therefore, the whole carbonate system is defined if two of the four observable quantities, e.g. $\sum CO_2$ and alkalinity, are measured.

$\sum CO_2$ is changed by CO_2 exchange with air and by formation or destruction of organic matter of solid carbonate particles. Alkalinity is not affected by gas exchange, but by dissolution of $CaCO_3$ particle and,

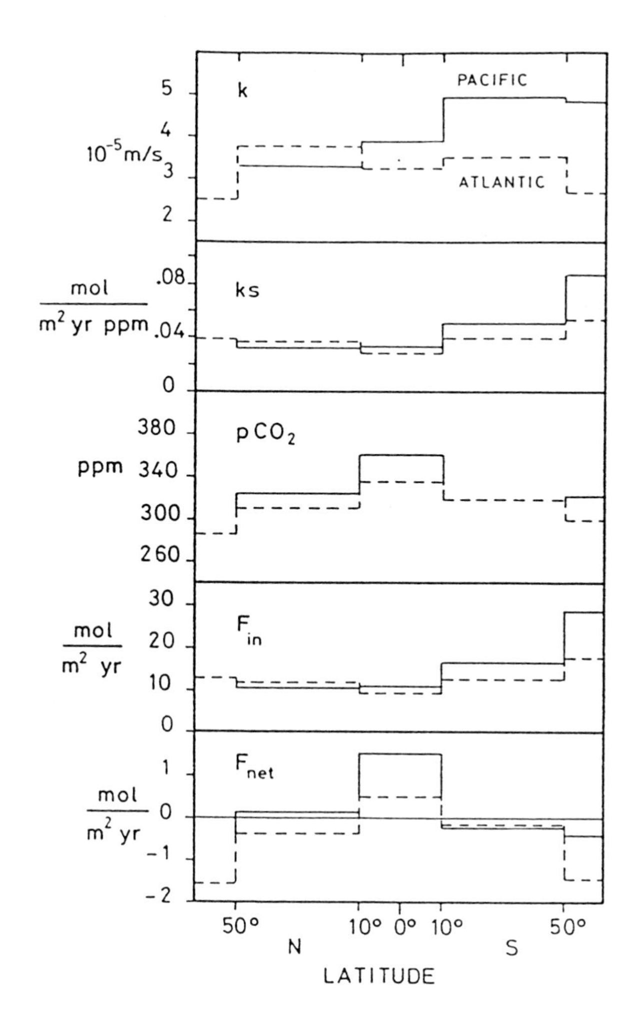

Figure 3. Estimated mean values for different latitude bands of CO_2 transfer velocities k and k s (related to liquid-phase and gas-phase concentrations, respectively), pCO_2, invasion flux of CO_2 and net flux from ocean to atmosphere. Solid line: Pacific Ocean, dashed line: Atlantic Ocean.

to a minor degree (via NO_3^- and PO_4^{3-}), of organic particles. Of special interest for air-sea exchange are processes affecting pCO_2. pCO_2 can be considered a function of ΣCO_2, alkalinity and temperature; other factors like salinity are of minor influence. In connection with anthropogenic CO_2, the relation between pCO_2 and ΣCO_2 (for constant A and T) is important. If the atmospheric CO_2 level varies, the equilibrium concentration of dissolved CO_2 in surface sea water varies proportionally, but not the concentrations of bicarbonate and carbonate ions because of chemical constraints. The resulting relative change of total dissolved inorganic carbon is smaller than that of CO_2 gas alone. This is taken into account by introducing a buffer factor ζ : if the CO_2 partial pressure is increased by p percent, then CO_2 increases by p/ζ percent only. This can be expressed as

$$\frac{\Delta \Sigma CO_2}{\Sigma CO_2} = \frac{1}{\zeta} \frac{\Delta pCO_2}{pCO_2} \qquad (10)$$

The buffer factor (also called Revelle factor) is a function of ΣCO_2 and alkalinity, to a lesser degree also of temperature (3), (71). On the average it is about 9 for temperate and equatorial, and about 14 for high-latitude surface water.

For constant ΣCO_2 and temperature, pCO_2 decreases as alkalinity increases. The analogue to the buffer factor for alkalinity, i.e. the ratio of relative pCO_2 change to relative alkalinity change, is about -8 to -13. This becomes of interest when e.g. considering carbonate sediment dissolution due to the acidification of sea-water by fossil CO_2.

For given ΣCO_2 and alkalinity, pCO_2 increases with temperature by 4 to 5 percent per degree Celsius.

2.5. The Oceanic Carbon Cycle

Cycling of carbon in the oceans is governed by transport by water and by formation and redissolution of organic and carbonate particles. Information on the time scales of water circulation is obtained from the distribution of natural ^{14}C in the ocean. The natural $^{14}C/C$ ratio of warm surface water was near 95 % (atmosphere: 100 %), and near 90 % or 93 % (Antarctic Ocean and northern North Atlantic, respectively) in surface water in regions of deep water formation. Values for deep water below 1500 m range from about 90 % in the North Atlantic to 76 % in the North Pacific, corresponding to mean ages (with respect to the corresponding cold surface waters) of 200 - 300 up to 1400 years. The distribution of natural and bomb-distributed ^{14}C has been described in detail by Broecker et al. (13) and by Siegenthaler (62).

Total CO_2 and alkalinity, but also the concentrations of elements like P and N, are depleted in surface water (Figure 4) because organic and carbonate particles carry them to depth, where organic particles are oxidized by microbial action and carbonate shells are redissolved.

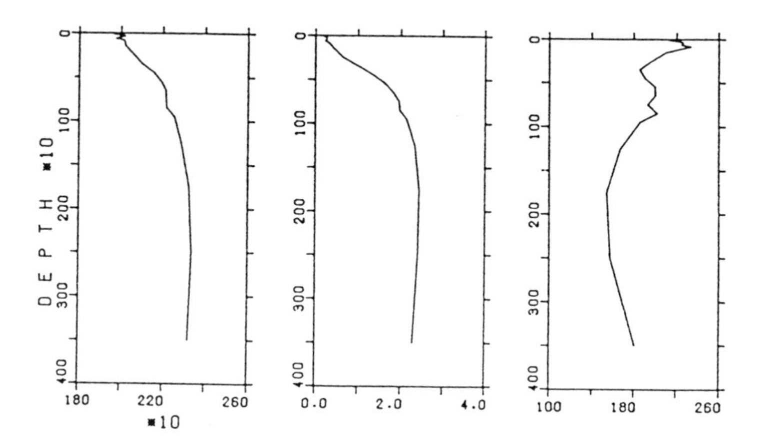

Figure 4. Vertical profiles in the Pacific Ocean, 10° - 50°S, from 0 - 4000 m depth, of (from left to right) $\sum CO_2$, phosphate and dissolved oxygen. Concentration units: µmol/kg (for $\sum CO_2$: unit = 10 µmol/kg); depth unit: 10 m.

The surface depletion of $\sum CO_2$ is about 10 to 20 percent of the deep water value, which can be interpreted in a first order estimate such that particle transport of carbon is about 10 to 20 percent of the transport by water circulation. To a minor degree, the surface - deep water difference is also due to the fact that the solubility of CO_2 is higher in the cold water of the deep-sea outcrop areas than in warm surface water.

The marine biological productivity depends on the availability of the elements N and P, for diatoms also of Si. These act as limiting nutrients in many oceanic regions. The mole ratios (Redfield ratios) of P, N and C are found to be relatively constant in biological material. A recent careful analysis of the variations of the composition of sea-water along isopycnal (equal density) surfaces by Takahashi (74) yielded best values for the ratios of concentration changes due to oxidation of organic matter of

$$P : N : C : -O_2 = 1 : 16 : (103 - 140) : 172 \qquad (11)$$

while the classical Redfield ratios are 1 : 16 : 106 : 138. The value for O_2 indicates how much oxygen is used for the oxidation of a given amount of organic matter; this includes the oxidation of phosphorus and nitrogen to phosphate and nitrate. The P : C ratio, as observed directly, was found to be 1 : 103, but this must be too large because of the influence of anthropogenic CO_2. The ratio 1 : 140 was obtained indirectly by assuming that the observed oxygen decrease was due to oxidation of organic carbon and NH_3; it would be larger (i.e. 1 : $<$ 140) if actually oxygen would be used also for oxidizing hydrogenated organic molecules. Thus the P : C ratio is not precisely known, but probably significantly lower than the classical value of 1 : 106.

Carbonate dissolution also adds total CO_2 to deep water; this contribution can be estimated from the concomitant alkalinity increase. On a global average, the ratio of carbon fluxes by organic and carbonate particles has been estimated to about 4 : 1 by Broecker et al. (13), while Takahashi et al. (74) found about 10 : 1 at the density horizons they considered. However, this ratio varies regionally; in areas of high productivity (equatorial upwelling zones, Antarctic ocean) diatoms, which produce opal, are predominant and the carbonate/organic particle ratio is correspondingly small.

2.6. The Cycle of Oxygen

The example of oxygen illustrates how cycles of different elements are coupled. Dry air contains 21 percent oxygen, and there are $1.2 \cdot 10^{21}$ g or $3.7 \cdot 10^{19}$ mol O_2 in the atmosphere. The photosynthetic reaction, which creates oxygen, can be written in a simplified way as

$$CO_2 + H_2O \longrightarrow CH_2O + O_2 \tag{12}$$

where CH_2O is a sum formula for organic material (neglecting minor elements like nitrogen and phosphorus). Thus one molecule of oxygen gas is created for every carbon dioxide molecule; because of this one-to-one relationship it is convenient to indicate masses in units of moles rather than grams.

Plant respiration (decomposition of organic matter) is just the reverse of photosynthesis:

$$CH_2O + O_2 \longrightarrow CO_2 + H_2O \tag{13}$$

Equations (12) and (13) show that the cycles of carbon and oxygen are directly coupled. Atmospheric oxygen must, therefore, exhibit variations of about the same absolute size but opposite sign as CO_2. They have not been directly observed so far because of their relative smallness - only a few ppm out of 210 000 ppm.

For calculating the mean atmospheric residence time of oxygen, gross, rather than net, primary productivity of terrestrial vegetation must be considered, since an oxygen molecule that has left a plant leaf is mixed into the bulk atmosphere within minutes. Assuming that global GPP on land is twice NPP yields a total terrestrial plus marine flux of about $175 \cdot 10^{15}$ g C yr^{-1} or $1.5 \cdot 10^{16}$ mol yr^{-1} (of C or O_2), which gives a mean atmospheric residence time for oxygen, with respect to cycling through the biosphere, of 2500 years. This is approximately the time-lag with which the isotopic composition of atmospheric oxygen follows global changes in $\delta^{18}O$ of the oceans: $\delta^{18}O$ of O_2 is linked via photosynthesis and respiration to the global water cycle and thus to the glacial-interglacial isotopic variations of sea-water (6).

The global rates of photosynthesis and plant respiration are nearly equal, so that the <u>next</u> oxygen flux is very small. According to equation (12), net production of oxygen occurs if organic matter is removed from the atmosphere-biosphere system, which happens during sedimentation. The corresponding sedimentation rate is about $6 \cdot 10^{12}$ mol/yr (68, Table 19). This yields a residence time of $6 \cdot 10^{6}$ yr for oxygen in the atmosphere-biosphere system. The total atmospheric mass of O_2 corresponds to only about 4 % of the estimated mass of sedimentary organic carbon. This indicates that most photosynthetic oxygen produced during geological time has been removed again, mostly as oxides of sulphur and iron (22).

According to Table 1, the total amount of recoverable fossil fuels plus biomass make up about $6.2 \cdot 10^{17}$ mol C. If <u>all</u> this carbon was burnt completely, this would use up about 2 percent of the atmospheric oxygen. For comparison, the barometric pressure - and with it the partial pressure of oxygen - decreases by 10 percent for an altitude increase of 1000 m. Thus, human activity does not create any problem of atmospheric oxygen. The impacts of energy use are different; problems on a local or regional scale due to production of pollutants like NO_x or SO_2, and the threat to global climate by the world-wide increase of CO_2 and infrared-active trace gases.

3. ANTHROPOGENIC INCREASE OF ATMOSPHERIC CO_2

3.1. Observations and Airborne Fraction

In 1958, C.D. Keeling started continuous measurements of the atmospheric carbon dioxide concentration by means of non-dispersive infrared spectroscopy at the observatory on Mauna Loa, Hawaii, and - using flask air samples - at the South Pole (28, 29). The results, and later observations at many other stations (35), have indicated a steady increase, presently at a rate of ca. 1.5 ppm/yr (Fig. 5). The increase is due to the production of CO_2 by the combustion of fossil fuels and, to a lesser degree, also to deforestation and land use (plowing leads to faster oxidation of the organic carbon stored in soils).

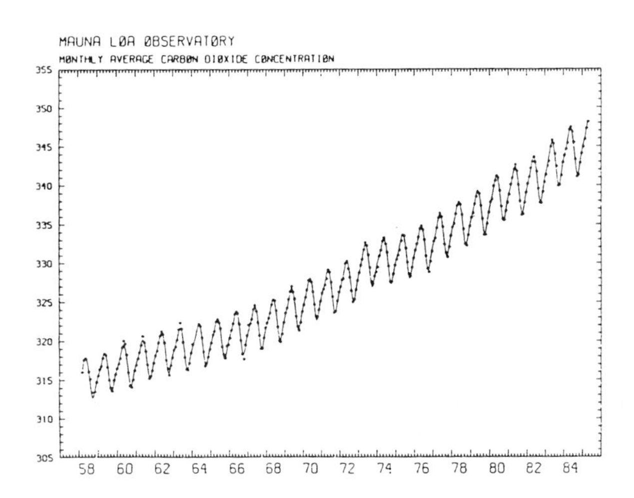

Figure 5. Atmospheric CO_2 concentration at Mauna Loa Observatory, Hawaii. Dots: monthly average values; smooth curve is a fit. Courtesy of C.D. Keeling.

Measurements of atmospheric CO_2 performed before 1958, generally by chemical techniques, were not reliable enough to accurately document the CO_2 increase. Different evaluations of 19th century measurements were undertaken (e.g. (15)). Based on a well-documented French series of data, an estimate for the 1880's is 285 to 290 ppm (61). However, there are uncertainties with all the old direct measurements, and it is questionable whether reliable absolute values can be reconstructed based on these data sets. The best method available at present for reconstructing the atmospheric concentration of CO_2 (and other trace gases) in the past is the analysis of air bubbles trapped in old polar ice (45), (52). For this purpose, only ice from cold regions (mean annual temperature below ca. -25°C), where no snow melting in summer occurs, is suited, because high CO_2 concentrations in melt layers, due to the high

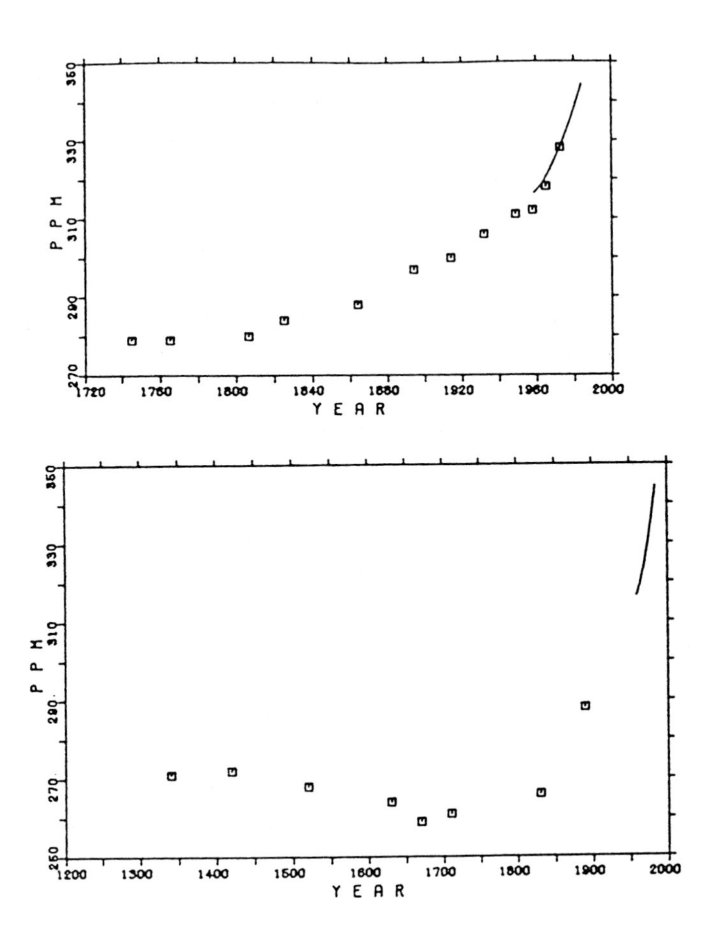

Figure 6. CO_2 concentration measured in air bubbles trapped in old polar ice from Antarctica (a) (top): Results from an ice core from Siple Station (45); (b) (bottom): Results from Station D57 (52). Solid lines: Mauna Loa results.

solubility of CO_2 in water, may lead to erroneous results. The results of the group at Bern (Fig. 6a) indicate a value of about 280 ppm before about 1800. The results of the Grenoble laboratory (Fig. 6b) are by about 10 to 20 ppm lower for the same time. The difference seems to be due, at least partly, to an experimental problem - selective gas-phase transport of CO_2 in the presence of water vapour; and based on a thorough recent intercomparison between the two laboratories the higher values of about 280 ppm appear to be nearer to the truth.

The age of the trapped air is smaller than that of the surrounding ice, because the pores in the firn are closed only at a depth of typically between 50 and 100 m. The age difference between ice and occluded air was for instance found to be 82 $\pm$ 10 yr for the Siple ice core (Fig. 6a); the range of $\pm$ 10 yr corresponds to the actual time interval of enclosure, different bubbles in the same ice layer having been closed at different depths.

Based on the ice core results, it can be concluded that the 18th century CO_2 concentration was near 280 ppm, possibly somewhat lower.

During the period 1860 to 1984, 183 $\pm$ $15 \cdot 10^{15}$ g C were released into the atmosphere due to fossil fuel combustion; the current rate of release is $5.2 \cdot 10^{15}$ g C. The amount of CO_2 produced by deforestation and land use in the same period is, according to the best current estimates, 150 $\pm$ $50 \cdot 10^{15}$ g C, and the corresponding current rate is 1.6 $\pm$ $0.8 \cdot 10^{15}$ g C (9). The atmospheric increase from 280 ppm to 344 ppm in 1984 corresponds to an airborne fraction of 41 percent of the total emission. The increase since the start of the Mauna Loa measurements corresponds to about 55 % of the <u>fossil</u> CO_2 released during that period; this value is often quoted because the actual net biospheric release is not well known.

3.2. Modelling the Oceanic Response to Carbon Cycle Perturbations

The major sink for excess CO_2 is the ocean. It contains about 63 times as much inorganic carbon as the pre-industrial atmosphere. Taking into account the buffer factor $\zeta \simeq 9$, about $63/\zeta = 7$ parts of all excess CO_2 will reside in the sea and 1 part in the air once the whole ocean has equilibrated with the atmosphere. Thus, after a long time the final airborne fraction will be roughly 12 percent. This value will, however, be reached only after about 10^3 years, when anthropogenic CO_2 has penetrated to the large volumes of the deep ocean. The time-dependent response to the anthropogenic input must be calculated using an atmosphere-ocean model, simulating the finite air-sea-gas exchange rate and the mixing of CO_2-laden surface water to depth. Since satisfactory ocean circulation models are not yet available for this purpose (see 55), simple perturbation models have been used in which the atmosphere is usually represented by one well-mixed box, the ocean by two or more boxes or by a diffusive reservoir.

The thermocline, extending from the bottom of the surface mixed layer to about 1 km depth, is the zone of the ocean most important in connection with the CO_2 increase. It has a characteristic mixing time-scale of a few decades. Few-box models have too coarse a resolution to simulate well the transport in the thermocline. A better approach is eddy diffusion. Rapid variations cause steeper concentration gradients and therefore larger diffusive fluxes, which cannot be simulated by first-order exchange between few boxes. This was the basis for the box-diffusion model of Oeschger et al. (47), in which the ocean consists of a mixed layer, assumed 75 m deep, and a deep sea in which transport occurs by vertical diffusion with constant eddy diffusion coefficient. The box-diffusion model has two adjustable transport parameters that have to be determined by means of a suitable tracer, usually natural or bomb-produced ^{14}C. These are (1) an air-sea gas exchange coefficient (see above), (2) a coefficient describing interior mixing, i.e. the vertical diffusion coefficient K of the deep sea.

The box-diffusion model is a perturbation model, not designed for simulating the whole steady state carbon cycle, but only the transport of excess CO_2 and of isotopic perturbations. Features supposed not to change with the anthropogenic CO_2 invasion are not considered. This can be illustrated by the carbon balance for some box i containing a carbon mass N_i at steady state and $(N_i + n_i(t))$ in a perturbed state; the flux from box i to box j is represented by $k_{ij} N_i$, and there is a carbon sources S_i into box i, e.g. due to particle dissolution:

Steady state: $$\frac{dN_i}{dt} = 0 = \sum k_{ji} N_j - \sum k_{ij} N_i + S_i$$

Perturbated state:

$$\frac{d(N_i+n_i)}{dt} = \sum k_{ji}(N_j+n_j) - \sum k_{ij}(N_i+n_i) + (S+s_i)$$

Perturbation only:

$$\frac{dn_i}{dt} = \sum k_{ji}\, n_j - \sum k_{ij}\, n_i + s_i$$

Thus, if the source term is constant with time ($s_i = 0$), it does not appear in the perturbation equation (= difference equation). This is specifically the case for the oceanic particle fluxes, since the biological productivity is governed by limiting nutrients like phosphate or by light, but not by the abundant carbon, and therefore does not respond to the CO_2 increase.

The concept of eddy diffusion is based on the assumption that the flux of matter can be described by $F = - K\, \partial C/\partial z$ (C = concentration). It is a simple parameterized way to describe the continuous character of mixing. The eddy diffusion coefficient K can be estimated based on the distribution of natural ^{14}C or of bomb-produced ^{14}C or tritium. Calibration with natural ^{14}C, by demanding that the surface value and the mean oceanic ^{14}C concentration be correctly simulated, yields $K = 1.3\ cm^2 s^{-1}$ (Oeschger et al. (47)). However, calibration with bomb-produced isotopes is preferred, since they have time-scale of change comparable to that of anthropogenic CO_2. A useful concept in connection with bomb-produced isotopes is the mean penetration depth, defined as the ratio of water-column inventory to surface concentration. Broecker et al. (14) estimated a global mean penetration depth of 375 m for tritium in about 1972, corresponding to $K = 1.7\ cm^2\ s^{-1}$. The same K value was obtained by Li et al. (37) from an evaluation of all GEOSECS tritium data. A calibration with bomb-^{14}C yielded $2.4\ cm^2\ s^{-1}$ or $7700\ m^2\ yr^{-1}$ (Siegenthaler (57)).

The step that mainly limits the oceanic uptake of anthropogenic CO_2 is vertical mixing. Thus, according to the box-diffusion model results, the present-day excess CO_2 concentration in surface water is about 80 % of its equilibrium value, i.e. gas exchange is rapid enough for chemical equilibrium to be nearly established. The fossil CO_2 production has increased in an approximately exponential way, with an e-folding time $\tau \approx 25$ yr, and the mean penetration depth of fossil CO_2 into the deep sea is $\sqrt{K\tau} = 438$ m (for $K = 2.4\ cm^2\ s^{-1}$), to which the depth of the mixed layer of 75 m has to be added. The pre-industrial atmosphere contained as much CO_2 as an ocean layer 67 m thick. Taking into account a degree of equilibration of surface water of 0.80 and a buffer factor of 9, the airborne fraction turns out to be

$$\frac{67\ \text{m}}{0.80\ 513\ \text{m}/9 + 67\ \text{m}}$$

The exact numerical calculation using the actual CO_2 production history yields a value of 0.61.

The ratio of observed CO_2 increase to fossil fuel input is about 0.55, and a lower airborne fraction is obtained if biospheric CO_2 emission is also considered. Thus, the box-diffusion model yields a higher atmospheric increase rate than actually observed. One possible reason for this discrepancy is that the horizontally averaged models do not appropriately include the rapid vertical exchange occuring at high latitude which may represent an additional channel for oceanic CO_2 uptake. Siegenthaler (57) extended the box-diffusion model to include a

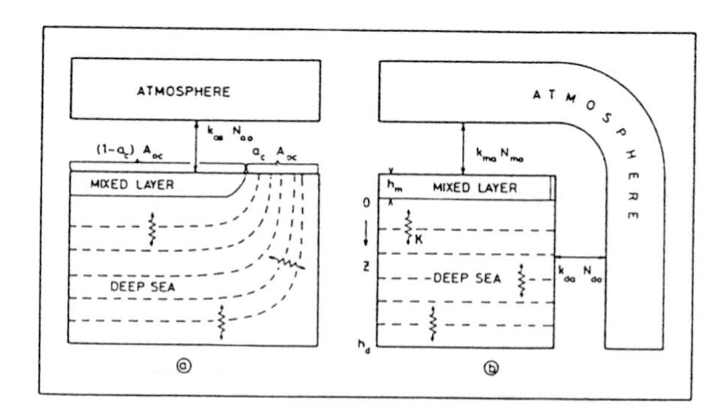

Figure 7. Schematic representation of outcrop-diffusion model of the ocean-atmosphere system (Siegenthaler (57)). Left: physical idea; right: mathematical representation.

high-latitude outcrop, so that deep water is partly directly ventilated from the atmosphere (outcrop-diffusion model, Fig. 7). With an outcrop covering 10 % of the ocean and a calibration using bomb-produced ^{14}C, the outcrop-diffusion model yields an airborne fraction of 0.52. In the outcrop region, the degree of equilibration with the atmospheric CO_2 excess is much lower than in warm surface water, i.e. the pCO_2 difference air-sea that drives net CO_2 invasion is larger. Therefore the outcrop is an efficient sink for excess CO_2.

For comparing different model versions, it is essential that they are always calibrated in a consistent way. Bomb-produced ^{14}C is often used for calibration, because its invasion into the ocean is partly analogous to that of excess CO_2. The analogue is, however, not perfect because the two perturbations behave differently with respect of air-sea exchange, since the buffer factor affects the uptake of CO_2, but not of ^{14}C (12). The equilibration time between a 100 m thick oceanic mixed layer for an isotopic perturbation is about 10 yr. For a CO_2 perturbation the equilibration time is only about 1 yr, because not all CO_2 must be exchanged between the two reservoirs as for ^{13}C or ^{14}C, but establishment of the new chemical equilibrium requires the transfer of only a fraction of the CO_2. The difference is illustrated by Figure 8, showing the atmospheric responses to pulse inputs of CO_2 and of ^{13}C (^{14}C would behave very similarly) according to the box-diffusion model. Note also that the decrease is first rapid and then slower and slower, i.e. it is not governed by one time constant, since after entering the upper ocean layers the perturbation penetrates more and more slowly into the thermocline and the deep sea.

3.3. CO_2 Release from the Terrestrial Biosphere and the "Missing CO_2 Sink"

The outcrop-diffusion model with an outcrop area of 5 to 10 percent can reproduce the atmospheric CO_2 increase assuming that fossil fuel burning is the only cause. However, ecologists' estimates of present CO_2 emission due to deforestation and land use are in the range $1.6 \pm 0.8 \cdot 10^{15}$ g C yr^{-1} (Bolin (9)) which is not negligible compared to fossil fuel consumption ($5 \cdot 10^{15}$ g C yr^{-1}). Thus, in the budget of excess CO_2,

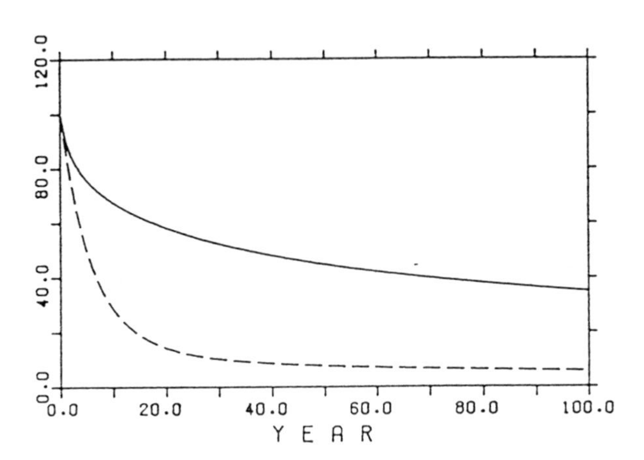

Figure 8. Model responses of the box-diffusion model of Oeschger et al. (47) calibrated using bomb-^{14}C, for a pulse input of isotopically labelled CO_2 into the atmosphere. Solid line: CO_2, dashed: $\delta^{13}C$. Initial values assumed as +100 ppm and +100 permil.

the total emissions seem not to be matched by the increase in atmosphere and ocean. Part of this problem may be non-existent, because the values of the airborne fraction mentioned above refer to simulation runs with fossil, i.e. quasi-exponentially increasing input. The results can be different if the input history deviated significantly from an exponential increase. If a large input of biospheric CO_2 occurred in the past, it is still influencing the present CO_2 concentration by providing a decreasing atmospheric baseline, since excess CO_2 is still being taken up by the ocean. This can be seen when considering the model response to a pulse input of atmospheric CO_2 (Figure 8). In this case, the ratio of atmospheric increase to present input rate becomes smaller.

Instead of determining the atmospheric concentration for a prescribed production history, we can calculate, by means of a carbon cycle model, the production rate from a prescribed concentration history. This has been done by Siegenthaler et al. (62) for the CO_2 concentrations

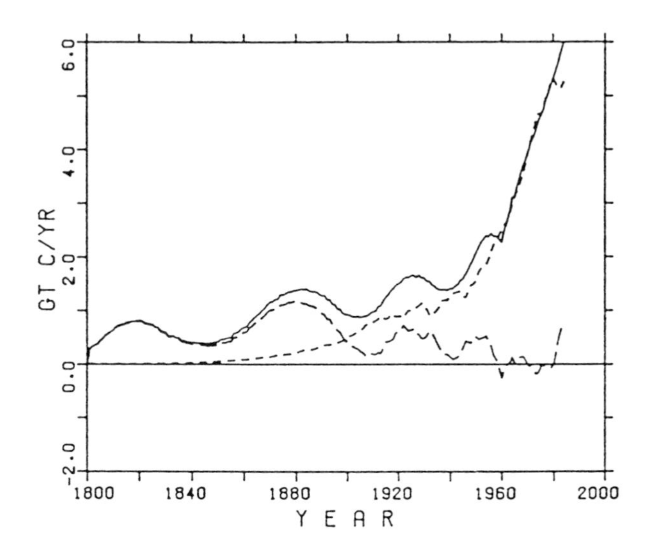

Figure 9. Total CO_2 production rate (solid line) calculated by deconvolving the CO_2 concentration history reconstructed from the Siple ice core results of Figure 6a. Short-dashed: fossil input rate (54); long-dashed: difference = non-fossil input rate.

as measured in the Siple ice core (cf. Figure 6a). Figure 9 shows the results. The calculated total production rate (solid line) minus the known fossil input (short-dashed) yields the non-fossil, i.e. probably biospheric CO_2 release rate (long-dashed). According to the box-diffusion model simulation, a major non-fossil input occurred in the last century, $66 \cdot 10^{15}$ g C until 1900 compared to $12 \cdot 10^{15}$ g C from fossil fuels; for the period 1900 to 1980, the total non-fossil input was $20 \cdot 10^{15}$ g C (fossil: $148 \cdot 10^{15}$ g C). Carrying out the deconvolution with the outcrop-diffusion model (a_c = 0.1) yields a higher non-fossil input ($156 \cdot 10^{15}$ g C til 1980) because of the higher flux into the ocean. The calculated cumulative non-fossil input by 1980 is compatible with the ecological estimate of

$150 \pm 50 \cdot 10^{15}$ g C. According to the box-diffusion and outcrop-diffusion model deconvolution, the net input from the biosphere during the past 20 - 30 years would have been 0 to $1 \cdot 10^{15}$ g C yr, compared to the direct estimate of $1.6 \pm 0.8 \cdot 10^{15}$ g C yr^{-1}. Thus, the "missing sink" is rather small and of the order of 0 to $2 \cdot 10^{15}$ g C yr^{-1}.

An additional effect that is possibly significant as a sink of excess CO_2 is the fact that plant productivity is stimulated by increased CO_2 concentration in the air. While this fertilizing effect has been clearly found in laboratory studies, it is difficult to assess its importance in nature where plant productivity is limited by other factors like space, light or nutrients. The seasonal CO_2 amplitude at Mauna Loa (and at other stations) has significantly increased from 1958 to 1982, by about 0.7 % per year (4). This probably indicates a strenghtened assimilation - respiration activity of the northern hemisphere land biota, which might be accompanied by a higher biospheric storage of carbon. Kohlmaier et al. (34) suggest that the additional annual storage in living biota and soil may amount to about 1 to 2 10^{15} g C yr^{-1}. With a sink of this size, the excess CO_2 budget would be perfectly balanced.

Other potential CO_2 sinks have been discussed, including enhanced carbonate sediment dissolution, enhanced production and sedimentation of particulate organic carbon (POC) due to fertilizer input, or carbon transport by rivers. The invasion of excess CO_2 leads to a decrease of the concentration of $CO_3^=$ ions in sea-water which consequently becomes more aggressive for $CaCO_3$ dissolution. Enhanced $CaCO_3$ dissolution leads to an increase of alkalinity and therefore to a higher CO_2 uptake capacity (lower buffer factor; see paragraph 2.4). Waters above 2 - 4 km depth are supersaturated with respect to calcite (the predominant form of $CaCO_3$ in sediments). Enhanced calcite dissolution will only be effective for CO_2 uptake when water containing excess CO_2 has been transported to depth, taken up additional alkalinity (i.e. $CO_3^=$ ions) and again returned to the surface. Thus, this process can only become active after a period corresponding to several ocean turnover times, i.e. about 10^3 years (Broecker et al. (13)).

Recent sediment trap observations in the North Pacific have indicated that the carbonate flux by aragonitic pteropods is considerably larger than previously assumed (Betzer et al. (7)). Aragonite is more soluble than calcite, and the particles dissolve rapidly at relatively shallow depth. Acidification of the water by anthropogenic CO_2 may lead to dissolution at even shallower depths and thus to an increase of alkalinity. It is not easy to assess the possible importance of such an effect without a model. Intuitively, one would judge it as not very significant, since no additional alkalinity is added to the ocean (as is the case for sediment dissolution), but the alkalinity is just injected somewhat nearer to the surface.

Due to man's activity, the amounts of phosphate and nitrate brought into the ocean by rivers has increased, which may act as fertilizers and lead to enhanced biological production in the sea. The additional phosphate input has been estimated as $3.2 \cdot 10^{10}$ mol yr^{-1}, the nitrate input as roughly $5 \cdot 10^{11}$ mol yr^{-1} (41). With the Redfield ratios P:N:C = 1:16:140, the phosphate and nitrate inputs would correspond to a maximum additional production of $5.4 \cdot 10^{13}$ g C or $1.0 \cdot 10^{14}$ g C. Only organic matter that escapes oxidation and is buried in the sediment corresponds to a carbon sink; this is the case for about 10 % of the total particulate flux (Emerson (20)). Thus, this mechanism provides for a sink of something like $0.5 - 1 \cdot 10^{13}$ g C yr^{-1}, which is negligible compared to the fossil CO_2 production rate of about $5 \cdot 10^{15}$ g C yr^{-1}. The hypothesis of Walsh et al. (75) that burial of organic carbon on the continental shelf is an effective carbon sink seems to be due to a confusion between anthropogenic fluxes and natural fluxes that have nothing to do with man's activity.

Kempe (32) has suggested that increased transport of POC to the ocean by rivers and increased deposition in lakes could provide a sink of up to $1 \cdot 10^{15}$ g C yr^{-1}. It is, however, uncertain what fraction of these fluxes actually corresponds to a sequestration of carbon, and how much is subject to recycling by oxidation. In addition, Kempe's estimate of up to $1 \cdot 10^{15}$ g C yr^{-1} appears to be biased towards high values.

3.4. Scenarios for Future CO_2 Concentrations

Future atmospheric CO_2 concentrations primarily depend on release rates by fossil energy consumption, but also on the uptake by the oceans. Figure 10 illustrates the influence of direct ventilation of the deep sea through an outcrop, covering 10 percent of the world ocean, for a high CO_2 production scenario (Siegenthaler (57)). The assumed final cumulative production corresponds to about 8 times the pre-industrial atmospheric

CO_2 mass and is probably unrealistically high; it is shown as a sensitivity study. The example demonstrates how slow the decrease will be even after production has stopped completely. There is a large difference between the cases with and without outcrop; this clearly shows that it is necessary to develop reliable ocean models for long-term estimates of future CO_2 concentrations that take into account in a realistic way the formation of thermocline and deep waters.

For the next several decades, however, existing models are sufficient for reasonably accurate predictions; uncertainties mainly stem from the scenarios of future CO_2 emissions. These have changed considerably in the past few years (see 43). According to the study of the National Research Council (43), the CO_2 concentration is expected to have doubled (reached 600 ppm) with 50 percent probability between 2050 and 2100. Bolin (9) concludes that a doubling will probably not occur before 2050, possibly even after 2100. The corresponding warming will be about 1° to 2°C due to CO_2 alone; anthropogenic trace gases will cause an additional warming of approximately the same size (see below).

3.5. Carbon Isotope Perturbations

Fossil fuel does not contain any ^{14}C, since it has been isolated from the atmosphere long enough that all ^{14}C has decayed. Thus, the emission of anthropogenic CO_2 caused a decrease of the $^{14}C/C$ and of the $^{13}C/^{12}C$ ratio in atmospheric CO_2 ("Suess effect").

By 1950 the cumulative fossil production corresponded to 10 % of the atmospheric CO_2 mass. If all this CO_2 had remained airborne, ^{14}C of atmospheric CO_2 would have decreased from 0 to -91 ‰. As a result of exchange with ocean and biosphere, the Suess effect was, however, considerably less. The decrease found by tree-ring studies is 20 ‰, and 17 ‰ if corrected for presumed natural fluctuations (Stuyver et al. (65)) (Figure 11). Calculations using the box-diffusion and the outcrop diffusion model yield a change of 18 to 20 ‰, in excellent agreement with observation.

The ^{14}C Suess effect reflects the CO_2 production from fossil fuel consumption, but not that from biomass changes. Both emissions, however cause a $\delta^{13}C$ decrease in atmospheric CO_2; air-CO_2 has $\delta^{13}C \approx -7$ ‰, fossil and biospheric carbon values near -25 ‰. A model simulation based on

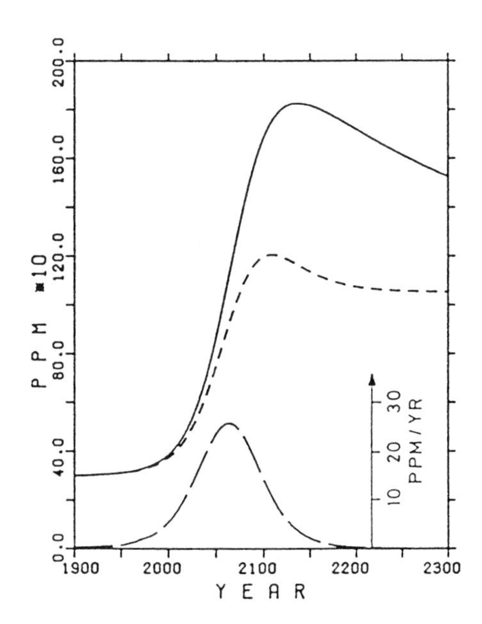

Figure 10. Model-calculated atmospheric CO_2 level (upper curves; scale from 0 to 2000 ppm) for an upper-limit production scenario (lower, long-dashed curve) according to which 4924 10^{15} g C ($\sim$ 8 times pre-industrial atmosphere) are finally burnt. Simulations by means of box-diffusion model (solid line) and outcrop-diffusion model with 10 percent deep-sea outcrop (dashed line), both bomb-^{14}C calibrated. Note the large influence of deep-sea ventilation by outcrop. From Siegenthaler (57).

assuming only fossil carbon release predicts a decrease of 1.1 ‰ until 1980. Additional biospheric CO_2 inputs must have led to a stronger $\delta^{13}C$ decrease. Several attempts have been made to reconstruct the atmospheric $\delta^{13}C$ history by means of tree-ring analyses and then to estimate past biospheric CO_2 releases (e.g. Peng et al. (50)). However, the $\delta^{13}C$ records of tree-rings exhibit much scatter; obviously, they do not reflect only atmospheric isotope variations, but are influenced by other environmental and plant-physiological factors. Stuiver et al. (67) attempted to correct the $\delta^{13}C$ values for such influences by using a correlation with tree-ring width; they obtained an average decrease of ca. 1.0 ‰ until 1980.

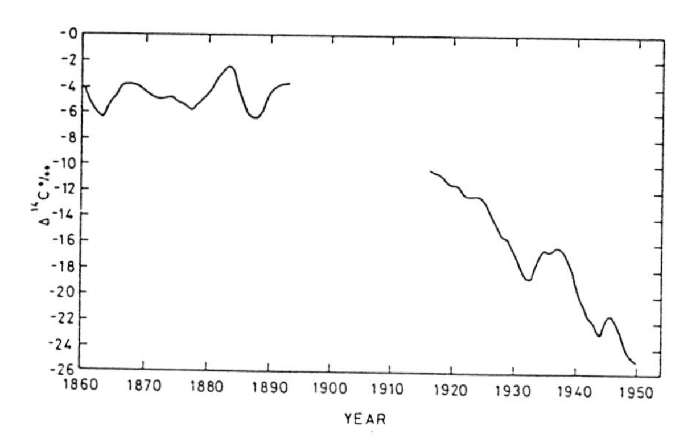

Figure 11. Decrease of ^{14}C in the atmosphere due to fossil fuel burning (Suess effect), reconstructed by high-precision measurements on tree-rings of known age. After Stuiver et al. (65).

A reliable $\delta^{13}C$ reconstruction is possible again by analyzing air trapped in old polar ice (21). First measurements performed at the University of Bern indicate a decrease of slightly more than 1 ‰ from 1800 to 1980, which is consistent with the CO_2 increase documented in Fig. 6a.

Tree-ring measurements have shown that the natural ^{14}C concentration was not quite constant in time (e.g. Damon et al. (18)). It was about 8 % higher 6000 to 8000 yr B.P. (before present) than in the 19th century. As a possible reason variations of the earth's magnetic field intensity have been discussed: radioisotope production in the atmosphere is due to cosmic rays, and the geomagnetic field partly shields the earth from cosmic radiation. Palaeomagnetic data seem to support this hypothesis, but the representativeness of these data for the actual geomagnetic field intensities has been challenged.

Short-period variations, "wiggles", of ca. 10 ‰ amplitude and ca. 200 yr quasi-period are a recurring feature of the tree-ring record. They are probably due to variations of the sun's magnetic field, as indicated by a good correlation with sunspot number (Stuyver et al. (64)). Model calculations show that these variations are attenuated, compared to the production changes, by a factor of about 20, because the perturbations are diluted by the carbon in the exchanging reservoirs.

The concentration of the cosmic-ray produced radioisotope ^{10}Be in ice cores also exhibits short-term fluctuations (Figure 12, top) that most probably are due to production rate changes. Taking these ^{10}Be concentrations as representative for radioisotope production rates and using the box-diffusion model of CO_2 exchange yields simulated ^{14}C variations well comparable to the observed ones (Figure 12, bottom) (5). The two isotopes have rather different geochemical behaviour - ^{10}Be is attached to aerosols and washed out from the atmosphere within weeks to a few years -, therefore the consistency of the two records is a strong indication that the fluctuations of ^{14}C and of ^{10}Be are indeed due to a common cause, i.e. modulation of the cosmic ray flux by solar activity. At the same time, the fact that the carbon cycle model seems to simulate well the attenuation of ^{14}C production variation provides a positive test of its validity.

4. CLIMATIC EFFECTS OF CO_2 INCREASE

There is a number of reviews of the carbon dioxide/climate issue, e.g. Clark (16), Hansen et al. (23) or Liss et al. (39).

Figure 13 shows the energy balance of the earth. The main factor determining global temperature is the solar radiation. Without re-emission of energy, the temperature of the surface would increase continuously, but the earth emits thermal radiation at a rate proportional to the fourth power of its absolute temperature.

Thus, the earth is heated up to a temperature for which, on a time-space average, the absorbed solar radiation is compensated by emitted infrared radiation. Based on this equilibrium and taking into

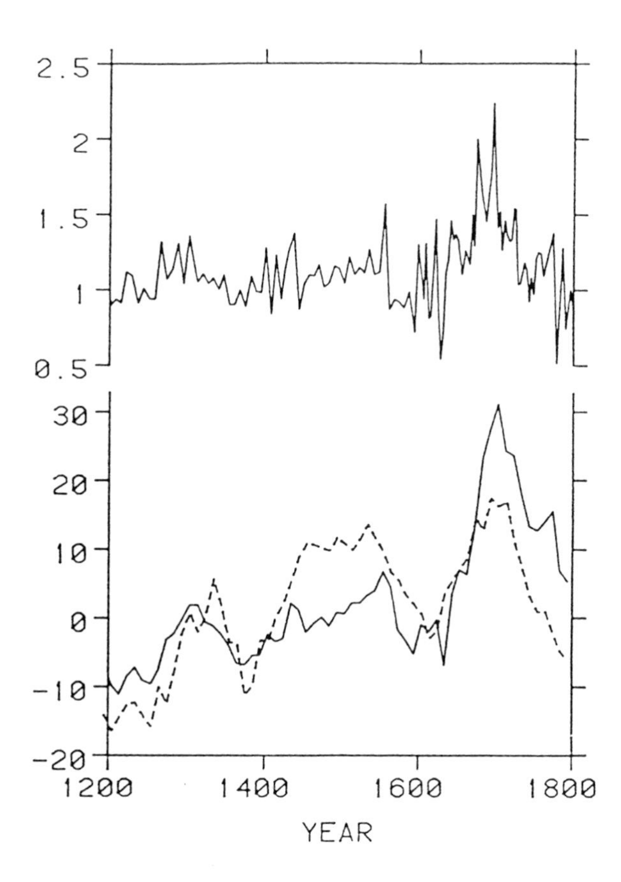

Figure 12. Top: ^{10}Be concentration (unit: 10^4 atoms/g ice) versus age in ice from Milcent, Greenland. Bottom: $\Delta^{14}C$ in the atmosphere, tree-ring measurements (dashed curve) (64) and and model-calculated (solid curve) based on the assumption that ^{10}Be variations are directly proportional to production rate. From Beer et al. (5).

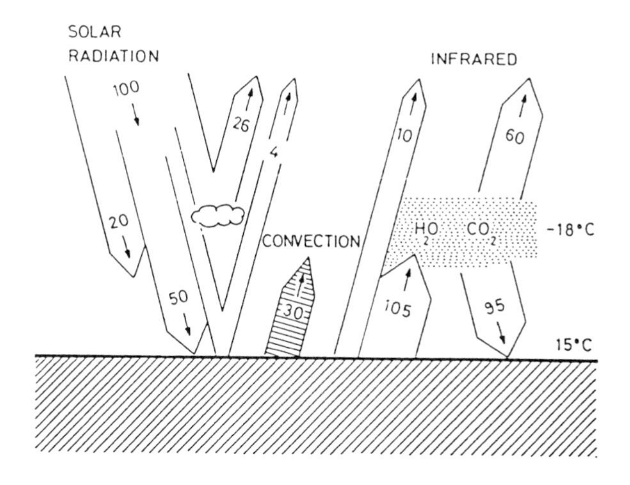

Figure 13. Energy balance of the earth. Incoming solar radiation intensity is set to 100 units. Left side: short-wavelength radiation; right side: infrared radiation. Infrared back-radiation from the atmosphere to the surface is of nearly equal size (95 units) as solar radiation at top of atmosphere. "Convection" also includes latent heat exchange.

account that about 30 % of the incident sunlight is not absorbed but reflected by clouds and the surface, an effective temperature (planetary temperature) of the earth can be calculated. This mean planetary temperature is 225 K or -18°C.

The actual mean surface temperature is, however, about 15°C and thus considerably higher. The difference is due to the fact that most of the infrared radiation to space is not emitted by the surface but by higher (and colder) atmospheric layers, because - in contrast to the visible sunlight which penetrates the cloud-free atmosphere without much loss - infrared radiation emitted by the surface is absorbed, mainly by water vapour, but also by carbon dioxide, ozone and other trace gases (right-hand side of Figure 13). The absorbed energy heats the lower atmosphere which re-emits infrared radiation partly back to the surface. In this way the surface receives, in addition to the solar energy, a considerable amount of thermal radiation and is heated up to an average temperature higher than the effective planetary temperature. This phenomenon, that the atmosphere is transparent to visible light but absorbs in the infrared an thus acts as a spectrally selective thermal isolation, is called greenhouse effect.

The infrared absorption by water vapour leaves a spectral window at wavelengths between 9 and 15 μm. In this range, absorption occurs by CO_2 and other trace gases but not by water vapour, and it is incomplete so that some radiation from the surface penetrates the atmosphere unaffected. An atmospheric CO_2 increase reduces the transmissivity in this spectral window and enhances the greenhouse effect. But also in other spectral regions, the greenhouse effect is amplified by a higher CO_2 concentration, because the radiation emitted by the surface is absorbed more strongly and the back-radiation originates from lower, and therefore warmer, atmospheric layers. In this way, an increase of CO_2 (and other infrared-active trace gases) leads to higher surface temperatures.

Climate models indicate that for a doubled CO_2 concentration, on a global average the downward infrared flux increases by 4 Wm^{-2} (Augustsson et al. (1)). For the whole earth this corresponds to an additional energy flux at the surface of $2000 \cdot 10^{12}$ W. This indirect heating due to the greenhouse effect can be compared with the direct anthropogenic energy input of $8 \cdot 10^{12}$ W at present, showing that on a global scale the direct warming by energy use is much weaker than the CO_2-induced warming.

To estimate temperature changes caused by increased atmospheric CO_2 concentrations, one has to consider, beside the changed radiation balance due to CO_2 alone, a number of feedback mechanisms. A strong positive feedback is the increase in absolute humidity going along with the warming of the earth's surface and the lower atmospheric layers. Another positive feedback effect results from changes in the earth's reflectivity for solar radiation (albedo) owing to a decrease of the snow and ice cover in polar areas caused by the warming. The combined temperature effect, including these feedback processes, is two to three times that caused by CO_2 alone.

The influence of clouds is difficult to model. Increasing cloud cover on the one hand leads to a higher albedo but on the other hand also to an increase of downward thermal radiation. The two effect partly compensate each other. At present it is not known if a higher surface temperature will lead to a change of cloud cover.

Calculations by means of a variety of climate models of different complexity indicate for a CO_2 doubling an average global temperature rise by 1.5° to 4°C, assuming that the energy fluxes at the earth's surface balance each other, i.e. that there is thermal equilibrium on a global scale.

While simple climate models provide estimates of CO_2-induced changes and permit to study the influence of specific processes and assumptions, reliable forecasts can finally only be expected from general circulation models (e.g. Manabe et al. (40)). These models predict that in high latitudes the warming is significantly larger than on average. This amplification in polar regions is partly due to the snow-albedo feedback mentioned above. At high latitudes there is also a seasonal asymmetry, with large temperature changes mainly in winter.

Temperature is not the only climate parameter of interest; changes in precipitation and evaporation are of equal significance and for some regions even more important. The results of Manabe et al. (40) indicate more precipitation in high latitudes. There are, however, still large problems with the simulation of the atmospheric water cycle which lead to uncertainties in the prediction.

The climate models studies discussed so far assume that the earth has adapted to the new radiative conditions and on average has reached thermal equilibrium. In reality, however, the world ocean has a considerable heat capacity, and while the temperature is rising, there is a relatively large flux of heat into the ocean needed to warm it up, so that the radiation budget is not balanced. Only recently has this aspect, which causes a delay in the global temperature increase, been included in the model discussion in a quantitative way (Hansen et al. (23); Hoffert et al. (25)). Since heat is transported in the ocean by the same processes as CO_2 - currents and turbulent water motions -, the box-diffusion ocean model can be also used for simulating the penetration of the temperature increase into the ocean (Siegenthaler et al. (60)). Figure 14 shows a model simulation of the CO_2-induced mean atmospheric warming for the CO_2 increase of Figure 6a, reconstructed from the Siple ice core (solid line in Figure 14) and for a model-predicted CO_2 increase due to fossil fuel emissions only (pre-industrial concentration 297 ppm; dashed line). In addition, observed mean northern hemisphere temperature anomalies are shown. The model simulations are for constant cloud-top temperature and a ratio of land area to ocean of 1:1; see Siegenthaler et al. (60) for technical details. For a reasonable range of climate model parameters, the calculated warming until 1980 is 0.39° to 0.54°C for the Siple CO_2 history, and 0.23° to 0.31°C for the

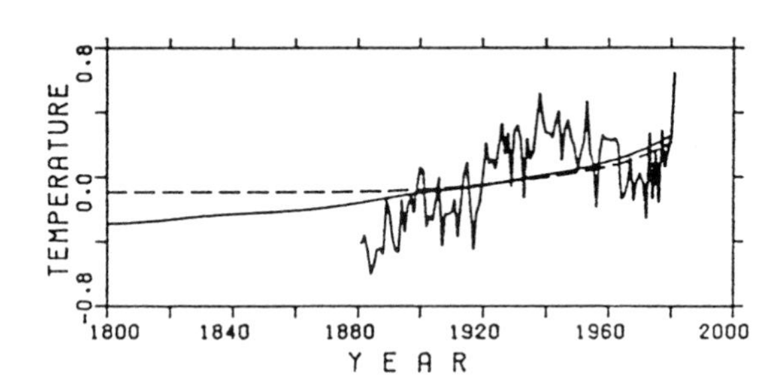

Figure 14. CO_2-induced global warming, calculated using the box-diffusion model of vertical oceanic heat transport of Siegenthaler et al. (60). Dashed line: for CO_2 increase due to fossil fuels only; smooth solid line: for CO_2 history according to the Siple ice core results of Figure 6a. Irregular curve: Observed mean Northern Hemisphere temperature anomalies (Jones et al. (27)).

case with fossil input only. These results were calculated taking the ocean's thermal inertia into account; they are about 30 to 50 percent lower than for radiative equilibrium. The observations do indicate a long-term temperature increase, but it is superimposed by relatively large noise, and it is not possible to attribute it with certainty to CO_2.

In addition to CO_2, also other anthropogenic trace gases (methane, CFMs, tropospheric ozone etc.) contribute to the greenhouse effect.

Ramanathan et al. (51) have estimated that for the period 1980 to 2030, the temperature change due to these other trace gases may be about equal to that due to CO_2. During the last 10^5 years, the mean global temperature was at most 1 to 2°C higher than at present. The probability is large that in 100 years this natural range will be exceeded due to anthropogenic emissions of CO_2 and trace gases. The problem is especially serious when considering that this heating cannot simply be turned off, since even after a complete production stop, atmospheric concentrations of CO_2 and trace gases decrease only very slowly. Thus, these anthropogenic climate changes appear as partly irreversible during human time-scales.

5. NATURAL CO_2 VARIATIONS

5.1. Seasonal Variations

Besides the continuous increase, atmospheric CO_2 records at many stations exhibit regular seasonal variations (Figures 5, 15). They are especially prominent in the northern hemisphere and reflect the biospheric cycle of growth and decay. In the growing season, the vegetation withdraws CO_2 from the atmosphere for the photosynthetic production of organic matter, and the atmospheric CO_2 is produced by soil respiration and the decay of plant material . On an annual basis, there is a nearly perfect balance between photsynthesis and decay, and the vegetation is neither a source nor a sink for atmospheric CO_2; this was at least the case before the biosphere was altered by human activities.

The biospheric origin of the seasonal variations is confirmed by a very good correlation of CO_2 with ^{13}C, which is due to the fact that the added or withdrawn biospheric CO_2 has a ^{13}C value near -25 ‰, i.e. considerably lower than that of air-Co_2 (Mook et al. (42)).

In equatorial regions and in the southern hemisphere, the seasonal amplitude becomes rather small. In the southern hemisphere, the continental area and correspondingly the vegetation biomass is relatively small. The seasonal amplitudes due to southern hemisphere vegetation, to interhemispheric air exchange and to air-sea exchange are comparable there and partly cancel each other. By means of two- and three-dimensional circulation models, the seasonal CO_2 variations can be used as tracers of atmospheric mixing (Pearman et al. (48); Keeling et al. (30); Keimann et al. (24)).

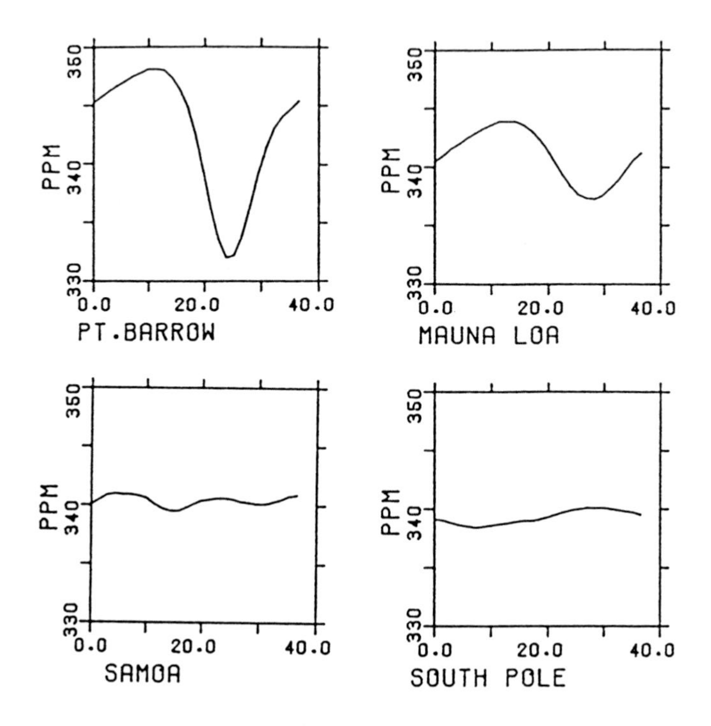

Figure 15. Seasonal CO_2 variations, 1982, for Pt. Barrow, Alaska (71.3°N), Mauna Loa, Hawaii (19.5°N), American Samoa (14.2°S), and South Pole (90°S). Data from Komhyr (35). Horizontal axis: day of the year (unit: 10 days).

5.2. Correlation with El Niño

The atmospheric CO_2 increase (Figure 5) is not quite regular from year to year. A correlation exists between growth rate and Southern Oscillation Index (SOI). As SOI, the difference of barometric pressure across the Pacific Ocean, between Darwin, Australia, and Easter Island, is taken. El Niño years coincide with periods of unusually low SOI. Bacastow (2) and Bacastow et al. (3) found that minima of SOI often coincide with a time of particularly high atmospheric CO_2 growth rate (Figure 16).

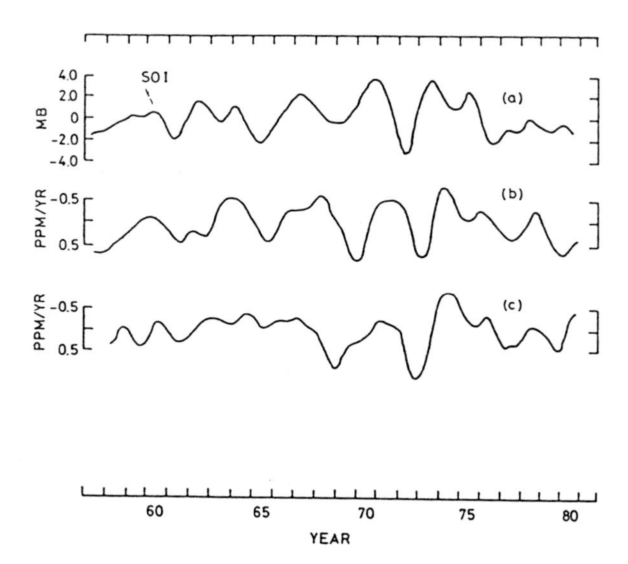

Figure 16. (a) Southern Oscillation Index, (b) and (c) time derivatives of seasonally adjusted and long-term detrended CO_2 concentrations at South Pole (b) and Mauna Loa (c). After Bacastow et al. (3).

Komhyr et al. (35) found, based on data from the NOAA network, that during the abnormally strong El Niño/Southern Oscillation event 1982/83 and also during the 1972 event, the CO_2 growth rate first decreased before increasing to higher than average values.

The phenomenon is not yet fully understood. There are various effects, partly acting in different directions. During El Nino years, equatorial upwelling of CO_2-rich water is weakened in the Pacific Ocean in response to changed atmospheric forcing. Presumably then the equatorial pCO_2 maximum (Figure 2) decays, and correspondingly the CO_2 source to the atmosphere is temporarily weakened (Keeling et al. (31)). A simple box-model, that also includes the warming of the equatorial surface water during El Niño, predicts a rapid atmospheric concentration drop, followed by a slow increase (Wenk et al. (78)), which, however, does not agree well with observation. On the other hand, $\delta^{13}C$ measurements seem to point to changes in the land biosphere affecting also the atmosphere (31).

The El Niño-related CO_2 oscillations are small, but of interest for the understanding of the factors that govern the atmospheric CO_2 concentration. In order to really understand them, it will be necessary to obtain detailed chemical and nutrient data from the Equatorial Pacific during an El Niño event.

5.3. Glacial/Interglacial Changes

Analyses of air bubbles trapped in old polar ice have shown that during the last ice age, the atmospheric CO_2 level was probably only about 200 ppm and that it increased at the transition to the Holocene to approximately its recent pre-industrial value (Neftel et al.(44); Delmas et al.(19)).

The atmospheric CO_2 concentration is strongly influenced by the chemical composition of ocean surface water, that in its turn depends on marine biological activity, as described in paragraph 2.5. Broecker (13) suggested that during glacial times, the amount of phosphate in the ocean was considerably larger than now, which would have led to a stronger depletion of total CO_2, and thus of pCO_2, in surface water. He suggested that the origin of this additional phosphate was erosion of nutrient-rich organic sediments on the continental shelves, deposited there during an earlier interglacial period of high sea level. This hypothesis can explain many of the observations, but the CO_2 level would change rather slowly which seems to be in contrast to the ice core results.

Another hypothesis relates the CO_2 variations to changes in the system of physical, chemical and biological processes in the ocean. In large oceanic areas, the biological productivity is limited by the availability of phosphate and nitrate. In these regions, a stronger vertical circulation would in a first approximation not change the surface concentrations of total CO_2 and of alkalinity. Increased upwelling of nutrients would lead to an enhanced flux of biogenic particles, but because of the fixed Redfield ratios between P, N and C, the additional particle flux of carbon would just be balanced by the increased upwelling rate of total CO_2. This is, however, not the case in the Southern Ocean around Antarctica where phosphate and nitrate are abundant in surface waters; biological activity is obviously governed by other factors there, like lack of light or rapid vertical mixing. This observation led to the hypothesis that CO_2 variations were caused by changes in ocean circulation that affected the distribution of chemical properties, including pCO_2, in the ocean surface (Siegenthaler et al. (61); Sarmiento et al. (56); Knose et al. (33)). A decrease in vertical mixing in the Southern Ocean, with constant biological productivity, would lead to lower surface concentrations of nutrients and of total CO_2 and consequently to a lower atmospheric CO_2 level. In spite of its relatively small area, the Southern Ocean turns out to be important for atmospheric CO_2, because it is in rapid exchange with the large volumes of the deep sea. This mechanism could lead to rather rapid CO_2 changes (within a few 100 years, Wenk et al. (77)), in contrast to Broecker's shelf sediment hypothesis.

It has not yet been possible to clearly decide which of the various hypothesis is the true explanation of the glacial/interglacial CO_2 variations. Besides this question, the studies have clearly demonstrated that physical, chemical and biological processes in the ocean do not act independently on the global carbon cycle, but that we have to consider them as a highly interactive system. On the experimental side, numerous studies on the history of the carbon cycle, e.g. on deep-sea sediments, have been induced by the detection of the natural CO_2 variations, as e.g. documented by the volume edited by Sundquist et al. (eds.) (69). Thus, the results of the ice core studies have led to large research activities that now go far beyond the issue of glacial/postglacial changes of atmospheric carbon dioxide.

ACKNOWLEDGMENTS

I thank S. Bohler for her skill and patience when typing the manuscript, and K. Hänni for drafting several figures.

REFERENCES

1. Augustsson, T. and V. Ramanathan, 1977: A radiative convective model study on the CO_2 climate problem. J. Atmos. Sci., 34, 448-451.
2. Bacastow, R.B., 1978: Modulation of atmospheric carbon dioxide by the Southern Oscillation. Nature, 261, 116-118.
3. Bacastow, R.B. and C.D. Keeling, 1981: Atmospheric CO_2 and the Southern Oscillation: effects associated with recent El Nino events. In: Proc. WMO/ICSU/UNEP Conference on Analysis and Interpretation of Atmospheric CO_2 Data, 109-112, World Meteorological Organization, Geneva.
4. Bacastow, R.B., C.D. Keeling, and T.P. Whorf, 1985: Seasonal amplitude increase in atmospheric CO_2 concentration at Mauna Loa, Hawaii, 1959-1982. J. Geophys. Res., 90, 10,529-10,540.
5. Beer, J., U. Siegenthaler, H. Oeschger, M. André, G. Bonani, M. Suter, W. Wölfli, R.C. Finkel, and C.C. Langway, 1984: Temporal ^{10}Be variations. In: Proc. 18th Internat. Cosmic Ray Conf., Bangalore, 9, 317-320.
6. Bender, M., L.D. Labeyrie, D. Raynaud, and C. Lorius, 1985: Isotopic composition of atmospheric O_2 in ice linked with deglaciation and global primary productivity. Nature, 318, 349-352.
7. Betzer, P.R., R.H. Byrne, J.G. Acker, C.S. Lewis, R.R. Jolley, and R.A. Feely, 1984: The oceanic carbonate system: a reassessment of biogenic controls. Science, 226, 1074-1077.
8. Bolin, B. (ed.), 1981: Carbon Cycle Modelling. SCOPE 16, Wiley.
9. Bolin, B., 1985: How much CO_2 will remain in the atmosphere? In: International Assessment of the Role of Carbon Dioxide and Other Greenhouse Gases in Climate Variations and Associated Impacts. World Meteorological Organization, Geneva, in press.
10. Bolin, B., E.T. Degens, S. Kempe, and P. Ketner (eds.), 1979: The Global Carbon Cycle. SCOPE 13, Wiley, New York.
11. Broecker, W.S., 1982: Ocean chemistry during glacial tiemes. Geochim. Cosmochim. Acta, 46, 1689-1705.

12. Broecker, W.S. and T.-H. Peng, 1974: Gas exchange rates between air and sea. Tellus, 26, 21-35.

13. Broecker, W.S. and T.-H. Peng, 1982: Tracers in the Sea. Eldigio Press, Lamont-Doherty Geological Observatory, Palisades, NY 10964.

14. Broecker, W.S., T.-H. Peng, and R. Engh, 1980: Modelling the carbon system. Radiocarbon, 22, 565-598.

15. Callendar, G.S., 1958: On the amount of carbon dioxide in the atmosphere. Tellus, 10, 243-248.

16. Clark, W.C. (ed.)., 1982: Carbon Dioxide Review 1982. Clarendon Press, Oxford.

17. Craig, H., 1957: Isotopic standards for carbon and correction factors for mass-spectrometric analysis of carbon dioxide. Geochim. Cosmochim. Acta, 12, 133-149.

18. Damon, P.E., J.C. Lerman, and A. Long, 1978: Temporal fluctuations of atmospheric ^{14}C: causal factors and implications. Ann. Rev. Earth Planet. Sci., 6, 457-494.

19. Delmas, R.J., J.-M Ascensio, and M. Legrand, 1980: Polar ice evidence that atmospheric CO_2 20,000 years BP was 50 % of present. Nature, 284, 155-157.

20. Emerson, S., 1985: Oceanic carbon preservation in marine sediments. In: Sundquist and Broecker, 1985, pp. 78-87.

21. Friedli, H., E. Moor, H. Oeschger, U. Siegenthaler, and B. Stauffer, 1984: $^{13}C/^{12}C$ ratios in CO_2 extracted from Antarctic ice. Geophys. Res. Lett., 11, 1145-1148.

22. Garrels, R.M. and E.A. Perry, 1974: Cycling of carbon, sulphur, and oxygen through geologic time. In: The Sea, vol. 5 (E.G. Goldberg, ed.), pp. 303-336. Wiley.

23. Hansen, J., A. Lacis, D. Rind, G. Russell, P. Stone, I. Fung, R. Ruedy, and J. Lerner, 1984: Climate sensitivity: analysis of feedback mechanisms. In: Climate Processes and Climate Sensitivity (J.E. Hansen and T. Takahashi, eds.), Geophysical Monograph 29, pp. 130-163, American Geophysical Union.

24. Heimann, M., C.D. Keeling, and C.J. Tucker, 1986b. A three dimensional model of atmospheric CO_2 transport based on observed winds. 2. Analyses of the seasonal cycle of CO_2. Tellus, in press.

25. Hoffert, M.I., A.J. Callegari, and C.T. Hsieh, 1980: The role of deep sea heat storage in the secular response to climatic forcing. J. Geophys. Res., 85, 6667-6679.

26. Jähne, B., W. Huber, A. Dutzi, T. Wais, and J. Ilmberger, 1984: Wind wave-tunnel experiments on the Schmidt number and wave field dependence of air/water exchange. In: Gas Exchange at Water Surface (W. Brutsaert and G.H. Jirka, eds.), pp; 303-309. Reidel.

27. Jones, P.D., T.M.L. Wigley, and P.M. Kelly, 1982: Variations in surface air temperatures. Part 1. Northern hemisphere, 1881-1980. Montly Weather Rev., 110, 59-69.

28. Keeling, C.D., J.A. Adams, C.A. Ekdahl, and P.R. Guenther, 1976: Atmospheric carbon dioxide variations at the South Pole. Tellus, 28, 552-563.

29. Keeling, C.D., R.B. Bacastow, and T.P. Whorf, 1982: Measurements on the concentration of carbon dioxide at Mauna Loa Observatory, Hawaii. In: Carbon Dioxide Review 1982, W.C. Clark (ed.), pp. 377-384.

30. Keeling, C.D. and M. Heimann, 1985: Meridional eddy diffusion model of the transport of atmospheric carbon dioxide. 2. The mean annual carbon cycle. J. Geophys. Res., in press.

31. Keeling, C.D. and R. Revelle, 1985: Effects of El Niño/Southern Oscillation on the atmospheric content of carbon dioxide. Meteoritics, 20, 437-450.

32. Kempe, S., 1984: Sinks of the anthropogenically enhanced carbon cycle in surface fresh waters. J. Geophys. Res., 89, 4657-4676.

33. Knox, F. and M. McElroy, 1984: Changes in atmospheric CO_2: influence of marine biota in high latitudes. J. Geophys. Res., 89, 4629-4637.

34. Kohlmaier, G.H., A. Janecek, C.D. Keeling, and R. Revelle, 1986: Analysis of the CO_2 stimulation effect of vegetation in connection with the atmospheric CO_2 amplitude increase. Tellus, in press.

35. Komhyr, W.D., R.H. Gammon, T.B. Harris, L.S. Waterman, T.J. Conway, W.R. Taylor, and K.W. Thoning, 1985: Global atmospheric CO_2 distribution and variations from 1968-1982 NOAA/GMCC CO_2 flask sample data. J. Geophys. Res., 90, 5567-5596.

36. Levitus, S., 1982: Climatological atlas of the World Ocean. NOAA Professional Paper 13.

37. Li, Y.-H., T.-H. Peng, W.S. Broecker, and H.G. Oestlund, 1984: The average vertical mixing coefficient for the oceanic thermocline. Tellus, 36B, 212-217.

38. Liss, P.S., 1983: Gas transfer: experiments and geochemical implications. In: Air-Sea Exchange of Gases and Particles (P.S. Liss and W.G.N. Slinn, eds.), pp. 241-298. Reidel.

39. Liss, P.S. and A.J. Crane, 1983: Man-made carbon dioxide and climatic change. Geo Books, Norwich.

40. Manabe, S. and R.J. Stouffer, 1980: Sensitivity of a global climate model to an increase of CO_2 concentration in the atmosphere. J. Geophys. Res., 85, 5529-5554.

41. Meybeck, M., 1982: Carbon, nitrogen and phosphorus transport by world rivers. Am. J. Sci., 282, 401-450.

42. Mook, W.G., M. Koopmans, A.F. Carter, and C.D. Keeling, 1983: Seasonal, latitudinal and secular variations in the abundance and isotopic ratios of atmospheric carbon dioxide. J. Geophys. Res., 88, 10,915-10,933.

43. National Research Council, 1983: Changing Climate. National Academy Press, Washington.

44. Neftel, A., H. Oeshger, J. Schwander, B. Stauffer, and R. Zumbrunn, 1982: Ice core measurements give atmospheric CO_2 content during the past 40,000 years. Nature, 295, 222-223.

45. Neftel, A., E. Moor, H. Oeschger, and B. Stauffer, 1985: Evidence from polar ice cores for the increase in atmospheric CO_2 in the past two centuries. Nature, 315, 45-47.

46. Nir, A. and S. Lewis, 1975: On tracer theory in geophysical systems in the steady and non-steady state. Part I. Tellus, 27, 372-383.

47. Oeschger, H., U. Siegenthaler, U. Schotterer, and A. Gugelmann, 1975: A box diffusion model to study the carbon dioxide exchange in nature. Tellus, 27, 168-192.

48. Pearman, G.I., P. Hyson, and P.J. Fraser, 1983: The global distribution of atmospheric carbon dioxide. Aspects of observations and modelling. J. Geophys. Res., 88, 3581-3590.

49. Peng, T.-H., W.S. Broecker, G.G. Mathieu, and Y.-H. Li, 1979: Radon evasion rates in the Atlantic and Pacific Oceans as determined during the GEOSECS program. J. Geophys. Res., 84, 2471-2486.

50. Peng, T.-H, W.S. Broecker, H.D. Freyer, and S. Trumbore, 1983: A deconvolution of the tree-ring based $\delta^{13}C$ record. J. Geophys. Res., 88, 3609-3620.

51. Ramanathan, V., R.J. Cicerone, H.B. Singh, and J.T. Kiehl, 1985: Trace gas trends and their potential role in climatic change. J. Geophys. Res., 90, 5547-5566.

52. Raynaud, D. and J.M. Barnola, 1985: An Antarctic ice core reveals atmospheric CO_2 variations over the past few centuries. Nature, 315, 309-311;

53. Roos, M. and G. Gravenhorst, 1984: The increase in oceanic carbon dioxide and the net CO_2 flux into the North Atlantic. J. Geophys. Res., 89, 8181-8193.

54. Rotty, R.M. and G. Marland, 1984: Production of CO_2 from fossil fuel burning by fuel type, 1860-1982. Report NDP-006, Carbon Dioxide Information Center, Oak Ridge National Laboratory, Oak Ridge.

55. Sarmiento, J.L., 1985: Three-dimensional ocean models for predicting the distribution of CO_2 between the ocean and atmosphere. In: The Global Carbon Cycle: Analysis of the Natural Cycle and Implications of Anthropogenic Alterations for the Next Century. Proc. 6th ORNL Life Sciences Symp. Springer (in press).

56. Sarmiento, J.L., and J.R. Toggweiler, 1984: A new model for the role of the oceans in determining atmospheric pCO_2. Nature, 308, 621-624.

57. Siegenthaler, U., 1983: Uptake of excess CO_2 by an outcrop-diffusion model of the ocean. J. Geophys. Res., 88, 3599-3608.

58. Siegenthaler, U., 1984: 19th century measurements of atmospheric CO_2 - a comment. Climatic Change, 6, 409-411.

59. Siegenthaler, U., 1986: ^{14}C in the oceans. In: Handbook of Environmental Isotope Geochemistry, vol. 3 (J.-Ch. Fontes and P. Fritz, eds.). Elsevier, in press.

60. Siegenthaler, U. and H. Oeschger, 1984. Transient temperature changes due to increasing CO_2 using simple models. Annals of Glaciology, 5, 153-159.

61. Siegenthaler, U. and T. Wenk, 1984: Rapid atmospheric CO_2 variations and ocean circulation. Nature, 308, 624-626.

62. Siegenthaler, U., H. Oeschger, and M. Heimann, 1986: Biospheric CO_2 sources since 1800 AD reconstructed by deconvolution of ice core CO_2 data. Tellus, in press.

63. Smethie, W.M., T. Takahashi, and D.W. Chipman, 1985: Gas exchange and CO_2 flux in the tropical Atlantic Ocean determined from ^{228}Rn and pCO_2 measurements. J. Geophys. Res., 90, 7005-7022.

64. Stuiver, M. and P.D. Quay, 1980: changes in atmospheric carbon-14 attributed to a variable sun. Science, 207, 11-19.

65. Stuiver, M. and P.D. Quay, 1981: Atmospheric ^{14}C changes resulting from fossil fuel CO_2 release and cosmic ray flux variability. Earth Planet. Sci. Lett., 53, 349-362.

66. Stuiver, M., H.G. Oestlund, and T.A. McConnaughey, 1981: GEOSECS Atlantic and Pacific ^{14}C distribution. In: B. Bolin (Editor), Carbon Cycle Modelling. Wiley, SCOPE 16, 201-221.

67. Stuiver, M., R.L. Burk, and P.D. Quay, 1984: $^{13}C/^{12}C$ ratios in tree-rings and the transfer of biospheric carbon to the atmosphere. J. Geophys. Res., 89, 11, 731-11, 748.

68. Sundquist, E.T., 1985: Geological perspectives on carbon dioxide and the carbon cycle. In: Sundquist and Broecker (1985), pp. 5-60.

69. Sundquist, T.T. and W.S. Broecker (eds.), 1985: The Carbon Cycle and Atmospheric CO_2: natural Variations Archean to Present. Geophys. Mongr. 32, American Geophysical Union.

70. Takahashi, T., 1985: Geographical and time variability of partial pressure of CO_2 in surface waters of the Atlantic Ocean. In: The Global Carbon Cycle: Analysis of the Natural Cycle and Implications of Anthropogenic Alterations for the Next Century. Proc. 6th ORNL Life Sciences Symposium. Springer, in press.

71. Takahashi, T., W.S. Broecker, S.R. Werner, and A.E. Bainbridge, 1980: Carbonate chemistry of the surface waters of the world ocean. In: Isotope Marine Chemistry. Uchida Rokakuho, Tokyo, pp; 291-326.

72. Takahashi, T., W.S. Broecker, and A.E. Bainbridge, 1981: Supplement to alkalinity and total carbon dioxide concentration in the world ocean. In: B. Bolin (ed.), Carbon Cycle Modelling. SCOPE 16, 159-199, Wiley.

73. Takahashi, T., D. Chipman, and T. Volk, 1983: Geographical, seasonal and secular variations of pCO_2 in surface waters of the North Atlantic Ocean. In: Carbon Dioxide Science and Consensus, pp. II.123 - II.145, CONF-820970, US Dept. of Commerce, Springfield, VA 22161.

74. Takahashi, T., W.S. Broecker, and S. Langer, 1985: Redfield ratio based on chemical data from isopycnal surfaces. J. Geophys. Res., 90, 6907-6924.

75. Walsh, J.J., R.L. Rowe, R.L. Iverson, and C.P. McRoy, 1981: Biological export of shelf carbon is a sink of the global CO_2 cycle. Nature, 291, 196-201.

76. Wanninkhof, R., J.R. Ledwell, and W.S. Broecker, 1985: Gas exchange - wind speed relation measured with sulphur hexafluoride on a lake. Science, 227, 1224-1226

77. Wenk, T. and U. Siegenthaler, 1985: The high-latitude ocean as a control of atmospheric CO_2. In: Sundquist and Broecker (1985), pp. 185-194.

78. Wenk, T., H. Oeschger, and U. Siegenthaler, 1986: Simulation of atmospheric CO_2 response to El Niño events by means of an ocean-atmosphere box model. Tellus, in press.

THE PHYSICS AND DYNAMICS OF THE CLIMATE SYSTEM

SIMULATION OF CLIMATE CHANGE

J F B Mitchell
Meteorological Office, Bracknell RG12 2SZ

Summary

The use of climate models to underfund and predict climate change is described, using examples from numerical studies of the effects of enhanced atmospheric CO_2.

1. INTRODUCTION, WHY MODEL CLIMATE?

The increases in atmospheric "Greenhouse" gases since 1860 have a radiative effect equivalent to a 40% increase in carbon dioxide concentrations, and by the middle of the next century, are expected to be equivalent to a doubling of carbon dioxide concentration. Simulations with detailed climate models indicate that this would produce a warming of 2 to 5 K in global mean surface temperature at equilibrium, with accompanying changes in precipitation, sea level and other parameters. The observed increase of 0.5 K since 1900 is consistent with the lower range of the estimated potential increase, allowing for a possible slowing of the global mean warming due to the ocean's large thermal inertia. Thus, man-made climate change is not just a possibility, it may already be occurring. Consequently, there is an ever pressing need to predict the likely changes in climate due to increases in trace gases in order to guide future energy policies and to minimize the possible climatic impacts. As discussed in Chapter 9, detailed 3-dimensional models of climate are the most promising method of providing the detailed information required for climatic impact assessment.

There is also considerable interest as to why the earth's climate has changed in the past (see Figure 1a). Climate models provide a means of testing theories of climate change. Conversely, periods from which reliable and wide-spread palaeoclimatic data are available may be used as a test of the fidelity of climate models, especially if the factors producing the climatic change are well known.

The rest of this paper is arranged as follows:

2. The "Greenhouse" effect.
3. The principal gases, past, present and future.
4. Climate feedbacks in CO_2 experiments.
5. Equilibrium climate change due to increased CO_2.
6. Modelling the transient response to increases in trace gases.
7. Uncertainties in the simulation and detection of the climati effect of increased trace gases.
8. Appeals to the past; simulations for 9000 years before present (9 K bp).

9. Concluding remarks.
10. Acknowledgements.

2. THE "GREENHOUSE" EFFECT

The globally averaged radiative heat balance of the earth-atmosphere system may be approximated by

$$(S_0/4)\,(1-\alpha) = \sigma T_e^{\,4} \qquad (1)$$

where S_0 is the solar "constant", α is the fraction of solar radiation reflected by the earth and atmosphere, σ is Stefan's constant and T_e is the mean or effective radiative temperature of the system. T_e is the radiative equilibrium temperature that the earth's surface would reach if the atmosphere was transparent to longwave (thermal or infra-red) radiation. With the current value of α (0.3), T_e = 255 K, some 33 K below the current observed global mean surface temperature.

Now consider a single layer atmosphere, longwave emissivity ε and shortwave albedo α in radiative equilibrium with the surface (Figure 1). For convenience, it will be assumed that the atmosphere does not absorb solar radiation. The surface temperature T_* is given by

$$T_*^{\,4} = T_e^{\,4}/(1-{}^{\varepsilon}/_{2}) \qquad (2)$$

and the atmospheric temperature T is given by

$$T^4 = \frac{T_*^{\,4}}{2} \qquad (3)$$

Thus, gases which absorb and emit longwave radiation (greenhouse gases) warm the surface and cool the atmosphere. As an exercise, the reader can show that in a two-level atmosphere in radiative equilibrium, temperature decreases with height.

In practice, the troposphere is not generally in radiative equilibrium, but is heated convectively from the surface. The global mean lapse rate, about 6 K/km, is slightly less than the dry adiabatic lapse rate due to latent heat released in regions of saturated ascent. The radiative equilibrium temperature of 255 K corresponds to a height of 5.5 km (point a on the solid profile in Figure 2). An increase in the concentration of greenhouse gases raises the effective emitting level, (point b in Figure 2) and reduces the effective emitting temperature T_e and outgoing radiance $\sigma T_e^{\,4}$. The system warms until the temperature at the new emitting level rises to the original value T_e (point c).

The stratosphere will cool if the concentration of greenhouse gases there is increased. The stratosphere is not convectively coupled to the surface but is in radiative equilibrium; an increase in longwave absorbers will produce enhanced emission to space which at equilibrium is compensated by a fall in emitting temperature.

3. THE PRINCIPAL ABSORBERS, PAST, PRESENT AND FUTURE

The concentration and radiative characteristics of the climatically most important longwave absorbers in the contemporary atmosphere are listed in Table 1, based on Dickinson and Cicerone (1986) and Ramanathan et al (1985). The greenhouse heating is defined as the change in net downward flux at the tropopause due to the presence of each gas. Since increasing the concentration of greenhouse gases produces a slight increase in the downward flux from the stratosphere, the change in net flux at the tropopause rather than at the top of the atmosphere is considered (Schneider, 1975). Note that the troposphere and surface are strongly coupled and hence can be regarded as a single system. Water vapour is by far the most important "Greenhouse" gas, followed by carbon dioxide. Although the concentrations of chlorofluorocarbons (CFCs) CFC11 and CFC12 are 10^{-6} that of carbon dioxide, their contribution to radiative heating is only a factor 10^3 less. These gases have absorption bands which are both very strong and lie in the atmospheric "window" from about 833 cm^{-1} to 1250 cm^{-1}, where there is little absorption by other gases.

Estimates of the pre-industrial (1860) concentrations of the main trace gases (other than water vapour) are shown in Table 2. The concentrations of carbon dioxide, methane and nitrous oxide are deduced from ice-core sampling. CFCs are produced solely by industrial activity and so their pre-industrial concentration is zero. Gases other than CO_2 produce 30% of the radiative perturbation to date, and the total effect is already equivalent to half that due to doubling carbon dioxide concentrations. There is considerable uncertainty in the estimated concentrations for 2035; the first three gases in Table 2 undergo natural cycles which are not completely understood, and for all the gases the industrial contribution is dependent on the world economy and the effectiveness of recent agreements to limit production such as those for carbon dioxide and CFCs. (The assumptions made to derive the concentrations in 2035 are listed in Mitchell (1989) and references therein.) If the estimates in Table 2 are not greatly in error, the radiative effect of the increase in all the gases since 1860 will be equivalent to a doubling of CO_2 before 2035, and this would occur even if CFCs do not increase any further.

The effect of alternative scenarios can be estimated from Table 2 by noting that the radiative effect of carbon dioxide varies approximately logarithmically with concentration, (Augustsson and Ramanathan, 1977), and that of methane and nitrous oxide as the square root of their concentration. The effect of CFC11 and CFC12 vary linearly with concentration because of their low concentrations and the fact that they absorb at wavelengths where there is little absorption by other gases.

4. CLIMATE FEEDBACKS IN CARBON DIOXIDE EXPERIMENTS

The radiative perturbation due to enhancing trace gases can (in principle) be calculated to any desired accuracy. Doubling carbon dioxide amounts increases the heating of the tropopause and surface by 4 Wm^{-2} (Lal and Ramanathan, 1984), which, assuming the current climate system behaves as a black body at mean temperatures near 255 K, would produce a warming of 1.1 K. However, the atmospheric response to changes in radiative heating is complex involving many processes which are not well understood. These may amplify the perturbation (positive feedbacks) or damp it (negative feedbacks).

Most equilibrium experiments to determine the climatic effects of increases in trace gases have assumed a doubling or quadrupling of carbon dioxide concentrations. The vertical profile of the radiative perturbation due to the other gases in Table 2 is qualitatively similar to that due to increasing carbon dioxide, (except near the tropopause where the minor gases produce a slight warming as opposed to a cooling (Ramanathan et al (1985)), so experiments with increased carbon dioxide only are probably adequate to represent the effect of all the gases. Here and in the next section, we will consider the equilibrium response to doubling carbon dioxide concentrations. First, we consider the main climate feedbacks which have been identified in numerical experiments.

A warmer atmosphere will hold more water vapour, and hence enhance the radiative perturbation due to carbon dioxide. Higher temperatures reduce the extent of highly reflective snow and ice, and so increase the absorption of solar radiation. In most simulations with increased carbon dioxide, cloud amount decreases, reducing the amount of insolation reflected back to space, and the mean cloud height increases, further enhancing the "Greenhouse" effect. It is also likely that cloud radiative properties may change. An increase in atmospheric moisture would tend to increase cloud water content and hence the reflectivity of cloud, producing a negative feedback. Thick clouds behave as black bodies at infra-red wavelengths, but thin high clouds may have an emissivity considerably less than unity. An increase in cloud water would tend to increase the longwave emissivity of high cloud and enhance the "greenhouse" effect, adding a further positive feedback.

The strength of individual feedbacks simulated in GCM experiments has been quantified using a simple zero-dimensional energy balance model

$$C_* \frac{d\Delta T}{dt} + \lambda \Delta T = \Delta Q \qquad (4)$$

where ΔQ is the radiative perturbation, ΔT is the change in surface temperature, C_* is the effective heat capacity of the system, and λ is defined as the feedback parameter. At equilibrium, $\Delta T_{eq} = \Delta Q/\lambda$. Thus the smaller λ, the greater ΔT_{eq}, implying a larger positive feedback. Assuming the feedbacks are linear and independent,

$$\lambda = \sum \lambda_i \ ; \ \lambda_i = \Delta Q_i/\Delta T_{eq} \qquad (5)$$

where i denotes the individual feedbacks, and ΔQ_i is the radiative perturbation due to the process i alone. For example, if i denotes changes in water vapour, then ΔQ_i is the radiative perturbation due to the increase in water vapour between the perturbed and control simulations, in the absence of changes in other parameters. Typical values for λ_i from GCM experiments are shown in Table 3. The water vapour feedback is easily the strongest, and there is good agreement between different models. The ice-snow albedo feedback is small, though it should be noted that the representation of sea-ice in current climate models is simplistic. The strength of the cloud feedback varies from model to model. Estimates of the effect of changes in cloud radiative properties range from a strong negative feedback (Somerville and Remer, 1984) to a weak positive feedback (Roeckner et al, 1987). The representation of clouds and their radiative properties are a source of major uncertainty in climate change experiments.

Other feedbacks have been analysed using one-dimensional radiative-convective models (for example, Schlesinger 1985). However, the reader is warned that the different modelling groups estimate the feedbacks in their experiments in different ways. The main value of feedback analysis is in identifying the processes which most affect climate sensitivity, and in helping to identify the reasons for discrepancies between models.

5. SIMULATED EQUILIBRIUM CLIMATE CHANGE DUE TO INCREASED CARBON DIOXIDE

The increases in carbon dioxide and other trace gases have been occurring on timescales of decades to centuries. It is generally assumed that one can ignore changes in the continental ice sheets on this timescale, but not changes in the ocean or sea-ice. Coupled ocean-atmosphere models are at present prohibitively expensive to run to equilibrium, so over the last decade atmospheric models coupled to a simple oceanic mixed layer of prescribed depth have been used to model the effects of increased CO_2. In order to allow for the advection of heat by the oceans, some models have included a prescribed oceanic heat flux C such that the mixed layer temperature T is given by

$$\rho h C_p \frac{dT}{dt} = S + C \qquad (6)$$

where ρ, h and C_p are respectively the density, depth and specific heat of the well mixed layer and S is the net downward heat flux from the atmosphere. C may be derived by prescribing the sea (surface) temperature T from climatology in the atmospheric model and integrating through several annual cycles to determine S as a function of time of year and location. Knowing S, h and $^{dT}/_{dt}$, one can derive C as a function of time and space, and prescribe it in equation 6 to determine the evolution of T in the coupled atmosphere mixed layer model.

In the remainder of this section, the effect on simulated climate of doubling carbon dioxide will be outlined using examples primarily from Wilson and Mitchell (1987a), henceforth referred to as WM. Schlesinger and Mitchell (1987) have reviewed three other comparable modelling studies; a fifth study is described by Schlesinger and Zhao (1987). Where appropriate, discrepancies between models are discussed.

Increasing carbon dioxide concentrations leads to a warming of the troposphere and a cooling of the stratosphere (Figure 3). In the tropics, the tropospheric warming increases with height, since in low latitudes the lapse rate adjusts towards the moist adiabatic which decreases with increasing temperature. The increase in warming with height is more pronounced in models with penetrative convection schemes as opposed to those with moist convective adjustment. Penetrative schemes are more efficient in transporting moisture upwards, and hence in warming upper levels.

In high latitudes in winter, the low-level inversion confines the warming near the surface. The reduction of snow and sea-ice leads to considerable enhancement of the warming in high latitudes in the winter hemisphere. Note that the removal of sea-ice in summer leads to increased absorption of solar radiation and storage of heat in the mixed layer, delaying the onset of freezing in autumn and early winter (Manabe and Stouffer (1980)). This leads to a pronounced warming, or more precisely, a reduced cooling relative to the control. In contrast, over sea-ice in summer, temperatures generally rise to the melting point of ice and are maintained there in both the control and anomaly simulation. Even where sea-ice is temporally removed in summer, the thermal inertia of the oceanic mixed layer substantially reduces the magnitude of the warming.

The influence of sea-ice is seen more clearly in the geographical distributions of seasonal mean changes in surface temperature (Figure 4). During December to February the maximum warming occurs in the Arctic, Hudson Bay and near Kamchatka where sea-ice is thinner or removed altogether in the $2xCO_2$ simulation, and the minimum occurs over Antarctic sea-ice in the Ross and Weddell Seas. In models in which sea-ice extents are excessive in the simulation of present day climate, the high latitude warming is exaggerated and the latitude of the maximum warming may be displaced equatorwards. The warming is generally less over the ocean than over land. Over land, evaporation may be limited by the dryness of the surface so that a greater proportion of the increase in radiative heating is used to raise surface temperature as opposed to increasing evaporation. In middle latitudes in winter, removal of snow in the $2xCO_2$ simulation also enhances the surface warming.

The warming is a minimum over the tropical oceans. The saturation vapour pressure for water increases exponentially with temperature, so assuming the relative humidity is approximately constant, potential evaporation also increases non-linearly with temperature. As a result more of the increase in radiative heating of the surface is converted to latent heat and less to raising surface temperatures and sensible heat as one goes from cold to warm regions. Increased evaporative cooling at the surface is linked to increased warming due to latent heat aloft, consistent with the reduction in lapse rate in the tropics noted above.

The changes in cloud are qualitatively similar from model to model (Figure 5). In general, there is an increase in high cloud and a reduction in upper and middle tropospheric cloud which in some models is most pronounced in the tropics and mid-latitudes. The changes in low cloud vary considerably from model to model, though increases in low cloud are generally restricted to middle and high latitudes. Thus in low latitudes, where insolation is greatest, there is a reduction in planetary albedo (high cloud is generally assigned a low albedo) leading to enhanced absorption of solar radiation, and in the extratropics, there are substantial increases in high cloud tending to reduce longwave cooling to space.

The warming of the atmosphere leads to an increase in specific humidity (41% in WM) and evaporation and precipitation (15% in WM). Precipitation does not increase everywhere, (Figure 6) and the spatial scale of the change is smaller than for temperature (compare Figure 4). Whereas there is substantial agreement between the temperature changes simulated by different models, there is little agreement between the regional details of the changes in precipitation, especially in the tropics. All models produce increased precipitation in high latitudes, particularly in winter. The increase in water vapour produces an increase in the transport of moisture into high latitudes, giving enhanced moisture convergence and precipitation. Most models simulate areas of decreased precipitation outside the intertropical convergence zone (ITCZ) and the subtropics, but there is little spatial correlation from model to model. In general, precipitation in the ITCZ is enhanced.

Changes in the variability as well as the mean climate are important in the assessment of the impacts of climatic change. Wilson and Mitchell (1987b) investigated changes in the variability over Western Europe in a

$4xCO_2$ sensitivity experiment. They found, for example, statistically significant reductions in the frequency and intensity of summer precipitation over Southern Italy (Figure 7).

There is a growing concern that increases in trace gases will lead to drier conditions in the northern and mid-latitude continents in summer. Even if mean summer precipitation increases (Figure 5) the soil may still become drier, as found by WM. Three out of five recent studies found a drying in mid-latitudes (Fig 8), whereas one model (Washington and Meehl, 1984) produced a wetter land surface on enhancing CO_2.

Mitchell and Warrilow (1987) have investigated the importance of land surface parametrizations on determining the simulated hydrological response to increased CO_2. In the standard formulation, snowmelt is used to augment soil moisture, or is run off if the ground is saturated. In the control simulation, snowmelt saturates the ground in late spring and the surface then dries rapidly through the summer (Figure 9a). On enhancing CO_2, precipitation increases in winter, raising soil moisture levels, but snowmelt and the associated summer drying occur earlier but from the same (saturated) level as in the control. The experiment was repeated, changing only the treatment of snowmelt such that all snowmelt is lost as run-off over frozen ground. The loss of snowmelt leads to a much drier control run (Figure 9b) and much less pronounced drying during summer. On enhancing CO_2, the enhanced precipitation in winter now enters in the soil, and though the summer drying starts earlier, it starts from a much higher level, and so the soil remains wetter for most of the summer. The model used by Washington and Meehl (1984) produces an excessively dry surface in the control simulation, apparently as a result of using too low a land surface albedo, and so behaves in a similar manner to the second version of the model described above.

Equilibrium simulations to date using models with a full dynamical representation of the ocean have used idealized geography and annual mean insolation (for example, Spelman and Manabe (1984); Manabe and Bryan (1985)). These studies suggest that the inclusion of the deep ocean does not substantially alter the equilibrium response. A faithful representation of the ocean is however essential if one is to model the time dependent response of climate to increases in greenhouse gases. This is considered in the following section.

6. MODELLING THE TRANSIENT RESPONSE TO INCREASES IN TRACE GASES

Although the concentration of "greenhouse" gases has and will continue to increase gradually, the temperature response of the atmosphere is likely to be slowed down by the large thermal inertia of the ocean. If the oceans are assumed to have a constant heat capacity C_* (equation 4), then the evolution of the temperature change ΔT due to an instantaneous increase in heating ΔQ at t=0 is given by

$$\Delta T = \frac{\Delta Q}{\lambda} (1 - \exp (- \lambda t/C_*)) \qquad (7)$$

The time τ taken to reach (1-1/e) of the equilibrium response is C_*/λ. Assuming $\lambda = 1\ Wm^{-2}\ K^{-1}$ and that the warming is mixed uniformly through the deep ocean, τ is about 500 years. On the other hand, if the warming is confined to the observed seasonal ocean mixed layer which has a depth of order 100 m, then τ is of order 10 years. These two estimates are probably upper and lower limits respectively, and the actual timescale for the oceans depends on the efficiency with which heat is mixed into the deep ocean and so varies between different parts of the ocean.

Rough estimates of the effect of oceanic inertia have been made using one-dimensional diffusive models of the ocean. The evolution of global mean temperatures assuming a doubling of effective carbon dioxide concentrations between 1850 and 2050 and vertical diffusion coefficients of 2.25 m^2 sec^{-1} are shown in Figure 10. Two cases are shown corresponding to an equilibrium increase of 4 K and 1.1 K due to doubling CO_2. For each case, the solid curve denotes the equilibrium response, and the dashed curves, the response of the diffusive ocean.

The diffusive model gives a very crude representation of vertical mixing in the ocean, and at equilibrium produces a temperature anomaly which is uniform with depth. Transient experiments have been carried out using coupled ocean atmosphere models in which carbon dioxide concentrations have been instantaneously doubled. The degree to which the warming penetrates the ocean varies with latitude, as can be seen in Figure 10 which shows the warming of an annually averaged ocean-atmosphere model with idealized geography 25 years after quadrupling carbon dioxide concentrations. Also shown in Figure 10 is the fraction of the equilibrium response achieved after 25 years, and the equilibrium response itself. Estimates of the response time τ derived from simple models at GCMs range from 10 to 100 years (Schlesinger, 1986) and possible reasons for the wide range of estimates have been investigated using simple diffusion models (Wigley and Schlesinger, 1985).

A numerical study carried out by Bryan et al (1988) indicates that the response in high latitudes of the southern hemisphere may be substantially slower than elsewhere because of the larger fractional ocean coverage and the existence of a meridional cell upwelling unmodified water from great depth. Hansen et al (1988) have estimated the time dependent response from 1958 to 2020, using a high, medium and low scenario for future increases in trace gases. In their model, the ocean is represented by a seasonally varying mixed layer. The mixed layer depth and advection of heat are assumed to remain unchanged as the concentration of trace gases is increased, and heat transfer into the deep ocean is represented by simple diffusion, the coefficient of diffusion being specified geographically on the basis of observations. As a result of their study, Hansen et al (1988) claim that the climatic effects should be clearly discernible in the 1990s. In contrast to Bryan et al (1988), they find that high southern latitudes is one of the regions where the warming should be detected earliest.

7. Uncertainties in the simulation and detection of the climatic effect of increases in trace gases.

(a) There are uncertainties in the pre-industrial concentration of carbon dioxide, and hence in the change in radiative forcing since then.

(b) The size of the equilibrium response to a given increase in gases is known only within a factor of 2 or so (Table 3), due mainly to uncertainties in the modelling of cloud and cloud radiative properties. These problems are currently an area of intensive research. The nature of the regional response, particularly in the hydrological cycle, varies from model to model. These discrepancies may be reduced by using higher horizontal resolution and improved physical paramatrizations.

(c) There is much to learn concerning the vertical mixing of heat in the oceans and its representation in numerical models. It is not known whether or not one has to resolve oceanic eddies in order to produce a reliable simulation of the changes in oceanic heat transport. Current global ocean GCMs have a horizontal resolution of 300-500 km. These require large horizontal diffusion coefficients which maintain numerical stability but also smooth out the more intense ocean currents.

(d) The climatic effects of enhanced trace gas concentrations have to be detected against the background of the natural variability of the atmosphere and ocean. The instrumental temperature record exhibits variability on timescales up to several decades. Some indication of the inherent variability of climate on the decadal timescales may be obtained from long (of order 100 years) simulations of coupled ocean atmosphere models Hence there remains some uncertainty in the observed warming.

(e) There remains the possibility that other factors are masking (or enhancing) the effects of increases in "greenhouse" gases. There has been much speculation concerning the possibility of changes in the solar "constant". Accurate measurements of the solar constant have only been available for the last decade or so, and these indicate a decrease of 0.1% between 1980 and 1985 (Willson et al, 1986). (The climatic effect of a 2% increase in solar constant is thought to be approximately equivalent to that due to a doubling of carbon dioxide concentrations, Wetherald and Manabe, 1975.) Variations in the concentration and composition of stratospheric aerosol resulting from volcanic eruptions have also been cited as a cause of climatic change, but the indices used to categorize volcanic dust do not allow a quantitative estimate of the associated radiative effects.

8. APPEALS TO THE PAST; A SIMULATION FOR 9000 BEFORE PRESENT (9 Kbp)

In order to try and limit some of the uncertainties discussed in the previous section, attempts have been made to model past climates and compare the simulated changes with the changes from the present deduced from palaeoclimatic data. Models have also been used to help understand past climates (for example Kutzbach and Guetter, 1986). The two periods chosen most frequently are the last glacial maximum (18 Kbp) and part of the Holocene around 9 Kbp. During the latter period, perihelion almost coincided with the summer as opposed to the winter solstice, enhancing insolation during the northern summer, especially in high northern latitudes. Here, a brief summary of some recent simulations for 9 Kbp (Mitchell et al, 1988) will be given. This period is of interest because

(a) The changes in boundary conditions are simple; the changes in orbital parameters are well known, CO_2 concentrations are believed to have been similar to present, and the ice sheet over North America was relatively small in extent.

(b) There is a large amount of palaeoclimatic data available from that period.

In the initial experiment, only the orbital parameters were altered. The enhanced insolation during the boreal summer produces an increase in surface temperature over the northern mid-latitude continents (Figure 12a). This enhances the land-sea contrast, lowering surface pressures over the continents (Figure 12b) and enhancing the monsoon precipitation over Venezuela, East Africa and Southern Asia (Figure 12c). The land surface becomes drier in mid-latitudes (Figure 13a) and wetter over most of the tropics. This is consistent with estimates of lake levels for the period (Figure 13b) which are lower than average in mid-latitudes and above average in the tropics. Although the broad pattern of simulated changes is in agreement with palaeoclimatic data, there are discrepancies, as over West Africa, where the model produces a drier surface at 9 Kbp, whereas the lake level data suggest that it was wetter then. This shortcoming has been the subject of further investigation.

9. CONCLUDING REMARKS

General circulation models have been used to investigate the climatic effects of a wide variety of other phenomena including sea surface temperature and sea-ice anomalies, changes in land surface characteristics due to desertification or deforestation, "nuclear winter" and waste heat from energy parks. Experiments have also been carried out to determine the influence of topography, land-sea contrasts and so forth on the general circulation of the atmosphere.

Although the use of climate models has become more frequent, each experiment should still be analysed carefully before one attempts to interpret the results. In doing so, one must bear in mind the limitations of the model and how such limitations will distort the simulated response. One can attempt to do this by isolating the mechanisms producing the changes in climate in the model, and assessing the physical realism of each step. Where there is some uncertainty concerning a parametrization on which the model's response is dependent, the experiment can be repeated using alternative forms of that parametrization to determine whether or not the uncertainties affect the results. One also has to consider whether or not all the physical processes relevant to the experiment have been included in the model. In short, one should always treat results from numerical models with a certain amount of cynicism until they are supported by a critical assessment of the model used and the simulated response.

10. ACKNOWLEDGEMENTS

The development, running and assessment of climate models is not a task that can be carried out by an individual. I am indebted to members of the Dynamical Climatology Branch of the Meteorological Office over the last

10 years for their help and support. I am also grateful to colleagues at other modelling centres who have provided additional information on their experiments. Peter Rowntree and Alayne Street-Perrott kindly provided Figures 10 and 13(b) respectively.

11. REFERENCES

Augustsson, T. and Ramanathan, V. 1977 A radiative — convective model study of the CO_2 climate problem. J. Atmos. Sci., 34, 448-51.

Bryan, K., Manabe, S. and Spelman, M.J. 1988 Interhemispheric Asymmetry in the transient response of a coupled ocean-atmosphere model to a CO_2 forcing. J. Phys. Oceanogr. 18, 851-867.

Dickinson, R.E. and Cicerone, R.J. 1986 Future global warming from atmospheric trace gases. Nature, 319, 109-115.

Hansen, J., Fung, I., Lacis, A., Rind, D., Lebedeff, S., Reudy, R. and Russell, G. 1988 Global climatic changes as forecast by Goddard Institute for Space Studies Three Dimensional Model. J. Geophys Res. (To appear).

Hansen, J. Lacis, A., Rind, D., Russell, L, Stone, P., Fung, I., Ruedy, R. and Lerner, J. 1984 Climate Sensitivity, Analysis of Feedback Mechanisms. In Climate Processes and Climate Sensitivity (ed J. Hansen and T Takahashi) Geophysical Monograph 29, 130-163. American Geophysical Union, Washington DC.

Kutzbach, J. E. and Guetter, P. J. 1986 The influence of changing orbital parameters and surface boundary conditions on climate simulations for the past 18,000 years. J. Atmos Sci. 43, 1726-1759.

Lal, M. and Ramanathan, V. 1984 The effects of moist convection and water vapour radiative processes on climate sensitivity. J. Atmos. Sci. 41, 2238-2249.

Manabe, S. and Bryan, K.Jnr. 1985 CO_2-Induced Change in a Coupled Ocean-Atmosphere Model and the Paleoclimatic Implications. J. Geophys. Res. 90, C11, 11,689-11,707.

Manabe, S. and Stouffer, R.J. 1980 Sensitivity of a global climate model to an increase in the CO_2 concentration in the atmosphere. J. Geophys. Res. 85, 5529-5554.

Mitchell, J.F.B. and Warrilow, D.A. 1987 Summer dryness in northern mid-latitudes due to increased CO_2. Nature 330, 238-240.

Mitchell, J.F.B., Grahame, N.S. and Needham, K.J. 1988 Climate simulations for 9000 years before present: Seasonal variations and the effect of the Laurentide ice sheet. J. Geophys Res. 93, 8283-8303.

Mitchell, J.F.B. 1989 The "greenhouse" effect and climate change. Submitted to Revs. of Geophysics.

Ramanathan, V., Cicerone, R.J., Singh, H.B. and Kiehl, J.T. 1985 Trace Gas and Trends and their potential role in climate change. J. Geophys. Res. 90, 5547-5566. J. Geophys. Res.

Roeckner, E., Schleses, U., Biercamp. J. and Loewe, P. 1987 Cloud optical depth feedbacks and climate modelling. Nature, 329, 138-139.

Schlesinger, M.E. 1985 Feedback analysis of results from energy balance and radiative-convective models. In "Projecting the climatic effects of increasing atmospheric carbon dioxide", edited by M.C. MacCracken and F.M. Luther, pp 280-319, US Department of Energy, Washington DC, 1985. (Available as NTIS, DOE (ER-2037 from Natl. Tech Inf.Serv., Springfield Va.)

Schlesinger, M.E. 1986 Equilibrium and transient climatic warming induced by increased atmospheric CO_2. Climate Dynamics 1, 35-51.

Schlesinger, M.E. and Mitchell, J.F.B. 1987 Climate model simulations of the equilibrium climatic response to increased carbon dioxide. Reviews of Geophysics 25, 760-798.

Schlesinger, M.E. and Zhao, Z. 1987 Seasonal climate changes induced by doubled CO_2 as simulated by the OSU atmospheric GCM/mixed layer model. Oregon State University Climatic Institute Report 70, 73pp.
Schneider, S.H. 1975 On the Carbon Dioxide Climate Confusion. J.Atmos.Sci. 32, 2060-2066.
Somerville, R.C.J. and Remer, L.A. 1984 Cloud Optical Thickness Feedbacks in the CO_2 Climate problem. J. Geophys. Res. 89, 9668-9672.
Spelman, M.J. and Manabe, S. 1984 Influence of Oceanic Heat Transport upon the sensitivity of a model climate. J. Geophys. Res. 89, 571-586.
Washington, W.M. and Meehl, G.A. 1984 Seasonal Cycle Experiment on the Climate Sensitivity Due to a Doubling of CO_2 with an Atmospheric General Circulation Model Coupled to a Simple Mixed Layer Ocean Model. J. Geophys. Res. 89, 9475-9503.
Wetherald, R.T. and Manabe, S. 1975 The effects of changing of the solar constant on the climate of a general circulation model. J. Atmos. Sci. 32, 2044-2059.
Wetherald, R.T. and Manabe, S. 1986 An investigation of cloud cover change in response to thermal forcing" Climate Change. 8, 5-24.
Wigley, T.M.L. and Schlesinger, M.E. 1985 Analytical solution for the effect of increasing CO_2 on global mean temperature. Nature 315, 649-652.
Willson, R.L., Hudson, H.S., Frohlich, C. and Brusa, R.W. 1986 Long term downward trend in solar radiance. Science, 234, 1114-1117.
Wilson, C.A. and Mitchell, J.F.B. 1987a A Doubled CO_2 Climate Sensitivity experiment with a GCM including a simple ocean. J. Geophys. Res, 92, 13315-13343.
Wilson, C.A. and Mitchell, J.F.B. 1987b Simulated climate and CO_2 induced climate change over Western Europe. Climatic Change, 10, 11-42.

12. BIBLIOGRAPHY

Physically-based modelling and simulation of climate and climate change
Vol 2 (1988) Schlesinger, M.E. (Ed) NATO Advanced Study Institute Series (Reidel, in Press). This volume covers applications of climate models as opposed to their development.

The "Greenhouse" Effect, Climatic Change and Ecosystems: 1987
Bolin, B., Doos, B.R., Jager, J. and Warrick, R.A. (Eds). SCOPE 29, Wiley, New York. Probably the best book to date covering the "greenhouse" effect and related topics.

Climate Model Simulations of the Equilibrium Response to Increased Carbon Dioxide: 1987
Schlesinger, M.E. and Mitchell, J.F.B., Rev. of Geophysics, 25, 760-798. A detailed review of CO_2 simulations up to 1986.

Issues in Atmospheric and Oceanic Modelling, Part A. Climate Dynamics: 1985
Manabe, S. (Ed). Advances in Geophysics Vol. 28, Academic Press (London). Chapters by Dickinson on Climatic sensitivity, Manabe and Wetherald on CO_2 and hydrology and Kutzbach on modeling of palaeoclimatics are of particular interest.

Palaeoclimatic Analysis and Modeling: 1985 Hecht, A.D. (Ed) John Wiley and Sons (New York). An excellent book describing both methods on analysis of palaeoclimatic data and some numerical experiments.

Table 1 Current concentrations and "greenhouse heating" due to trace gases. The main absorption bands and band strengths (a measure of the probability of a molecule absorbing a photon at the band wavelength) are shown for the less abundant gases.

Gas	Concentration (ppm)	Principal absorption bands Position (cm^{-1})	Strength $cm^{-1}(atm\ cm)^{-1}STP$	Greenhouse heating Wm^{-2}
Water Vapour	~3000			~100
Carbon Dioxide	345	667	(many bands)	~50
Methane	1.7	1306	185	1.7
Nitrous Oxide	0.30	1285	235	1.3
Ozone	$10\text{-}100\times10^{-3}$	1041	376	1.3
CFC11	0.22×10^{-3}	846 1085	1965 736	0.06
CFC12	0.38×10^{-3}	915 1095 1152	1568 1239 836	0.12

Table 2. Past and projected greenhouse gas concentrations and associated changes in "greenhouse" heating ΔQ

Gas	Assumed 1860 Concentration	ΔQ 1860 to 1985	Estimated 2035 Concentration (ppm)	Estimated ΔQ 1985 to 2035 (Wm^{-2}) Range
Carbon Dioxide	275.	1.3	475	1.8
Methane	1.1	0.4	2.8	0.5
Nitrous Oxide	0.28	0.05	0.38	0.15
CFC11	0	0.06	1.6×10^{-3}	0.35
CFC12	0	0.12	2.8×10^{-3}	0.69
TOTAL		1.9		3.5

Table 3. ANALYSIS OF FEEDBACKS AND TEMPERATURE CHANGES DUE TO DOUBLING CO_2, BASED ON TYPICAL VALUES FROM GCM EXPERIMENTS

FEEDBACK	STRENGTH (WM^{-2} K^{-1})	TEMP. CHANGE (Cumulative)
NONE	+3.7	1.1
WATER VAPOUR	-1.4	1.7
ICE/SNOW	-0.3	2.1
CLOUD AMOUNT	-0.9	4.0
CLOUD RADIATIVE PROPERTIES	-0.1 to 0.5	2.7 - 4.5

FIGURE CAPTIONS

1. In idealized single layer atmosphere, reflectivity α, longwave emissivity ε and temperature T in radiative equilibrium with the surface at temperature T_*.

2. The effect on vertical temperature profile of increasing atmospheric CO_2 concentrations (schematic). Increasing CO_2 raises the emitting level from a to b, and the profile then warms from b to c to restore the effective emitting temperature to its original value.

3. The height/latitude distribution of simulated temperature changes due to doubling CO_2 concentrations. Areas of decrease are stippled (a) Northern winter (December to February) (b) Northern Summer (June to August). From Wilson and Mitchell (1987a).

4. Simulated changes in surface temperature due to doubling CO_2 concentrations. (a) Northern winter (b) Northern Summer.

5. The height/latitude distribution of simulated annual mean changes in cloud from three different models. Areas of decrease are stippled, and the units are percent of total cover. Top, from Wetherald and Manabe (1986); Centre, from Hansen et al (1984); bottom, from Washington and Meehl (1984).

6. Simulated changes in precipitation rate (mm/day) due to doubling CO_2 concentrations. Areas of decrease are stippled (a) Northern winter (b) Northern Summer. From Wilson and Mitchell (1987a).

7. Simulated daily precipitation totals (mm) at a grid point over Southern Italy. Top panel, with quadripled CO_2 amounts and enhanced sea surface temperatures: bottom panel, with present CO_2 amounts and sea surface temperatures. From Wilson and Mitchell (1987b).

8. Simulated changes in soil moisture (cm) due to doubling CO_2 amounts. (a) From Manabe and Wetherald (1987) (b) Hansen et al (1984) (c) Washington and Meehl (1984) (d) Schlesinger and Zhao (1987) (e) Wilson and Mitchell (1987a).

9. Simulated monthly mean soil moisture averaged over land between 45° and 60°N for $1xCO_2$ (solid line) and $2xCO_2$ (dashed) line. (a) Allowing snowmelt to augment soil moisture (b) Running off snowmelt if the ground is frozen. From Mitchell and Warrilow (1987).

10. Estimates of the time dependent response to doubling effective CO_2 concentrations between 1850 and 2050, made using a one-dimensional diffusive ocean model with a vertical diffusion of 2.25 cm^2 sec^{-1}. The solid curves represent the instantaneous equilibrium values, the dashed curves the actual response allowing for oceanic thermal inertia. The upper curves allow for climate feedbacks, the lower curve neglect feedbacks. (Rowntree 1988, personal communication).

11. Height (depth)/latitude diagrams of the simulated changes in atmospheric (oceanic) temperatures due to quadrupling CO_2 amounts in an annually averaged coupled ocean-atmosphere model using idealized topography. Upper panel, the changes 25 years after instantaneously quadrupling CO_2 amounts. Middle panel, the percentage of the equilibrium response achieved after 25 years. Lower panel, the equilibrium response. From Spelman and Manabe (1984).

12. Simulated changes during northern summer due to change in the earth's orbital parameters to those appropriate to 9 Kbp. (a) Surface temperature (K), (b) Sea-level pressure (mb), (c) Precipitation rate (mm/day). Areas of decrease are stippled. From Mitchell et al (1988).

13. (a) Simulated changes in soil moisture (9 Kbp-present). Contours at 0, ±2, ±5 cm. Areas of decrease are stippled, and reductions greater than 2 cm are stippled heavily. From Mitchell et al (1988).
(b) Changes in lake level status between 9 Kbp and present. (Street-Perrott, 1988, personal communication.)

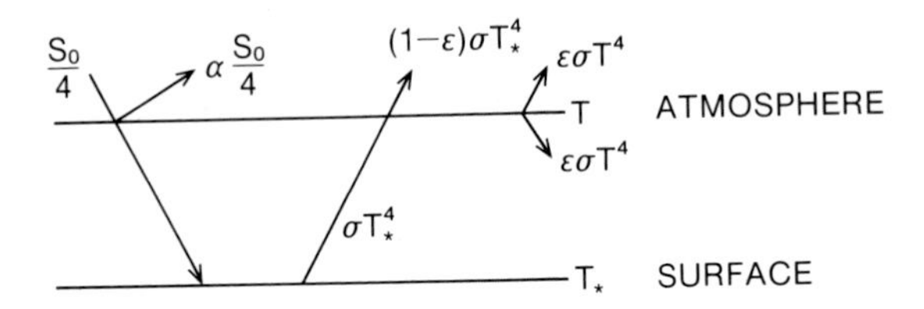

Figure 1

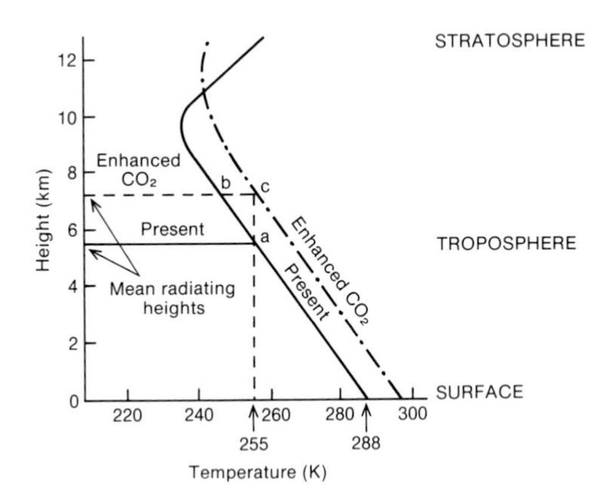

Figure 2

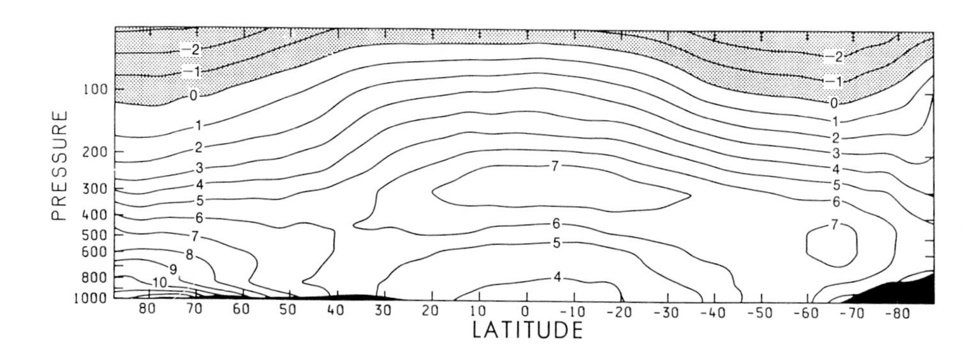
PRESSURE
100
200
300
400
500
600
800
1000
-2
-1
0
1
2
3
4
5
6
7
8
9
10
7
6
5
4
-2
-1
0
1
2
3
4
5
6
7
80 70 60 50 40 30 20 10 0 -10 -20 -30 -40 -50 -60 -70 -80
LATITUDE

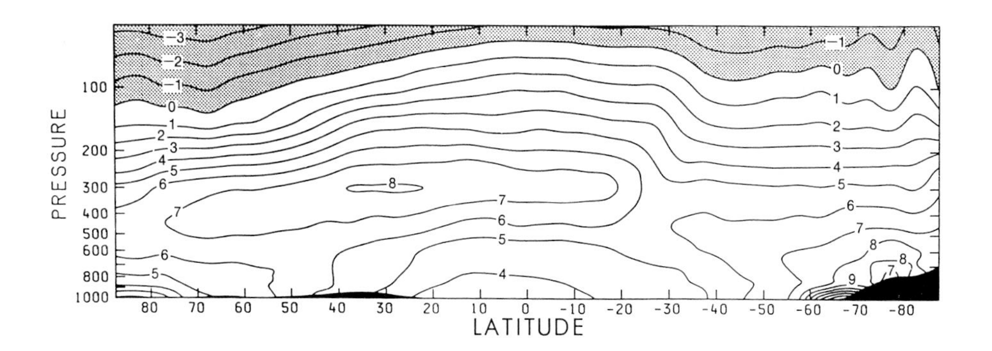
PRESSURE
100
200
300
400
500
600
800
1000
-3
-2
-1
0
1
2
3
4
5
6
7
6
5
8
7
6
5
4
-1
0
1
2
3
4
5
6
7
8
8
7
9
80 70 60 50 40 30 20 10 0 -10 -20 -30 -40 -50 -60 -70 -80
LATITUDE

Figure 3

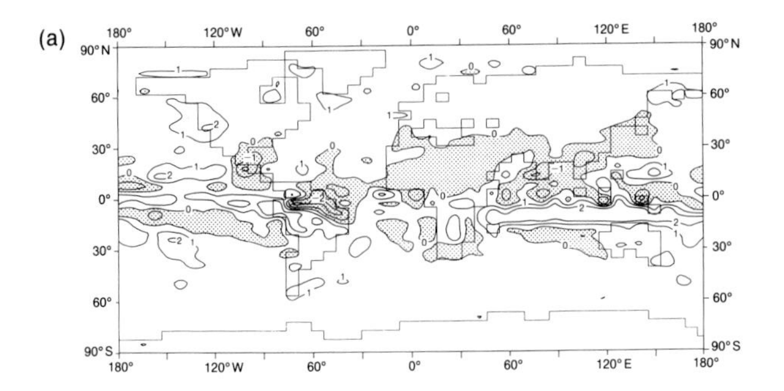
(a)
180°
120°W
60°
0°
60°
120°E
180°
90°N
60°
30°
0°
30°
60°
90°S

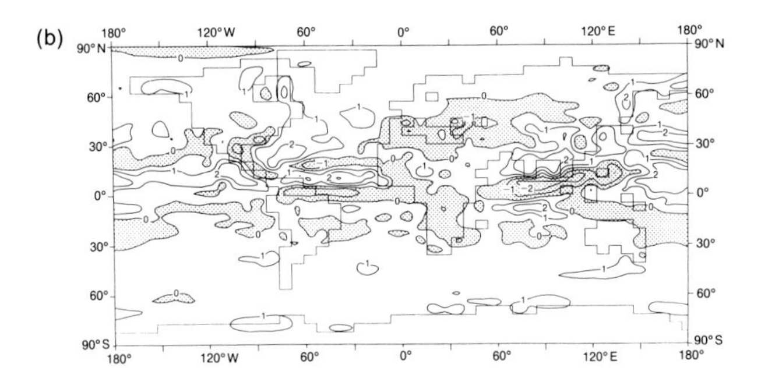
(b)
180°
120°W
60°
0°
60°
120°E
180°
90°N
60°
30°
0°
30°
60°
90°S

Figure 4

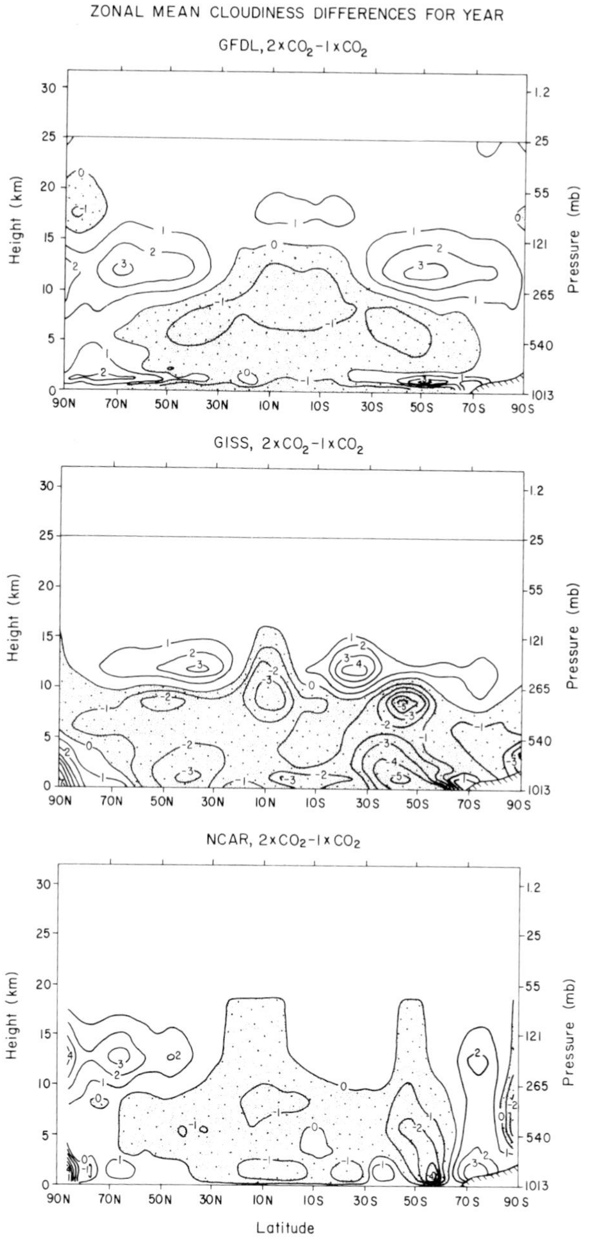
ZONAL MEAN CLOUDINESS DIFFERENCES FOR YEAR
GFDL, 2xCO2-1xCO2
GISS, 2xCO2-1xCO2
NCAR, 2xCO2-1xCO2
Height (km)
Pressure (mb)
90N 70N 50N 30N 10N 10S 30S 50S 70S 90S
Latitude

Figure 5

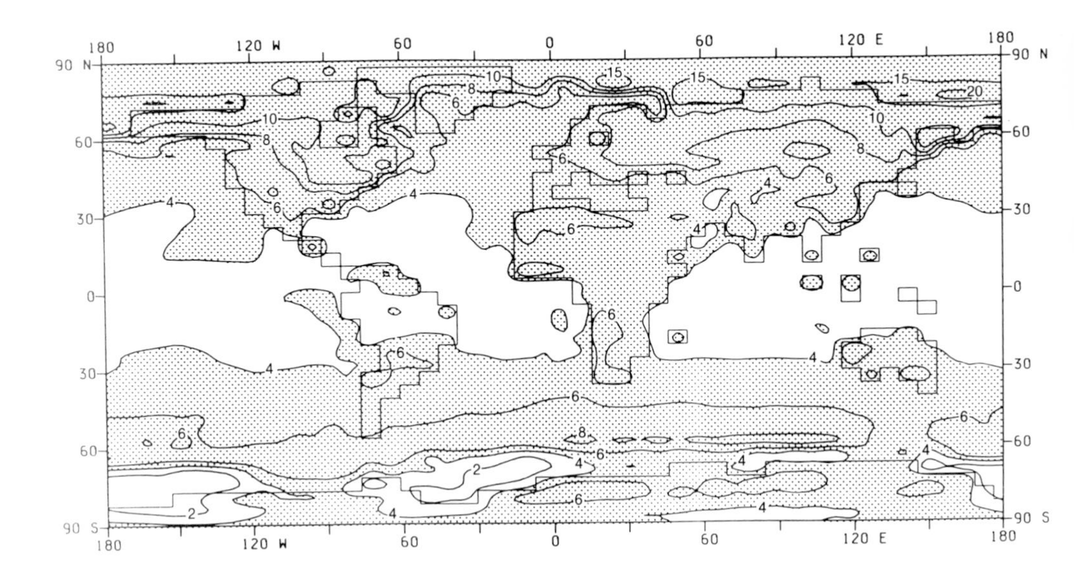
180
120 W
60
0
60
120 E
180
90 N
60
30
0
30
60
90 S

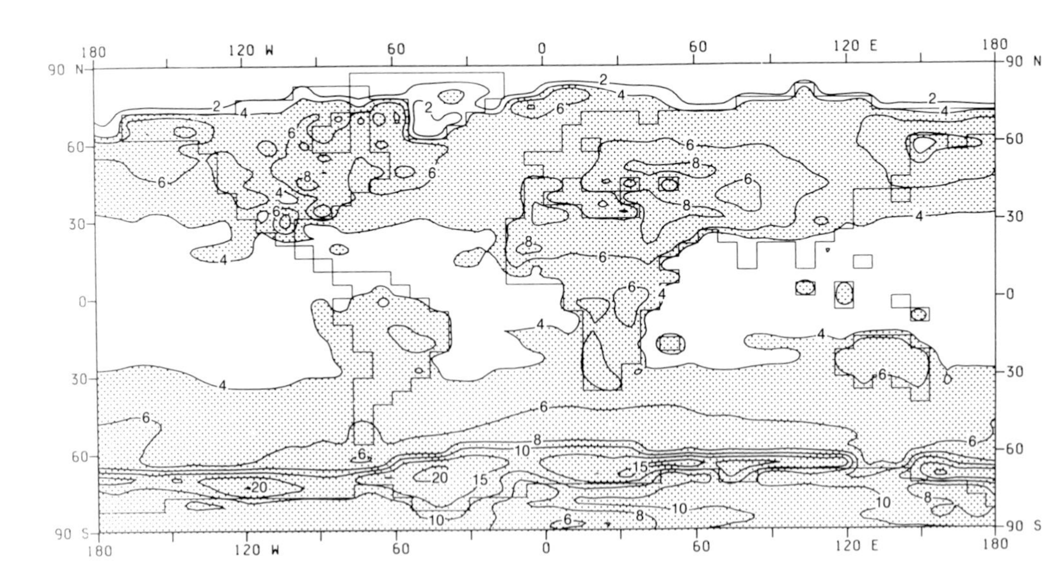
180
120 W
60
0
60
120 E
180
90 N
60
30
0
30
60
90 S

Figure 6

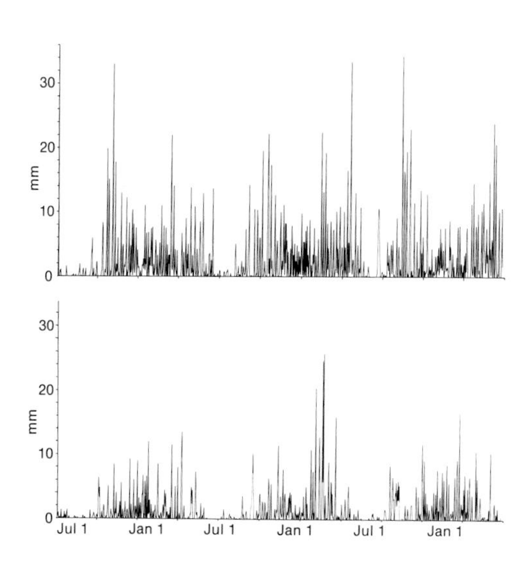

Figure 7

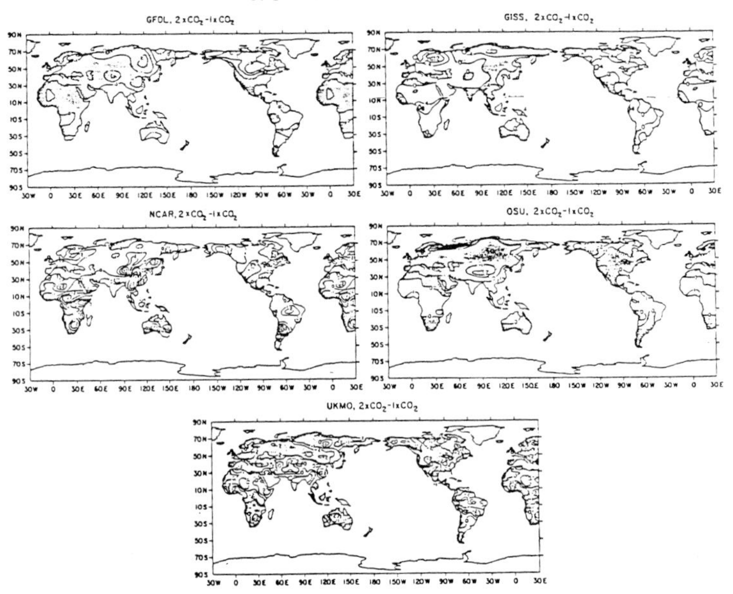

SOIL WATER DIFFERENCES FOR JJA
GFDL, 2xCO2-1xCO2
GISS, 2xCO2-1xCO2
NCAR, 2xCO2-1xCO2
OSU, 2xCO2-1xCO2
UKMO, 2xCO2-1xCO2

Figure 8

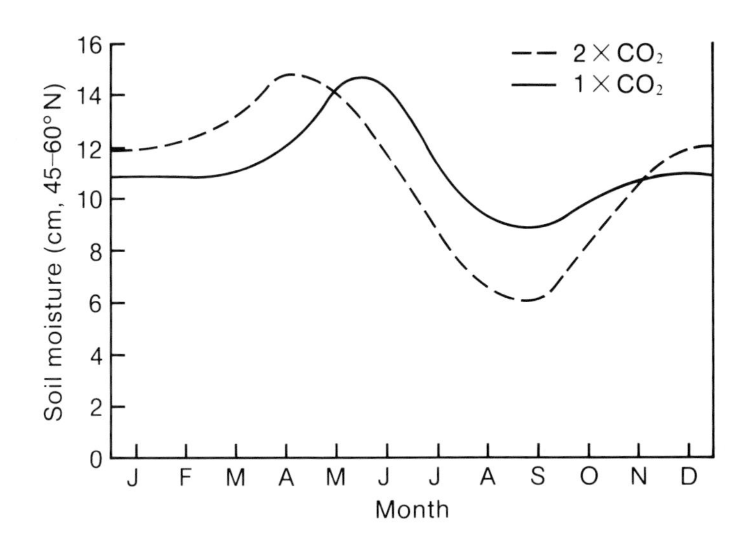

Figure 9 (a)

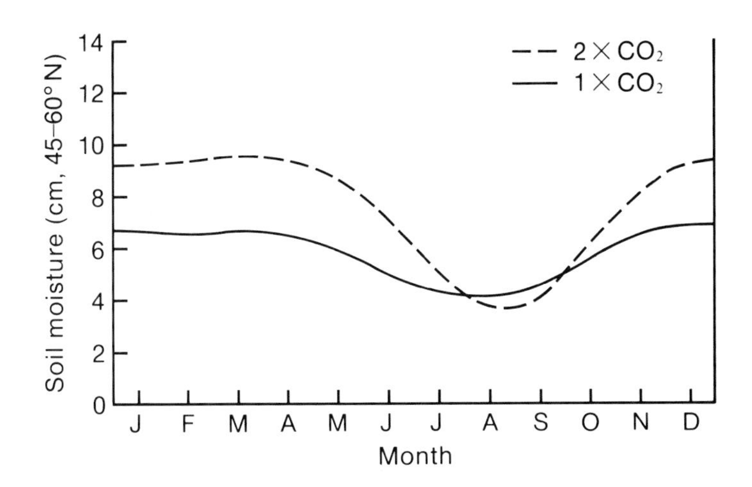

Figure 9 (b)

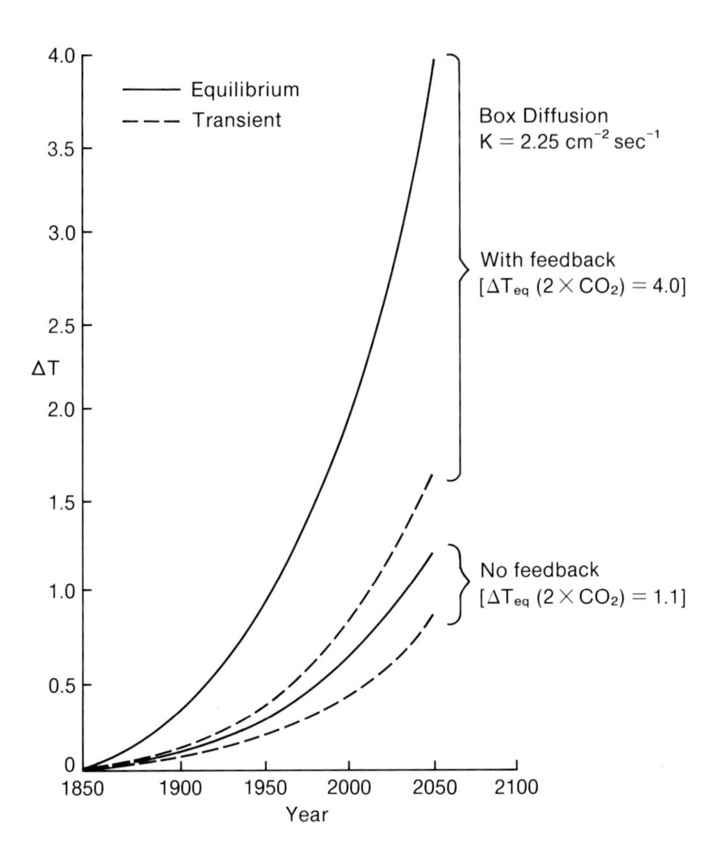
4.0
3.5
3.0
2.5
ΔT
2.0
1.5
1.0
0.5
0
1850
1900
1950
2000
2050
2100
Year
Equilibrium
Transient
Box Diffusion
K = 2.25 cm^{-2} sec^{-1}
With feedback
[ΔT_{eq} (2 × CO_2) = 4.0]
No feedback
[ΔT_{eq} (2 × CO_2) = 1.1]

Figure 10

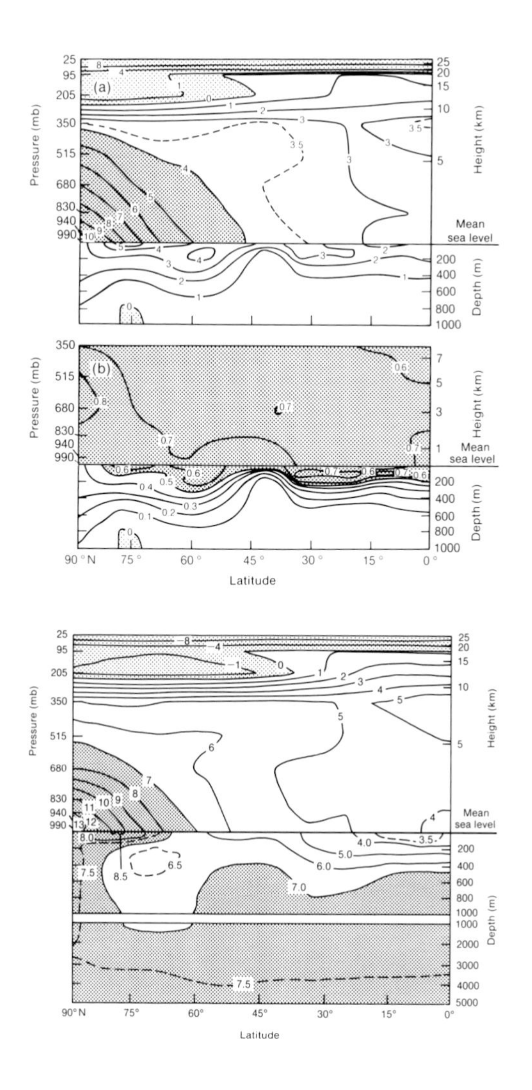
(a)
Pressure (mb)
Height (km)
Mean sea level
Depth (m)
(b)
Latitude
90°N
75°
60°
45°
30°
15°
0°

Figure 11

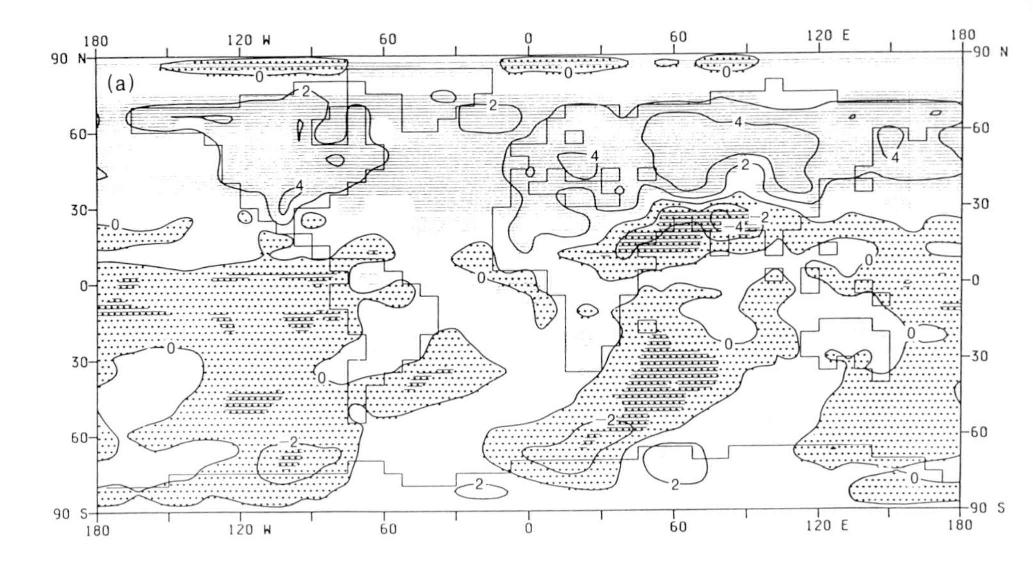

Figure 12 (a)

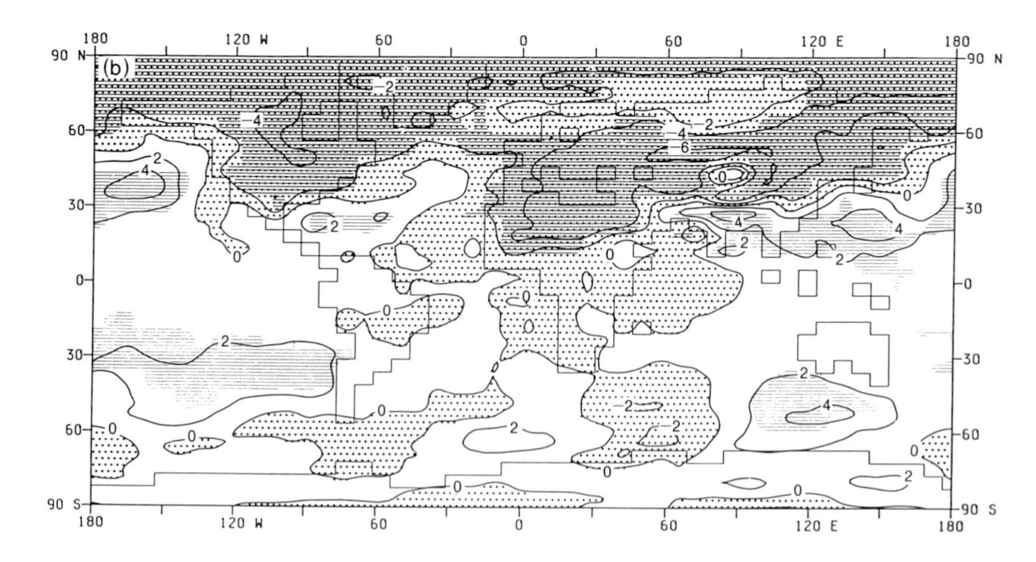

Figure 12 (b)

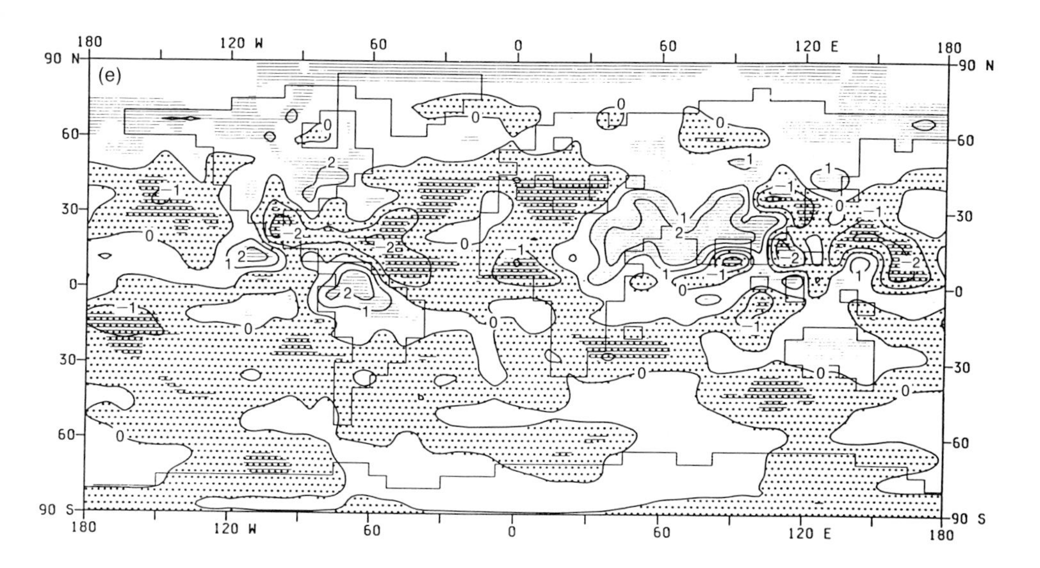

Figure 12 (c)

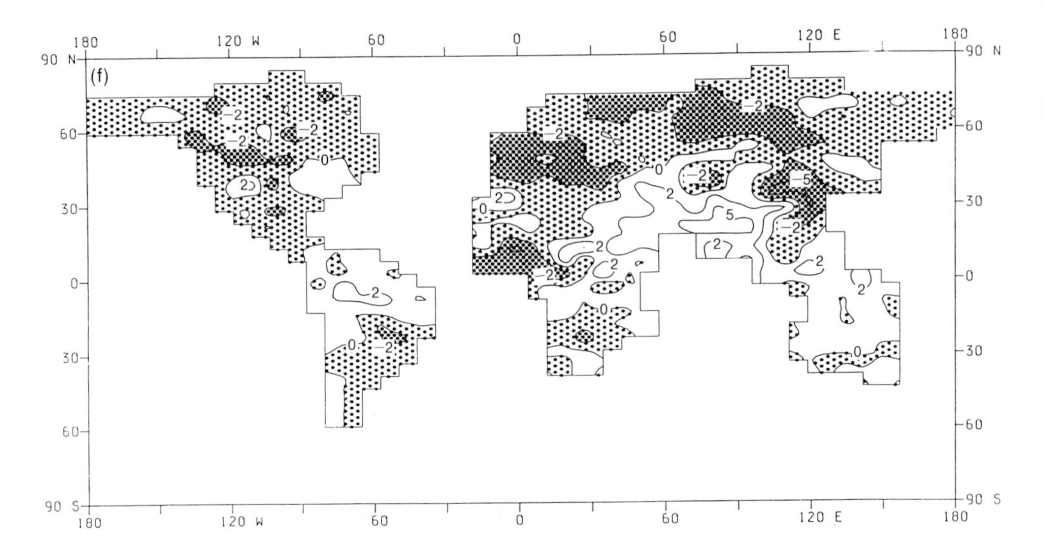

Figure 13 (a)

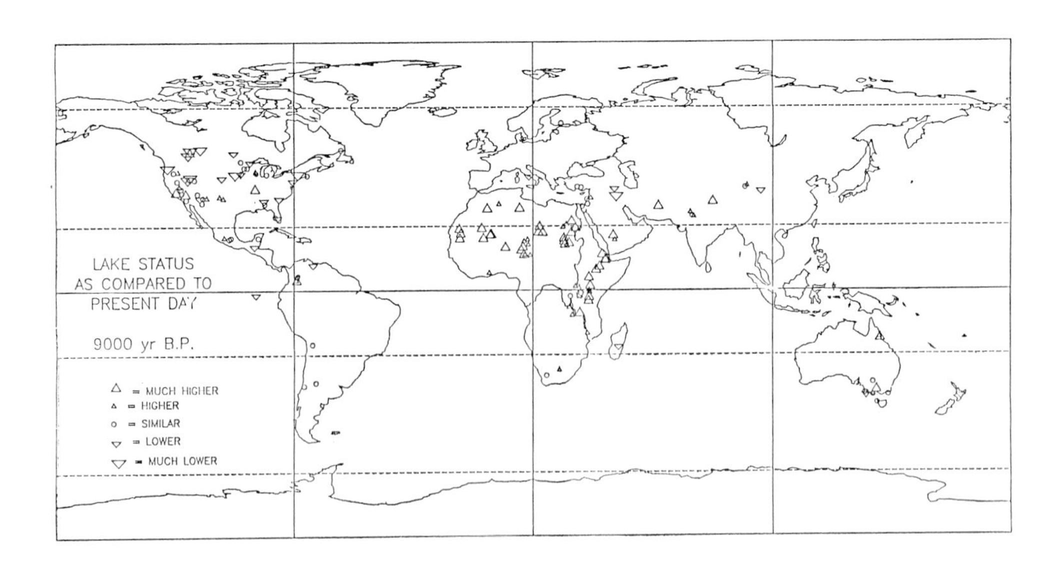

Figure 13 (b)

EUROEPAN SCHOOL OF CLIMATOLOGY AND NATURAL HAZARDS

Course on

"Climatic Change and Impacts: A General Introduction"

5. CLIMATIC IMPACTS

CLIMATIC CHANGES AND FUTURE LAND USE POTENTIAL IN EUROPE

M.L. PARRY and T.R. CARTER

Atmospheric Impacts Research Group
School of Geography
University of Birmingham
P.O. Box 363
Birmingham B15 2TT
U.K.

Summary

Future climatic changes due to increasing atmospheric concentrations of radiatively active trace gases are likely to induce significant regional changes in land use potential in Europe. Three approaches to assessing the effects of climatic change on land use potential are identified: descriptive, correlational and physiological. These are illustrated (with particular reference to agriculture) using examples from assessments already completed both within Europe and elsewhere in the world. Potential technological and management responses to climatic change are also briefly outlined.

An important conclusion is that our present imprecise understanding of future changes in climate or of possible trends in a range of socio-economic factors, does not allow for firm "predictions" of the likely state of European agriculture under an altered climate. However, assessments of such effects do provide useful indications of the "sensitivity" of land use to a range of possible future conditions.

INTRODUCTION

Background and objectives

There are growing indications that the earth is experiencing a long-timescale warming as a result of increasing concentrations of carbon dioxide and other radiatively active gases in the atmosphere. By the middle of the next century, global mean air temperatures could be 1.5°C to 5.5°C above what they are now (1). In Europe, model estimates suggest that mean temperatures may be between 2°C and 4°C above the present norm in summer and between 4°C and 14°C above in winter, with the greatest increases occurring at high latitudes. There are indications that precipitation may be greater in winter than today while in summer some regions may be drier, others wetter (2).

The purpose of this paper is to consider what might be the effects of these possible changes of climate on land-use potential in Europe. Our focus is on agriculture, and the emphasis is on identifying approaches that

* A modified version of this paper is being published in Brouwer, F.M. and Chadwick, M.J. (eds) (1989) Land Use Changes in Europe, Kluwer, Dordrecht.

are useful in the development of appropriate policies for managing the so-called "greenhouse effect" or for responding to it.

In the following sections we illustrate, with the aid of examples, some methods that have been employed in previous studies to evaluate the effect of climatic changes on agricultural potential. Our conclusion is that there are advantages in investigating the effects of climatic variations using a range of models, from simple descriptive indices (that tell something about the character of the climatic resources for agricultural production), through statistical models that facilitate the prediction of plant responses to climatic variations (though with little explanation of the mechanisms) to more complex physiological models that consider the processes of plant growth and development. Similarly, it is helpful to consider the effects of a range of climatic scenarios providing information both on the magnitude and on the rate of climatic change. Finally, in order to make the results more relevant for policy-makers, they should be interpreted in the light of a range of adaptations that could be implemented to limit, to anticipate, or to react to climatic change.

SPECIFIC APPROACHES

In this section, we review previous studies which have related climate and plant response, both for the present-day climate and under conditions of a changed climate. The scope of the review extends beyond agricultural crops to include other plants, as a number of studies have also considered the potential pattern and productivity of natural vegetation. Three broad approaches can be identified: descriptive, correlational and physiological.

The descriptive approach

In essence, this approach involves describing and comparing the spatial patterns of a plant attribute and the spatial patterns of those environmental phenomena (including climate) thought to influence this attribute. The approach, can be thought of as comprising five elements:

(1) The spatial pattern of the plant attribute (for example, plant growth rate, productivity or spatial distribution) is mapped based on measurements and observations from ground survey, aerial photographs or satellite imagery.

(2) The spatial patterns of important environmental phenomena are mapped (for example, soils on the basis of survey information, and long-term average climate).

(3) The mapped distributions in (1) and (2) are compared visually to identify any qualitative similarities.

(4) Those environmental factors that display a spatial pattern that corresponds to a particular plant attribute are selected as potential proxy indicators of that attribute.

(5) Where the proxy indicators are climatic, the effects of a climatic change can be simulated by adjusting the observed or derived climatic variables according to the "scenarios" of climatic change. The resulting pattern of climate can then be used qualitatively to infer changes in the spatial distribution of the plant attribute under study.

On their own, however, such direct measures of climate are seldom appropriate for describing the important climatic effects on plants. It more useful to combine these variables into agroclimatic indices that bear

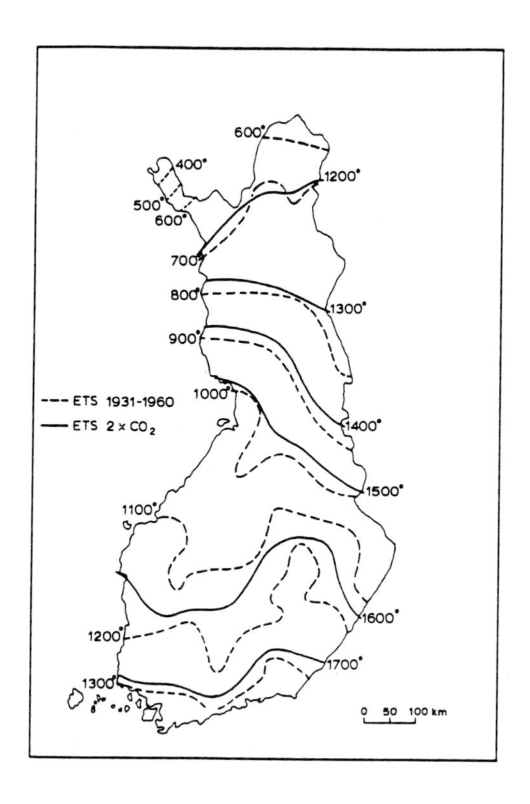

Figure 1 Effective temperature sums (degree-days) in Finland for the period 1931-1960 (after 33) and estimates for the GISS 2 x CO_2 climatic scenario. Source: (3).

a closer relation to the plant response observed. For example, effective temperature sum (the sum of air temperatures throughout the year above a threshold temperature) can be related to the growth requirements of agricultural crops and forests, and has been mapped for Finland by (3) under both the present-day (1931-1960) and GISS 2 x CO_2 scenario climates (Figure 1). Similarly, the length of the potential growing period under present-day and 2 x CO_2 climatic conditions has been computed for the whole of Europe (4). A third example is the hydrothermal coefficient (a measure combining information on growing season temperatures and precipitation), which has been used by (5) to characterise agroclimatic regions according to the degree of aridity in the Stavropol territory of the European USSR (Figure 2).

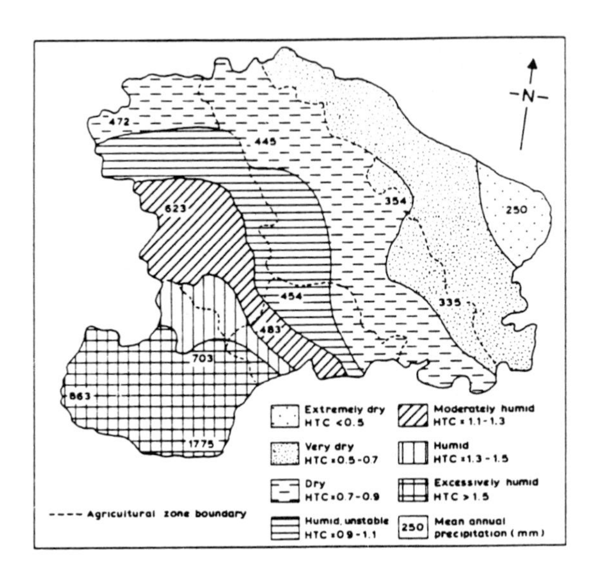

Figure 2 Agroclimatic regions of the Stavropol Territory, southern European USSR, defined according to the hydrothermal coefficient (HTC). Source: (5).

The correlational approach

The descriptive approach (above) can be used to identify those climatic variables that appear, qualitatively, to be related to certain plant attributes. In order to quantify these relationships, the conventional method employed is to use statistical techniques of correlation and regression. In traditional ecology this type of analysis involves relating certain climatic elements (e.g., air temperature, effective temperature sum, precipitation, potential evapotranspiration, etc.) to observed plant attributes (e.g., presence/absence, productivity, growth rate, yield, etc.) across a global or regional set of locations to produce a single prognostic model. Validation of the predictive performance of the model is conducted by using climatic data from an additional site (not used in the model construction) to "predict" the plant attributes at that site, which are then compared with actual (observed) plant data. Following validation, global sets of long-term mean climate can be input to the model to predict plant attributes at as many sites as there are meteorological stations supplying data, hence providing a basis for mapping. Similarly, the mean climate can be adjusted according to specified climatic scenarios, and these altered values input to the model.

There are many examples of this approach, including relationships that have been identified between vegetation zones and climate (e.g., the Holdridge Life Zone system - 6,7), between net primary productivity (NPP) of natural vegetation and climate (e.g., the "Miami Model" of NPP - 8) and between agricultural potential and climate (e.g., the Climatic Index of Agricultural Potential - 9). The correlational approach has also been employed in estimating the effects of climatic change. For example, a semi-empirical model of biomass potential (10) - combining elements of the methods to compute potential evapotranspiration (9), and to relate evapotranspiration to NPP (11) - was used to estimate the percentage changes in theoretically achievable biomass production of a mixed grass sward (relative to the present-day) under two 2 x CO_2 scenario climates in the European Community (12, 13 - Figure 3). A second example is the distribution of vegetation zones or "zonobiomes" as defined by Walter (14), which has been mapped for Europe (15) under both the present-day (1931-1960) climate and a hypothetical 2 x CO_2 climate (Figure 4).

There are a number of significant drawbacks in using the correlational approach, however, the first being that it does not attempt to explain the mechanisms and processes whereby climate affects plant response. The relationships are statistically derived, presenting a black box with climate as an input and plant attributes as outputs. Secondly, results are strongly dependent upon the implicit assumption that it is valid to employ a statistical relationship developed on the basis of spatial variations in climate to estimate plant responses to temporal variations in climate at individual sites. Thirdly, the large changes in climate implied in some future projections are well outside the observed range of climatic variations at individual sites, and may well either lie outside the range or not resemble the combination of conditions to be found anywhere on the earth at the present day. Thus, the predictions of future climatically-determined plant attributes rely strongly on uncertain model extrapolations.

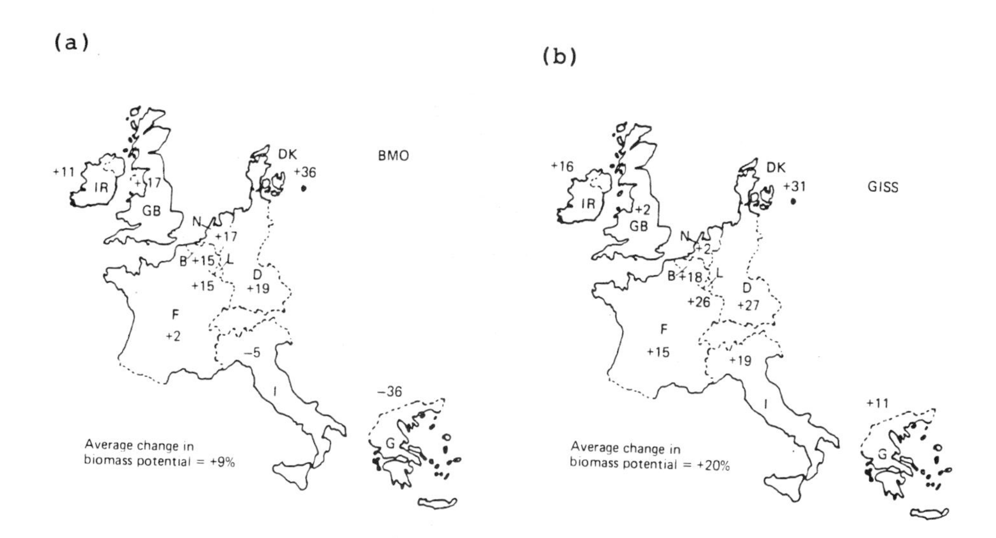

Figure 3 Modelled changes in average biomass potential of ten European Community countries relative to the present-day under: (a) the BMO 2 x CO_2 climatic scenario and (b) the GISS 2 x CO_2 climatic scenario (after 10). Source: (13).

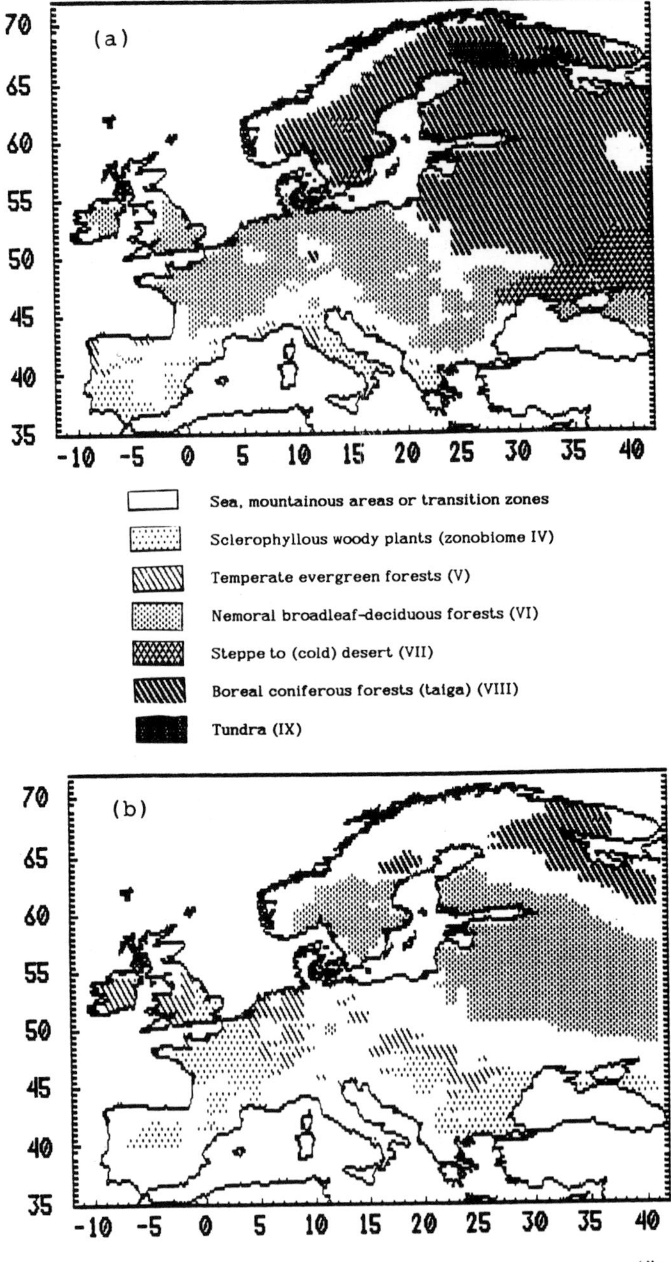

Figure 4 The potential distribution of natural vegetation ("zonobiomes" - 14) based on average temperature and precipitation for: (a) the 1931-1960 period and (b) hypothetical increases in mean annual temperature and annual precipitation amount of 5°C and 10%, respectively. Source: (15).

Partial solutions can be found, however, to the above problems. They involve the use of temporally- (rather than spatially-) based correlation models. These models are commonly employed to estimate regional crop yields, and are constructed by relating time series of observed crop yields to time series of one or more observed climatic variables over an equivalent period, using statistical techniques such as regression or principal components analysis. Such models should be tested in regions other than that for which they were constructed to indicate their general applicability, but such global relationships are rarely obtainable for single crops. Alternatively, separate models can be constructed for different regions and operated in parallel to provide a large-area assessment of crop responses to climatic variations. For example 294 regression equations were developed by Hanus to relate climatic variables to wheat yields at 42 sites in the European Community (16), and subsequently used to estimate the possible effect of two 2 x CO_2 scenario climates on yields (12, 13). The results were averaged across individual countries, and are presented as percentage changes in yield for the period January to July, and as monthly-averaged changes during the same period, both relative to recorded national average yields (Table I).

A second technique, which has some potential for testing the validity of extrapolating temporally-derived statistical relation-ships outside the range of observed regional conditions, is the use of regional analogues (17). This involves seeking those regions that have a climatic regime at the present day which resembles the future climate projected for a region of interest. Examination of present-day crop responses to climatic conditions in the "analogue region" may provide useful information about the robustness of model relationships in the study area.

To illustrate this, mean monthly temperatures in Iceland estimated for the GISS 2 x CO_2 scenario (see above) were found to be markedly warmer than the warmest months observed in the instrumental record in Iceland, resembling instead those of northern Britain at the present-day (18). In studies of the likely effects of the 2 x CO_2 climate on grass yields in Iceland, information on temperatures and grass growth in northern Britain were thus used to justify the extrapolation of relationships developed between temperature and grass growth under present-day conditions in Iceland.

Table I Mean monthly and cumulative changes in wheat yield estimated for eight European Community countries in response to two scenarios based on global climate model projections of the future climate under doubled concentrations of atmospheric carbon dioxide. Values are expressed as percentages of 1975-1979 average yields (after 13)

Country	Average yield of wheat* and spelt, 1975-1979 (t/ha)	Mean monthly yield changes (%)		Cumulative yield changes (%)	
		GISS 2 x CO_2 scenario	BMO 2 x CO_2 scenario	GISS 2 x CO_2 scenario	BMO 2 x CO_2 scenario
Ireland	4.79	-1.3	-2.5	-8.8	-17.5
Denmark	5.25	+2.7	+0.2	+18.7	+1.1
Netherlands	5.82	+0.2	0.0	+1.2	+0.3
Belgium	4.72	-1.5	-1.1	-9.5	-6.8
Luxembourg	3.09	+1.0	+1.0	+7.8	+6.1
France	4.38	-1.4	-1.8	-9.6	-12.3
F.R.G.	4.66	-0.2	-1.1	-1.1	-8.6
Italy	2.58	0.0	-0.4	-0.8	-1.2

* Crop is winter wheat in all countries except Ireland, where not specified.

The physiological approach

In contrast to the correlational approach, the physiological approach attempts to incorporate our understanding of plant physiological processes within a model framework. Some of these processes are represented as deterministic functions that are well enough understood to be regarded as accepted biophysical laws. Others that are either poorly understood or of secondary interest are frequently represented by empirical functions. Thus, no physiological approach can be described as truly deterministic; all incorporate at least some empirical (black-box) elements. The different plant processes are brought together into a model framework that is usually dynamic and frequently non-linear. For example, this approach is often capable of representing critical levels of plant tolerance to climate. If these levels are exceeded only slightly, this may have a disproportionate effect on plant response. Such non-linearities and discontinuities of response are rarely incorporated in correlation-based schemes.

There have been few simulation studies in Europe which employ physiological models to evaluate the potential effects of climatic change. Models have been used, however, to evaluate the climatic control on plant attributes in Europe under the present-day climate, and several are potentially applicable to climatic change experiments. For example, a semi-empirical grass yield model was developed for Europe by Hume et al. (19). This requires daily climatic data as input, and has been used to evaluate and map grass yields for much of western Europe on the basis of data from the 1951-1980 period.

Elsewhere in the world, there are a number of illustrations of physiological models being used in climatic change experiments. For example, the so-called "Chikugo model" has been used to estimate net primary productivity of natural vegetation, both at a global scale (20) for the present-day (1931-1960) climate, and at a national scale (21) for scenarios of changed climate in Japan (Figure 5). In form, the Chikugo

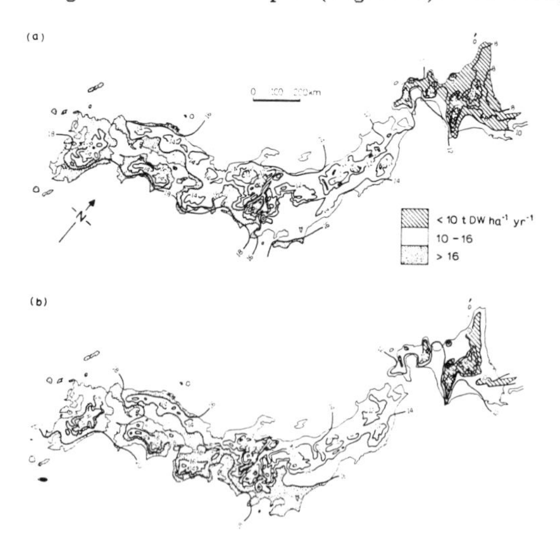

Figure 5 Distribution maps of net primary productivity of natural vegetation estimated using the Chikugo model for: (a) the baseline climate (1951-1980) and (b) the GISS 2 x CO_2 climatic scenario. Source: (21).

model closely resembles correlational models of NPP, but it differs from these in that the simple model-equation is derived from physiological theory. Use of this model in Europe would provide a useful comparison with the results of the Miami model (see above).

In the United States, an intensive study by the Environmental Protection Agency (EPA) is employing generic crop-climate models to assess the possible effects of future long-term climatic change on crop yields in four regions of the United States (22). These include the CERES models for wheat and maize (23, 24), which have been tested across a wide range of environmental conditions and thus appear to offer promising possibilities for adoption in climatic change experiments in regions such as Europe, providing that appropriate input data are available to run the models.

A SUMMARY OF POSSIBLE EFFECTS OF CLIMATIC CHANGE ON AGRICULTURAL POTENTIAL IN EUROPE

In this section we note briefly the results of some recent studies of effects of climatic change on agricultural potential in Europe. It should be emphasised that a high degree of uncertainty is attached to all such results. They are included here for illustrative purposes only. To facilitate comparison of the results, only those that relate to the same scenario of climatic change are discussed here. This is the GISS 2 x CO_2 scenario (see above), which has been employed in two studies of European agriculture: an EC-funded study of the European Community countries, completed in 1983 (12) and an IIASA/UNEP-funded study, which included four regional case studies in Europe (among eleven studies worldwide), completed in 1987 (25, 26).

The performance of the GISS model in simulating present-day climate in Europe has been tested by comparing the GISS 1 x CO_2 model estimates with observed 30-year averaged data (Figure 6). It may be noted here that while the GISS model reproduces the observed values of mean monthly air temperature reasonably well, the fit for mean monthly precipitation is poor. The deviations between estimated and observed precipitation in certain areas may exceed 100 percent in some months. It follows that our confidence in the GISS 2 x CO_2 estimates is also subject to large uncertainties. It should therefore be emphasised that they are regarded in all the studies described below not as predictions, but as scenarios of future climate that are useful in extending our understanding of the methods of estimating biophysical and economic effects of climatic change.

Estimated effects in the European USSR

These results are summarised from two IIASA/UNEP case studies in the European USSR. Most results are from the first study, of three regions in the northern half of the area [the Leningrad, Cherdyn and Central (Moscow) regions - 27], while the remainder are for a second study in the Saratov region, in the south-east of the area (28).

The GISS model-derived data were utilised in two ways in the USSR study: (i) as a discrete, step-like perturbation between the present climate and the doubled CO_2 climate (ii) as the basis of a scenario of transient changes of climate, when combined with an empirical approach developed by Vinnikov and Groisman (29). Assuming a doubling of CO_2 to occur in the year 2050, smooth linear changes in climate were assumed between the present-day and the year 1995 and between 1995 and 2005,

(a)

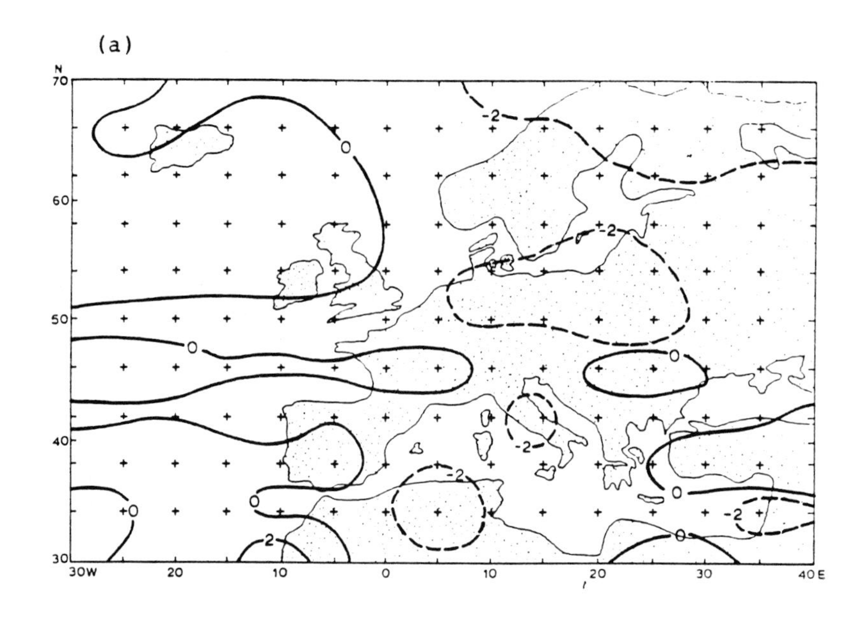

(b)

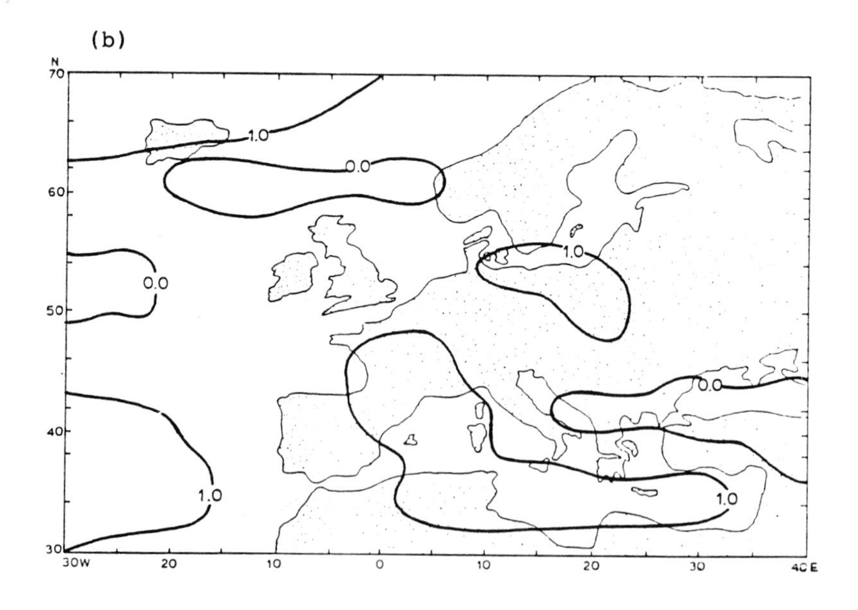

Figure 6 Differences between the GISS GCM control climate and the observed climate (1931-1960) over Europe for: (a) mean annual temperature (K), and (b) mean annual precipitation rate (mm/day). Source: (34).

according to the Vinnikov and Groisman estimates, followed by a third linear trend from 2005 to 2050, to reach the levels of change estimated by the GISS model. The period 1951-1980 was adopted as a reference against which to compare the scenario results.

Several arbitrary scenarios of climatic change (+1.0°C and +1.5°C relative to the baseline) were also considered in the northern USSR study. These are summarised here but not discussed in detail.

Estimates of effects are summarised in Figure 7. In the Leningrad Region winter rye yields (estimated using the semi-empirical VNIISI-Obukhov environmental model) are reduced by 13 percent as a result of higher temperatures and increased rates of evapotranspiration, assuming present-day technology (27). The response of winter rye yield and some environmental parameters were also considered in relation to the transient climatic scenario over 1980-2035. Winter rye yields (relative to the upward trend estimated for improved technology and management) increase up to 2010, but are 23 percent below trend by 2035. An explanation for this is that relatively small increases in temperature appear to be beneficial, but these are subsequently countervailed by large increases in precipitation, leading to increased soil degradation (due to increased waterlogging and erosion) and increased nitrate leaching with concomitant increases in surface water pollution and reduced soil fertility (Table II).

In the Cherdyn region, spring wheat yields (estimated using a process-based crop-growth simulation model) decreased slightly relative to yields simulated for the 1951-1980 baseline period, owing to water stress and to the premature development of the crop under the increased temperatures. When the "direct" effects of doubled CO_2 concentrations on crop photosynthesis were combined with the climatic effects, however, a 17 percent increase in yields was estimated (27).

Similar experiments using the same model, but in the warmer and drier Saratov region (some 1000km to the south of Cherdyn), showed a yield increase of about 14 percent relative to the baseline (80 percent if direct CO_2 effects are included), mainly due to the 22 percent increase in growing season precipitation implied by the GISS 2 x CO_2 scenario for this region (28).

Table II Changes relative to trend of winter rye yield (Y), an index of soil fertility (SI), groundwater level (GW) and surface water pollution by nitrogen from agricultural watersheds (POL) in response to a scenario of "transient" changes in growing season air temperature (T) and precipitation (P) relative to the 1951-1980 baseline period (adapted from 27).

			Changes relative to trend			
Year	T (°C)	P (mm)	Y (%)	SI (%)	GW (%)	POL (%)
1990	+0.4	+8	+4	-1	+2	+18
2000	+0.8	+16	+5	-4	+4	+30
2010	+1.2	+38	+5	-11	+9	+75
2020	+1.6	+74	-4	-22	+14	+213
2035	+2.2	+127	-23	-42	+38	+325

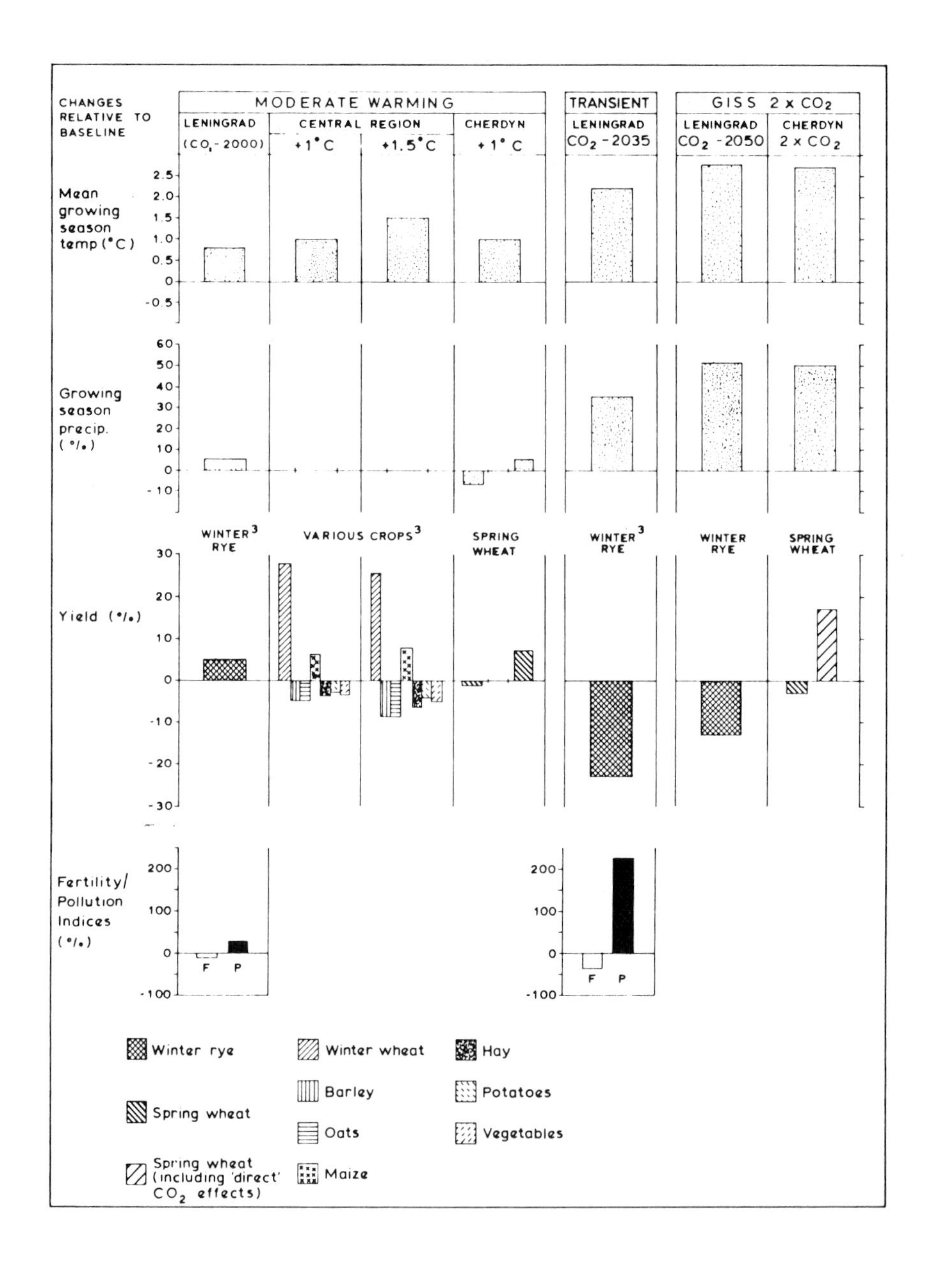

Figure 7 Estimated effects of climatic variations on agricultural production in the northern European USSR. Baseline climate is 1951-1980 for the Leningrad and Cherdyn regions and 1931-1960 for the Central Region. 1 = climatic change estimated using an empirical method (29); 2 = assumed date of CO_2-doubling; 3 = relative to technology trend. Data from (27); source: (17).

Further estimates for arbitrary changes in temperature are summarised for the Central Region (Moscow area) in Figure 7. In general, temperature increases of 1.0°C and 1.5°C lead to increases in yields (estimated using a process-based crop production model) of winter wheat and maize for silage, but decreases in yields of barley, oats, potatoes, hay and vegetables (27).

Estimated effects in western Europe

The majority of these results are summarised from the 1983 EC study (12). Two crop-climate models (introduced in Section 3, above) were used to assess the effects of changes in temperature and precipitation estimated under both the GISS and the UKMO 2 x CO_2 scenarios: an empirical/statistical (correlational) model for estimating winter wheat yields (16), and a semi-empirical model of biomass production (10). Changes in average annual yields of winter wheat and spelt by country were presented above in Table I. Results for biomass potential are given in Figure 3. Comparison between the results is difficult for such aggregated (national-level) data, and no clear conclusions can be drawn. There are indications that the effect of the GISS scenario on wheat yields is more negative than that of the BMO scenario, while the effects on biomass potential appear to be more conservative for the GISS than for the BMO scenario. However, perhaps the most important lesson to be drawn from the discrepancies between results, is that they emphasis the uncertainties surrounding GCM projections of future climate.

In a further study, a crop-climate simulation model developed by Hough (28) was employed to estimate the response of winter wheat yields to the GISS 2 x CO_2 scenario temperature changes in northern England (31). Assuming no change in crop varieties, the mean annual temperature increase of about 3.7°C is estimated to shorten the growing period of crops in lowland areas by about two months and to reduce mean yields of well-watered crops by up to one-third. In contrast, the temperatures at upland sites above about 500 metres (where no crops are grown today) would be close to optimum for winter wheat growth. Furthermore, the effects of water stress on crops (not modelled here) could be expected to intensify this implied upland-lowland reversal in relative crop potential. Similar results have recently been obtained using another winter wheat productivity model for a range of hypothetical temperature increases at sites in the United Kingdom (32).

Estimated effects in northern Europe

These results are summarised from two of the IIASA/UNEP case studies: of Iceland (18) and of Finland (3).

In both countries, the estimated increase in both mean annual and growing season temperatures under the GISS 2 x CO_2 scenario is greater than further south in Europe (exceeding 4°C in nearly all months). Precipitation is also estimated to increase throughout the year in both countries.

In Iceland, effects under the GISS 2 x CO_2 scenario exceed those estimated for anomalously warm years in the instrumental record (Figure 8). This is due both to the large degree of warming and to the relatively low present-day variability inherent in Iceland's highly maritime climate.

Under the GISS 2 x CO_2 scenario, hay yields are estimated (using an empirical statistical model) to increase by 66 percent relative to the

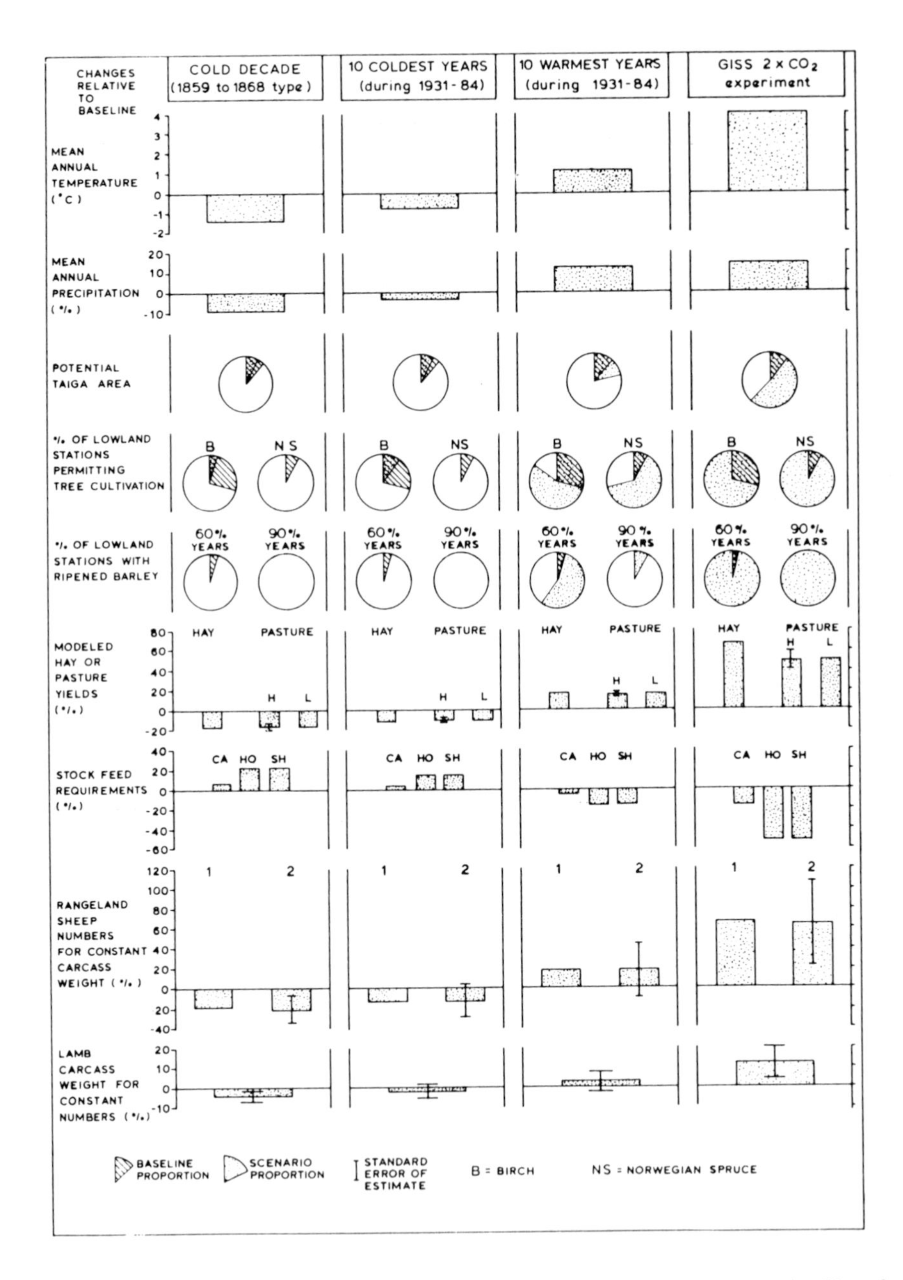

Figure 8 Estimated effects of climatic variations on agricultural production in Iceland. Baseline climate is 1951-1980. H = high input (120 kg/ha nitrogen); L = low input (80 kg/ha); Ca = cattle; Ho = horses; Sh = sheep; 1 = from national model; 2 = from refined national model. Data from (18); source: (17).

present-day (compared with a range of -13 to +18 percent for the 10 coldest years and the 10 warmest years in the period 1931-1984, respectively). The respective estimates for rangeland carrying capacity and sheep carcass weight are increases of 64 percent and 12 percent (compared with ranges of -13 to +18 percent and -2 to +3 percent). Under this scenario, barley would ripen 6 years in 10 at all lowland locations (compared with only 4 percent of lowland locations at present), and the theoretical forested area would increase sixfold (18).

In Finland, effects under the GISS 2 x CO_2 scenario exceed those for anomalously warm periods under the present-day climate (Figure 9), though the difference is less marked than in Iceland. The latter is due partly to a weaker temperature/yield relationship for grain crops in Finland than for grass crops in Iceland, partly to the sensitivity of Finnish grain yield to precipitation changes, and partly to the greater inter-annual variability of temperatures in Finland than in Iceland.

Yields of grain crops were estimated using empirical-statistical models in this study. Under the GISS 2 x CO_2 scenario, barley yields are estimated to increase by between 9 percent and 21 percent (depending on location and on the model employed). This compares with an equivalent range of +1 to +12 percent in an anomalously warm period like the 1930s. Spring wheat yields (assuming that longer-season varieties are introduced under the warmer climate) are estimated to increase by about 10 percent in southern Finland and by about 20 percent in central Finland (compared with +5 percent and +15 percent, respectively, for present varieties during a recent anomalously warm period (3).

POTENTIAL TECHNOLOGICAL AND MANAGEMENT RESPONSES

Changes in management practices or technology, which are assumed to be fixed in the majority of impact experiments, are the natural farm-level responses to cope with climatic changes. These can be simulated by some models in so-called "adjustment experiments", a number of which were employed in the IIASA/UNEP studies (17):

(1) Changes in crop variety:

- Changes from spring-sown to winter-sown cereal varieties to take advantage of warmer winters and withstand better the increased frequency of moisture stress in some regions resulting from higher temperatures.

- Changes to varieties with higher thermal requirements, to exploit longer and warmer growing seasons.

- Changes to varieties giving less variable yields.

(2) Changes in fertilizing and drainage:

- Altered fertilizer applications, to optimise yields under the changed climate.

- Improved field drainage, to reduce the risk of water-erosion, offset the risks of waterlogging from increased precipitation and dispose more efficiently of nitrate pollutants.

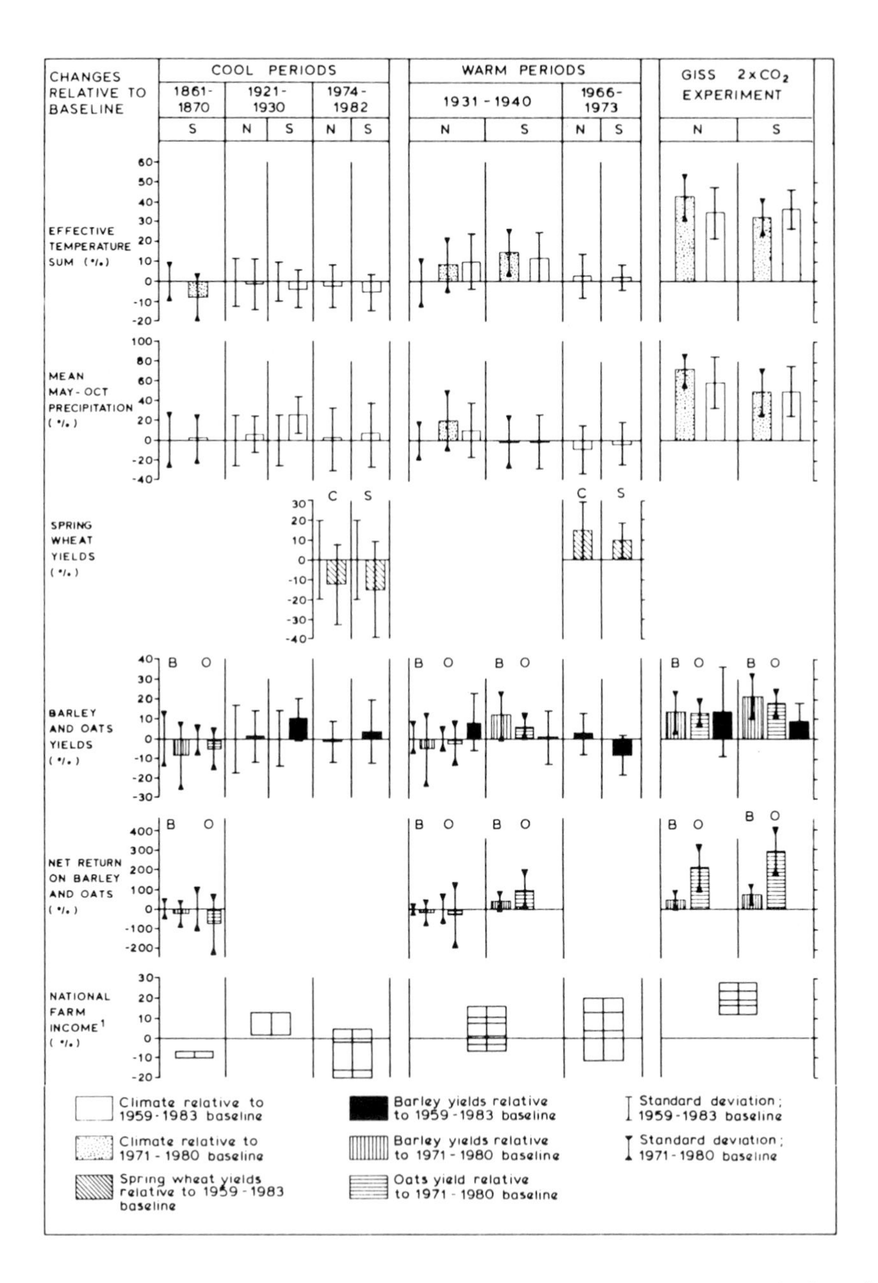

Figure 9 Estimated effects of climatic variations on agricultural production in northern (N), central (C) and southern (S) Finland. Baseline climate is 1959-1983 or 1971-1980, as shown. 1 = estimated farm income based on yield estimates from all regions. Data from (3); source (17).

(3) Changes in land allocation:

* Changes of land use to optimise production.
* Changes of land use to stabilise production.

CONCLUSIONS

Extrapolating assessments of effects

A necessary conclusion from the above findings is that it is at present inappropriate to attempt any "predictions" of the state of agriculture in Europe under a future altered climate. Rather, the results so far obtained represent a useful sensitivity analysis of present-day agriculture, enabling the identification of those aspects and areas that may be especially vulnerable to climatic variations. However, when viewed as a whole, the findings do indicate the following:

(1) The varying degrees of absolute change in climate at different locations are not necessarily reflected in the spatial pattern of their impacts. Instead, they tend to be a function of the change in climate relative to the existing (baseline) conditions.

(2) Yield-climate relationships for different crops are almost always non-linear, differ between locations, and are closely tied to levels of technology and management (e.g., fertilizer application levels, planting dates, etc.), which also vary spatially.

(3) Future changes in climate will lead to a spatial shift of crop productivity potential requiring a relocation of cropping patterns from the present situation.

(4) Changes in mean climate (even assuming no changes in interannual variability) can lead to changes in crop yield variability. For example, while mean grain yields are estimated to increase in all parts of Finland under the GISS 2 x CO_2 scenario, interannual yield variability decreases in the south but increases in the north of the country, probably due to the large precipitation increases projected for the north (3).

(5) Most of the assessments summarised above assume that factors such as technology and management remain fixed at the present-day levels.. This assumption is unrealistic in the context of the long-term nature of the projected climatic changes.

Specific research priorities

Given the uncertainties described above, the following items (which are not necessarily specific to Europe) merit further research:

(1) Provision of improved scenarios of future climatic change, incorporating regional (sub-grid-scale) detail and information on anomalous climatic events that can have a major effect on agricultural output.

(2) The refinement and validation of impact models, with further

experimentation for a wider range of agricultural activities, reflecting not only biophysical effects but also "downstream" economic and social effects of climatic variations.

(3) Consideration of the effects of climatic changes in combination with other related or unrelated effects occurring simultaneously (e.g., the direct effects of CO_2 on crop yields, the effects on crop yields of climatically-related changes in the incidence of pests and diseases).

(4) Further experiments with a range of potential farm-level adjustments in agriculture, to explore their efficacy in mitigating impacts or exploiting options.

(5) More exploration of the potential economic and political responses to climatic change (e.g., restructuring of price support systems, insurance mechanisms, etc.).

REFERENCES

1. WMO (1986). Report of the International Conference on the Assessment of the Role of Carbon Dioxide and of Other Greenhouse Gases in Climate Variations and Associated Impacts, Villach, Austria, 9-15 October 1985. WMO No. 661, World Meteorological Organization, Geneva, 78 pp.

2. SCHLESINGER, M.E. and MITCHELL, J.F.B. (1985). Model projections of the equilibrium climatic response to increased carbon dioxide. In M.C. MACCRACKEN and F.M. LUTHER (eds) Projecting the Climatic Effects of Increasing Carbon Dioxide, United States Department of Energy, DOE/ER-0237, Washington, D.C., pp. 81-147.

3. KETTUNEN, L., MUKULA, J., POHJONEN, V., RANTANEN, O. and VARJO, U. (1988). The effects of climatic variations on agriculture in Finland. In M.L. PARRY, T.R. CARTER and N.T. KONIJN (eds) The Impact of Climatic Variations on Agriculture Volume 1 Assessments in Cool Temperate and Cold Regions, Kluwer, Dordrecht, The Netherlands, pp. 513-614.

4. BROUWER, F.M. (1988). Determination of broad-scale landuse changes by climate and soils. Working Paper WP-88-007, International Institute for Applied Systems Analysis, Laxenburg, Austria, 21 pp.

5. NIKONOV, A.A., PETROVA, L.N., STOLYAROVA, H.M., LEBEDEV, V.Yu., SIPTITS, S.O., MILYUTIN, N.N. and KONIJN, N.T. (1988). The effect of climatic variations on agriculture in the semi-arid zone of the European USSR. A. The Stavropol Territory. In M.L. PARRY, T.R. CARTER and N.T. KONIJN (eds) The Impact of Climatic Variations on Agriculture Volume 2. Assessments in Semi-Arid Regions, Kluwer, Dordrecht, The Netherlands, pp. 579-627.

6. HOLDRIDGE, L.R. (1947). Determination of world plant formations from simple climatic data. Science, 105, 367-368.

7. HOLDRIDGE, L.R. (1964). Life Zone Ecology. Tropical Science Center, San Jose, Costa Rica.

8. LIETH, H. (1975). Modeling the primary productivity of the World. In H. LIETH and R.H. WHITTAKER (eds) Primary Productivity of the Biosphere. Springer-Verlag, New York, pp. 237-263.

9. TURC, L. and LECERF, H. (1972). Indice climatique de potentialite agricole. Science du Sol, 2, 81-102.

10. BRIGGS, D.J. (1983). Biomass potential of the European Community. Report for the Environment and Consumer Protections Service of the European Community, Sheffield, U.K.

11. LIETH, H. and BOX, E.O. (1972). Evapotranspiration and Primary Productivity; C.W. Thornthwaite Memorial Model. Publications in Climatology, Volume 25 (2), C.W. Thornthwaite Associates, Centerton/Elmer, New Jersey, pp. 37-46.

12. MEINL, H., BACH, W., JAEGER, J., JUNG, H-J., KNOTTENBERG, H., MARR, G., SANTER, B. and SCHWIEREN, G. (1984). Socio-economic impacts of climatic changes due to a doubling of atmospheric CO_2 content. Commission of the European Communities, Contract No. CLI-063-D, 642 pp.

13. SANTER, B. (1985). The use of general circulation models in climate impact analysis - A preliminary study of the impact of a CO_2-induced climatic change in west European agriculture. Climatic Change, 7, 71-93.

14. WALTER, H. (1985). Vegetation of the Earth and Ecological Systems of the Geobiosphere (3rd ed). Springer-Verlag, New York.

15. DE GROOT, R.S. (1987). Assessment of the potential shifts in Europe's natural vegetation due to climatic changes and the implications for conservation. Young Scientists Summer Program 1987, Final Report. International Institute for Applied Systems Analysis, Laxenburg, Austria, 34 pp.

16. HANUS, H. (1978). Forecasting of crop yields from meteorological data in the EC countries. Statistical Office of the European Communities, Agricultural Statistical Studies, No. 21.

17. PARRY, M.L. and CARTER, T.R. (1988). The assessment of effects of climatic variations on agriculture: aims, methods and summary of results. In M.L. PARRY, T.R. CARTER and N.T. KONIJN (eds) The Impact of Climatic Variations on Agriculture. Volume 1. Assessments in Cool Temperate and Cold Regions, Kluwer, Dordrecht, The Netherlands, pp. 11-95.

18. BERGTHORSSON, P., BJORNSSON, H., DYRMUNDSSON, O., GUDMUNDSSON, B., HELGADOTTIR, A. and JONMUNDSSON, J.V. (1988). The effect of climatic variations on agriculture in Iceland. In M.L. PARRY, T.R. CARTER and N.T. KONIJN (eds) The Impact of Climatic Variations on Agriculture Volume 1. Assessments in Cool Temperate and Cold Regions, Kluwer, Dordrecht, The Netherlands, pp. 383-509.

19. HUME, C.J. ET AL. (1986). A quantitative study of the effects of climate variability and climatic change on herbage production from

intensively managed grassland in western European countries. Commission of the European Communities, Contract No. CLI-058081-UK (H).

20. UCHIJIMA, Z. and SEINO, H. (1988). Probable effects of CO_2-induced climatic change on agroclimatic resources and net primary productivity in Japan. Bull. Natl. Inst. Agro-Environ. Sci., 4, 67-88.

21. YOSHINO, M., HORIE, T., SEINO, H., TSUJII, H., UCHIJIMA, T. and UCHIJIMA, Z. (1988). The effect of climatic variations on agriculture in Japan. In M.L. PARRY, T.R. CARTER and N.T. KONIJN (eds) The Impact of Climatic Variations on Agriculture. Volume 1. Assessments in Cool Temperate and Cold Regions, Kluwer, Dordrecht, The Netherlands, pp. 725-868.

22. EPA (1988). Report to Congress on the Effects of Global Climate Change. Office of Policy, Planning and Evaluation, United States Environmental Protection Agency, Washington D.C. (in preparation).

23. JONES, C.A. and KINIRY, J.R. (eds) (1986). CERES - maize: a simulation model of maize growth and development. Texas A and M University Press, College Station, 198 pp.

24. RITCHIE, J.T. and OTTER, S. (1985). Description and performance of CERES-wheat: a user-orientated wheat yield model. In W.O. WILLIS (ed) ARS Wheat Yield Project. USDA-ARS. ARS-38 pp. 159-175.

25. PARRY, M.L., CARTER, T.R. and KONIJN, N.T. (eds) (1988a) The Impact of Climatic Variations on Agriculture. Volume 1. Assessments in Cool Temperate and Cold Regions, Kluwer, Dordrecht, The Netherlands, 876 pp.

26. PARRY, M.L., CARTER, T.R. and KONIJN, N.T. (eds) (1988b) The Impact of Climatic Variations on Agriculture. Volume 2. Assessments in Semi-Arid Regions, Kluwer, Dordrecht, The Netherlands, 764 pp.

27. PITOVRANOV, S.E., IAKIMETS, V., KISILEV, V.I. and SIROTENKO, O.D. (1988a). The effects of climatic variations on agriculture in the subarctic zone of the USSR. In M.L. Parry, T.R. Carter and N.T. Konijn (eds) The Impact of Climatic Variations on Agriculture. Volume 1. Assessments in Cool Temperate and Cold Regions, Kluwer, Dordrecht, The Netherlands, pp. 615-722.

28. PITOVRANOV, S.E., MAXIMOV, A.D., SIROTENKO, S.E., ABASHINA, E.V., PAVLOVA, V.N. and CARTER, T.R. (1988b). The effects of climatic variations on agriculture in the semi-arid zone of the European USSR. B. The Saratov Region. In M.L. PARRY, T.R. CARTER and N.T. KONIJN (eds) The Impact of Climatic Variations on Agriculture Volume 2. Assessments in Semi-Arid Regions, Kluwer, Dordrecht, The Netherlands, pp. 629-664.

29. VINNIKOV, K. Ya. and GROISMAN, P. Ya. (1979). An empirical model of present-day climatic changes. Meteorologia i Gidrologia, 3, 25-28 (in Russian). English translation available in: Soviet Meteorol. Hydrol., 3.

30. HOUGH, M.L. (1980). A proposed cereal yield simulation model. Agricultural Memorandum No. 892, Meteorological Office, Bracknell, Mimeo, 14pp.

31. CARTER, T.R. (1988). Climatic change and cropping margins in upland Britain. Unpublished Ph.D. thesis, University of Birmingham, U.K., 370 pp.

32. SQUIRE, G.R. and UNSWORTH, M.H. (1988). Effects of CO_2 and climatic change on agriculture. Contract Report to the Department of the Environment, University of Nottingham, Sutton Bonington, U.K., 31 pp.

33. KOLKKI, O. (1969). Katsaus Suomen ilmastoon. Ilmatieteen Laitoksen Tiedonnantoja, 18, 1-18, Helsinki, Finland.

34. BACH, W. (1988). Development of climatic scenarios: A. From general circulation models. In M.L. PARRY, T.R. CARTER and N.T. KONIJN (eds) The Impact of Climatic Variations on Agriculture Volume 1. Assessments in Cool Temperate and Cold Regions, Kluwer, Dordrecht, The Netherlands, pp. 125-157.

G. Maracchi
Università di Firenze
Istituto di Analisi Ambientale e telerilevamento applicati all'agricoltura - Consiglio Nazionale delle Ricerche

Impacts of climatic change on crops

Summary

The response of crops to climate depends on their physiological characteristics.

In order to assess this response in a way suitable to forecast a climatic change as derived from climate models, the identification of the relevant parameters for crop production is needed.

We summarize the criteria for choosing these parameters, and an agroclimatic prodedure - AGROCLIMET - is proposed.

Introduction

Climatic change could affect crops both in gross production and geographic distribution.

The degree of flexibility of the response of crops to the environment is related to their genetic variability which can be naturally or artificially induced. As an example, the area of cultivation of wheat extends from high latitudes to North Africa as a result of breeding techniques done in the past.

On the other hand, the population characteristics and flexibility of trees and natural vegetation being the result of natural mechanisms are more restricted.

The physiological characteristics of plants depend upon their genetics and biochemical processes. The whole determines the diversity of plants in terms of morphology, leaf size and geometry, roots size and length, height, and as a response to the environment, the length of the growing season, the growing temperature threshold, the functioning of the photosynthetic system, stomata resistence among other factors.

Table I summarizes the main processes and parameters determining production.

TABLE I. Main processes and parameters of plant production.

Growth		Development	
Photosynthesis	Radiation Temperature	Phenology	Daylength Temperature
Transpiration	Water availability Saturation deficit Radiation Wind Speed Temperature		Water
Nutrition	Soil pH, E_7C.C Nutrients		

One of the aims of the Climatology Research Programme of the European Community is to evaluate climate change. A change on crops response to geographical distribution or production is to take into consideration.

The prediction of the effect of a climatic change, or of the increase of CO_2 concentration directly on crops production and their economic consequences is to be attempted.

Factors of crop production

Photosynthesis is on the basis of crop production. It is a light dependent process (Table I). When the other factors are at the best level there is a linear relationship between the biomass accumulation and the summation of daily amount of radiation intercepted by the canopy during the growing period. The architecture of foliage strongly affects the interception of photosynthetically solar radiation, which is a fraction of the global solar radiation. The repartition of the light in the canopy can be computed by means of the extinction coefficient.

The capacity of the canopy to accumulate dry matter as a function of environmental parameters can be measured by the Water Use Efficiency index. This is the ratio between the amount of biomass produced and the amount of water transpire by the canopy. There is a linear relationship between the amount of transpired water and the total biomass. Water availability to plants is highly variable in space and time, as it primarily depends on rainfall distribution, topography, soil characteristics. As a result, plants adopt many strategies in order to face aridity. Among the means to adapt to climatic constraints are: depth of roots, size and cupping of leaves, stomata control, cuticolar characteristics, length of growing period. A similar response to temperature relates often to the strategy of plants to best use the rainfall season. As an example, winter cereals can germinate, survive and photosynthesize at lower temperatures than those of cereals from tropical regions like maize, millet, or sorghum.

In this way winter cereals can use spring rainfall, being at maturity right at the beginninhg of the dry period. Breeding was done in the past for achieving cereal characteristics which correspond to a better use of the climatic resources. This is the reason why the most ancien crops like cereals are spread out over a large area of the world. A greater difference between varieties exists as compared to differences among species which poses difficulties to using cereal crops' behaviour as indicators of a climatic change. In order to achieve that, the degree of flexibility or adaptability of different species has to be taken into consideration. Otherwise an agroclimatic procedure based on the analysis of the relevant parameters affecting agriculture production is defined and the change in these parameters due to climatic change is computed.

Climate, microclimate and crops

Climate could in a way be defined as the statistical distribution of air temperature, air humidity, solar radiation, rainfall, wind speed in a period of time varying from instantaneous values to year average values without any interference with the local topography and surrounding objects. A climatic classification tends to group timewise meteorological parameters in order to allow for a comparison among different stations which otherwise could be very difficult with a large number of data. Daily data are filtered by means of statistical techniques (like taking the average over a certain period). The most common time intervals adopted to statistically process meteorological data are five days, one week, ten days, one month. In the case of discrete events the original set of data referring to one year should be the summation instead of the average taken over the period in question. For the description of the basic processes of agricultural production this reveals not to be the best procedure. An alternative procedure will be described in the sequence.

In the light of the definition of climate above given, its variability relates to the main land features which include hills or mountains, lakes, and also the distance from the sea, mean sea level values, latitude, longitude, general circulation features. For a flat surface variability would be of the order of many tenths of kilometers while in rough areas it would be of the order of kilometers.

Microclimate is the result of the interaction between climate, topography and vegetation. Microclimate is of the order of meters to some tenths of meters and strongly affect the response of vegetation. While climate can be studied on the basis of a standard meteorological network, microclimate should be investigated by making use of local measurements or models taking into account local characteristics for each parameter. Knowledge of microclimate is an important and necessary instrument for crop management as for assessing the effect of perturbations introduced by roads, buildings, artificial lakes, industries, small urban centres.

The numerical acquisition of conventional topographic maps is a way to interpolate the meteorological standard data to compute the effect of topography. For instance computing slope, structure and horizon of a surface we can compute solar radiation on a hill side and on the basis of altitude of a valley, the distribution of temperature, in particular during the night, and that of water vapour can be obtained.

In view of a possible climatic change affecting agricultural productivity, the study of microclimate may be a little disregarded for the moment.

On the other hand, an interpolation of data collected by the network stations (which reflect orographic features) can be made, in order to define a grid to superimpose to the grids used for computing meteorological data which are output of atmospheric general circulation models.

In order to assess a possible effect of climatic change on agriculture production the following procedure may be adopted:

- Definition of the size of the grid on which the model is based;
- Description of the type of agriculture done at the moment;
- Description of the relevant parameters taken from an ecophysiology of crops point of view;
- Identification of the most relevant agroclimatic parameters;
- Processing of such parameters on a basis of historical data to put in evidence their trends.

AGROCLIMET procedure

On the basis of the above a procedure has been set up jointly by the Institute for Environmental Analysis and Remote Sensing for Agriculture (I.A.T.A. - C.N.R. - Florence) and the C.E.S.I.A. - Centre for Informatics in Agriculture (Academy of Georgofili - Florence).

It is the AGROCLIMET (Agroclimatic Methodology) which we propose to use to assess the effect of climatic change on agriculture.

The size of the grid at regional or national level corresponds to the density of the available meteorological stations where a data series of at least thirty years exists. Once the positioning of the stations is often irregular depending on the topography an interpollation programme based on the numerical description of topography has been prepared. At European scale the grid size should correspond to the density of stations used at the computer center of the European Centre for Medium-Range Weather Forecasts (ECMRWF) at Reading.

The description of a region in agricultural terms is done on the basis of the following information:

- main existing crops
- growing season

The boundary of a region is that adopted by the National bureau of Census which combines districts which have main crops similar characteristics in elementary regions.

The parameters describing the crop's ecophysiological response to environment are:

Thermal parameters

- Growing temperature threshold
- Optimum temperature
- Minimum bearable temperature
- Stop growth temperature
- Degree days summation from emergence to maturity

Radiative parameters

- Coefficient of the function biomass v/s summation of radiation
- Light extinction coefficient in the canopy

Water parameters

- Water use efficiency
- Depth of roots
- Crop coefficients

Pheonological parameters

- Growing period length
- Length of interphases
- Photoperiod sensitivity

Morphological parameters

- Leaf area index
- Height
- Harvest index
- Relative ratio of leaves, stems, roots
- Total biomass at maturity

By taking into consideration the relative role of the above mentioned parameters and processes involved an agroclimatic procedure can be divised in order to characterize a territory from an agroclimatic viewpoint.

For AGROCLIMET the agroclimatic indices are the following:

- Beginning of the growing period computed at 5°C for winter crops and 10°C for summer crops. The values of 5°C and 10°C are computed as on mobile averages to filter the interdaily variability of mean daily temperature.

- Last Spring day with temperature below 0°C.

- Beginning of dry period. This is computed by means of calculating the effective rainfall and the balance between rainfall summation and potential evapotranspiration.

- Total rainfall during the growing period.
- Length of period with the greatest number of dry days.
- Length of growing period for short root crops.
- Length of growing period for long root crops.
- Total number of dry days.
- Total potential evapotranspiration.
- Absolute minimum temperature.

These paramters which are subdivided into two groups, including on one hand those related to the radiative-thermal processes and on the other those related to the water balance, are computed for a thirty years period. A significant deviation of these parameters from a normal distribution could mean a severe change in agricultural production in the region. By considering the output of general circulation models in the case of a doubling of CO_2 scenario, it is possible to forecast which change to foresee. The output of climatic models is at the moment available at a scale of at least 100 km x 100 km, while AGROCLIMET runs at a scale of 30 km x 30 km or smaller. Timewise climatic models give outputs as monthly averages while AGROCLIMET requires daily data in order to fullfill its requirements. A model is then needed to bridge the scales for making possible some reasoning on the effect of climatic change on agriculture.

If on one hand crops and cultivations in general have some degree of flexibility of responding to climatic factors as a consequence of breeding and husbandry technics, wood trees and natural vegetation might be good indictiors of shift of climatic conditions. Their sensitivity to environmental conditions is better than that of cultivated plants. Bearing this in mind, a cooperative programme at European scale could be devised by choosing appropriate species with the aim of detecting a climatic change.

A scheme like AGROCLIMET could be used possibly by introducing some refinements determined by natural vegetation specific characteristics.

Conclusions

During the last thirty years many international collaborative programmes were sponsored by international agencies including the European Community in the field of crop physiology, ecophysiology, agrometeorology, modelling. Much information and experimental data have been collected on the agricultural production main processes; several models have been devised for describing in some detail the phenomena which link physiology with climate. A comparison of such models is required for choosing among them the best to use for detecting a climatic change and forecast its economical effect possibly at global scale.

In order to achieve that two different steps are proposed:

- the first concerns the comparison of the existing models for evaluating the feasibility of their use bearing in mind that the numbers of inputs and their time and spatial resolution should be compatible with the data available from the European Community standard meteorological network. Therefore an inventory of models available in institutes participating to the European Programme on Climatology would constitute a very useful exercise for testing the validity and comparing their results with the experimental data available at many locations.

- having hosen a best fitting model, the second step could be the linking of the European Programme on Climatology information to the data bank at ECMRWF in a way to make it possible of being used by a large community of scientists.

A narrower cooperation among the scientists working on crop models and those working on general circulation models is of envisaging.

BIBLIOGRAPHY

- Belmans C., Wesseling J.G., Fedders R.A. (1983) - Simulation model of the water balance of a cropped soil: SWATRE-J. of hydrology 63, 3/4, 1983.
- C. Conese, G. Maracchi, F. Miglietta, L. Bacci, M. Romani (1988) Metodologia di formazione e gestione di una banca dati territoriali e modelli agroclimatici di produttività. Quaderno IPRA n. 20, 1988.
- Doorenbos J., Kassam A.H. (1979) - Yield response to water - FAO irrigation and drainage paper n. 33 - Food and Agriculture Organization of the United Nations, Rome.
- Feedes R.A., Kowalik P.J., Zaradny H. (1978) - Simulation of field water use and crop yield - Center for agriculture publishing and documentation, Wageningen, The Netherlands.
- Field J.A., Parker J.C., Powell N.L. (1984) - Comparison of field and laboratory measured and predicted hydraulic properties of a soil with macropores - Soil Sciences 138, 6: 385-397.
- Floyd R.B., Braddock R.D. (1984) - A simple method for fitting average diurnal temperature curves - Agricultural and Forest Meteorology, 32, 2: 107-121.
- Gaastra P. (1959) - Photosynthesis of crop plants as influenced by light, carbon dioxide, temperature, and stomatal diffusion resistance - Mededelingen van de Landbouwhogeschool te Wageningen, Netherlands, 59: 1-68.
- Goudriaan P.J., van Laar H.H. (1978) - Calculation of daily totals of the gross CO_2 assimilation of leaf canopies - Netherlands. Journal Agricultural Sciences, 26: 373-382.
- Hackett C., Carolane J. (1982) - Edible horticultural crops - Academic Press, New York.
- Hillel D., Talpaz H. (1976) - Simulation of soil water dynamics in layered soils - Soil Science 122, 54-62.
- Jarvis P.G. (1976) - The interpretation of the variations in leaf water potential and stomatal conductance found in canopies in the field. - Philosophical Transactions of the Royal Society of London, B. 273: 593-610.
- Maracchi G. (1982) - Long-term consequences of technological development: Italian case study - International Institute for Applied Systems analysis - Austria, CP-82-72.
- Maracchi G., Miglietta F. (1984) - Agroclimatic classification of central Italy - Proceedings EEC Meeting of Climatology Project - Sophia Antipolis. "Current Issue in Climate Research". Commission of the European Communities, 1986, p. 309-322.
- Maracchi G., Miglietta F., Raschi A., Vazzana C., Bacci L. (1985) - A crop model based on simple meteorological inputs for large scale agricultural productivity simulation. - Proceedings of the IASTED International Symposium "Modelling and Simulation", Lugano, June 24-26, 1985.
- Miglietta F., Bacci L., Cervo R., Maracchi G. (1986) - Indici di produttività potenziale Climatica - Le procedure di calcolo - C.E.S.I.A. Manuali Tecnici - N. 1, 1986.

INTERACTION BETWEEN BIOSPHERE AND CLIMATE

H. Lieth
University of Osnabrück, FB 5, D-4500 Osnabrück

Summary

The following paper discusses the interaction between climate and biosphere, especially vegetation. The key climatic parameters influencing vegetation performance are discussed as well as the key vegetation parameters influencing weather and climate. The vegetation variables impacting climate are divided into physical and chemical parameters, and from the latter the CO_2-exchange between biosphere and atmosphere is especially elaborated because of the role it plays for the greenhouse effect. The discussion is based on the Osnabrück Biosphere Model describing the global carbon flux and predicting the CO_2-level in the atmosphere from all relevant inputs and sinks between 1860 and present. The use of satellite remote sensing for climate relevant functional and structural vegetation analyses and its use for monitoring land use and vegetation changes is briefly discussed. The need for climate impact studies for agriculture, forestry, natural vegetation, limnology, and fishery is stressed. The paper is meant as an introduction to the literature cited through which a meaningful research work on interactions between climate and biosphere can be started.

1. INTRODUCTION

It is common knowledge that climate and vegetation are interdependent. Certain climate types are so closely correlated with certain vegetation features that noticeable, structural vegetation features were used in the past to delineate climate types (Köppen (1), Troll and Paffen (2). Similarly were climate types used to construct vegetation maps (Walter and Box (3). Even more elaborate was the attempt by Box (4) to correlate vegetation properties with climate variables.

If such a causal correlation between climate and vegetation exists, one can logically assume that vegetation will change locally if the climate changes. Vegetation responses will include changes for agriculture and forestry and will, therefore, be of great importance for the organic production and management of landscapes.

The impact of vegetation upon climatic variables is most prominently felt in the water-, energy-, and heat-balance analyses. The basic relations and the vegetation parameters of concern are explained by Geiger (5), Thornthwaite (6), Penman (7), and others, the correlation between climate and vegetation based on climate diagrams was extensively done by Walter and Lieth (8). Most of the cited publications deal with the direct influences of vegetation structure and function upon local climate. The discussion of the global climate change probability has surfaced another dimension of vegetation-related impact to the climate: The release of water vapour and trace gases to the atmosphere and the uptake and storage capacity of water and CO_2 in the vegetation body. The importance of these vegetation functions has been the main reason for extensive global budgeting and modelling attempts, many of which are summerized and evaluated in Lieth, Fantechi and Schnitzler ed. (9), Lieth (10) Esser (11), Esser et al. (12), DOE Report TR 040 (13), Trabalka ed. (14) among numerous other treatments.

The high probability of a man-induced climate change requires careful monitoring of relevant environmental parameters. It also requires a monitoring program of biospheric changes which may be caused by presently ongoing climate changes. The vast majority of individually existing numerical relations between plant and animal responses to climatic variables is unknown, not to speak of secondary responses caused by soil properties changing with climate.

The three aspects of climate/biosphere interactions mentioned above require therefore an intensive dialogue between climatologists and geoecologists. This dialogue needs the implementation of large computer-based models and requires therefore the participation of information specialists.

This paper will primarily demonstrate the input of the geoecologist to the climate change discussion. With the help of a set of tables and figures it will demonstrate the working philosophies of the vegetation scientists. It needs to be said, however, that serious work on climate/vegetation relations has to be based upon large data sets like we have accumulated in Osnabrück in the "VW file"[1]. Similar data sets are available from several authors cited in the reference list.

2. THE IMPACT OF CLIMATE UPON VEGETATION

The presently discussed climate change probability will have a severe impact upon the entire biosphere, if it materializes. We restrict our discussion in this paper to vegetation only, because the immediate feedback from a climate induced change in the biosphere to the climate will come from vegetation changes.

We can expect a change for free living animals as well. The large mammals in savannah regions may equally be forced to respond to climate changes as well as the numerous small soil animals and the insects. This response of animal groups to a change of climate will have a feedback to the climate via the vegetation. We cannot deal with this parameter, however. The feedback relations between vegetation and climate are complicated already and the basic parameters of this relation need to be known before secondary variables can be incorporated into feedback models.

-----

1) The DATA VW file containing for many locations around the world the major climate relevant parameters of ecosystems structure and function together with the environmental constraints.

The climate has an overwhelming influence on the vegetation. It is responsible for the physiognomic stand structure, which remains very similar under the same climate, even if the species composition is very different because of different soil conditions. The species composition, however, changes also with climate changes. The most influential climatic parameters in this respect are temperature, precipitation, and seasonality. The species composition on the land areas changes drastically with a change of 5°C annual average temperature between -5°C and 25°C and also with a doubling of the annual precipitation starting with about 250 mm to 500, 1000 and over 2000 mm rainfall. The most prominent climate components responsible for vegetation structure and functions are listed in table 1.

The parameters listed in table 1 are not only relevant for wild plants and animals but also for crops and forests, for the probability of contagious deseases, for the kinds of housings, the organization of settlements, the road conditions, and the basic functions of the human society in general. If the climate variables of table 1 change drastically, we can expect serious responses of all cultivated plants and animals. It is therefore essential for the comfort of the coming generation that global changes in climate are accurately predicted and sufficient knowledge about biospheric response probabilities is accumulated.

Table 1: Climatic variables, important for the structure, function, and species composition of natural vegetation.

1. Radiation
   - short wave radiation
   - long wave radiation
   - intensity, extreme values
   - day length
   - seasonality

2. Temperature
   - mean value
   - extreme values
   - daily fluctuations
   - seasonality

3. Precipitation
   - mean annual sum total
   - seasonality
   - physical state: rain, mist, snow, ice

4. Atmospheric properties
   - air moisture, saturation deficit
   - CO_2-level
   - other trace gases
   - aerosol level
   - air movement
     - mean values,
     - extreme values, hurricanes

3. THE IMPACT OF VEGETATION ON CLIMATE.

As indicated in the introduction it is necessary to discuss the biospheric properties relevant to climate in two categories: 1st influences due to biospheric in- and outputs upon the chemical composition of the atmosphere, and 2nd influences due to physical, structural properties of vegetation stands. Table 2 shows the principal components of the impact possibilities. From the items listed in table 2 the physical properties are directly used by climatologists interested in global circulation models. The physical structure of the ecosystem has a strong influence on temperature and flow patterns of the atmosphere near the ground. This is especially important because physiognomic properties of natural ecosystems occur similarly over vast areas but are significantly different in different climatic regions. From the functional properties of ecosystems the water usage is also directly important for the climatologist. The evaporation of water has an impact upon the energy balance, the heat balance, and the probability of cloud formation. The importance of these parameters for the global circulation modellers and for climate change probabilities has been demonstrated in

---

Table 2: Biospheric properties important for climate changes

1. Physical properties (mainly structural properties)

1.1 Stand structure of ecosystems
- hight
- density
- canopy structure
- foliage geometry

1.2
- surface properties
- spectral properties
- surface roughness

2. Chemical properties (Mainly functional properties)
These properties are usually connected to soil properties in which the respective ecosystem roots.

2.1 Water usage
- evapotranspiration
- water use efficiency
- water storage

2.2 Metabolism
- gas exchange (=CO_2)
- emission of airborne compounds (=trace gases and aerosols
- chemical compounds of the vegetation body
  - herbaceous material
  - woody material
  - humus compounds (including peat, raw humus, and soil organic compounds).

several papers cited in the reference list and is dealt with in other contributions to this volume.

The vegetation exchanges annually a large amount of chemicals with the atmosphere and the soil. The main components of relevance to climate are the CO_2, the water vapour, the nitrogene compounds, and a variety of compounds released as small particulate matter, the socalled aerosols.

Metabolic properties of ecosystems, except the water- evaporation, are not directly felt by the local climate. The turnover rates of some metabolic substances are so large, however, that they can change the composition of other geophysically important bodies, e.g. atmosphere and fresh water. With respect to climate impacts the uptake of released CO_2 and the production of CH_4 are the most important substances for the interaction between climate and biosphere. It is therefore important to understand the global carbon cycle and the role the biosphere plays in it. The key fluxes and poolsizes are listed in table 3.

As shown in table 3, the biosphere is a prime pool and flux channel for carbon. Photosynthesis and respiration are the largest movers of carbon. The terrestrial biomass is the largest pool into which the assimilation can allocate major portions of carbon, mainly into humus, wood, and litter produced by higher plants. Large amounts of CH_4 are released by bacteria from rice fields, swamp vegetation, ungulate herds (incl. cattle) and termites.

Comparing the numbers for pools and rates, one can see that the biosphere can play an enormous role for the composition of the atmosphere, if human activities or natural disasters change the size or the function of the biosphere. The climate-relevant impact on the atmosphere is mainly due to the changes of spectral properties of the atmosphere by trace gases. Of prime importance for any climate/biosphere feedback assessment is therefore the calculation of carbon uptake, storage, and release of the biosphere in relation to fossil fuel burning and the ocean/atmosphere gas exchange. In order to elaborate the necessary carbon balance one needs a detailed analysis of the contribution from the different vegetation types. This fact is known since many years and has been discussed in earlier papers by Junge and Czeplak (15) and Lieth (16). Our present task is, therefore, to determine quantitatively as accurate as possible the climate-relevant functions of vegetation types and the changes from human interference. The relation between biosphere functions and climate/atmosphere properties need to be cast into mathematical equations which can be entered into a computer model so that force functions, fluxes and feedback mechanisms can be tested. An attempt to do that is described in the following chapter.

4. COUPLING ATMOSPHERE AND VEGETATION IN THE OSNABRÜCK BIOSPHERE MODEL

The coupling of global matter exchanges between biosphere and atmosphere requires for the stationary balance an elaborate set of tables for vegetation properties and differential equations. If one attempts to model annual changes of climate sensitive to individual portions, the problem becomes even more complicated.

Table 3: Reservoir sizes and flow rates of the global carbon cycle. Acc. to Lieth ed. (29).

Table 3a: Exchange and transport rates of the global carbon cycle the decades 1970-1980 from Lieth ed (29).

Exchange and Transport Rates	Authors	Rate/Year 10^{15} gC
Gross exchange Atmosphere/Ocean	Broecker *et al.* (1979) (30) Peng *et al.* (1979) (31)	70-87
Gross Primary Productivity	Aselmann & Lieth (1983) (32)	110-120
Net Primary Productivity	Aselmann & Lieth (1983) (32)	57
Gaseous Primary Productivity		1-3
Marine Primary Productivity	de Vooys (1979) (33)	43.5
Fresh water Primary Productivity	de Vooys (1979) (33)	2.3
Vulcanism	Degens (1989) (34)	0.2
Rivers anorg. dissolved	Kempe (1979a) (35)	0.45
Rivers org.	Kempe (1984) (36) Meybeck (1982) (37) Schlesinger & Melack (1981) (38)	0.2-0.8
Net Silicate weathering	Kempe (1979b) (39)	0.01-0.02
Marine sedimentation	Kempe (1979b) (39)	0.26-0-12
Anthropogenic disturbances:		
Fossil fuels and cement (1979,80)	Rotty (1983) (40)	5.3
Destruction of biosphere	Moore *et al.* (1981) (41) Houghton *et al.* (1982) (42)	1.8-4.7
Sum total of anthropogenic contributions:		
Increase in the atmosphere (57% Air-borne-fract.)	Bacastow & Keeling (1981) (28)	3.0±0.2
Ocean sink	Broecker *et al.* (1979) (30)	1.9±0.4
Disturbances in the fresh water cycle	Kempe (1984) (36)	0.8±0.5
Sum total:		5.7±0.6
Biospheric remaining sink: (CO_2-fertilization, deposits on land, reforestation)	Olson (1982) (43)	0.6-4.3

Table 3b: Sizes by weight of the major carbon reservoirs on earth

Reservoir	Author	C-concentration	C-mass 10^{15}g (if no exponent is given)
Atmosphere (total mass 5.14×10^{21}g)			
Preindustrial C-content	Barnolla et al. (1983) (44)	260 ppmv	550
C-content for 1982	Fraser et al. (45)	340 ppmv	717
Ocean (total mass 1.384×10^{24}g)			
Inorganic C-content	Bolin et al. (1979) (46)	29g/m^3	39000
Organic C-content	Mycke and Kempe (1983) (47)	07g/m^3	1000
C in particulate organic matter	Mopper and Degens (1979) (48)		30
Biosphere			
living biomass	Ajtay et al. (1979) (49)		560
standing dead			30
litter			69
humus			$1\text{-}6\text{-}2.0 \times 10^{18}$
marine plankton	Mopper and Degens (1979) (48)		3
Lithosphere (total mass of crust 24×10^{24}g)			
Inorganic C	Kempe 1979 (39)	.20%	
Organic C		.07%	17.5×10^{21}
All burnable fossil sources	Laurmann and Rotty (1983) (50)	best estimate	upper limit
		4131	16655
Oil		230	380
Gas		143	230
Coal		3510	6315
Oil shists and bituminous sands		75	200
Methanclathrat	Bell 1981 (51)		
under permafrost			2×10^{18}
in marine sediments			100×10^{18}

In order to solve this problem for the global carbon flux we have constructed the Osnabrück Biosphere Model (OBM), Esser (11), Lieth (10) (17). This model incorporates the natural ecosystemic functions as well as human activities impinging on the biosphere type distribution, composition, and function. The model operates globally on a matrix of 2.5 x 2.5 degrees. For each element the following carbon flux relevant components are quantitatively assessed on a per annum time lapse:

Table 4: Mean stand ages and factors for separating NPP into herbaceous and woody portions for 31 formations of the potential natural vegetation. Values were derived using our DATAVW files, gaps were filled by ranking. Ranked units carry an *. The vegetation types were taken from Schmidthüsen (18).

Formation	Mean stand age (years)	Herbaceous factor
Tropical moist lowland forest	200	0.37
Tropical dry lowland forest	80*	0.4 *
Tropical mountain forest	80*	0.37*
Tropical savanna	5	0.98
Tropical paramo woodland	10*	0.95*
Tropical paramo grassland	1*	1.0 *
Puna formation	2*	1.0 *
Subtropical evergreen forest	200	0.37
Subtropical deciduous forest	150	0.44
Subtropical savanna	5	0.90
Subtropical halophytic formation	5	0.9
Subtropical steppe and grassland	1	1.0
Temperate steppe and meadow	1	1.0
Subtropical semidesert	16	0.85
Xeromorphic formation	20	0.4
Desert (tropical, subtropical, cold)	5*	0.85*
Mediterranean sclerophyllous forest	100*	0.4 *
Mediterranean shrub and woodland	15	0.47
Temperate evergreen (coniferous) forest	130	0.29
Temperate deciduous forest	150	0.38
Temperate woodland	25	0.53
Temperate shrub formation	10*	0.85
Temperate bog and tundra	5	0.48
Boreal evergreen coniferous forest	100	0.34
Boreal deciduous forest	100	0.38*
Boreal woodland	15	0.6 *
Boreal shrub formation	10*	0.85*
Woody tundra	10	0.7
Herbaceous tundra	2	1.0
Azonal formation	5*	0.60*
Mangrove	50	0.29

percentage of vegetation type as listed in table 4
area changed into agriculture by man
primary productivity of the vegetation
proportion of assimilates stored in wood
amount of litter produced
amount of litter decomposed.

The storage of wood in the standing biomass and the amount of litter decomposed in relation to litter production together with the amount of biomass removed by people for agricultural usage are the main budget items responsible for the CO_2-exchange balance between biosphere and atmosphere. In the OBM the total biomass is achieved at saturation age calculated from the net primary productivity rate which is predicted from the climate parameters annual mean temperature in centigrades, annual mean sum total for precipitation and from soil fertility estimates (see table 5). Table 4 contains therefore only the mean saturation age

---

Table 5: The set of equations to calculate the production and depletion of dry matter in the vegetation based on empirical correlations to environmental parameters and the distribution of assimilates.

Table 5.1: Net primary productivity

$$(NPP/g \times m^{-2} \times a^{-1} = NPP\ (\text{Hamburg Model}) \times F(CO_2))$$

$$NPP\ (\text{Hamburg Model}) = \min\ (F(T),\ G\ (N)) \times F_0\ (\text{Soil fertility})$$

$$F(T) = \frac{3000}{1 + e^{(1.315-0.119 \times T/°C)}}$$

$$G(N) = 3000 \times (1-e^{(-0.000664 \times Pp/mm)})$$

$T/°C$ = mean yearly temperature

Pp/mm = mean yearly sum total precipitation

F_0 = (soil fertility factor) see table 5.11

Plant productivity factor for CO_2-concentration in the atmosphere

$$F\ (CO_2) = A \times (1-\exp(-R(CO_2-80)))$$

$$A = 1 + \frac{F_0}{4},\ R = -\ln \frac{A-1}{A} \times \frac{1}{240}$$

Table 5.11: Factors for F_O (soil fertility) for the most frequent soils of the world (Esser) (11). The soil classifications are taken from the FAO/UNESCO World soil map (52).

FAO Soil-unit	Soil-factor	FAO-Soil-unit	Soil-factor	FAO Soil-unit	Soil-factor
Acrisols		Lithosols		Regosols	
Ag	0.87	I	0.52	Rc	1.61
Ah	0.22	Iy	1.14	Re	1.14
Ao	0.70			Rx	0.91
other A	0.60	Fluvisols		other R	1.20
Cambisols		J	0.49	Solonetz	
		Je	0.61		
Bd	0.94	other J	0.55	So	0.59
Be	1.69				
Bh	1.58	Kastanozems		Andosols	
Bx	0.76				
		Kh	1.96	Tv	1.65
Chernosems		Kl	1.61		
		other K	1.80	Xerosols	
Cl	0.99				
		Luvisols		Xh	0.42
Podsoluvisols					
		La	0.34	Yermosols	
Dd	0.83	Lc	1.04		
		Lf	1.65	Y	0.30
Ferralsols		Lg	2.78	Yh	0.66
		Lo	0.85	Yt	0.09
Fx	0.55			Yl	0.23
		Histosols			
Gleysols				Solonchak	
		Od	1.39		
Gh	0.47			Zo	0.44
Gx	0.57	Podsols		Zt	0.03
other G	0.50			other Z	0.20
		Ph	0.56		
		Po	0.61		
		other P	0.55		

Table 5.2: Biomass pool (P) and litter (LP) - "steady state" phytomass at stand age A and mean NPP

$$P/kg = 0.5918 \times 10^{-3} \times A^{(0.79216 \times NPP)}$$

Litter (LP)

$$LP/g \times m^{-2} \times a^{-1} = NPP - P - E$$

E = Yield in exploited ecosystems $/g \times m^{-2} \times a^{-1}$
P = Standing crop
The litter amount is split into woody and non woody according to table 5.4

Table 5.3: Litter depletion rate in % of litter pool for herbaceous litter and woody litter per year in dependence of temperature and precipitation

Herbaceous litter

$$Dh/\% \times A^{-1} = \min(F1(T), G1(N))$$

$$F1(T) = 7.67 \times e^{(0.0926 \times (T/°C + 6.41))} + 17.06$$

$$G1(N) = \left(\frac{50}{0.0215 + e^{(4.2 - 0.0053 \times Pp/mm)}} + 670\right)$$

$$\times \left(\frac{0.094}{0.7 + e^{(0.0023 \times Pp/mm - 5.05)}} + 0.076\right)$$

$$\times \quad 0.64 \times (1 - e^{(0.001 \times Pp/mm)})$$

Woody litter

$$Dw/\% \times A^{-1} = \min(F2(T), G2(N))$$

$$F2(T) = 2.67 \times e^{(0.0522 \times (T/248\ °C + 31.63))} - 2.51$$

$$G2(N) = \left(\frac{27.8}{0.021 + e^{(8.53 - 0.0095 \times Pp/mm)}} + 712\right)$$

$$\times \left(\frac{0.126}{1.51 + e^{(9.003 \times Pp/mm - 4.65)}} + 0.05\right)$$

$$\times \quad 0.5 \times (1 - e^{(-0.001 \times Pp/mm)})$$

T/°C = Mean annual temperature
Pp/mm = Mean annual sum total precipitation
Dh = Decomposition rate herbaceous
Dw = Decomposition rate woody

values for the maximum biomass amount of the vegetation types and the percentage of woody material of the annual net primary production. Table 3 contains only values for a contracted type list. The model used actually more than 100 vegetation types according to the atlas of Schmidthüsen (18). The vegetation boundaries of this atlas were planimetered for each geographical matrix element, and the respective values,percentage for each vegetation type, entered into a total matter balance for each grid element.

The global biospheric functions are balanced against the human technological CO_2-output, the atmosphere/ocean CO_2-exchange volume, the CO_2-concentration change in the atmosphere, and the biosphere to

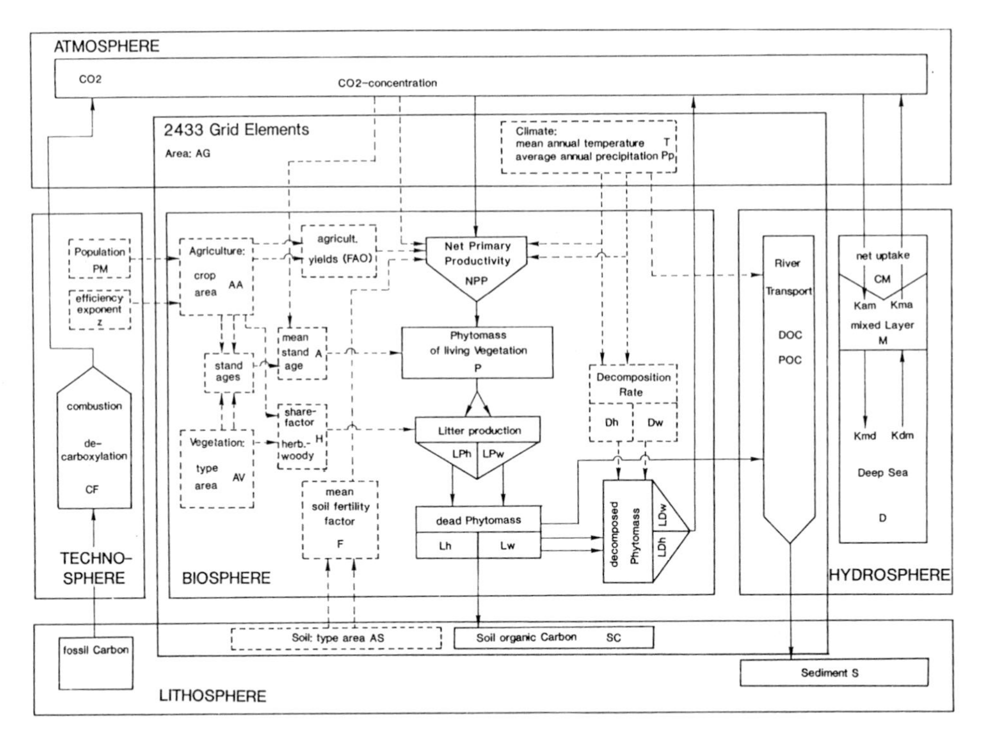

Fig. 1. Flow chart of carbon in the Osnabrück Biosphere Model. The model describes the carbon flow from the lithosphere through the technosphere into the atmosphere, its exchange between atmosphere and ocean as well as atmosphere and biosphere, and its eventual redeposit into the lithosphere. Main part of our model is the contribution of the biosphere to the carbon flux, either by anthropogenic changes of land use as well as climatic changes. The carbon flows are indicated with solid arrows, flow control variables are shown as dashed lines. The quantification of the flux through the biosphere and its dependence on natural variables are shown in tables 4 and 5. Other constraints are summerized in the text. The model is fully described by Esser (11) and Lieth (10). Results of a model run are shown in figures 2a and 2c.

hydrosphere flux of carbon. This latter was assessed separately during several SCOPE symposia which were published by Degens (19), Degens, Kempe, Soliman (20) and Degens, Kempe, Herrera (21). The flow chart of the total system is shown in fig. 1. A detailed explanation of this flowchart is given by Esser (11) and in Lieth, Fantechi and Schnitzler (9) together with some applications of the model for the time span between 1860 and 1980. In the meantime the model has been used and modified to incorporate the Ocean model of Maier-Reimer and Hasselmann (22) which yields results of the carbon flow pattern very similar to the diffusion Box model by Siegenthaler and Oeschger (23) which we used in the beginning.

Fig. 2a shows the CO_2-amount in the atmosphere from preindustrial times until recent predicted with the OBM using the Maier-Reimer and Hasselmann (22) regionalized ocean model. Fig. 2b shows the distribution of the total anthropogenic CO_2-input over the compartments living biomass, litter, ocean and atmosphere between 1820 and 1980. Fig. 2c shows the CO_2-changes in the atmosphere during the last 30 years, when we use a box diffusion model for the ocean. For more details we suggest to consult the papers by Esser (11) or Lieth (10).

The main impacts of the atmospheric CO_2-changes to the biosphere are noticed in the balance between overall CO_2-assimilation and community respiration. If we base our balance on net primary productivity and the decay rate of the accumulated matter, we are able to calculate the biospheric contribution to the global atmospheric CO_2-changes. If we correlate these vegetation functions to climatic and edaphic parameters we can use the same models to calculate the impact of possible climate changes. For this reason we have established the sequence of equations which correlate the net primary productivity and the litter decay rate to the climatic variables mean annual temperature and precipitation, to the CO_2-concentration of the atmosphere, and to the fertility of soil. The latter is used as a table function for net primary productivity calculation only. The differentiation in woody and nonwoody biomass, as shown in table 4, is needed to implement the litter decay rate equations. The set of equations is listed in table 5. The entire model, together with a set of data, may be obtained from the author. A more detailed explanation is given by Esser (11) and Lieth (10).

5. FUTURE NEEDS IN MODELLING CLIMATE/BIOSPHERE INTERACTIONS

As shown in figs. 2a and c, the Osnabrück Biosphere Model simulates quite well the CO_2-level changes during the industrial development of mankind. This, together with a realistic account for biomass development and litter amounts, provides already some confidence in the accuracy of the model. The fact that the model uses climatic parameters to predict vegetation responses makes it useful to forecast changes of vegetation functions, if climatic changes will occur.

During several years of discussions the model has gradually been improved. It requires still more improvement, if it is to be used on a regional basis. The most important improvement of the model appears to be the better time resolution. Most of the world`s vegetation functions under seasonal changes of either temperature or precipitation, together with solar radiation and daylength. All these are decisive forces for the vegetation, as shown in table 1. The initial OBM uses annual averages to

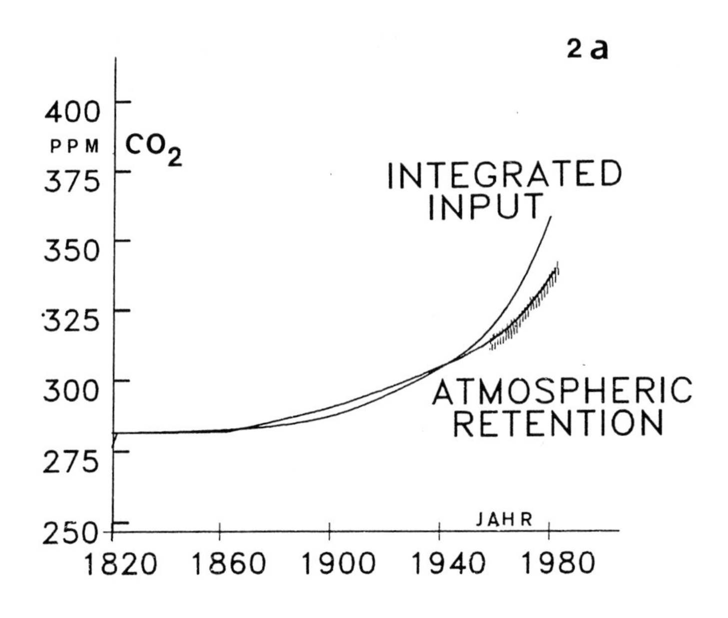
2 a
400
PPM
CO_2
375
350
325
300
275
250
1820
1860
1900
1940
1980
JAHR
INTEGRATED
INPUT
ATMOSPHERIC
RETENTION

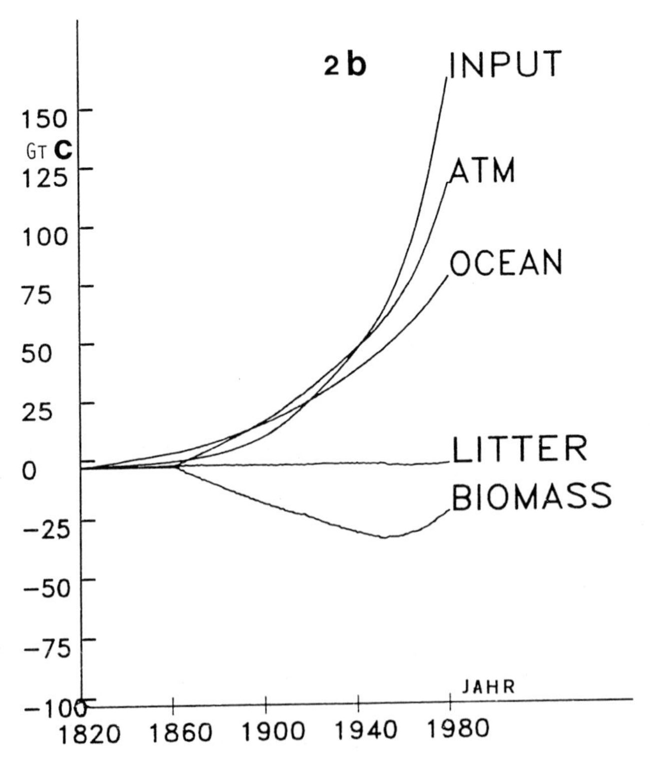
2 b
150
GT C
125
100
75
50
25
0
−25
−50
−75
−100
1820
1860
1900
1940
1980
JAHR
INPUT
ATM
OCEAN
LITTER
BIOMASS

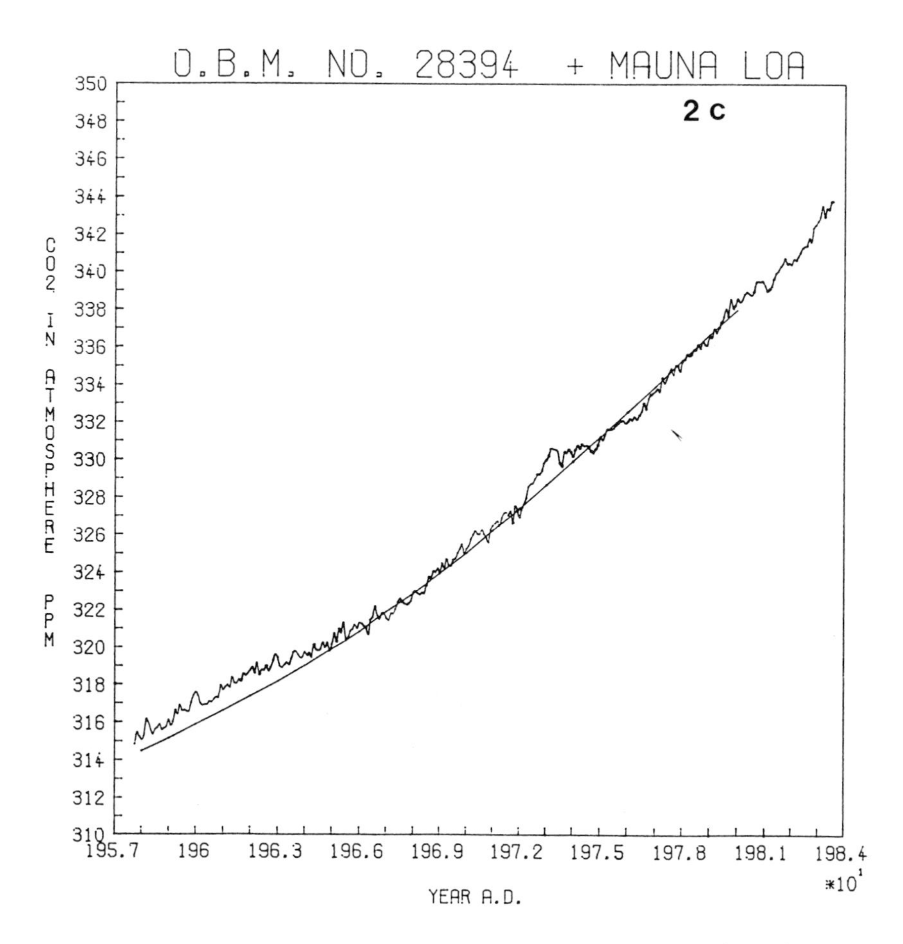

Fig. 2. Results of simulation runs for the global carbon flux since the start of industrialization. Figures 2a and b after Hasselmann and Maier-Reimer (22), figure 2c after Esser (11).

Figs. 2a and b. Simulation of the global carbon flow using the Osnabrück Biosphere Model with the Hamburg Ocean Model (22). 2a shows the increase of CO_2 in the atmosphere predicted by the simulation under the assumption the preindustrial CO_2-level was close to 280 ppm. The dashed band at the right hand side of the curve depicts the values measured at Mauna Loa (28). Fig. 2b shows the contributions of the major carbon pools to the global carbon flux through time under the assumption of the simulation yielding the results shown in fig. 2a.

Fig. 2c shows an earlier simulation of the atmospheric CO_2-level between 1957 and 1984 using the OBM together with the box model for the ocean by Siegenthaler and Oeschger (23). The curvilinear line represents the simulated annual CO_2-increase, the wavy line depicts quarterly averages of the Mauna Loa monitoring.

calculate global averages. This assumes a vegetation that can cope wholesale with such changes, which, however, is usually not the case, as shown by Wigley (24) for many crops. This problem is especially important for agriculture, and here especially in regions where drought or low temperature limit the growth of plants already. The recently published books by Parry et al. (25), Decker and Achutuni (26) give some insight. It is therefore necessary to change in the OBM the productivity and decay calculations to monthly averages and at a later state to daily or weekly values. This needs to be done similar to the approach offered by King et al. (27), where seasonality of production and decomposition are balanced in the way done by Lieth (16), so that the seasonal changes of the CO_2-flux between atmosphere and biosphere for each vegetation type can be simulated. This would enable us to validate our model along the seasonal fluctuations of atmospheric CO_2-levels shown by the curves of Bacastow and Keeling (28).

The applicability of our model for regional climate impacts requires also that we exchange the standard climate in our model to a regionally variable climate. Since we assume that future climate changes will vary greatly in different regions, we must provide in our model the possibility to use real meteorological data to reconstruct past situations as well as weather generators for regional climate simulations to predict vegetation responses. Presently we work on subroutines necessary to do this.

6. THE USE OF SATELLITE REMOTE SENSING TO DETECT AND MONITOR VEGETATION FUNCTIONS AND CHANGES

In recent years remote sensing techniques have been greatly improved. It is now possible to detect changes in biospheric functions and in land uses on a global scale conveniently by satellite remote sensing. Detailed studies have been made by many researchers, see evaluation of literature by Kawosa (53). Our own contributions to this field are the detection of land use changes in South America, as shown in table 6 from Esser and Lieth (54), and the correlation of microwave signals with primary productivity and water usage in figure 3 from Choudhury (55), Choudhury and Tucker (56) and Lieth (57).

Landsat images are available since the early seventies when the ERTS satellites observed the spectral radiation of the earth surface over the visible and infrared wavelengths with an accuracy of 70x70m per pixel and lateron with smaller pixel sizes. The ratio between visible and near infrared reflection of sunlight from the earth surface, the socalled vegetation index, turned out to be useful for detecting large-scale deforestation and the length of vegetation period of annual crops. The normalized vegetation index was also correlated with primary productivity with some success by Tucker et al. (58).

The analysis of vegetation changes during the seventies of this century with Landsat images by Esser and Lieth (54) and Lieth (57) is shown in table 6. The values were compared with statistical assessments by Richards et al. (59) and found to be in reasonable agreement. This method appears therefore useful to monitor land use changes annually. Since this change can be converted to carbon release or to potential CO_2-uptake by the vegetation we can use this method to test carbon flux and balance models.

Satellite imaging in the microwave frequencies is being used in various ways. Its advantage lies in the fact that clouds do not obscure

Country	Change in area		Balance	Evaluated area	Related phytomass changes	Richards et al. 1983 Mean 1958-1978
	Decrease	Increase	(km^2/a)	(%)	(t/a)	(km^2/a)
	(1)	(2)	(3)	(4)	(5)	(6)
Brasil	6889	199	6689	76	$80 \times 10^6 C$	7708
Argentina	1533	17	1516	85	12×10^6	1880
Paraguay	791	0.4	790	80	9.5×10^6	227
Venezuela	360	3	357	65	3×10^6	960
Bolivia	76	0	76	40	1×10^6	713
Chile	69	3	66	90	0.3×10^6	104
Columbia	3	0	3	30	0.02×10^6	280
Peru	0	0	0	34	-	649
Ecuador	0	0	0	6	-	187
Uruguay	0	0	0	95	-	-48
South America	9721	233	9497	72	106×10^6	12660

Table 6. Areal changes in the natural vegetation of South American countries, as derived from the evaluation of 934 landsat scenes. Column 4 gives the percentage area of each country covered by overlapped parts of scene pairs. Related phytomass changes (column 5) estimated using the Osnabrück biosphere model. Column 6 gives a recent country-based estimate prepared by Richards et al. (59). (Table from Esser and Lieth (54)).

the signals wanted. Furthermore, polarizations are possible in various ways which enhance the possibility to discern surface properties of the earth otherwise not detectable. One example useful for the analysis of climate/vegetation interaction was recently shown by Choudhury (55) who used the polarization difference of the 37 Ghz emission of the global surface measured continuously by the Nimbus satellite to monitor the ice cover on the ocean. In analysing the polarization difference on land he discovered a correlation between the polarization difference and a variety of vegetation borne parameters like primary productivity, evapotranspiration, amount of rainfall, seasonal CO_2-variation in the atmosphere, and normalized vegetation index. While the real reason for the polarization remains to be analyzed it seems possible to use the method for monitoring water usage by the vegetation or its net primary productivity. The latter has been discussed in Lieth (57). The water usage expressed in units of evapotranspiration is shown in fig. 3. The figure shows a correlation for crude averages and it remains to be seen to which detail such analyses can be carried and what use can be made out of it to monitor regionally relevant vegetation functions.

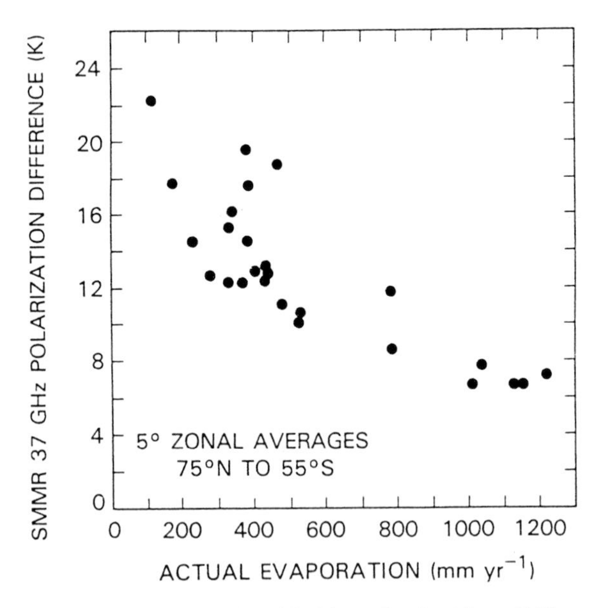

Fig. 3. Relation between the 37 GHz polarization difference expressed in K° temperature difference and evaporation. The ΔT values are derived from the Nimbus monitoring, while the evaporation values are from Baumgartner and Reichel (60). These evaporation values are calculated from surface observations and on empirical formulae with some subjective adjustments.

One of the first applications by Choudhury (55) was the prediction of mean average net primary productivity which was compared in (57) against earlier assessments by Lieth. From this map an enlarged section is given in fig. 4. The figure shows Africa, Arabia, India, Southeast Asia, and Australia. The scaling of the NPP averages ranges from below .25 till 2.00 $kgxm^{-2}xyear^{-1}$. The regression line between NPP and 37 GhPD levels out at around 2. $kgxm^{-2x}yr^{-1}$. But the map shows correctly the increasing NPP from desert over grassland to various forest types. Across this sequence the map shows that the water availability is the limiting factor for NPP. The majority of the entries in fig. 3 stem from that area.

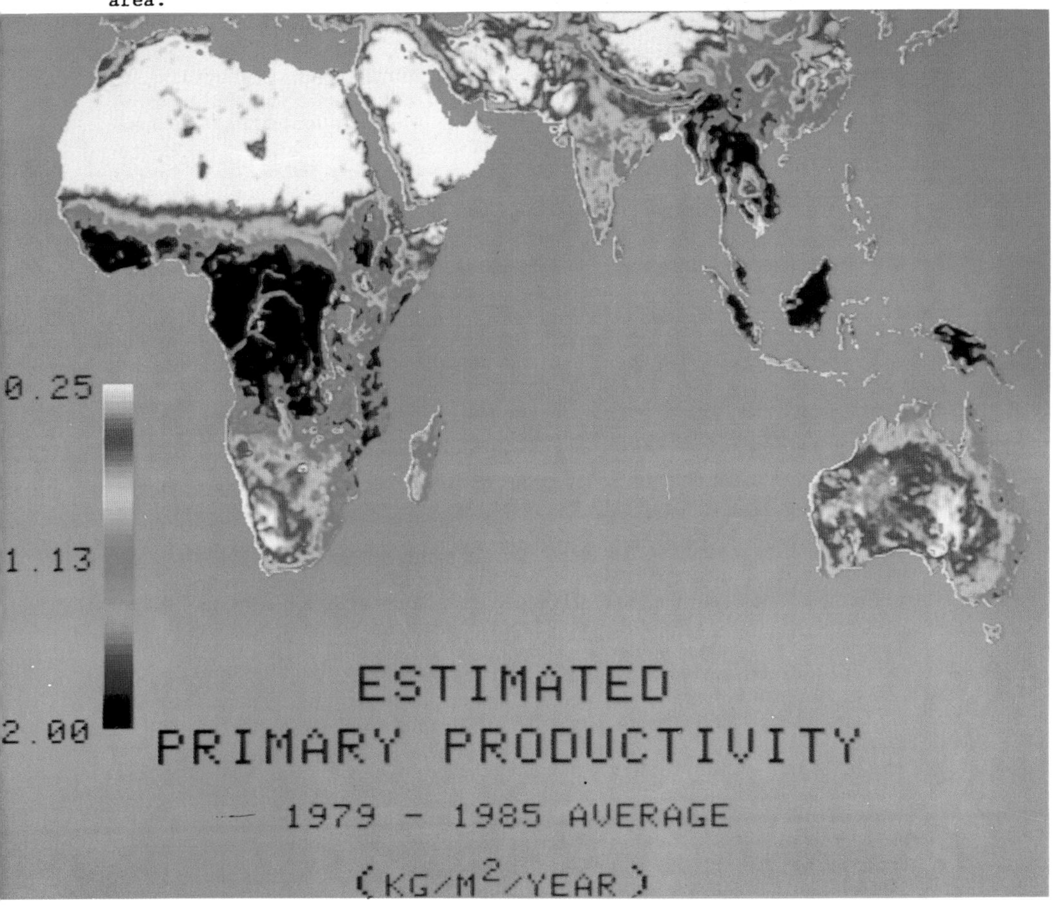

Fig. 4: Average net primary productivity of the land vegetation of the south eastern part of the globe calculated from the 37 GHz polarization difference by Choudhury (55). The section is an enlargement from the map shown and discussed in (55) and (57). The reason for the correlation between the polarization difference and annual mean NPP or actual evaporation as shown in fig. remains to be investigated. The correlation between actual evaporation and annual NPP has been discussed or cited in several of the papers by Lieth (19) (17) (57). Map courtesy of Dr. Choudhury NASA Goddard Space Flight Center.

REFERENCES

(1) Köppen, W. 1931: Die Klimate der Erde, Grundriß der Klimakunde, 2nd Ed., Berlin und Leipzig.

(2) Troll, C. & K.H. Pfaffen 1980: Jahreszeitenklimate der Erde. - Berlin.

(3) Walter, H. & E.O. Box 1976: Global classification of natural terrestrial ecosystems. Vegetatio, Vol. 32, 2: 76-81.

(4) Box, E.O. 1981: Macroclimate and plant forms: An introduction to predictive modeling in phytogeography. T:VS Vol. 1, H. Lieth (ed.). Dr. W. Junk Publishers, The Hague, Boston, London.

(5) Geiger, T. 1961: Das Klima der bodennahen Luftschicht. 4. Auflage. Viehweg, Braunschweig.

(6) Thornthwaite, C.W. 1931: The climates of North America according to a new classification. Geogr. Rev. 21: 633-655.

(7) Penman, H.L. 1956: Evapotranspiration - an Introductory Survey. Neth. Journ. of Agr. Sciences, 4: 9-29.

(8) Walter, H. & H. Lieth 1960-1967: Klimadiagramm Weltatlas. Gustav Fischer Verlag, Jena.

(9) Lieth, H., R. Fantechi & H. Schnitzler (eds.) 1984: Progress in Biometeorology, 3. Lisse, Swets and Zeitlinger.

(10) Lieth, H. 1985: A dynamic model of the global carbon flux through the biosphere and its relation to climatic and soil parameters. Int. J. Biometeorol. 29 (2): 17-31.

(11) Esser, G. 1985: Der Kohlenstoff-Haushalt der Atmosphäre - Struktur und erste Ergebnisse des Osnabrücker Biosphären Modells. 160 p. Veröff. Naturf. Ges. zu Emden von 1814. Vol. 8, Emden.

(12) Esser, G., I. Aselmann & H. Lieth 1982: Modelling the Carbon Reservoir in the System Compartment Litter. S. 39-59. In: Mitt. aus dem Geol.-Paläontol. Inst. der Univ. Hamburg. Proceedings of a workshop at Hamburg Univ., 8-12 March 1982. Transport of Carbon and Minerals in Major World Rivers, Part 1, E.T. Degens (ed.) SCOPE/UNEP Sonderbd., Vol. 52.

(13) DOE report TR 040 1988: A primer on greenhouse gases. DOE/NBB 0083. March 1988. US Dept. of Energy, Washington DC.

(14) Trabalka, J.R. (ed.) 1987: Atmospheric Carbon Dioxide and the Global Carbon Cycle. 316. p. DOE/ER 0239, Washington DC.

(15) Junge, C.E. & Czeplak 1968: Some aspects of seasonal variation of carbon dioxide and ozone. Tellus 20:422-433.

(16) Lieth, H. 1963: The role of the vegetation in the carbon dioxide content of the atmosphere. Journal of Geophysical Research 68: 3887-3898.

(17) Lieth, H. 1986: Das Osnabrücker Biosphären Modell als Simulationsmodell zur Beschreibung der globalen Änderung des Kohlenstoffkreislaufs. In: Advances in System Analysis, D.P.F. Möller (Ed.) Erwin-Riesch-Workshop. System Analysis of Biol. Processes, 2nd Ebernburger Working Conference, Vol. 2, p. 29-45. Braunschweig/Wiesbaden, Vieweg & Sohn.

(18) Schmithüsen, J. 1976: Atlas zur Biogeographie. Mayers großer physischer Weltatlas, Bd. 3, Mannheim.

(19) Degens, E.T. (ed.) 1982: Transport of carbon and minerals in major world rivers. Part 1., Mittl. aus dem Geol.-Paläontol. Inst. der Univ. Hamburg SCOPE/UNEP Sonderbd., Vol. 52.

(20) Degens, E.T., S. Kempe & H. Soliman (eds.) 1983. Transport of carbon and minerals in major world rivers. Part 2, Mittl. aus dem Geol.-Paläontol. Inst. der Univ. Hamburg SCOPE/UNEP Sonderbd.,Vol. 55.

(21) Degens, E.T., S. Kempe & R. Herrera (eds.) 1985: Transport of carbon and minerals in major world rivers. Part 3, Mittl. aus dem Geol.-Paläontol. Inst. der Univ. Hamburg SCOPE/UNEP Sonderbd., Vol. 58.

(22) Maier-Reimer, E. & K. Hasselmann 1987: Transport and Storage of CO_2 in the ocean - an inorganic ocean circulation carbon cycle model. Climate Dynamics 2: 63-90.

(23) Siegenthaler U., & H. Oeschger 1978: Predicting future atmospheric carbon dioxide levels. Science 199, 388-395.

(24) Wigley, T.M.L. and Jones, P.D. 1981: Detecting CO_2-induced climage change.

(25) Parry, M.L., T.R. Carter, N.T. Konijn, eds. 1988: The Impact of Climatic Variations on Agriculture. Vol. 2, Assessment in semi-arid regions. 764 p., Kluver Dordrecht.

(26) Decker, W.L., R. Achutuni 1988: The use of statistical climate-crop models for simulating yield to project the impacts of CO_2-induced climate change. 42 p. TR043 NTIS, USDC, Springfield VA.

(27) King, A.W., D.L. De Angelis, W.M. Post 1987: The seasonal exchange of carbon dioxide between the atmosphere and the terrestrial biosphere: Extrapolation from site specific models to regional models. (281 p. ORNL/TM-10570 Oak Ridge TN.

(28) Bacastow, R. & C.D. Keeling 1981: Atmospheric carbon dioxide concentration and the observed airborne fraction. In: B. Bolin, ed., "Carbon Cycle Modelling", SCOPE Rep. 16, 103-112. Chichester, Wiley & Sons.

(29) Lieth, H.(Ed.) 1985: Das Kohlendioxid in der Atmosphäre als Teil des globalen Kohlenstoffkreislaufs und seine Wechselwirkungen zu Klima und Pflanzenwachstum. Veröff. d. Naturforschenden Ges. z. Emden von 1814, Band 1-1.1 Serie 1. Emden.

(30) Broecker, W.S., Takahashi, H.J. Simpson & T.-H. Peng 1979: Fate of fossil fuel carbon dioxide and the global carbon budget. Science 206, 409-417.

(31) Peng, T.H., W.S. Broecker, G.G. Mathieu & Y.-H.Li 1979: Radon evasion rates in the Atlantic and Pacific Oceans as determined during the GEOSECS program. J. Geophys. Res. 84, 2471-2486.

(32) Aselmann, I. und H. Lieth 1983: The implementation of agricultural productivity into existing global models of primary productivity. In: E.T. Degens, S. Kempe and H. Soliman, eds. "Transport of Carbon and Minerals in Major World Rivers" Vol. 2, Mitt. Geol.-Paläont. Inst. Univ. Hamburg, SCOPE/UNEP Sonderbd.

(33) de Vooys, C.G. 1979: Primary production in aquatic environments. In: B Bolin, E.T. Degens, S. Kempe & P. Ketner, eds."The Global Carbon Cycle" SCOPE Rep. 13, 259-292. Chichester, J. Wiley & Sons.

(34) Degens, E. 1989: Annual volcanic carbon dioxide emission: An estimate from eruption chronologies. Environm. Geo. 4: 15-21.

(35) Kempe, S. 1979a: Carbon in the freshwater cycle. In: B. Bolin, E.T. Degens, S. Kempe & P. Ketner, eds. "The Global Carbon cycle" SCOPE Rep. 13, 317-342, Chichester, J. Wiley & Sons.

(36) Kempe, S. 1984: Sinks of the anthropogenically enhanced carbon cycle in surface fresh waters. J. Geophys. Res. 89, 4657-4676.

(37) Meybeck, M 1982: Carbon, nitrogen, and phosphorus transport by world rivers. Amer. J. Sci. 282, 401-450.

(38) Schlesinger, W.H. & J.M. Melack 1981: Transport of organic carbon in the worlds rivers. Tellus 33:172-187.

(39) Kempe, S. 1979b: Carbon in the rock cycle. In: B. Bolin, E.T. Degens, S. Kempe & P. Ketner, eds." The Global Carbon Cycle" SCOPE Rep. 13, 343-377, Chichester, J. Wiley & Sons.

(40) Rotty, R. 1983: Distribution of and changes in industrial carbon dioxide production. J. Geophys. Res. 88(C2), 1301-1308.

(41) Moore, B., R.D. Boone, J.E. Hobbie, R.A. Houghton, J.M. Melillo, B.J. Peterson, G.R. Shaver, C.I. Vörösmarty & G. Woodwell 1981: A simple model for analysis of the role of terrestrial ecosystems in the global carbon budget. In: (B. Bolin, ed.) "Carbon Cycle Modelling", SCOPE Rep. 16, 365-385, Chichester, Wiley & Sons.

(42) Houghton, R.A., J.E. Hobbie, J.M. Milillo, B. Moore, B.J. Peterson, G.R. Shaver & W. Woodwell 1983: Changes in the carbon content of terrestrial biota and soils between 1860 and 1980: a net release of CO_2 to the atmosphere. Ecological monographs 53: 235-262.

(43) Olson, J. 1982: Earth`s vegetation and atmospheric carbon dioxide. In: W.C. Clark, ed., "Carbon Dioxide Review: 1982", 388-398, New York, Oxford Univ. Press.

(44) Barnolla, J.M., D. Raynaud, A. Neftel & H. Oeschger 1983: Comparison of CO_2-measurements by two laboratories on air from bubbles in polar ice. Nature 303:410-413.

(45) Fraser, P.J., P. Hyson & G. I., Pearman 1981: Some considerations of the global measurements of background atmospheric carbon dioxide. Proc. WMO/ICSU/UNEP Scientific Conference on Analysis and Interpretation of Atmospheric CO_2-Data, Bern 14-18 Sept. 1981, 179-186.

(46) Bolin, B., E.T. Degens, S. Kempe & P. Ketner (eds.) 1979: The global carbon cycle. SCOPE Rep. 13, 1-491, J. Wiley & Sons, Chichester.

(47) Mycke, B. & S. Kempe 1983: Dissolved organic carbon in the Atlantic and Pacific. In: S. Kempe (ed.). "RV, SONNE IV Cruise, May, June 1978, Cruise, Bremerhaven-Panama-Hawaii", Final. Rep. to the German Res. Council, unpublished.

(48) Mopper, K. & E.T. Degens 1979: Organic carbon in the ocean: Nature and cycling. In: B. Bolin, E.T. Degens, S. Kempe & P. Ketner (eds.). "The global carbon cycle" SCOPE Rep. 13, 293-316, J. Wiley & Sons, Chichester.

(49) Ajtay, G. L., P. Ketner & P. Duvigneaud 1979: Terrestrial primary production and phytomass. In: B. Bolin, E.T. Degens, S. Kempe & P. Ketner (eds.). "The global carbon cycle" SCOPE Rep. 13, 129-181, J. Wiley & Sons, Chichester.

(50) Laurmann, J.A. & R.M. Rotty 1983: Exponential growth and atmospheric carbon dioxide. J. Geophys. Res. 88 (C2), 1295-1299.

(51) Bell, P.R. 1981:Methane hydrate and the carbon dioxide question. In: W.C. Clark (ed.). "Carbon Dioxide Review: 1982", 401-406, Oxford Univ. Press, New York.

(52) FAO/UNESCO: Soil map of the world 1:5000000, Paris and Rome 1974.

(53) Kawosa, M.A. 1985: The use of satellite imagery for vegetation oriented land use mapping and forest systems dynamics of the Himalaya countries. Veröffentl. d. Naturforschenden Ges. zu Emden von 1814. Vol. 2 2-3 Dl. 138 p. Emden.

(54) Esser, G. & H. Lieth 1986: Die globale Waldflächenänderung in den siebziger Jahren erfaßt mit Landsataufnahmen. BMFT-Statusseminar: Die Nutzung von Fernerkundungsdaten in der BR Deutschland. Garmisch-Partenkirchen 20.21.1.1986. Publications of the DGLR 1986.

(55) Choudhury, B.J. 1988: Estimates of primary productivity over the Thar Desert based upon Nimbus-7 37 GHz data: 1979-1985. Intl. J. of Remote Sensing (in press).

(56) Choudhury, B.J. & C.J. Tucker 1987: Monitoring global vegetation using Nimbus-7 37 GHz data: Some empirical relations. Intl. J. of Remote Sensing, 8: 1085-1090

(57) Lieth, H. 1988: Starting a new term for the International Journal of Biometeorology. Intl. J. Biometeorol. 32: 1-10.
(58) Tucker, C.J., C.L. Vanpraet, M.J. Sharmon & G von Ittersum 1985: Satellite remote sensing of total herbaceous biomass production in the Senegalese Sahel: 1980-1984. Remote Sens. Environ. 17: 233-248.
(59) Richards, J.R., J.S. Olson & R.M. Rotty 1983: Development of a data base for carbon dioxide releases resulting from conversion of land to agricultural uses. Institute for Energy Analysis, Oak Ridge Associated University. ORAU/IEA-82-10(M), ORNL/TM-8801.
(60) Baumgartner, A. & E. Reichel 1975: The water balance, New York, Elsevier.

SEA LEVEL CHANGE AND THE CLIMATE CONNECTION: PAST AND FUTURE

R.A. Warrick
Climatic Research Unit
University of East Anglia
Norwich, UK

Summary
The prospect of global warming from increasing concentrations of "greenhouse gases" has generated considerable concern over the possibility of a rise in sea level. A significant portion of the world's population lives on low-lying river deltas, islands and coastal plains and could be potentially threatened. What are the factors - climate-related and otherwise - that could cause global sea level to rise? Has sea level been rising? Can we find support for a climate-sea level connection from past observations? Will sea level rise in the future, and, if so, by how much? This chapter addresses such questions. The intention is to present an overview of the problems and complexities involved, and to provide some estimates of both past and future sea level change. The emphasis is on global change on the decade-to-century timescale.

1. FACTORS INFLUENCING SEA LEVEL

Attempts to measure global sea level changes caused by ocean volume - or "eustatic" - changes have relied on tide gauge data, in particular those data maintained by the Permanent Service for Mean Sea Level at the Bidston Observatory, UK. Aside from potential difficulties such as geographical bias, insufficient record length and the like, tide gauge data are all subject to a major complicating factor: vertical land movement. Since tide measurements are made in relation to a fixed bench-mark, the data reflect both land and ocean surface movements - i.e. "relative" sea level. Attempts to measure the eustatic component must subtract the movement of land.

Vertical Land Movements There are many factors affecting vertical land movements. These vary in scale from the local to the global. In local situations the pumping of groundwater or oil can lead to land subsidence, as in Bangkok where in places land is currently sinking at rates up to 10-15cm/yr, making the city increasingly vulnerable to flooding (Nutulaya, 1989). In Hong Kong the sheer weight of urban development is compacting the foundation of marine mud and increasing the threat of floods from storm surges (Yim, 1990).

Regionally, tectonic movements and deltaic sedimentation can contribute to vertical land movement. The marshlands around the Mississippi Delta, for instance, are disappearing at the rate of as much as 100km^2 per year due to subsidence of the Delta (at the rate of a metre per century) and to the sediment starvation of the marshlands as a result of flood control engineering works on the Mississippi

Channel (Boesch, 1982; Day et al. 1990). In tectonically active areas, like the eastern Mediterranean basin, relative sea level changes can be large - and sometimes sudden. One portion of the Greek coast dropped about 1.5m after three earthquakes in 1981 (Pugh, 1987). Throughout the world, one can also find instances of land emergence as well as subsidence at these scales.

Accounting for land movements from geological processes operating on the regional-global scale is particularly crucial in attempts to identify secular changes in eustatic sea level from tide gauge data. The most important of these are the lingering crustal movements - "isostatic" effects - from the retreat of the Laurentide and Fennoscandian ice sheets following the last glaciation. In areas previously compressed by the large ice masses the land is still "rebounding" thousands of years later. Tide gauge measurements in the Baltic Sea region show that in some parts of Fennoscandia sea level is dropping by more than 1m per century. Similar effects are observed in parts of Alaska.

Such glacio-isostatic effects are not limited to the previously glaciated areas. In areas peripheral to the continental glaciers - the so-called "forebulge" - the land was elevated and is now subsiding. This is probably why sea level in the southeast UK shows such a rapid rate of rise relative to northern Britain, for example (Woodworth, 1987). In fact the Earth's visco-elastic adjustment produces sea level effects that are world-wide (Peltier, 1985). Removing such long-term vertical land movements is critical to estimating eustatic sea level change (see below).

Eustatic Factors. There are many factors that influence eustatic sea level as well.* A number of these have been conveniently categorised by Mörner (1987) (see Figure 1). Various horizontal and vertical land movements can, in addition to affecting relative sea level measurements by tide gauges, change the ocean basin volume and the shape of the "geoid" surface (a gravitational equipotential) and therefore eustatic sea level. For example, even in the mid-ocean basins of the Pacific and Atlantic, sea level is probably very slowly dropping. There, the slow global re-distribution of mass within the Earth following de-glaciation has been lowering the geoid surface, with the result that seawater is slightly decanting poleward (Peltier, pers. comm.). The in-flow of sediments alters the ocean basin volume, as do plate tectonics and other earth movements. However, in general the effects of such long-term geological processes on ocean basin volume (and thus sea level) are small on the century time-scale, probably less than several mm per century.

*The term "eustatic" as used by Mörner (1987) describes any absolute change in sea-level, regardless of causation. Thus eustatic variables include changes in ocean basin volume, as shown in Figure 1. However, the common usage of the term is narrower and refers strictly to ocean volume changes, whether due to changes in mass of the ocean (e.g. by land ice melting), or to changes in volume at constant mass (i.e. steric" changes).

EUSTASY	OCEAN LEVEL CHANGES	VERTICAL AND HORIZONTAL GEOID CHANGES	OCEAN BASIN VOLUME	TECTONO-EUSTASY	EARTH-VOLUME CHANGES
				TECTONICS	OROGENY
					MID-OCEANIC RIDGE GROWTH
					PLATE TECTONICS
					SEA FLOOR SUBSIDENCE
					OTHER EARTH MOVEMENTS
					SEDIMENT IN-FILL
				ISOSTASY	LOCAL ISOSTASY
					HYDRO - ISOSTASY
					INTERNAL LOADING ADJUSTMENT
			OCEAN WATER VOLUME	GLACIAL EUSTASY	
					WATER IN SEDIMENT, LAKES AND CLOUDS, EVAPORATION, JUVENILE WATER
			OCEAN MASS/LEVEL DISTRIBUTION	GEOIDAL EUSTASY	GRAVITATIONAL WAVES
					TILTING OF THE EARTH
					EARTH'S RATE OF ROTATION
					DEFORMATION OF GEOID RELIEF (DIFFERENT HARMONICS)
		DYNAMIC CHANGES	DYNAMIC SEA LEVEL CHANGES		METEOROLOGICAL
					HYDROLOGICAL
					OCEANOGRAPHIC

Figure 1. An illustration to show the eustatic variables (from Mörner, 1987).

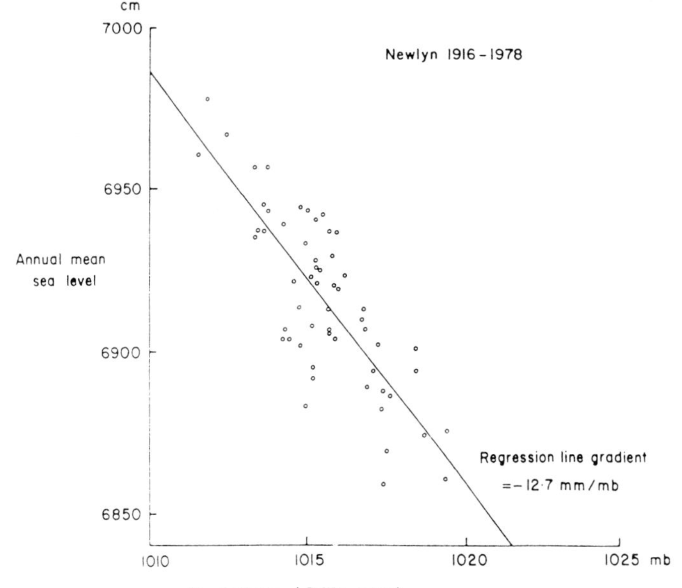

Figure 2. The relationship between annual mean levels and annual mean air pressures at Newlyn (from Pugh, 1987).

Importantly, meteorological and oceanographic factors can have quite pronounced effects on the level of the sea in relation to the geoid. Atmospheric pressure changes directly affect sea level, theoretically amounting to a 1cm rise for every millibar decrease in pressure - the "inverted barometer" effect (Figure 2) (see Pugh, 1987, for discussion). Typically, this effect alone might produce a range of sea level variation of around 50cm over the course of a year in the mid-latitudes. Winds and ocean currents, too, alter the sea surface in relation to the geoid. For instance, years of El Niño occurrence result in sea levels off South America that are about 20cm higher than during Anti-El Niño years (Nicholls, 1987; Wyrtki, 1982).

In terms of future climatic change, such oceanographic and meteorological factors could, regionally, have marked effects on sea level. However, changes in regional pressure patterns and ocean currents cannot yet be predicted reliably with current climate models (MacCracken and Luther, 1985; Wigley and Santer, 1990), so the associated regional sea level changes cannot yet be estimated reliably either. In terms of investigating past (last 100 years) trends in global-mean sea level, such factors operate on short-time scales and at regional levels, so it is usually assumed that they average out.

What are the climate-related factors operating on a decade-to-century timescale that could affect global-mean sea level? Principally, they are: thermal expansion of the oceans, and the increased melting of land ice.

At a constant mass, the volume of a column of water changes with changes in density - a "steric" change in sea level. Density of seawater is a function of temperature and salinity. At temperatures close to 0°C (at 35°/oo salinity), seawater expands as it warms. This thermal expansion is larger in warmer water than in colder water. Thus, for a given degree of warming one would expect more expansion in low latitudes than in high latitudes, and in the surface layer as compared to deeper layers of the oceans. A change in salinity, too, affects ocean density and causes expansion, but given the range of possible concentration changes its effect is minor compared to that of temperature. With global warming from increasing concentrations of greenhouse gases, the ocean surface warms and heat is transferred to the deeper layers of the oceans. The rate of heat transfer determines the rate of thermal expansion.

Potential changes in ocean volume from land-ice come from three sources: alpine glaciers, the Greenland ice sheet and the Antarctic ice sheet. (The melting of floating sea ice - e.g. Arctic sea ice - of course, has a negligible effect on sea level.) The alpine glaciers contain a relatively small amount of water (30-60cm in equivalent sea level) but a comparatively short response times with respect to changes in climate (of the order of 10-100 years). In contrast, the sea level equivalents of the Greenland and Antarctic ice sheets are large (about 7m and 70m respectively) but with dynamic response times on the order of thousands of years (Oerlemans, 1989). Thus, on the decadal timescale, one might expect little sea level contribution from the dynamic response (e.g. changes in ice flow velocities and iceberg discharge) of the polar ice sheets, but in the long-term the contribution could be large indeed. For instance, the during the last interglacial (the Eem, 120 000 years ago), when the global temperature was 1-2°C warmer than today, sea level was probably about 5m higher, perhaps due to drastic reduction in the West Antarctic Ice Sheet

(Mercer, 1978)); this rise in sea level probably took thousands of years to occur. (For further discussion see Oerlemans this volume).

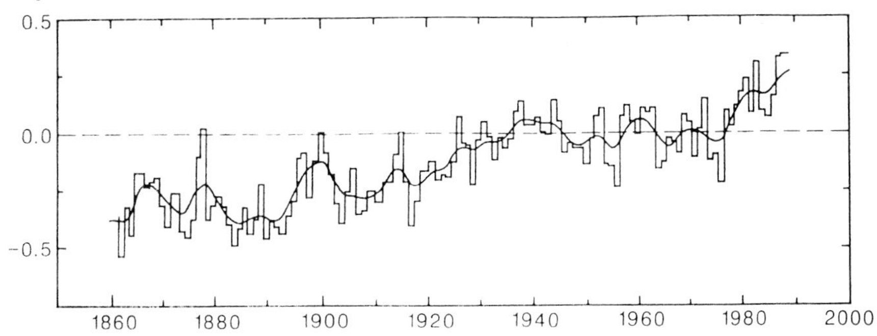

Figure 3. The global temperature record 1860-1988. The values are anomalies from the 1950 to 1979 reference period (updated from Jones et al., 1988).

2. THE LAST ONE-HUNDRED YEARS

During the last 100 years, the world has, on average, been getting warmer. The global-mean temperature curve in Figure 3 indicates a rather strong warming trend during 1910-1940, a subsequent period of little change to about the mid-1970s, followed by a sharp warming continuing through the 1980s. Overall, the world has warmed by about 0.5±0.1°C (Jones et al 1988). Has global-mean sea level also risen? And, if so, is there supporting evidence of a climate-sea level connection?

Analyses of tide-gauge data. Table 1 lists the results of the major studies of global-mean sea level trends. The striking aspect of these analyses is their general consensus: sea level has been rising. There are some discrepancies regarding the rates of rise, but even these are not as serious as one might initially suppose. For instance, the 3.0mm/yr obtained by Emery (1980) is based on the period 1935-1975 when rates of rise were apparently higher (also see Barnett, 1984); in addition, the particular statistical techniques used may have inadvertantly eliminated stations with low rates of rise (Aubrey, 1985). Similarly, the recent study by Peltier and Tushingham (1989), which obtains 2.4±0.9mm/yr, applies to the period of faster rise, 1920-1970; when calculated for a longer period (1900-1970), their estimated rate drops to 1.2mm/yr (Peltier and Tushingham, 1990), in close agreement with other estimates.

A major methodological issue is how to account for the vertical land movements and other factors, noted above, that contaminate the data. Various methods are used. Barnett (1983, 1984) simply eliminated stations with obvious isostatic effects (e.g. Baltic region) and assumed that other effects would average out in a global-scale analysis. Gornitz and Lebedeff (1987; also Gornitz et al., 1982, Gornitz, 1990) first eliminated data on a station-by-station basis, and then filtered the remaining station data using late Holocene sea level indicators to compensate for long-term isostatic and neo-tectonic effects. A comparison of Figures 4A,B and C shows how the spatial variance is reduced markedly. Peltier and Tushingham (1989) used a geophysical model for the same purpose, which has the advantage of global coverage. In large part, these differences in approach account for residual discrepancies in estimated rates of sea level rise.

In accord with recent assessments, it can be concluded from analyses of tide gauge data that, over the last 100 years, sea level has rising at the average rate of about 1.0-1.5mm/yr, as a best estimate (for recent reviews see Robin, 1986; Barnett, 1988; Woodworth, 1990; PRB, 1985). Can this rise be explained invoking the climate-related factors mentioned above? If so, this would lend credence to the notion of future sea level rise from greenhouse warming.

Table 1. Estimates of mean "global" sea level increase (modified and updated from Robin, 1986).

Author	Rate (mm/year)
Kuenen (1950)	1.2-1.4
Fairbridge and Krebs (1962)	1.2
Lisitzin (1974)	1.1±0.4
Emery (1980)	3.0
Gornitz et al. (1982)	1.2
Klige (1982)	1.5
Barnett (1983)	1.4±0.1
Barnett (1984)	2.3±0.2
Gornitz and Lebedeff (1987)	1.2±0.3 1.0±0.1
Peltier and Tushingham (1989)	2.4±0.9

Thermal Expansion. The possible effects of thermal expansion on past sea level rise has been investigated using relatively simple one-dimensional energy-balance climate models (e.g. Gornitz et al., 1982; Wigley and Raper, 1987). In this type of model, the land and oceans in both hemispheres are represented by "boxes" and heat transfer is parameterised as a diffusive process (see Figure 5). In the model used by Wigley and Raper (1987), the other important tunable parameter, besides the diffusivity, is the "climate sensitivity"; this is the equilibrium global-mean temperature response for a CO_2 doubling, thought to lie within the range 1.5-4.5°C (for resent assessments see WMO, 1986; MacCracken and Luther, 1985). The model also includes upwelling in order to maintain a realistic temperature profile, and is forced by changes in greenhouse gas concentrations. This, and other similar models, are used primarily to estimate the time-dependent ("transient", as opposed to "equilibrium") temperature response to changes in greenhouse forcing (for review see Hoffert and Flannery, 1985). The inclusion of expansion coefficients (which vary according to initial temperature, salinity and presssure, and therefore to latitude and depth) allows thermal expansion to be calculated at the same time.

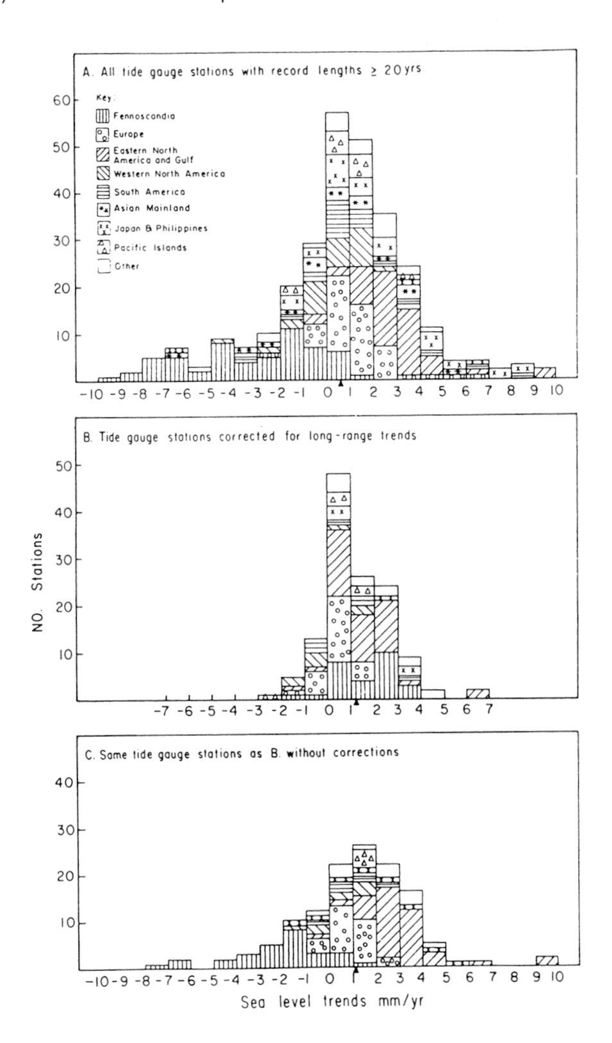

Figure 4. Histogram of the number of stations versus sea-level trends. (A) All tide-gauge stations with record lengths $\geq$20 years. (B) The same for tide-gauge stations from which long-range trends have been subtracted. (C) Same subset of stations as (B), long-range trends included (from Gornitz and Lebedeff, 1987).

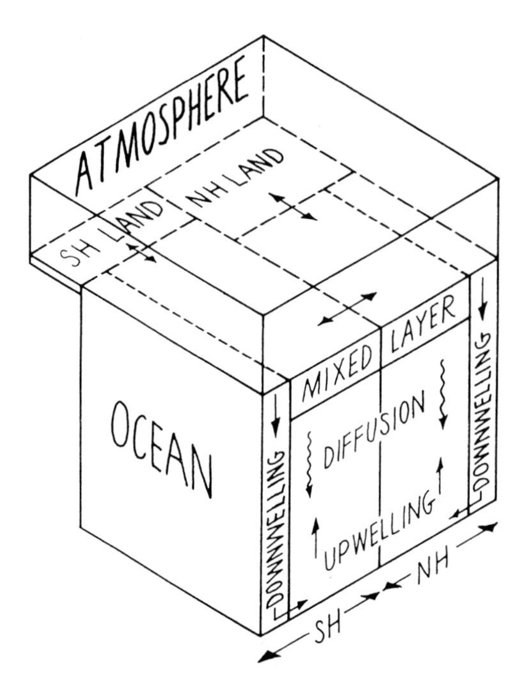

Figure 5. Schema of box-upwelling-diffusion energy-balance climate model used by Wigley and Raper (1987).

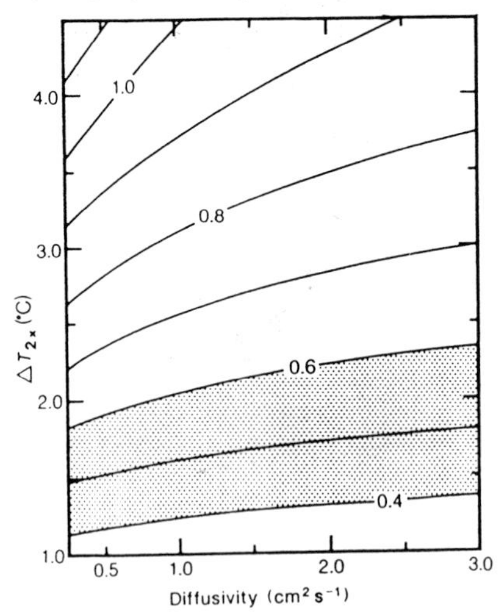

Figure 6. 1880-1985 warming (°C) due to observed increases in greenhouse gas concentrations for different diffusivities (κ) and equilibrium CO_2-doubling temperature changes (ΔT_{2x}). The observed warming range of 0.4-0.6°C is shown stippled (from Wigley and Raper, 1987).

In the study by Wigley and Raper (1987), the observed changes in greenhouse gas concentration to 1985 are used to force their model. Through sensitivity analyses, they first determined the combinations of climate sensitivity and diffusivities which were "allowable" within the constraints of the observed past warming (i.e. 0.4-0.6°C) (Figure 6). The associated range of thermal expansion, similarly constrained, was also calculated (Figure 7). The finding was that, on average, the oceans expanded by about 2-5cm over the period 1880-1985, assuming greenhouse warming (that is, assuming no other climate forcing other than greenhouse gases).

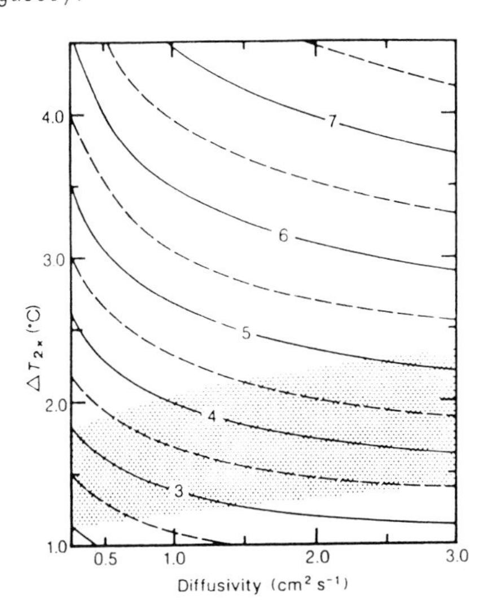

Figure 7. 1880-1985 thermal-expansion-induced sea-level rise (cm) due to observed increases in greenhouse-gas concentrations for different diffusivities (κ) and equilibrium CO_2-doubling temperature changes (ΔT_{2x}). The stippled area gives the range of κ and ΔT_{2x} values compatible with the observed 1880-1985 global warming (from Wigley and Raper, 1987).

Land Ice Contributions. In general, the contribution of changes in land ice is hampered by the paucity of long-term measurements of volume and mass balance changes. It is apparent that, on average, most of the world's *alpine glaciers* have been shrinking during this century (Grove, 1988). The only existing, published attempt to arrive at a global estimate of alpine glacier contributions to sea level is that of Meier (1984). Meier assumed that the magnitude of the long-term balance (for which data are scant) may be related to the seasonal mass balance amplitude (for which data are more widely available), and that this may be used as a scaling factor in order to derive a global, long-term estimate. From his analysis, it can be concluded that glaciers contributed roughly 2-6cm to sea level over the last 100 years.

It should be borne in mind that such analyses are crude at best. For both past and future glacier contributions to sea level, careful region-by-region analyses are required. Combinations of data analysis and energy-balance and mass-balance models, in conjunction with detailed climate data and climate models, should be used. This sort of research is critical to predicting sea level rise but has yet to be undertaken in a systematic fashion.

The *Greenland ice sheet*, too, suffers from lack of measurements. Various estimates of the mass balance components (ablation, accumulation, calving) have been made, but these are of limited time spans and largely restricted to West Greenland only. There is no firm evidence to suggest that, currently, the mass balance is negative (Robin, 1986; PRB, 1985). In fact, using satellite radar altimetry, Zwally et al. (1989; also Zwally, 1989) concluded that, over the period 1978-86, there was net growth of the Greenland ice sheet.

However, at the century timescale, there is circumstantial evidence to suggest that there has been an average negative net balance in Greenland, for two reasons. First, the outlet glaciers have retreated markedly. Second, the mean-summer temperature has warmed substantially from the pre-1900 average within the Arctic in general (Kelly et al., 1982) and in Greenland in particular. This warming reached a peak about mid-1930s with a cooling trend thereafter, but still with a positive accumulated temperature anomaly over the entire time period. A net warming would be expected to produce a negative net mass balance since the ablation rate is directly related to temperature, and since ablation is likely to dominate over the accumulation rate (also directly related to temperature) (Oerlemans, 1990).

As a first approximation, the past sea level contribution of Greenland can be estimated as the net accumulated temperature deviations multiplied by the "sensitivity", taken as about 0.4mm/yr per degree C in sea level equivalent (based on Bindschadler, 1985; Ambach, 1985; Oerlemans, 1990). This "static" response gives a Greenland contribution to sea level over 1880-1980 of approximately +3cm; the uncertainty is probably ±3cm (Warrick and Oerlemans et al., 1990). The dynamic response can be effectively ignored at this timescale.

The mass balance of the *Antarctic ice sheet* is very uncertain. In contrast to Greenland, there is very little melt runoff in Antarctica. Ice is "lost" primarily through calving (iceberg discharge) and basal melting of floating ice shelves (which does not contribute appreciably to sea level). In terms of sea level, it is the total ice discharge across the grounding line - the separation between grounded and floating ice - that matters. Direct measurements of calving rates are sparse, and data on ice stream velocities are generally limited to surface measurements with little empirical evidence for estimating flow at depth. Accumulation data are better, but estimates of total Antarctic-wide accumulation rates vary considerably. Most assessments (e.g. Robin, 1986; PRB, 1985) conclude that Antarctic mass balance is either zero or slightly positive. But since the mass balance components are very large, small imbalances can make a significant difference to estimates of sea level effects. The Antarctic factor is perhaps the largest uncertainty in the estimation of past and future sea level change.

Total Contributions. If, for the moment, we assume that Antarctica is in balance, the sum of the "best estimate" values for the other contributing factors compares quite well with the observed sea level rise (see Table 2).

Table 2: Estimated values of the contributing factors to observed sea level rise.

Source	cm/century
Thermal expansion	3.5
Alpine glaciers	4.0
Greenland	3.0
Antarctica	0.0 ?
TOTAL	10.5
OBSERVED	10-15

If, however, we are to accept a positive Antarctic mass balance of 0.6mm/yr (with the possibility of up to 1.2mm/yr; PRB, 1985), the observed rise would remain largely (or even totally) unexplained. This would suggest either an (as yet) unidentified source, or would perhaps invoke the high estimates from within the ranges of uncertainty associated with thermal expansion, alpine glaciers and the Greenland ice sheet. The latter would give approximately 17cm and would compensate for a positive mass balance in Antarctica.

3. FUTURE SEA LEVEL RISE

This section briefly recounts one particular set of estimates of future climate and sea level changes. For further details, see Raper et al. (1990), Wigley (1989), Warrick and Farmer (1990) and Warrick and Wigley (1990). For additional discussion of future sea level rise, see Oerlemans (this volume).

In broad terms, the rate of future sea level partly depends on the rate of global warming, which in turn depends on three major factors: future changes in radiative forcing from increasing greenhouse gas concentrations; the sensitivity of climate to such radiative forcing changes; and the rate of oceanic heat uptake (which also determines the rate of thermal expansion). By taking ranges of values for these factors, scenarios of future warming can be constructed, as shown in Table 3.

A "best estimate" is that the equivalent concentration of CO_2 (taking into account other major greenhouse gases) will be double the pre-industrial level (i.e. 560ppmv) sometime around the year 2030; that the climate sensitivity to such a doubling is 2.5-3.5°C; and that the diffusivity is $1.0 cm^2 sec^{-1}$. For the period 1985-2030, this set of assumptions produces a global warming of 1-2°C. With credible "high" and "low" scenarios, the warming could be as little as 0.5°C or as much as 2.5°C. The model used to make the temperature projection is that of Wigley and Raper (1987), described earlier. The associated thermal expansion effects from the model are noted in Table 3.

The estimated contributions to sea level from land ice depend on both future climatic change and ice sensitivities. For alpine glaciers a simple global glacier model was used that accounts for uncertainties

Table 3. Projected climate and sea level changes, 1985-2030 (from Warrick and Farmer, 1990).

	MAJOR ASSUMPTIONS		WARMING (°C)	SEA LEVEL CHANGE (cm)				
	Equiv. CO_2 conc. (ppmv)	Clim. sens. (°C)	Surface air temp. change (from 1985)	Thermal expansion	Alpine glaciers	Greenland	Antarctica	TOTAL
LOW	505	1.5	0.5	4.3	2.2	0.8	-2.3	5.0
BEST ESTIMATE RANGE	544	2.5	1.1	8.4	8.3	1.8	-1.9	16.6
	603	3.5	1.9	14.0	12.6	3.1	-3.3	26.4
HIGH	649	4.5	2.5	17.8	19.2	3.9	3.0	43.9

in response time, initial ice mass, and sensitivities to climate warming (see Raper et al. 1990). Ignoring the dynamic responses, the best-estimate sensitivities for Greenland and Antarctica are, in terms of sea level equivalent, approximately 0.4mm/yr per degree C and -0.5mm/yr per degree C (after Bindschadler, 1985 and Oerlemans, 1989, respectively). The reason that the Antarctic sensitivity is negative is that increasing temperatures are likely to increase precipitation and thus accumulation, while iceberg discharge is related to the long-term dynamic response and is likely to remain unchanged in the short-term.

The best estimate is that global-mean sea level will be about 17-26cm higher than today by the year 2030. This rise is caused primarily by oceanic thermal expansion and increased melting of alpine glaciers. The small positive and negative contributions from Greenland and Antarctica roughly cancel each other out.

There is an additional point about future sea level rise that is often not recognised. First, although we speak of "global-mean" sea level, this does not necessarily imply that sea level will rise uniformly around the world. Just as sea level varies spatially because of the "inverse barometer" effect (see above), regional differences in changes in ocean density (as from warming) and thus thermal expansion could produce differences in steric sea levels that could persist. One estimate suggests that with global warming the differences can be of the order of ±20% of the global-mean (Raper et al., 1990).

In addition, changes in the global re-distribution of mass from ice melting will affect vertical surface movement (from the Earth's visco-elastic response) and the surface geoid. For example, using a geophysical model, Clark and Primus (1987) calculated the spatial differences in sea level change from a global eustatic rise of 100cm due to a hypothetical partial melting of the Greenland ice sheet (see Figure 8). They found, for instance, that in some places (e.g. Iceland) sea level could actually fall. In short, in case of future sea level rise, the adage "water finds its own level" is not necessarily appropriate.

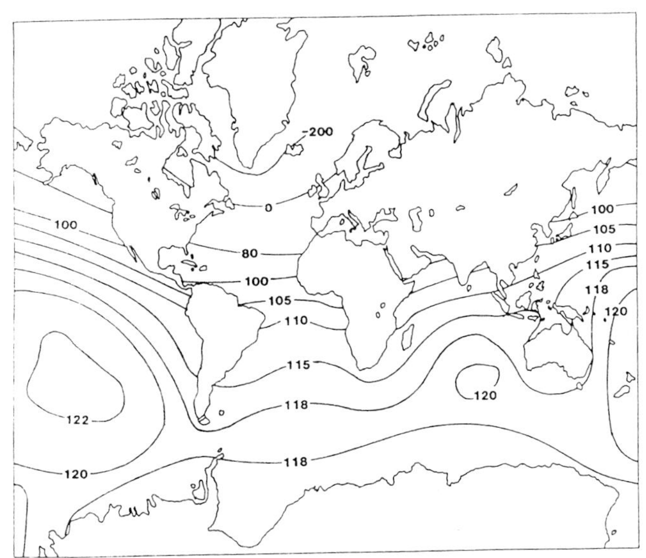

Figure 8. A map to show the amount of sea-level rise caused by the melting of part of the Greenland ice sheet. Contours are in centimetres if the eustatic sea-level rise is 100 centimetres. Alternatively, contours are in percent of eustatic sea-level rise (from Clark and Primus, 1987).

4. SUMMARY

There are many factors that influence sea level - on short- and long-term time scales and at local to global space scales - and that complicate the identification of secular changes in eustatic sea level from tide gauge data. Nevertheless, there is fairly strong evidence to conclude that global sea level has been rising over the last one hundred years. After accounting for vertical land movements, the average rate of rise of eustatic sea level appears to lie within the range 1.0-1.5mm/yr. Furthermore, there are grounds for believing that oceanic thermal expansion and increased melting of alpine glaciers and the Greenland ice sheet, possibly associated with global warming, may have been responsible. This gives cautious support to the notion that future global warming may indeed lead to an acceleration in the rate of sea level rise.

Future sea level rise will depend in part on the rate of global warming, which in turn depends on future changes in greenhouse gas concentrations, the sensitivity of climate and the thermal inertia of the oceans. A set of "best-estimate" assumptions produces a world that is about 1-2°C warmer than today on average, with an associated sea level rise of 17-26cm. This would represent a rate of rise approximately 3-6 times faster than that experienced over the last 100 years.

ACKNOWLEDGEMENT

The help of Miss Elaine Barrow in the preparation of this document is greatly appreciated.

REFERENCES

Ambach, W. (1985): Characteristics of the heat balance of the Greenland ice sheet for modelling. *Journal of Glaciology* **30**, 3-12.

Aubrey, D.G. (1985): Recent sea levels from tide gauges: problems and prognosis. In: *Glaciers, Ice Sheets, and Sea Level: Effect of a CO_2-induced climatic change.* DOE/ER/60235-1 (U.S. Department of Energy, Carbon Dioxide Research Division, Washington) pp. 73-91.

Barnett, T.P. (1983): Recent changes in sea level and their possible causes. *Clim. Change* **5**, 15-38.

Barnett, T.P. (1984): Estimation of "global" sea level change: a problem of uniqueness. *J. Geophys. Res.* **89**, 7980-7988.

Barnett, T.P. (1988): Global sea level change. In: NCPO, *Climate variations over the past century and the greenhouse effect.* A report based on the First Climate Trends Workshop, 7-9 September 1988, Washington D.C. National Climate Program Office/NOAA, Rockville, Maryland.

Bindschadler, R.A. (1985): Contribution of the Greenland ice cap to changing sea level: present and future. In: *Glaciers, ice sheets, and sea level: effect of a CO_2-induced climatic change.* DOE/ER-0235 (U.S. Department of Energy, Carbon Dioxide Research Division, Washington), pp.258-266.

Boesch, D.F. (ed) (1982): *Proceedings of the conference on coastal erosion and wetland modification in Louisiana: causes, consequences and options.* FWS-OBS-82/59 (Fish and Wildlife Service, Biological Services Program, Washington).

Clark, J.A. and J.A. Primus (1987): Sea level changes resulting from future retreat of ice sheets: an effect of CO_2-warming of the climate. In: M.J. Tooley and I. Shennan (eds.) *Sea Level Changes.* Basil Blackwell Ltd, Oxford.

Day, J.W., W.H. Conner, R. Constanza, G.P. Kemp and I.A. Mendelssohn (1990): Impact of sea level rise on coastal ecosystems. In: R.A. Warrick and T.M.L. Wigley (eds.) *Climate and sea level change: observations, projections and implications.* Cambridge University Press, Cambridge (in press).

Emery, K.O. (1980): Relative sea levels from tide gauge records. *Proceedings of the National Academy of Sciences* **77**(12) 6968-6972.

Fairbridge, R. and O. Krebs (1962): Sea level and the Southern Oscillation. *Geophys. J.R. Astron. Soc.*, **6**, 532-545.

Gornitz, V. (1990): Mean sea level changes in the recent past. In: R.A. Warrick and T.M.L. Wigley (eds.) *Climate and sea level change: observations, projections and implcations.* Cambridge University Press, Cambridge (in press).

Gornitz, V. and S. Lebedeff (1987): Global sea-level changes during the past century. In: D. Nummedal, O.H. Pilkey and J.D. Howard (eds.) *Sea-level fluctuation and coastal evolution* (SEPM Special Publication No. 41).

Gornitz, V., S. Lebedeff and J. Hansen (1982): Global sea level trend in the past century. *Science* **215** pp.1611-1614.

Grove, J.M. (1988): *The Little Ice Age.* (Methuen, London).

Hoffert, I.M. and B.P. Flannery (1985): Model projections of the time-dependent response to increasing carbon dioxide. In: M.C. MacCracken and F.M. Luther (eds.) *Projecting the climatic effects of*

increasing carbon dioxide. DOE/ER-0237 (U.S. Department of Energy, Carbon Dioxide Research Division, Washington) pp.149-190.
Jones, P.D., T.M.L. Wigley, C.K. Folland, D.E. Parker, J.K. Angell, S. Lebedeff and J.E. Hansen (1988): Evidence for global warming in the past decade. *Nature* **332**, p.790.
Kelly, P.M., P.D. Jones, C.B. Sears, B.S.G. Cherry and R.K. Tavakol (1982): Variations in surface air temperatures: Part 2. Arctic regions, 1881-1980. *Monthly Weather Review* **110**, pp.71-83.
Klige, R.K. (1982): Oceanic level fluctuations in the history of the earth. In: *Sea and Oceanic Level Fluctuations for 15,000 years*, 11-22. Acad. Sc. USSR, Institute of Geography, Moscow.
Kuenen, Ph. H. (1950): *Marine Geology*. New York, Wiley.
Lisitzin, E. (1974): *Sea-level Changes*. Elsevier Oceanography Series, 8, Amsterdam-Oxford-New York, Elsevier Scientific.
MacCracken, M.C. and F.M. Luther (eds.) (1985): *Projecting the climatic effects of increasing carbon dioxide*. DOE/ER-0237 (US Department of Energy, Carbon Dioxide Research Division, Washington).
Meier, M.F. (1984): Contributions of small glaciers to global sea level. *Science* **226**, 1418-1421.
Mercer, J.H. (1978): West Antarctic ice sheet and CO_2 greenhouse effect: a threat of disaster. *Nature* **271**, 321-325.
Mörner, N.A. (1987): Models of global sea-level changes. In: M.J. Tooley and I. Shennan (eds.). *Sea Level Changes*. Basil Blackwell Ltd., Oxford.
Nicholls, N. (1987): The El Niño/Southern Oscillation phenomenon. In: M. Glantz, R. Katz and M. Krenz (eds.) *The Societal Impacts Associated with the 1982-83 Worldwide Climate Anomalies*. National Center for Atmospheric Research, Boulder, Colorado.
Nutulaya, P. (1989): Investigation of land subsidence in Bangkok during 1978-1988. In: J.D. Milliman (ed.) *Proceedings of the SCOPE Workshop on rising sea level and subsiding coastal areas*. Bangkok, 1988. John Wiley, Chichester.
Oerlemans, J. (1989): A projection of future sea level. *Climatic Change*.
Oerlemans, J. (1990): Possible changes in the mass balance of the Greenland and Antarctic ice sheets and their effects on sea level. In: R.A. Warrick and T.M.L. Wigley (eds.) *Climate and sea level change: observations, projections and implications*. Cambridge University Press, Cambridge (in press).
Peltier, W.R. (1985): Climatic implications of isostatic adjustment constraints on current variations of eustatic sea level. In: *Glaciers, ice sheets, and sea level: effect of a CO_2-induced climatic change*. DOE/ER/60235-1 (U.S. Department of Energy, Carbon Dioxide Research Division, Washington), pp.258-266.
Peltier, W.R. and A.M. Tushingham (1989): Global sea level rise and the greenhouse effect: might they be connected? *Science* **244**, 806-810.
Peltier, W.R. and A.M. Tushingham (1990): The influence of glacial isostatic adjustment on tide gauge measurements of secular sea level. *J. Geophys. Res.* (in press).
Polar Research Board (PRB) (1985): *Glaciers, ice sheets, and sea level: effect of a CO_2-induced climatic change*. DOE/ER/60235-1 (U.S. Department of Energy, Carbon Dioxide Research Division, Washington), pp.258-266.
Pugh, D.T. (1987): *Tides, Surges and Mean Sea-Level*. John Wiley and Sons

(Chichester), 472 pp.

Raper, S.C.B., R.A. Warrick and T.M.L. Wigley (1990): Global sea level rise: past and future. In: J.D. Milliman (ed.) *Proceedings of the SCOPE Workshop on rising sea level and subsiding coastal areas.* Bangkok, 1988. John Wiley, Chichester (in press).

Robin, G. De Q. (1986): Changing the sea level. In: B. Bolin, B.R. Döös, J. Jäger and R.A. Warrick (eds.) *The greenhouse effect, climatic change and ecosystems.* John Wiley and Sons, Chichester, pp. 323-359.

Warrick, R.A. and G. Farmer (1990): The greenhouse effect, climatic change and rising sea level:implications for development. *Trans. Inst. Br. Geogr.*, (in press).

Warrick, R.A. and J. Oerlemans et al. (1990): Global-mean sea level. Chapter for Working Group 1. Report of the Intergovernmental Panel on Climate Change (Draft document).

Warrick, R.A. and T.M.L. Wigley (eds.) (1990): *Climate and Sea Level Change: Observations, Projections and Implications.* Cambridge University Press, Cambridge (in press).

Wigley, T.M.L. (1989): Scientific assessment of climate change and its impacts. In: *Seminar on climatic change.* Presentations of a seminar at Downing Street, 26 April 1989 (U.K. Department of the Environment, London).

Wigley, T.M.L. and S.C.B. Raper (1987): Thermal expansion of sea water associated with global warming. *Nature* **330**, 127-131.

Wigley, T.M.L. and B.D. Santer (1990): Statistical comparison of spatial fields in model validation, perturbation and predictability experiments. *J. of Geophys. Res.* (in press).

Woodworth, P.L. (1990): Workshop conclusions on sea level changes. In: R.A. Warrick and T.M.L. Wigley (eds.) *Climate and sea level change: observations, projections and implications.* Cambridge University Press, Cambridge (in press).

Woodworth, P.L. (1987): Trends in U.K. mean sea level. *Marine Geodesy* **11**, pp.57-87.

World Meteorological Organization (WMO) (1986): Report of the International conference on the assessment of the role of carbon dioxide and of other greenhouse gases in climate variations and associated impacts, Villach, Austria, 9-15 October 1985, WMO-No. 661 (WMO Geneva).

Wyrtki, K. (1982): The Southern Oscillation, ocean-atmosphere interaction and El Niño. *Marine Technology Society Journal 16*, 3-10 (U.S.A).

Yim, W. W-S. (1990): Future sea level rise in Hong Kong and possible environmental effects. In: R.A. Warrick and T.M.L. Wigley (eds.) *Climate and sea level change: observations, projections and implications.* Cambridge University Press, Cambridge (in press).

Zwally, H.J. (1989): Growth of Greenland ice sheet: Interpretation. *Science 246*, 1589-1591.

Zwally, H.J., A.C. Brenner, J.A. Major, R.A. Bindschadler and J.G. Marsh (1989): Growth of Greenland ice sheet: Measurement. *Science 246*, 1587-1589.

EUROPEAN SCHOOL OF CLIMATOLOGY AND NATURAL HAZARDS

Course on

"Climatic Change and Impacts: A General Introduction"

STUDENTS' PAPERS

DRY MATTER ACCUMULATION, CO_2 GAS EXCHANGE, AND WATER UPTAKE OF PLANTS AND PLANT COMMUNITIES UNDER INCREASED ATMOSPHERIC CO_2 CONCENTRATION

M. Forstreuter
University of Osnabrueck
General Group of Ecology
Barbarastr. 11
D-4500 Osnabrueck

SUMMARY
The effect of long term exposure to elevated CO_2 levels on biomass production at 18 herbaceous plant species was studied. The grain yield, CO_2 gas exchange, and water uptake on plant communities of *Trifolium pratense* and *Festuca pratensis* were examined at 340, represent ambient conditions, 450, 600 and 800 ppm atmospheric CO_2 concentrations in mini-glasshouses under field conditions. Plant species grown at nearly doubling CO_2 differed widely in the response to dry weight accumulation, all species together show an average increase of about 30 % dry weight. The grain yield of *Trifolium pratense* and *Festuca pratensis* was increased with CO_2 enrichment. A positive effect (+30%) on the CO_2 gas exchange of a whole plant community was measured at 450 ppm compared to 340 ppm CO_2 at high photon flux densities. The daily balance of net CO_2 uptake showed that 33 % more carbon dioxide was incorporated in a plant stand at 450 ppm compared with 340 ppm CO_2 in the air. The soil water storage was reduced in the plant community at 340 ppm CO_2 concentration compared to that at the enhanced CO_2 levels. Parallel measurements of matric soil potentials indicated an increase in water use efficiency under CO_2 enrichment.

1. INTRODUCTION

It is now well established that carbon dioxide is increasing in the global atmosphere, see Figure 1 (1). This is principally due to the increased use of fossil fuels, but also to other human activities, such as deforestation, erosion and oxidation of agriculturally used soils (2,3).

Rising levels of atmospheric carbon dioxide may have a significant effect on the growth of individual plants and plant communities. The phytomass of plants contains about 45 % of carbon in the dry matter. Carbon is fixed by green plants through the reduction of atmospheric CO_2 in the process of photosynthesis.

How can plants respond to increased carbon dioxide supply?
An assemblage and analysis of 430 already existing observations on growth and yield of agricultural and non-agricultural plants under enriched CO_2 concentrations are available (4). Other authors reported about the direct effects of increased atmospheric CO_2 concentration on vegetation (5,6,7). There are several reports about the influence of atmospheric CO_2 enrichment on photosynthesis and short-term growth, however rarely have any crops been grown from seed to maturity under different CO_2 concentrations in their atmosphere. Only a few field experiments were done on whole plant communities in an enriched CO_2 concentration (8,9,10).

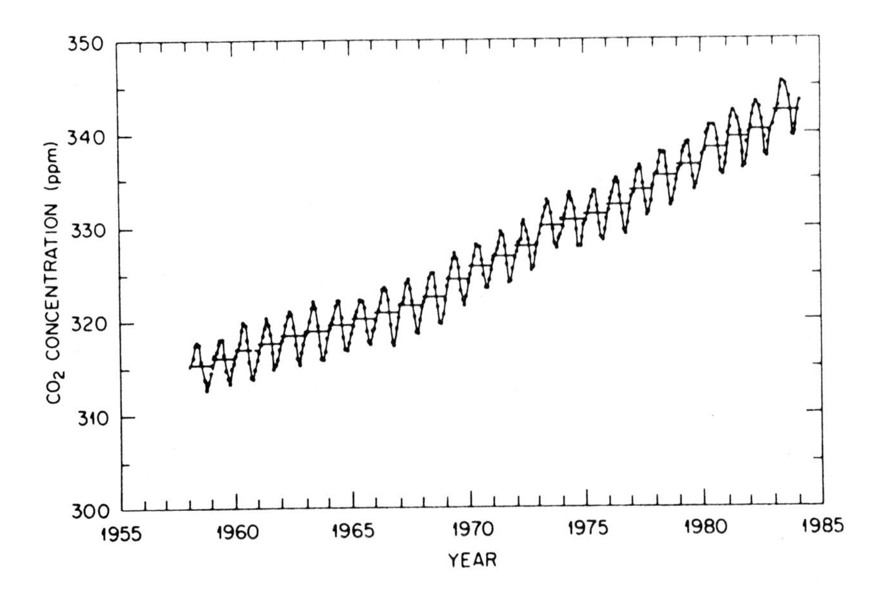

Figure 1: Concentration of atmospheric CO_2 at Mauna Loa Observatory, Hawaii, expressed as a mole fraction in parts per million of dry air. The dots depict monthly averages of visually selected data adjusted to the center of each month. The horizontal bars represent annual averages. Data were obtained by C.D. KEELING, Scripps Institution of Oceanography, University of California, La Jolla, California.

The present studies were undertaken to analyse the long term effects of CO_2 enrichment on different component processes, such as dry matter accumulation, CO_2 gas exchange, and water uptake during a whole vegetation period under nearly natural environments.

2. MATERIALS AND METHODS

Experiments in enriched atmospheric CO_2 concentrations on dry matter production of 18 different plant species were made in 1986 in CO_2 controlled greenhouses (11,12). The species were grown in plastic pots over the vegetation period, in nearly 340 ppm and 600 ppm (fluctuating between 380 and 700 ppm), plant materials were harvested at different growth stages.

A study of a whole plant community under field conditions was done with a mixture (1:1) of red clover (*Trifolium pratense* L. var. Lero) and meadow fescue (*Festuca pratensis* Huds. var. Cosmos II). The plants were grown simultaneously from seedlings in acrylic mini-greenhouses (prototype: 13,10) under field conditions. Four of these model-ecosystems containing 60 cm homogenized garden soil were supplied with 340 ppm, representing ambient, 450, 600 and 800 ppm CO_2 in the air for long term measurements of the net carbon dioxide flux. The air temperature in the mini-greenhouses was adjusted to the air outside (±0.5 C°). The windspeed was measured outside and a relative airstream inside the mini-greenhouses was fan produced . Beyond that, the relative air humidity, the photon flux density (photosynthetically active radiation) outside and inside the mini-greenhouses, and the CO_2 gas exchange rates of the sytems were continuously registered with the climatic parameters, measuring the intervals of 48 sec and computed to half hour means with the standard deviation. The precipitation was measured every day and an equal amount of water was added in the mini-greenhouses.

Direct measurements of the soil water content were made on soil samples (0 - 40 cm) of known weight by oven drying at 105 C° to constant weight. The bulk density of the soil was determined separately (14). The conversion of the water content per weight unit into water content per volume unit was done by multiplying the weight percentage by the bulk density of soil under study.

Field measurements of the matric potential of soil were made with tensiometers in each plant stand grown under different CO_2 concentrations. They consisted of a porous ceramic cup filled with water which was buried in the soil at 40 cm depth and connected by a water filled tube to a manometer near the top. The manometer indicates the pressure drop on the water in the porous cup, which is in equilibrium with the matric potential of the water in the soil (15).

3. RESULTS

3.1 DRY MATTER ACCUMULATION

The possible impact of elevated CO_2 concentrations on the biomass production of annual and perennial plants was studied in a greenhouse experiment in detail. Table 1 shows the results of the experiment during the 1986 vegetation period (11,12). The results of all herbaceous species indicate that plants with reserve tissue increase their productivity with elevated CO_2 concentration and those species without storage organs did not show significant increases. Individual plant species differ widely in the response to enriched CO_2 concentration. The mean response of the total biomass production summarizing all species showed a positve CO_2 effect of 30 % under nearly doubling CO_2.

The reproductive development rates at concentrations up to 800 ppm CO_2 were investigated in whole plant stands under field environments during the vegetation period of 1988. The grain yield was increased by CO_2 enrichment (Fig. 2) and became noticeable as a weight increase of the individual seeds.

3.2 CO_2 GAS EXCHANCE

Additionnally to the technique of harvesting plant material to obtain information about the dry matter accumulation of enriched CO_2-treatment in greenhouses, measurements were made of the CO_2 gas exchange on whole plant communities, roots and soil included. One dayplot of the field experiment with red clover (*Trifolium pratense*) and meadow fescue (*Festuca pratensis*) is shown in Fig. 4. The curves of CO_2 flux were closely correlated with solar radiation. The respiration during the night was enhanced under enriched CO_2 concentration, because more phytomass was accumulated. At the beginning of the day the CO_2 net uptake in both plant stands became positive and below full sun-light the CO2 net uptake was nearly the same. At high light levels much higher rates of gas exchange were maintained in the enriched CO_2 conditions than in 340 ppm treatment. The plant stand under the CO_2 concentration of 450 ppm fixed up to 34 % more carbon dioxide at the level of 1200 µE m^{-2} s^{-1} of photon flux density against the 340 ppm treatment.

The daily balance of both model ecosystems, 30 g CO_2 m^{-2} d^{-1} at 340 ppm CO_2 and 40 g CO_2 m^{-2} d^{-1} in 450 ppm CO_2 treatment, showed that 33 %

		TS
1. Grasses		
Dactylis glomerata	Above soil line	0.94
	Root	1.35
Festuca pratensis	Above soil line	0.56
	Root	0.84
Festuca rubra	Above soil line	1.08
	Root	1.35
Lolium perenne	Above soil line	2.04
	Root	1.83
	Leaf	0.59
Hordeum vulgare	Shoot	1.07
	Root	1.92
	Leaf	1.2
Secale cereale	Shoot	0.9
	Root	1.09
Triticum sativum	Leaf	1.08
	Shoot	1.27
	Root	1.34
2. Herbaceous plants with storage organs		
Beta vulgaris	Leaf	1.07
ssp. vulgaris var. crassa	Shoot	0.77
	Root	1.72
Beta vulgaris	Leaf Shoot	1.26
	Root	1.60
Plantago major	Leaf	1.38
	Shoot	1.18
	Root	2.03
Rumex obtusifolius	Leaf	1.12
	Shoot	1.16
	Root	0.94
Taraxacum officinale	Above soil line	0.77
	Root	0.84
3. Herbaceous plants without storage organs		
Capsella bursa pastoris	Above stoil line	0.10
	Root	0.88
Chenopodium album	Above soil line	2.24
	Root	0.67
Polygonum aviculare	Above soil line	1.52
	Root	1.74
Trifolium pratense	Leaf Shoot	2.28
	Root	2.55
Trifolium repens	Leaf	0.86
	Shoot	3.3
	Root	2.35
Urtica dioica	Leaf	0.75
	Shoot	0.73
	Root	1.13

Table 1: Relative response on dry matter accumulation (TS) to elevated CO_2 level by 18 different herbaceous plants growing under present (340 ppm) and under elevated (fluctuating between 380 and 700 ppm) CO_2 levels in a greenhouse experiment. Plants are divided into grasses, herbaceous plants with storage tissue and those without storage organs. These are results from an experiment conducted during the vegetation period 1986. Data were collected by (11,12) from (23).

more carbon dioxide was fixed under the enriched ambient CO_2 concentration.

The fertilization factor of ambient CO_2 concentration on the net primary production used for the OBM (Osnabrueck Biosphere Model) is shown in Figure 3. The curves are calculated with the equation from (16,17).

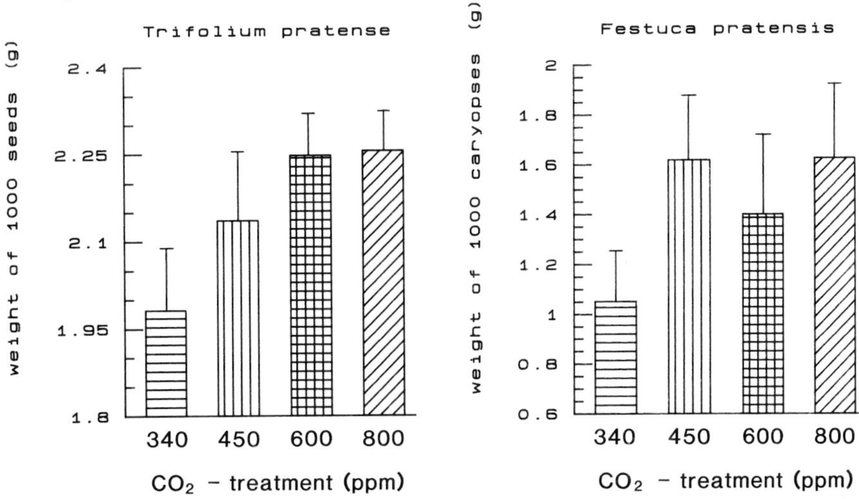

Figure 2: Long term effects of 340, 450, 600 and 800 ppm CO_2 in the air on the weight of 1000 seeds and caryopses of *Trifolium pratense* and *Festuca pratensis* in an field experiment (n=10).

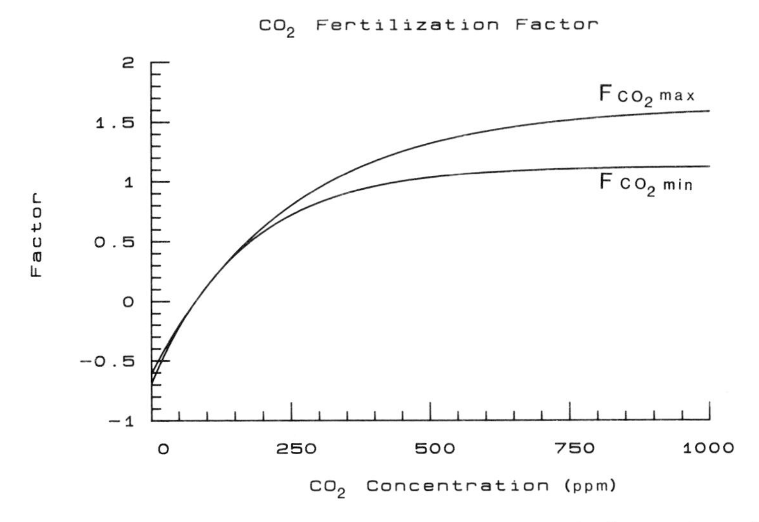

Figure 3: The curves show the minimum (F_{CO2} min) and the maximum (F_{CO2} max) CO_2 fertilazation factor over a range of atmospheric CO_2 concentrations (ppm) in the air, calculated with the equation from ESSER and LIETH (17) with a minimal (0.5) and maximal soil factor (2.5).

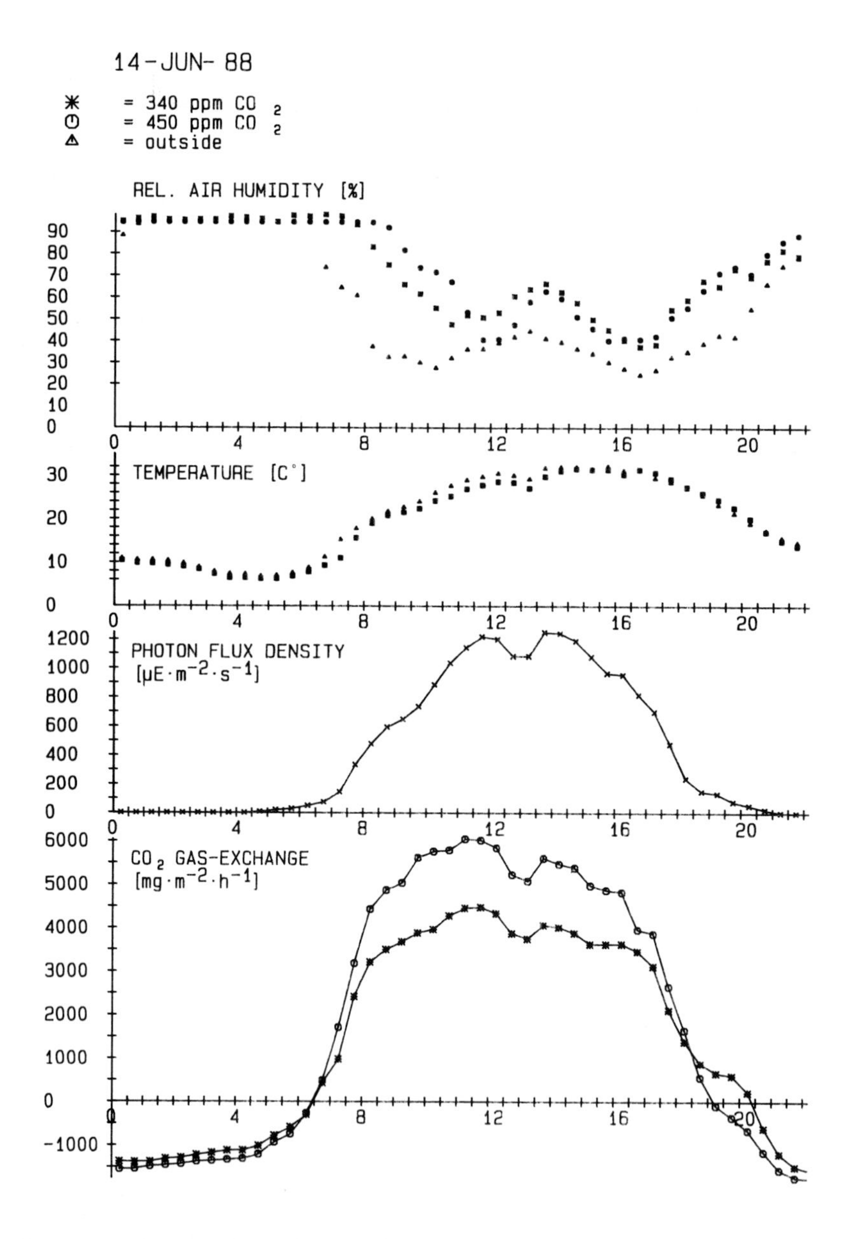

Figure 4: One dayplot of relative air humidity (%), temperatures (C°), photon flux density (μE m^{-2} s^{-1}) and CO_2 gas exchange rates in model ecosystems of *Trifolium pratense* and *Festuca pratensis* (mixture 1:1) in 340 ppm CO_2, representing ambient, and 450 ppm CO_2 in the air.

3.3 WATER UPTAKE

During the vegetation period of 1987 and 1988 the water uptake of plant communities of *Trifolium pratense* and *Festuca pratensis* under 4 different CO_2 concentrations was studied. Measurements of change in soil water content were done by two techniques: measure of soil moisture on soil samples and direct field measurements of matric potential with tensiometers. Table 2 shows the results of soil moisture content as % of soil dry weight during the experiment at 4 different CO_2 treatments. The experiment started on 18th June 1987 with a soil water content of 13.7 % in each plant stand. During the growth season the soil water content in the 340 ppm was reduced compared to that in the enriched CO_2 treatments. The greatest difference of 3.9 % of water content occurred in May 1988 in the plant stand under 450 ppm CO_2. The 600 ppm and 800 ppm treatments show higher water content levels of 2.6 % and 2.4 % than plant stands under the 340 ppm CO_2 concentration at that time.

Continued measurements of the matric potential of the soil with tensiometers, as shown in Figure 5, indicate that the amount of water use by evapotranspiration differs during the growth period and was reduced in the plant stands under enriched CO_2 concentrations.

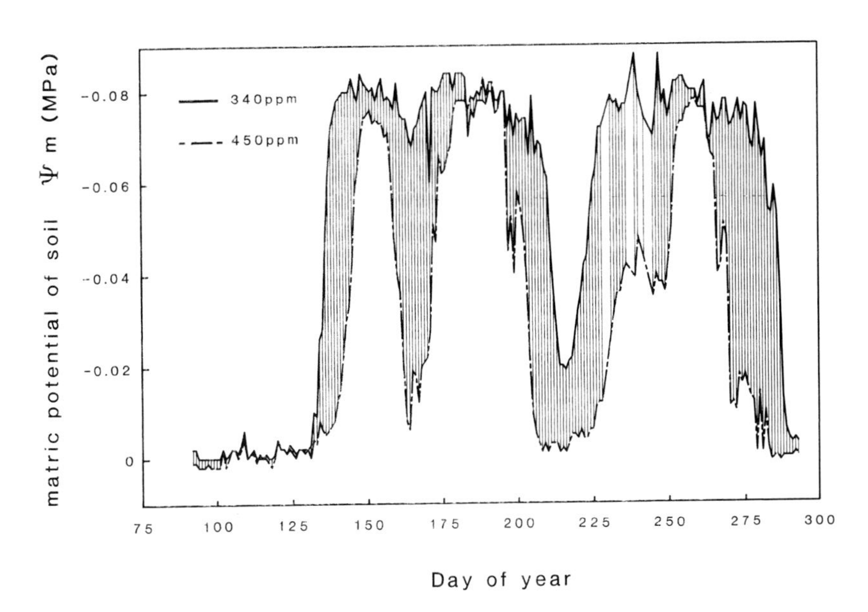

Figure 5: Variation in matric potential of soil in a plant community of *Trifolium pratense* and *Festuca pratensis*, as measured with tensiometers at 40 cm depth in two different atmospheric CO_2 concentrations in a field experiment during 1988.

CO_2-Treatment	Soil moisture content in % of dry weight					
Date:	18/7/	19/11/87	19/1	18/5	25/5	5/7/88
340 ppm	13.7 ±0.7	14.2 ±0.8	18.2 ±0.3	8.3 ±2.0	7.2 ±1.0	8.1 ±1.7
450 ppm	13.7 ±0.7	15.9 ±0.7	20.4 ±1.0	12.2 ±2.1	10.4 ±1.3	11.0 ±3.0
600 ppm	13.7 ±0.7	15.1 ±0.7	20.3 ±0.2	10.9 ±2.9	9.5 ±1.4	9.6 ±1.4
800 ppm	13.7 ±0.7	13.8 ±1.2	19.5 ±0.8	10.7 ±1.6	9.0 ±1.0	9.3 ±1.5

Table 2: Soil water content in % of soil dry weight in a model ecosystem of *Trifolium pratense* and *Festuca pratensis* (mixture 1:1) at four different CO_2 concentrations in the air (soil: sand;bulk density 1.28 g cm^{-3}) during a long term field experiment (n=10).

4. DISCUSSION

The present results indicate a positive fertilization effect of increasing atmospheric CO_2 level on dry matter production and net CO_2 uptake in the early stage of the vegetation period. The effect of 30 % more dry matter accumulation by doubling CO_2 corresponds with other investigations described in the literature (5). Greenland ecosystems are a sink for carbon dioxide in the early vegetation period, but at the end of the growth season the expected higher decomposition rate of the biomass reduces this effect (10). This change into a higher accumulation of biomass in sommertime and a higher respiration of CO_2 at the end of the year may affect the annual amplitude of the Mauna Loa curve (Figure 1).

The research has shown that the transpiration ratio (18) (somestimes called water requirement), the amount of water lost per unit of dry weight produced and is reciprocal of water use efficiency, decreased with increasing CO_2 concentration in the atmosphere. Short-term leaf responses of plants to increasing CO_2 have been well studied (19,20). In contrast with well-studied leaf level CO_2 responses, the ecosystem level, long term responses are rarely available. This study shows that increasing CO_2 concentration enhanced the dry matter accumulation and net CO_2 uptake; additionally, the soil water storage was on a higher level although more dry matter was produced. This indicates an increase in the water use efficiency (WUE), defined in terms of unit of water used per unit of dry matter produced, often expressing water loss as both evaporation and transpiration (21) under enriched CO2 concentrations. It has been predicted that higher WUE will cause sites to become more mesic and for the vegetation to change (22). The effects of enriched CO_2 on evapotranspiration and plant water requirements of whole plant communities need considerably more attention in the future.

REFERENCES

1. KIMBALL, C.D. (1985). Concentration of atmospheric CO_2 at Mauna Loa observation. In WHITE, M.R. (ed.), Characterization of Information Requirements for Studies of CO_2 Effects: Water Resources, Agriculture, Fisheries, Forests and Human Health. DOE/ER-0236, Washington.

2. ESSER, G. (1987). Sensitivity of global carbon pools and fluxes to human and potential climatic impacts. Tellus 39B, p.245-260.

3. MOORE, B., BOONE R.D., HOBBIE, J.E., HOUGHTON, R.A., MELILLO, J.M., PETERSON, B.J., SHAVER, G.R.,VOEROESMARTY, C.I. and WOODWELL, G. (1981). A simple model for analysis of the role of terresrial ecosystems in the global carbon budget. In: BOLIN,B.(ed.). Carbon Cycle Modelling, SCOPE Rep. 16, Chichester, New York, 365-385.

4. KIMBALL, B.A. (1983). Carbon dioxide and agricultural yield: An assemblage and analysis of 430 prior observations. Agron J 75,779-788.

5. STRAIN, B. and CURE, J. (1986). Direct effects of carbon dioxide on plants and ecosystems. A bibliography with abtracts. ORNL/CDIC-13, Nat. Techn. Inform. Ser., Springfield, VA, 22161. pp. 1-197.

6. OVERDIECK, D. and LIETH, H. (1986). Long-term effects of an increased atmospheric CO_2 concentration level on terrestrial plants in model-ecosystems, Universitaet Osnabrueck, Final Report of the project, CLI-075-D, Osnabrück.

7.OVERDIECK, D. (ed.) (1987). Untersuchungen über die voraussichtlichen Langzeiteffekte einer CO_2-bedingten Klimaveränderung auf die einheimische Vegetation, Universitaet Osnabrueck, Abschlußbericht des Vorhabens KF 20096, Osnabrück.

8. OVERDIECK, D. and BOSSEMEYER, D. (1985). Long time effects of elevated CO_2 concentration on CO_2 gas exchange of a model ecosystem (in German). Angew. Bot. 59, 179-198.

9. PRUDHOMME, T.I., OECHEL, W.C. and HASTINGS, S.J. (1984) Net ecosystem gas exchange at ambient and elevated carbon dioxide concentrations in Tussock Tundra at Toolik Lake, Alaska: An evaluation of methods and initial results. In: McBeath, J.H.(ed.). The Potential Effects of Carbon Dioxide-Induced Climatic Changes in Alaska: Proceedings of a Conference. School of Agricultural and Land Resources Management, Univ. of Alaska, Fairbanks, AK. Misc. Publ. 83-1.pp.155-162.

10. OVERDIECK, D. and FORSTREUTER, M. (1987). Langzeit-Effekte eines erhöhten CO_2-Angebotes bei Rotklee-Wiesenschwingelgemeinschaften, Verhandlungen Gesellschaft für Ökologie, Band XVI, Göttingen, 197-206.

11. KRUSE, M. (1987). Trockensubstanzproduktion und Mineralstoffaufnahme von krautigen Kultur- und Wildpflanzen besonders Getreide- und Hackfruchtarten, unter verschiedenen CO_2-Gehalten in der Luft. Diplomarbeit an der Universitaet Osnabrueck, FB 5 Biologie/Chemie, Osnabrueck.

12. STEGMANN, S. (1987). Trockensubstanzproduktion und Mineralstoffaufnahme von krautigen Kultur- und Wiesenpflanzen, besonders Wiesen- und Trittpflanzen, Diplomarbeit an der Universitaet Osnabrueck, FB 5 Biologie/Chemie, Osnabrueck.

13. FORTREUTER, M. (1986). CO_2-Gaswechsel und Entwicklung von Model-Oekosystemen mit Rotklee- und Wiesenschwingel-Gemeinschaften bei erhöhtem CO_2-Angebot. Diplomarbeit an der Universitaet Osnabrueck, FB 5 Biologie/Chemie, Osnabrueck.

14. SCHLICHTING, E. and BLUME, H.P. (1966). Bodenkundliches Praktikum, Paul Parey, Hamburg, 209 Seiten.

15. KRAMER, P.J. (1983). Water Relations of Plants. Academic Press, Inc., London. 489p.

16. LIETH, H. (1985). A dynamic model of the global carbon flux through the biosphere and its relation to climatic and soil parameters. Int J Biometeorol 29 (Suppl 2): 17-31.

17. ESSER, G. and LIETH, H. (1989). Atmosphere as a part of the biosphere. In DIEMINGER, HARTMANN, LABITZKE, LEITINGER, HERBST (eds.). Physics of the upper atmosphere. Landolt-Börnstein, Group V, Vol. 5, Kap. 1.1.9. Springer, Berlin.

18. LARCHER, W. (1980). Oekologie der Pflanzen. 3.Aufl., Stuttgart.

19. BJORKMANN, O. and PEARCY, R.W. (1983). "Physiological Effects". 65-106. In LEMON, E.R. (ed.). CO_2 and the response of plants to rising levels of atmospheric carbon dioxide. American Association for the Advancement of Science. Selected Symposium 84. West view Press, Boulder, Colorado.

20. KIMBALL, B.A. and IDSO, S.B. (1983). Increasing atmospheric CO_2: Effects on crop yield, water use and climate. Agric Water Management 7, 55-72.

21. TEARE, I.D., KANEMASU, E.T., POWERS, W.L. and JACOBS, B.S. (1973). Water use efficiency and its relation to crop canopy area, stomatal regulation and root distribution. Agron J 65, 207-211.

22. IDSO, S.B. and QUINN, J.A. (1983). Vegetational redistribution in Arizona and New Mexico. In: Response to a doubling of the atmosphere CO_2 concentration. Climatological Publications, Scientific Paper No. 17, Laboratory of Climatology, Ariz. State Univ., Tempe, AZ, 52 pp.

23. LIETH, H. (1988). Starting a new term for the International Journal of Biometeorology, Int J Biometeorol 32: 1-10.

INCLUSION OF DOWNDRAUGHT PROCESSES IN THE UK METEOROLOGICAL OFFICE MASS FLUX CONVECTION SCHEME

D. Gregory
Dynamical Climatology Branch
UK Meteorological Office

Work is underway to include downdraught processes in the mass-flux convection scheme used in the UK Meteorological Office 11-layer climate model (Slingo et al (1989)). Development of the revised scheme is being carried out using a single column version of the model initialised with easterly wave data from the GATE experiment (Thompson et al (1979)). The GATE data sets provide information concerning surface fluxes and large-scale forcing of convection. These are supplied to the model which is integrated through several cycles of the easterly wave (each 80 hours in duration). The downdraught is based upon an inverted entraining plume starting from the model layer in which an equal mixture of updraught and environmental air, brought to saturation, is negatively buoyant on descent to the next model layer after evaporation and freezing processes have been accounted for. The initial downdraught mass-flux is set to one tenth of the initial updraught mass-flux. On reaching the surface the downdraught air is detrained over the two lowest model layers. Model results are compared with the GATE data and also results from a study of cloud ensembles using a detailed cloud model and the same GATE data set (Gregory and Miller (1989)).

Results are presented from a 20 day integration, 6 wave cycles in all, and are composites over the last four cycles. Figure 1 compares the heating profiles for the updraught and downdraught between 20 and 30 hours of the wave cycle for the present scheme representing updraughts only, the new version including the downdraught and from a detailed cloud model. The present scheme provides a poor representation of the updraught heating profile being flat in the lower half of the convective layer and peaking around 600mb, reflecting the shape of the large-scale cooling which it balances. The detailed cloud model indicates a peak in updraught heating at 700mb. Inclusion of downdraught processes brings better agreement with the detailed cloud model. Figure 2 shows the evolution of 10 hour averages of rainfall and mixing ratio for the lowest model layer. Peak values of rainfall are well simulated by the revised scheme. The structure of the boundary layer is much improved, the stability of the two lowest model layers being retained whereas the present scheme over stabilises the boundary layer due to excessive evaporation of rain below cloud base. The magnitude of the moisture content of the lowest model layer and its temporal evolution over the wave cycle are improved by the inclusion of the downdraught so improving the evolution of the surface latent heat flux.

Several problems remain to be resolved. The upper troposphere is too warm (by 2-3°C above the freezing level (around 600mb)) and the atmosphere is too moist above the boundary layer. These errors may be caused by the downdraught being too active in the later half of the wave cycle, destabilising the atmosphere and so causing convection to exceed the supplied large-scale forcing during this period. A more complex specification of initial downdraught mass flux may bring improvement.

REFERENCES

Gregory, D., and Miller, M. J. 1989 : A numerical study of the parameterization of deep tropical convection. Submitted to the Quart. J. Roy. Met. Soc..

Slingo, A., Wilderspin R. C., and Smith, R. N. B. 1989 : The effect of improved physical parametrizations on simulations of cloudiness and the earth's radiation budget. Submitted to the J. Geophys. Res..

Thompson, R. M., Payne, S. W., Recker, E. E., and Reed, R. J. 1979 : Structure and properties of synoptic-scale wave disturbances in the ITCZ of the eastern Atlantic. J. Atm. Sci., 36, p53-72.

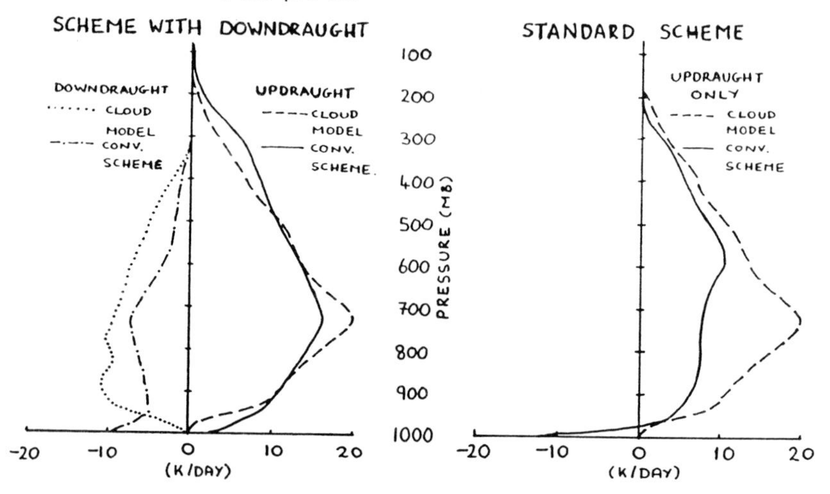

Figure 1. Comparison of updraught heating and downdraught cooling profiles from standard and downdraught convection scheme and detailed cloud model.

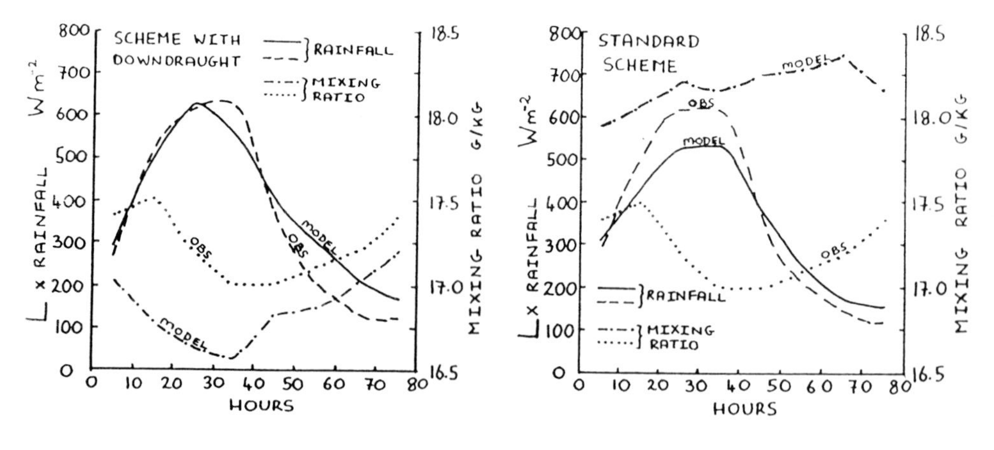

Figure 2. Evolution of rainfall and mixing ratio of lowest model layer over a composite wave cycle.

A 14O,OOO-YEAR CLIMATIC RECONSTRUCTION OF THE CONTINENTAL CLIMATE FROM TWO EUROPEAN POLLEN RECORDS

J. GUIOT, J.L. de BEAULIEU, A. PONS and M. REILLE
Laboratoire de Botanique historique et Palynologie, UA CNRS 1152, Faculté des Sciences et Techniques St-Jérôme, 13397 Marseille Cedex, France

On continents, pollen analysis provides the most reliable climatic proxy data (1). The pollen spectrum, given by the percentages of the different taxa present in a given site provides a reliable indication of the regional vegetation, and hence of the regional climate. Quantitative analysis of the spectra recorded in the past enables theoritically to provide numerical reconstructions. An original method already described (2) and improved (3) has enabled us to use the pollen-stratigraphical records of La Grande Pile (4) and Les Echets (5), in eastern France, each of which contains a sediment record extending over the last 14O millennia. The modern data, covering Europe, Northern Africa and Siberia, are represented by 227 spectra, composed by 52 taxa, selected so as to represent the greatest variety of present-day vegetation types and climates. Seasonal temperatures and precipitation were interpolated for the localities from which spectra were collected (2).

1. METHOD

The method is based on the assumption that the different climates, their corresponding vegetation types and contemporaneous pollen spectra that have succeeded each other in the course of time in the study area can be found to-day in very similar forms. It follows, therefore, that for climates to be represented that are quite different from those of the studied locality at the present time, a very extensive study area is required, which can provide a wide range of analogues. The method involves the following steps:

a) the establishment of representative modern pollen spectra from a large area with wide variations in present-day climatic conditions and vegetation cover;

b) the definition of precise climatic data for all the sites from which pollen spectra have been obtained;

c) the identification of modern spectra closest in similarity (the "best analogues") for each fossil spectra;

d) the utilisation of the present day climatic parameters corresponding to these best analogues in order to infer past climatic conditions.

The anthropogenic disturbance of the modern vegetation has been minimized, in the reconstructions, to some extent by using a particular method for calculating the best analogue spectra. The percentages corresponding to each plant have been attributed a "loading factor" (which we term "Palaeobioclimatic Operator") which does not express its present correlation with climate but the role it has played throughout the period covered by the reconstruction.

The reconstructions are primarily obtained and drawn as a function of the depths. The reconstructions in the two sites are presented in a time-scale and averaged in Figure 1. This time-scale is based on ^{14}C dates after 30,000 year B.P. (5,6) and on ^{18}O stages generally admitted (7) before.

2.RESULTS

Over the whole reconstructed period, there appears two distinctly warm and humid episodes : the last Interglacial (Eemian) and the Holocene. Both of them succeed to a complex period marked by an important rise in temperature and in precipitation.

A clear succession of two relatively warm and humid periods occurred after the Interglacial. The first, *i.e.* the St-Germain 1 Interstadial, appears as a climatically complex period. The two periods which succeeded to the Interglacial and the St-Germain 1 stadial (respectively the Melisey 1 and the Melisey 2 stadials) were, as expected, cold and dry. The second temperate period, *i.e.* the St-Germain 2 interstadial, suggests overall warm and humid conditions, but it ends with a tendency towards low temperatures and high precipitation. Thus, the following period, the Lower Würmian Pleniglacial, was first humid and cold, then very cold and very dry, as were the "late Eemian-Melisey 1" and the "late St-Germain 1-Melisey 2" periods.

If one agrees with Ruddiman and Mc Intyre (8) that an accumulation of continental ice implies a cold and humid continent as opposed to a hot Atlantic Ocean between latitudes 5O° and 6O° North, the results presented here suggest three major ice accretion periods in Europe.

The first corresponds to the very humid and markedly cold climate of the final part of the Eemian, forming a prelude to the even colder and dry Melisey 1 stadial. The second is the end of the St-Germain 1 interstadial (very humid but moderately cold) which was succeeded by the cold and dry Melisey 2 stadial. The third ice accretion period corresponds to the end of St Germain 2 interstadial and to the beginning (markedly cold but moderately humid) of the Lower Pleniglacial, before the second very cold and dry part of this major stadial.

REFERENCES

(1) Birks, H.J.B. and Birks, H.H. (1980). Quaternary Palaeoecology, E.Arnold, London, 289pp.
(2) Guiot, J. (1987). Late Quaternary climatic change in France estimated from multivariate pollen time-series. Quaternary Research, 28, 100-118.
(3) Guiot, J., Beaulieu, J.L. de, Pons, A., and Reille, M., A 140,000 yr climatic reconstruction from two european pollen records. Nature, to be published.
(4) Woillard, G. (1979). The last interglacial-glacial cycle at Grande Pile in Northeastern France. Bull. Soc. Belge Géol., 88, 51-69.
(5) Beaulieu, J.L. de, Reille, M. (1984). A long upper Pleistocene pollen record from Les Echets, near Lyon, France. Boreas, 13, 111-132.
(6) Woillard, G.M., Mook, W.G. (1982). Sciences, 215, 159-161.
(7) Imbrie, J., Hays, J.D., Martinson, D.G., McIntyre, A., Mix, A.C., Morley,J.J., Pisias, N.G., Prell, W.L., Shackleton, N.J. (1984). The orbital theory of Pleistocene climate: support from a revised chronology of the marine ^{18}O record. In A.L. Berger et al. (eds) Milankovitch and Climate, Reidel, Dordrecht, part I, 269-305.
(8) Ruddiman, W.F., McIntyre, A., Niebler-Hunt, V., Durazzi, J.T. (1980). Quaternary Research, 13, 33-64.

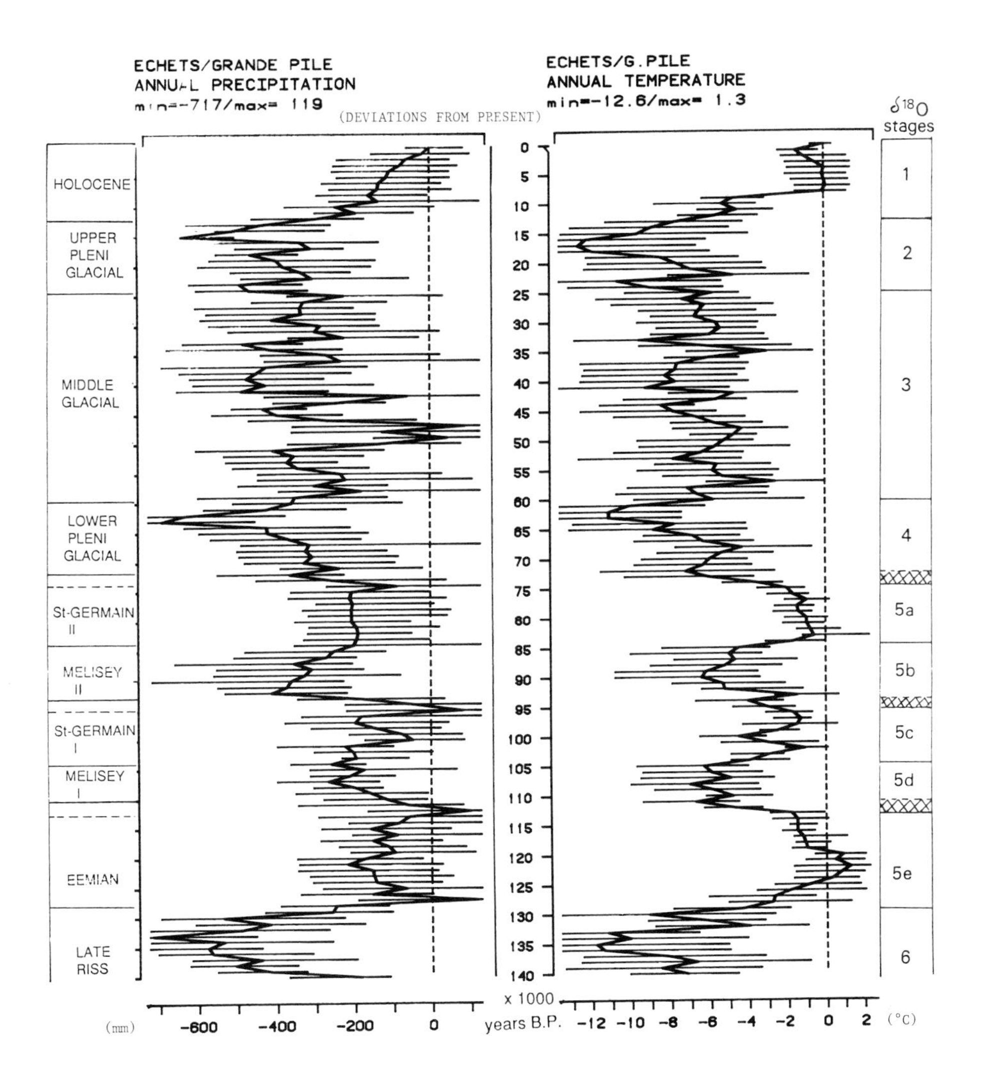
ECHETS/GRANDE PILE
ANNUAL PRECIPITATION
min=-717/max= 119
(DEVIATIONS FROM PRESENT)
ECHETS/G.PILE
ANNUAL TEMPERATURE
min=-12.6/max= 1.3
δ18O
stages
HOLOCENE
UPPER
PLENI
GLACIAL
MIDDLE
GLACIAL
LOWER
PLENI
GLACIAL
St-GERMAIN
II
MELISEY
II
St-GERMAIN
I
MELISEY
I
EEMIAN
LATE
RISS
1
2
3
4
5a
5b
5c
5d
5e
6
x 1000
years B.P.
(mm)
-600
-400
-200
0
-12
-10
-8
-6
-4
-2
0
2
(°C)

HYDROMETEOROLOGICAL ASSESSMENT PROGRAM IN EAST AFRICA.

J.E.Hossell*
Department of Meteorology,
University of Reading, U.K.

INTRODUCTION

An hydrometeorological assessment project, funded by the World Bank, is currently being undertaken with the aim of evaluating the water resources available in six East African countries - Sudan, Uganda, Djibouti, Kenya, Somalia, and Ethiopia. One element of this study involves the analysis of the raingauge and meteorological networks in these countries, to determine the extent of their coverage, the quality of the data gathered and to produce updated mean annual isohyetal maps to aid future hydrological resource planning.

Thus far information has only been obtained from three of the nations, Sudan, Djibouti, and Uganda. The methods described below are those used to analyse the Sudanese data set.

BACKGROUND

The planning of hydrological schemes necessitates a reliable description of the rainfall climatology in the area concerned. This should include long term records from a network of raingauges, which adequately define any spatial and temporal fluctuations in

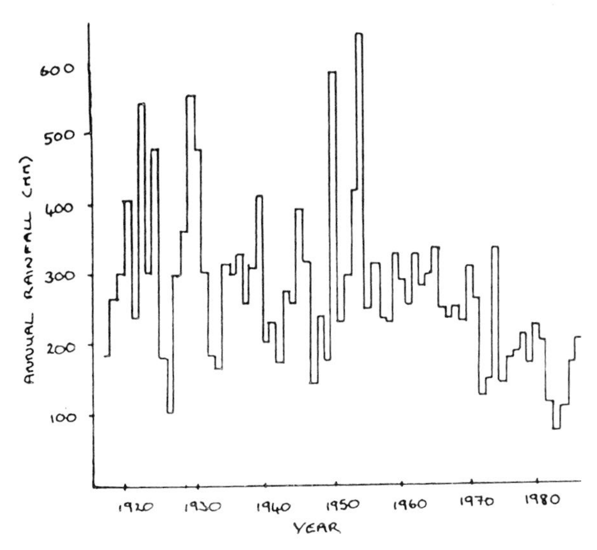

FIGURE 1. Annual Precipitation at El Fasher, Sudan.

*Currently a Ph.D. student at the Department of Geography, University of Birmingham.

the precipitaion patterns. These criteria prove difficult to satisfy in areas, such as semi-arid Africa, which show persistant dry periods.

Evidence of the decline in annual rainfall over Africa, in general, and the Sudan, in particular, is well documented. Figure 1 shows the pattern of rainfall totals for El Fasher, since the beginning of the century. It is clear that there has been a decline in amounts from the end of the 1960 s and that this trend has intensified in the present decade.

The last editon of the country's mean annual isohyetal map was in 1965, when the 1931-60 raingauge station averages were prodcued (see Figure 2).

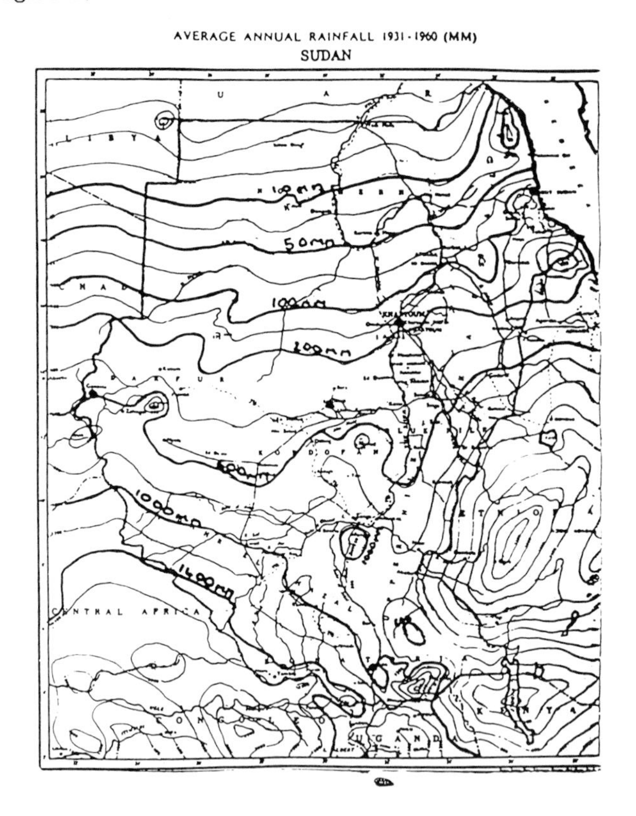

FIGURE 2. Sudan Met. Department 1931-60 Isohyetal Map.

It is important to update this map to enable the precipitation resources of the country to be accurately assessed.

Climatic means are generally calculated over 30 year periods. However, since the decrease in rainfall totals is only evident for the last 20 years, any averaging over a longer time will tend to obscure the severity of the recent decline. As data for the Sudan were available from the Sudan Met. Department (SMD.) Annual Meteorological and Rainfall Reports for 1950-86, it was decided to divide the 37 years of information into to two data sets, 1950-67 and 1968-86, in an attempt to show the mean rainfall pattern of both the earlier 'wet' and the later 'dry' periods.

DATA ANALYSIS

There were 840 raingauge stations operating in the Sudan between 1950 and 1986, though some ran for less than 5 years and with the problems of inadequate quality control and, due to the civil war in the south of the country, with data collection, many more have rather fragmented records. The data were in the form of monthly and annual totals and only those stations with no more than 10% of the years missing in each period were considered for use in the production of the isohyetal map. In addition, to avoid over representation around the Gezeira region, south of Khartoum, numbers were further restricted to 4 gauges per 2 x 2 degree grid square. However, stations operating in both periods were included regardless of location.

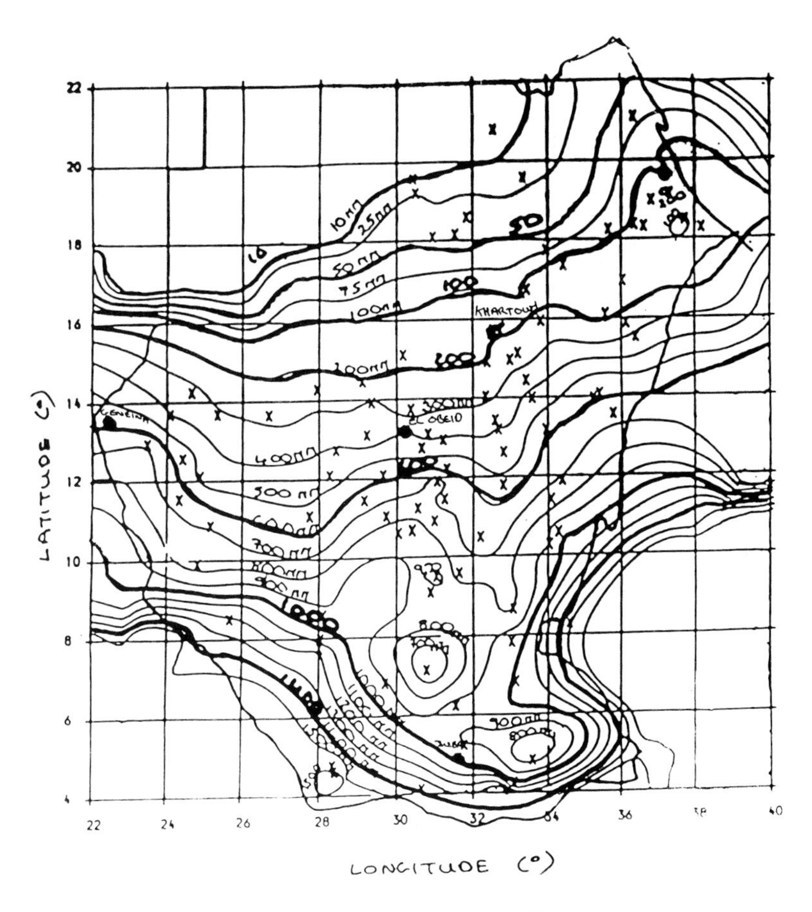

FIGURE 3. Mean Annual Isohyetal Map 1950-67.

There were 104 stations in the 1950-67 data set that fitted the above criteria, but only 45 in the later period. 37 of the stations were in operation throughout both time intervals. Despite the smaller number of stations used in this current analysis the pattern of rainfall is similar to that shown by the SMD. 1931-60 map seen in Figure 2.

Missing monthly totals needed to complete data years were replaced by the median monthly value at the appropriate station. These figures were considered more representative than the mean because of the skewed nature of the rainfall distribution. Interpolation of missing values from data at neighbouring stations was not regarded as being sufficiently accurate, as a result of the high spatial variability of the precipitation across the country and the large distances between gauges in some areas.

Figure 4 is the 1968-86 map. Although the pattern of isohyets across the country is similar for both periods, the rainfall values are lower in the more recent one.

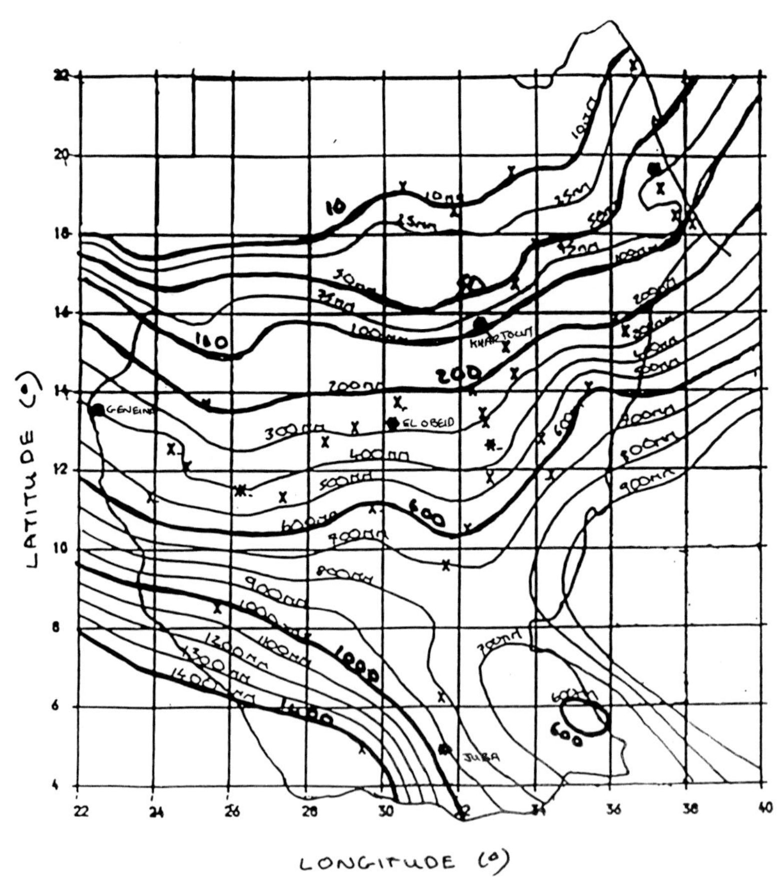

FIGURE 4. Mean Annual Isohyetal Map 1968-86.

Most stations experienced an average decrease in annual rainfall of approximately 100mm. For example the mean yearly precipitation at Khartoum airport between 1950-67 was 192mm but between 1968 and 1986 it was 102mm. Rainfall over the majority of the country has decreased, on average, by more than 10%.

CONCLUSIONS

It is clear that for areas where precipitation has been declining for the past 2 decades, such as Sudan, the use of the 30 year mean is misleading and tends to lead to an overestimation of rainfall totals. For the purpose at least of water resource planning, it is safer to rely on a pessimistic rather than optimistic assessment of the expected precipitation amount.

CLOUD RADIATIVE PROPERTIES AND CLIMATE SENSITIVITY

W.J. Ingram
UK Meteorological Office

Clouds have an important effect on atmospheric radiative fluxes in both the solar and longwave regions of the spectrum. They increase the planetary albedo, cooling the surface/troposphere system, but also absorb outgoing longwave radiation and re-radiate it to space at a colder temperature, a "greenhouse" effect that warms the surface/troposphere system. The strength of these effects can be estimated using satellite data or atmospheric general circulation models (AGCMs). Probably both are of the order of 50 Wm^{-2} in the annual global mean, with the albedo effect dominating to give a net cooling of order 10 Wm^{-2}. This is of similar size to the 4 Wm^{-2} annual global mean forcing of the surface/troposphere system due to a doubling of CO_2 levels, the standard case for studying trace gas influence on climate. This might suggest that cloud feedbacks cannot play a significant role in climate change. However, the two cloud forcing effects are largely due to different clouds. The greatest contribution to the albedo effect is from thick water clouds at low latitudes, but the "greenhouse" effect is greatest for high ice clouds. Since the physical processes associated with the clouds are quite different, we cannot expect them to react similarly to climate change. Thus the comparison above should not use the magnitude of their sum, but their individual magnitudes. These are large compared with the 4 Wm^{-2} CO_2 forcing, so that fractionally small changes in cloud forcing in response to climate change could provide a major feedback.

References (1) and (2) discuss the uncertainties in the prediction of climate change which are due to uncertainties in the strength of feedbacks, including cloud feedback. The feedback on climate change due to changes in cloud distribution (sometimes split into cloud height and cloud amount, but the distinction is often unnatural) has been studies using various AGCMs. Major discrepancies still exist between models or different versions of the same model (e.g. reference (2), and the results reported below). Research therefore continues to identify the reasons for these disagreements and the physical mechanisms underlying the responses, and hence the likely response of the real climate system (e.g. reference (3)).

However, the possible feedbacks due to changes in cloud radiative properties have been much less studies. CO_2-induced warming increases both atmospheric water vapour content (because saturation specific humidity increases rapidly with temperature while relative humidity changes little with climate) and precipitation (because increases in the overall radiative cooling of the atmosphere and heating of the surface are balanced by increased upward latent heat transports). Thus one might also expect increases in the intervening stage, cloud water content. This would increase both the "greenhouse" and albedo effects, giving a positive and negative contribution to the overall feedback, and one cannot say a priori which would be the larger. In fact cloud radiative properties depend on the cloud microphysics as well as the total condensed water path, but our knowledge of this is not adequate for proper parametrization in AGCMs, especially for ice clouds. This also precludes the representation of the potentially important feedbacks discussed by reference (4).

The experiments of Somerville and Remer (5) with a radiative-convective (globally averaged) model suggested that changes in cloud radiative properties due to changes in cloud water content might provide a very strong negative feedback on climate change. However, they omitted the "greenhouse" feedback entirely, and the cloud they prescribed for their model probably exaggerated the albedo effect. Also, in such a model it is necessary to prescribe arbitrarily a dependence of cloud water content on climate, and estimates derived from its variation with time and place within the present climate are not necessarily appropriate.

There is thus a need to study cloud radiative property feedbacks using AGCMs, where we can not only represent all these effects, but can also base the parametrizations more closely on the physical processes that we believe to act in the real atmosphere. One such study has been published (6), using an AGCM that predicted cloud water paths, and calculated cloud radiative properties from them following Stephens (7). They found a fairly strong positive feedback (λ =-0.7 $Wm^{-2}K^{-1}$, in the notation of (1)).

However, as references (1) and (2) show, different AGCMs can give very different responses to the same forcing, so that a result from one should not be accepted as representing the response of the real atmosphere without careful checking and comparison. One should use observational data if possible, and if not, other AGCMs of different formulation, and should examine the nature of the physical processes involved rather than just their overall effects.

An experiment is therefore being carried out to investigate the impact of interactive cloud radiative properties on climate sensitivity in the UK Meteorological Office climate model. A prognositc scheme to calculate layer cloud water content has been incorporated (8). This has been coupled to radiative fields using a simple exponential formula for the emissivity and a parametrization based on the work of Liou and Wittman (9) for the solar region. (It is intended to replace the latter with the scheme of Slingo (10) in the near future. Among the advantages this has over the schemes of (7) and (9) and are that its assumptions about cloud microphysics are explicit.) Otherwise the atmospheric model is similar to the improved model of (11), with some modifications to the land surface scheme. It is coupled to a "slab" ocean model as in the experiment of (12).

The experiment has not yet been run to equilibrium, but preliminary indications are that the fairly strong positive cloud feedback found by (12) ($\lambda \simeq -0.7$ $Wm^{-2}K^{-1}$) has become strongly negative ($\lambda \simeq +1.2$ $Wm^{-2}K^{-1}$). This reduces the overall climate sensitivity by more than a factor of three. However, cloud radiative properties seem to be responsible for only a part of this negative feedback, probably in the range 0.9 to 1.2 $Wm^{-2}K^{-1}$. (This compares with 2.4 $Wm^{-2}K^{-1}$ for the work of (5) with their parameter f set to what they suggested was the most likely value.) The rest of the change seems to be due to change in the cloud distribution feedback with the new cloud scheme. In the experiment of (12), the amount of low-level cloud at mid latitudes and of middle-level cloud at low latitudes deceased when the climate warmed, giving a positive albedo feedback. In the new experiment, they increase, giving a negative albedo feedback. This seems to be due to the fact that in the new cloud scheme (8) precipitation of cloud water is much faster for ice cloud than for water cloud. Thus cloud cover in regions just above the freezing level in the control run becomes much more persistent in the warmed climate. This feedback seems not to have been discussed in the literature to date. Its presence in the model is probably realistic, but its strength may well be exaggerated.

It is intended to publish a more detailed description when the experiment has converged and been fully analysed, and also to study the sensitivity of the cloud feedback to the parametrization of both the conversion of cloud water to precipitation and the radiative properties of clouds, and possibly to other factors which affect the distribution of moisture in the atmosphere, such as the parametrization of convection.

Acknowledgements

This work is part of a collective research programme in the UK Meteorological Office and many colleagues have contributed, especially Catherine Senior, John Mitchell and Paul Whitfield.

References

(1) Mitchell, J.F.B., 1988. Simulation of climatic change. This volume.

(2) Schlesinger, M.E. and Mitchell J.F.B., 1987. Climate model simulations of the equilibrium climatic response to increased carbon dioxide. Rev. Geophys., 25, 4, pp. 760-798.

(3) Mitchell, J.F.B. and Ingram W.J., 1989. On CO_2 and climate : a diagnostic study of mechanisms of change in cloud and precipitation. In preparation.

(4) Charlson, R.J., Levelock, J.E., Andreae, M.O. and Warren, S.G., 1987. Oceanic phytoplankton, atmospheric sulphur, cloud albedo and climate. Nature, 326, pp. 655-661

(5) Sommerville, R.C.J. and Remer, L.A., 1984. Cloud optical thickness feedbacks in the CO_2 climate problem. J. Geophys. Res., 89, D6, pp. 9668-9672.

(6) Roeckner, E., Schlese, U., Biercamp, J. and Loewe, P., 1987. Cloud optical properties and climate sensitivity. Nature, 329, 6135, pp. 138-140.

(7) Stephens, G.L., 1978. Radiation profiles in extended water clouds 2. Parametrization schemes J. Amos. Sci., 35, pp. 2123-2132.

(8) Smith, R.N.B., 1989. A scheme for prediction layer clouds and their water content in a general circulation model. Submitted to Quart. J. Roy. Met. Soc.

(9) Liou, K.-N. and Wittman, G.D., 1979. Parametrization of the radiative properties of clouds. J. Atmos. Sci., 36, pp. 1261-1273.

(10) Slingo, A., 1989. A GCM parametrization for the shortwave radiative properties of water clouds. Submitted to J. Climate.

(11) Slingo, A., Wilderspin, R.C. and Smith, R.N.B., 1989. The effect of improved physical parametrizations on simulations of cloudiness and the earth's radiation budget. To appear in J. Geophys. Res.

(12) Wilson, C.A. and Mitchell, J.F.B., 1987. A doubled CO_2 climate senstivity experiment with a global climate model including a simple ocean. J. Geophys. Res., 92, D11, pp. 13315-13343.

DETERMINATION OF SURFACE ALBEDO FOR AGROCLIMATOLOGICAL USES

E. Lopez-Baeza
Remote Sensing Unit. University of Valencia
Spain

Summary

The spatial, temporal and spectral resolutions of meteorological satellites make them suitable for near-real-time albedo mapping over large areas, particularly at a regional scale. This work refers to an application of the "minimum-radiance" method to METEOSAT WEFAX-format images directly received at a SECONDARY DATA USER STATION (SDUS) for the obtention of semiquantitative albedo maps in the agricultural region of Valencia (Spain).

1. INTRODUCTION

The region of Valencia in the Spanish Mediterranean coast economically relies mainly on agriculture. Citrics, rice and vineyards are probably the most representative products of the region and Remote Sensing techniques are increasingly been used in some agricultural applications. There are a number of projects so far supported by the CONSELLERIA DE AGRICULTURA of the local government which are related to inventories and evaluations of citric and rice parcels (1), definition of climatic zones with agricultural purposes (2), estimation of surface covered by greenhouses (3), etc. There are also some studies going on which are addressed to the analysis of the effects caused by some impact phenomena which take place in the region and damage agricultural productions. These mainly are night cooling radiation frosts and forest fires.

The climate system is driven primarily by solar energy absorbed at the surface (4). Solar radiation is, therefore, of considerable interest, particularly in agroclimatological applications. Surface albedo is also a fundamental climatological parameter because in some ways it integrates the effects of more variable quantities. Albedo variations are symptomatic of climatic variations and at the same time, changes in surface properties due to albedo changes may influence the climatology of a region in the long term.

Satellites can play an important rôle in resolving the difficult problem of properly representing the average surface albedo over large areas. The spatial resolution of the visible channel of METEOSAT (2.5 x 2.5 km^2 at subsatellite point) and its frequent time coverage (an image of the whole Earth disc every 30 min), make this satellite ideal for near-real-time monitoring of the environment.

2. RADIATIVE TRANSFER MODEL

One of the aims of this work has been the evaluation of the performance of a highly-implemented SECONDARY DATA USER STATION (SDUS) when carrying out albedo zonations at a regional scale. We have made use of a very simple and well-known model (5) to relate the solar radiation reaching the ground with the satellite measurement when we assume a clear-sky atmosphere. We have simplified it further in order to investigate the validity of its application to METEOSAT secondary data (WEFAX format).

The irradiance E_0 proceeding from the sun and getting to the top of the atmosphere undergoes a set of reflection, absorption and scattering processes within an earth-atmosphere column which extends from the top of the atmosphere to the ground. If we consider the atmosphere as a whole, no matter whether it contains clouds or not, and with albedo A_c, E_0A_c represents the fraction of E_0 reflected by the atmosphere (assuming it Lambertian and neglecting absorption) and $E_0(1-A_c)$ represents the transmitted fraction to the ground surface of albedo A_s. A part of the radiation reaching the surface is absorbed by the ground and the rest is reflected back to the atmosphere from where it undergoes an infinite set of reflections and transmissions as represented in Figure 1.

By adding up all the fluxes reaching the ground, it may be shown that the total irradiance is

$$E_s \approx E_0 \ (1-A_c)/(1-A_sA_c)$$

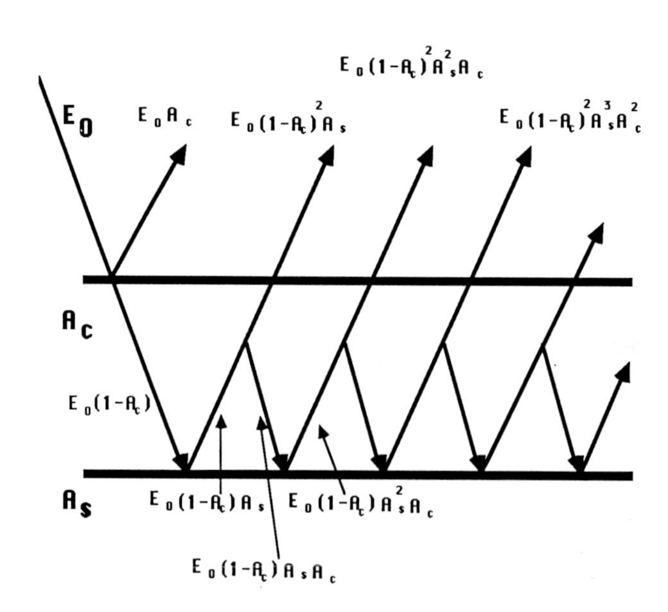

Figure 1. Diagram of the radiative transfer model used

and by doing so with all the fluxes coming out of the atmosphere, it may also be shown that the radiance measured by the satellite is

$$E_{sat} \approx E_0 \left\{ A_C + [(1-A_C)^2 A_S]/(1-A_S A_C) \right\}$$

This radiance may also be expressed as

$$E_{sat} = E_0 A_p$$

where A_p is the planetary albedo. All the above equations lead to the expression of planetary albedo as a function of the albedos A_C and A_S,

$$A_p = A_C + [(1-A_C)^2 A_S]/(1-A_S A_C)$$

from where the surface albedo, A_S, may be considered as the minimum of a measurement set taken by the satellite in a period of time, that is,

$$A_S \approx (A_p)_{min}$$

Besides assuming an isotropic field for the ground surface, we have also assumed a clear-sky atmosphere so that its influence is minimum. The measurements were all taken at about noon time in order to avoid the solar zenith angle dependence. We have consider the visible channel of METEOSAT (0.4 - 1.1 μm) as a good estimate of the full spectral reflectance (0.2 - 4.0 μm).

3. METEOSAT SDUS DATA

For this work we recorded one week images (9-16 July, 1988) of METEOSAT at 12.30 in format C2D which corresponds to the visible image for Europe in analogue WEFAX format. This is a format originated at the EUROPEAN SPACE OPERATIONS CENTER (ESOC) with 2.5 km ground resolution at subsatellite point and which may be received at SECONDARY DATA USER STATIONS (SDUS) (6). In those images we concentrated on a window of 256 x 256 pixels which covered Spain. We had previously checked that the images were cloud-free by visual inspection of the visible and the corresponding infrared images. An advantage of SDUS images is that they may be easily acquired for the visible, infrared and water vapour channels of METEOSAT and at a low cost. However, they are not calibrated and may be altered at ESOC to optimise visualization.

We are presently undertaking some sort of "calibration" by making use of the so called TEST PATTERN 4. This is a fixed electronic signal generated at ESOC and disseminated to SDUS which is composed of 32 grey levels in steps of 8 values, corresponding 0 to black and 255 to white. We have carried out several pixel by pixel correlations between different tests for different days or times and always have obtained correlation coefficients of the order of 0.98 - 0.99. This only means that our receiving equipment does not sensitively disturbs the original signal.

We have also compared a C2D analogue image with its corresponding B-format image in primary data. The correlation was studied between small homogeneous zones corresponding to both images and again the correlation coefficient was of the order of 0.99.

From the temporal set of images, a composite one can finally be obtained where each pixel radiance corresponds to the minimum visible value of the corresponding pixel of each individual image. This technique allows the elimination of clouds without using thresholds to distinguish them. The method permits a simple and quick treatment of satellite data and their application in near-real-time studies over large areas with high spatial and temporal resolutions.

Figure 2 shows an albedo image resulting from the application of the method to the images from 9th to 16th July, 1988 and Figures 3 and 4 show the zoom of the image for Spain and the Region of Valencia, respectively. The values given in the scales are digital counts and are not yet normalized.

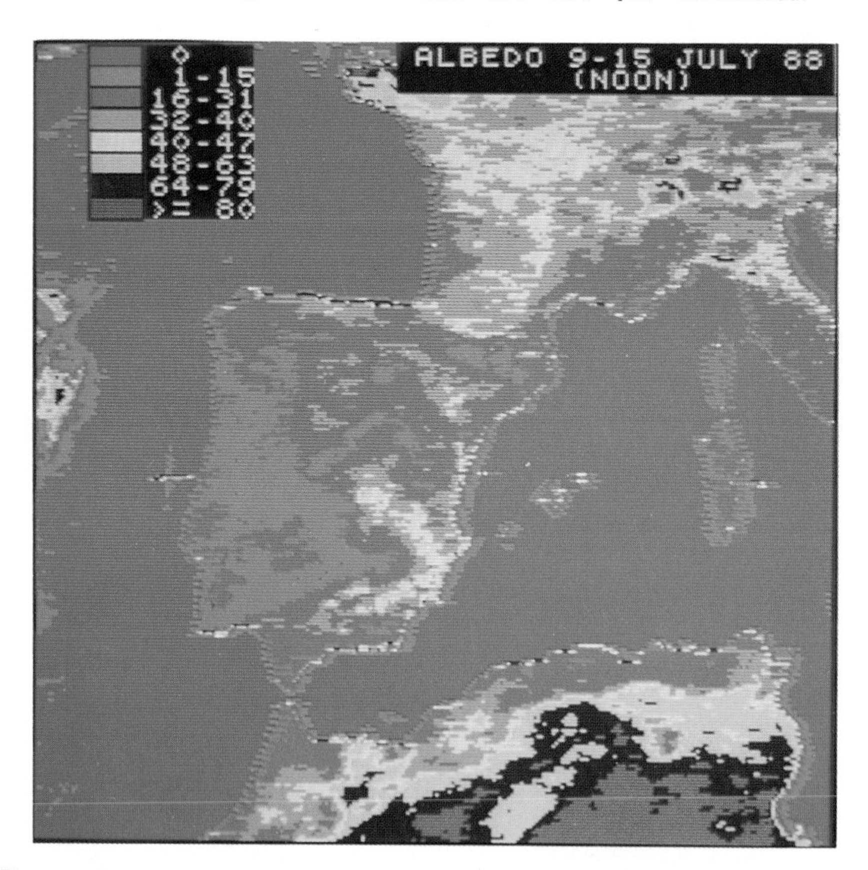

Figure 2. Composite map of minimum counts of the METEOSAT image set for July, 1988

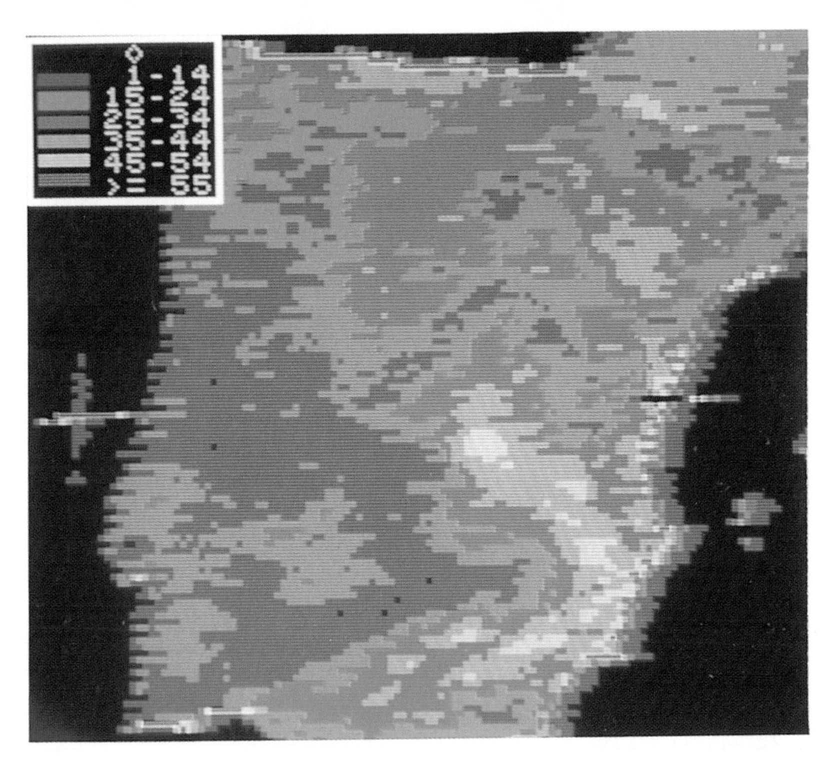

Figure 3. Zoom image for Spain

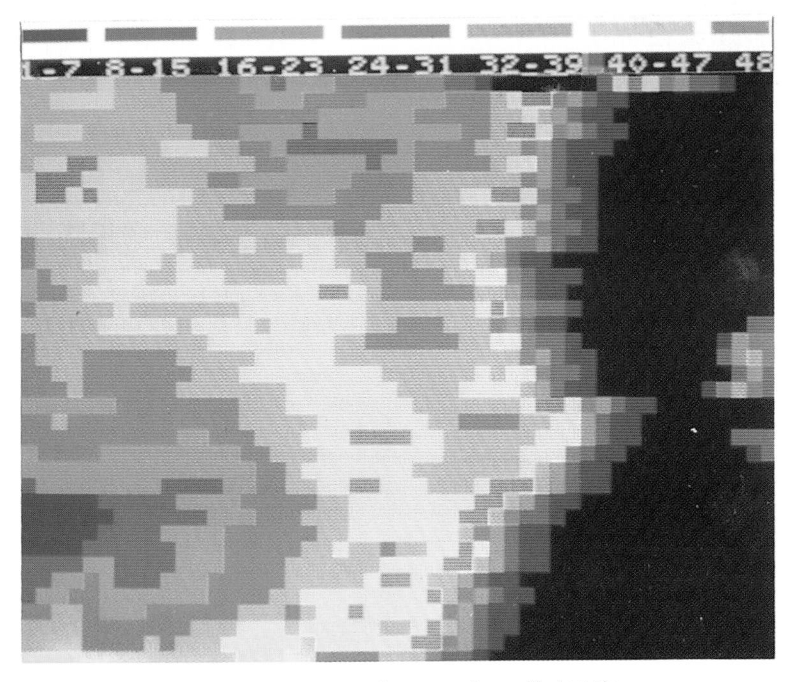

Figure 4. Zoom image for Valencia

4. CONCLUSIONS

We are trying to develop a methodology for near-real-time mapping of albedo and solar radiation of a semiquantitative character using METEOSAT secondary data received in WEFAX format. The model is of a preliminar nature because of the drastic assumptions so far imposed on an accepted simplified method. A primary albedo zonation may be obtained which should be improved using METEOSAT high-resolution images if a more accurate work is needed.

6. ACKNOWLEDGEMENTS

This work has been financily supported through the Project PA 85/172 of the CAICYT (Spain) and the author would also like to express his gratitude to Professor J. Meliá and Dr. D. Segarra with whom it has been carried out.

REFERENCES

(1) GILABERT,A. and MELIA,J. (1987). Clasificación de una zona agrícola utilizando imágenes LANDSAT-5TM. II Reunión Nacional del Grupo de Trabajo de Teledetección. Valencia 17 y 18 Diciembre 1987

(2) UNIDAD DE INVESTIGACION EN TELEDETECCION. Universitat de València (1987). Localización de Zonas Climáticas y Estudio de un Censo Hortícola Mediante Técnicas de Teledetección. Informe Final del Convenio de Colaboración entre la Consellería de Agricultura y Pesca de la Generalitat Valenciana y la Universitat de València. Noviembre 1987.

(3) MORENO,J. and GANDIA,S. (1988). The influence of the high spatial resolution of SPOT-HRV data on the separability of spectral classes. Proceedings of IGARSS'88 Symposium, Edimburgh, Scotland 13-16 September 1988. Ref ESA SP-284, vol 1, 597.

(4) WILSON,M.F. and HENDERSON-SELLERS,A. (1982). Construction of a Global Surface Albedo Map from Satellite Data. Proceedings on the Int. Confe. on Remote Sensing and the Atmosphere. Liverpool 15-17 Dec. 1982.

(5) DEDIEU,G., DESCHAMPS,P.Y. and KERR,J.H. (1987). Satellite estimation of solar irradiance at the surface of the Earth and of surface albedo using a physical model applied to METEOSAT data. J. Clim. Appl. Met. 26, 79-87.

(6) EUROPEAN SPACE OPERATIONS CENTER (ESOC) (1981?). METEOSAT WEFAX Transmissions. ESA.

MODELLING RADIATIVE TRANSFER IN THE MIDDLE ATMOSPHERE

A. J. L. MANNING
Department of Atmospheric, Oceanic and Planetary Physics
University of Oxford, U.K.

Summary

In the middle atmosphere, there is strong coupling between radiative, dynamical and chemical processes. The emitted terrestrial radiation field and the absorbed solar radiation field are modified by the presence of clouds and gaseous constituents such as water vapour, carbon dioxide, ozone, nitrogen dioxide and methane, which exhibit intense and complex absorption in this region of the spectrum. In order to further our understanding of the role of radiation in the stratosphere and mesosphere, a flexible numerical model has been developed for calculating radiative fluxes and heating rates. Particular attention has been paid to water vapour, with the intention of including other minor constituents and also some minor absorption bands of major constituents. The scheme produces sets of latitude- and height-dependent heating rates.

1. INTRODUCTION

In recent times much attention has been paid to phenomena such as the depletion of the ozone layer and greenhouse warming, in effect, to the consequences of changing the concentrations of the various constituents in the earth's atmosphere. The influx into the middle atmosphere of unnatural quantities of, in particular, chlorine and anthropogenically-produced carbon dioxide is likely to affect the earth's climate both adversely and irreversibly. In order to model accurately the response of the atmosphere to these changes, and to evaluate the importance of various gases in the complex interlinked processes taking place, it is important to understand as fully as possible the rôle played by radiative transfer. Indeed, any long-term calculations require a detailed radiation scheme if they are to represent the behaviour of the atmosphere correctly.

To this end, a flexible numerical model has been developed to investigate radiation in the middle atmosphere. Given atmospheric profiles of temperature and water vapour concentration, and spectral data for the bands under consideration, the scheme calculates the radiative fluxes at specified atmospheric levels, and hence latitude- and height-dependent heating rates. Water vapour has been singled out for detailed study, and the inclusion of other minor constituents and also some minor absorption bands of constituents such as ozone, carbon dioxide, methane, nitrogen dioxide and the chlorofluorocarbons has been planned. Water vapour is considered to be of particular interest since although it causes atmospheric cooling which is smaller in magnitude than that due to carbon dioxide or ozone, it is a very important participant in the interlinked dynamical, chemical and radiative processes. It is also a valuable tracer, as its chemical lifetime in these regions is relatively long, and is a major greenhouse gas, with a powerful feedback mechanism. However, in spite of its major rôle, water vapour has received little attention in previous studies of the middle atmosphere.

Most current models indicate that water vapour is responsible for approximately 10% of the total cooling in the stratosphere, with its contribution increasing in the lower stratosphere. Published model cooling rates, especially those in the middle atmosphere, vary considerably due to uncertainties in mixing ratio and in the treatment of the transmittance; these uncertainties may affect the middle atmospheric heating rates considerably. In addition, Doppler and Voigt effects on transmittance have yet to be fully investigated. The suitability of the various band and line-by-line models available for the calculation of transmittance has been evaluated, and schemes for the accurate vertical interpolation of temperature and absorber amount have been assessed. Together with an analysis of the methods used within the radiation scheme, these studies are described in section 3, while section 2 details the atmospheric and spectral data used by the model.

2. INPUT DATA FOR THE MODEL

The model uses temperature and H_2O profiles from the recent LIMS (**L**imb **I**nfrared **M**onitor of the **S**tratosphere) global data set, thus enabling heating rates to be calculated at a wide range of latitudes and heights. The LIMS infrared radiometer operated for seven and a half months between late October 1978 and May 1979, with the observations extending latitudinally from 64°S to 84°N. The vertical range of the measurements was between the 100mb ($\sim$ 15km) level and the 0.1mb (64km) level for temperature and O_3, and between the 100mb and the 1mb (48km) level for H_2O (7).

The scheme incorporates temperature and H_2O data previously tabulated at seventeen heights, between the 1000mb level and the 1mb level, for each of nineteen latitudes between $-\pi/2$ and $+\pi/2$. For every latitude there are points at which no valid measurements of temperature or H_2O concentration were made; generally, these appear at the very lowest tropospheric levels, as well as at high southern latitudes. At bad data points in the troposphere, the Barnett and Corney proposed-CIRA (**C**ospar **I**nternational **R**eference **A**tmosphere) temperatures were used (1). This reference atmosphere encompasses the 0–80km height region between the latitudes of 80°S and 80°N, and draws on measurements of atmospheric parameters made during various satellite experiments.

Tropospheric H_2O values, of which there were no LIMS measurements made, were derived from the corresponding temperatures using the following method: the saturation vapour pressure at each point, and hence the H_2O concentration, were calculated as a function of temperature and relative humidity from the empirical expression of Murray, with the relative humidity being derived from the pressure as described in Manabe and Wetherald (3 and 4). Since the LIMS data extends only to 64°S, complications arise at high latitudes in the southern hemisphere, where there are no other available H_2O measurements. This lack of data would be a problem if the radiation scheme were to be incorporated into a global circulation model, but for the self-contained radiation studies the southern polar region has been neglected.

The initial spectral data used in the model were in the form of band parameters designed for use with the Goody model, and covered nineteen spectral intervals in the 0–2200 cm^{-1} spectral region. The data were originally used by Rodgers and Walshaw, and the parameters listed are combinations of mean line intensity, mean line spacing and mean half width (5). For the first set of ten bands, (0–1100 cm^{-1}), the temperature dependence of line width is taken into account; in the model, the temperature dependence coefficients modify the amount of H_2O calculated to be present in a particular atmospheric layer. Cooling from lines beyond 1200 cm^{-1} is slight and confined to the lower atmosphere, rendering unimportant the absence of similar parameters to describe the temperature dependence for these bands.

Several line data compilations were also incorporated, including the AFGL (**A**ir **F**orce **G**eophysics **L**aboratory), HITRAN (**HI**gh Resolution **TRAN**smission Molecular Absorption) and the GEISA (**G**estion et **E**tude des **I**nformation **S**pectroscopiques **A**tmosphérique) databases. The 1986 edition of the HITRAN database is the most recent update of the AFGL line compilation and lists for each line parameters such as wavenumber, line intensity, air- and self-broadened halfwidth, energy of the lower state, temperature dependence of the air-broadened halfwidth and transition probability (6). Information is provided for 28 species, in the spectral region 0–17900 cm^{-1}. At present the temperature dependence of line width and intensity are not included in calculations of transmittance, but such an effect would be relatively simple to account for.

3. THE RADIATION SCHEME

The form of the radiative transfer equations used in the model is the $\int \tau dB$ form, in which $F_\nu^\uparrow(z)$ and $F_\nu^\downarrow(z)$, the upward and downward radiative fluxes at wavenumber ν, evaluated at the level a distance z above the lower boundary of the model, are calculated as follows:

$$F_\nu^\uparrow(z) = [F_\nu^\uparrow(0) - \pi B(T_0)]\tau(0,z) + \pi B(T_z) - \int_0^z \pi\tau(z',z)\frac{dB(T_{z'})}{dz'}dz'$$

$$F_\nu^\downarrow(z) = \pi B(T_z) - \pi B(T_\infty)\tau(z,\infty) + \int_z^\infty \pi\tau(z',z)\frac{dB(T_{z'})}{dz'}dz'$$

where,

$T_z =$ temperature at level z,
$B(T_z) =$ Planck function evaluated at temperature T_z and wavenumber ν,
$\tau(z',z) =$ transmittance between layers z and z'.

This formulation is the most suitable for numerical computation because of the cancellation of the $\pi B(T_z)$ terms which occurs when the net flux across level z is calculated. In addition, it is preferable to the $\int B d\tau$ form since it is more straightforward to compute dB/dz than $d\tau/dz$. The temperature structure of the atmosphere is known explicitly, and the dependence of B on T is simple, so dB/dz is most easily calculated as $(dB/dT)(dT/dz)$.

The net flux $F_\nu^{net}(z)$ is given by

$$F_\nu^{net}(z) = F_\nu^\downarrow(z) - F_\nu^\uparrow(z)$$

and the heating rate in the layer between z_i and z_{i+1} of

$$\frac{dT}{dt} = -\frac{1}{\rho c_p}\frac{\Delta F}{\Delta z} = \frac{g}{c_p}\frac{\Delta F}{\Delta p}$$

where,

$\Delta F = F_\nu^{net}(z_i) - F_\nu^{net}(z_{i+1})$,
$\Delta p = p(z_{i+1}) - p(z_i)$,
$\Delta z = z_{i+1} - z_i$,
$\rho =$ total air density,
$c_p =$ specific heat capacity of dry air at constant pressure.

Given a model atmosphere comprising temperature and absorber concentration values specified at points on a latitude-height grid, the radiation scheme will derive the heating rate dT/dt at each defined level. Intermediate processes include:

- modification and extrapolation of the atmospheric data,
- selection and manipulation of the spectral data,
- interpolation of temperature,
- interpolation of absorber amount,
- evaluation of the Planck function, B, and its derivative, dB/dT,
- calculation of the inter-level transmittances, and
- computation of the upward, downward and net flux at each level.

The interpolation of temperature is carried out by using cubic splines rather than the less cumbersome methods such as linear interpolation, as the former produce a smoother and more accurate temperature function which can still be easily differentiated. Interpolation by cubic splines is performed in the troposphere, followed if necessary by linear interpolation over a small region where the atmosphere is effectively isothermal, and then by a return to the use of splines over the remaining atmospheric layers.

The form of the radiative transfer equation employed in the model demands that the integral of the Planck function $B(\nu,T)$ and of its derivative $dB(\nu,T)/dT$ be calculated over each spectral interval. This integration is performed numerically.

In order to calculate the transmittance between any two atmospheric layers, it is necessary to know the amount of absorbing material lying between them. This is accomplished in the model by a subroutine based on the method described by Rodgers and Walshaw (5). The Curtis-Godson scaling approximation is used to simulate absorption in an atmospheric path where temperature and pressure, and hence such parameters as line width and line strength, are no longer constant, and the temperature dependence coefficients of the Goody band data modify the calculated absorber amount Δm.

At present, to approximate the integration of transmittance over zenith angle θ, Δm is multiplied by a diffusivity factor of 1.66. This approximate factor of 1.66 is frequently employed where errors of up to 1.5% are acceptable, although for optically thin gases a more appropriate value may be nearer 2.0. Evidently, it may be preferable in certain circumstances to integrate explicitly over zenith angle, especially when the model has been extended to include gases other than water vapour.

The choice between a band or a line-by-line model for calculating the transmittances between atmospheric layers is of major importance, especially in the case of H_2O. The most accurate method for calculating the transmittance over a certain spectral interval is to use line-by-line calculations, which resolve the structure of each individual line, but which are lengthy and computationally expensive. They require detailed knowledge of the line positions, shapes, intensities and widths, and of their dependence on temperature, pressure and absorber density. Band models, such as the Goody and Malkmus random band models, make certain assumptions about the distribution of spectral lines, and generally reproduce the main features of the transmittance spectrum. In their standard form these particular models are far from ideal outside the Lorentz regime, and require considerable modification to take account of Doppler and Voigt effects.

When using a band model for H_2O, the choice of bandwidth may strongly affect the resulting transmittances. A 5 cm^{-1} bandwidth is often used, but this may not necessarily be universally applicable; in parts of the spectrum such an interval may contain too few lines for the band model statistics to be valid. The optimum band width is chosen after comparison of the model transmittances either with line-by-line results or with experimental data, if the latter are available. The transmittances may also depend strongly on the distribution of the bands in the spectral interval under consideration; if the bands are shifted along, very different transmittances may result. The H_2O transmittances produced by the Goody and Malkmus models are quite similar over the 0–2200 cm^{-1} spectral region.

The GENLN2 line-by-line model developed by Edwards computes the transmittances of user-specified atmospheric paths for any band within the spectral region 0–17900 cm^{-1}, and for any of the 28 gases included in the HITRAN compilation (2). Although at present the radiation scheme uses random band models to calculate the transmittances of water vapour, it would be preferable to incorporate the GENLN2 line-by-line values for reasons of accuracy. The calculations would be executed for a large number of temperatures, pressures and absorber concentrations, and the results tabulated and extracted when needed. Ideally, there would be many such tables, one for each gas to be included in the scheme, plus others for the overlap regions, where each overlapping gas would necessitate an additional table.

Such a process would also provide a simple yet accurate means of incorporating the Voigt and Doppler line shapes, in addition to the Lorentz line shape, whereas inclusion of the Voigt profile into the band model calculations demands that various approximations be made which necessarily cause a decrease in accuracy. Incorporation of the line-by-line transmittances would be the ideal solution to the problem of inaccuracy, although the large tables which this would cause to be generated would be rather unwieldy.

There are two options available within the radiation scheme for calculating the vertical integral term of the radiative transfer equation: one may use either Gaussian integration, in which the points at which the integral is to be calculated are specified beforehand, or adaptive integration, whereby the number of points at which the integral is calculated varies according to the shape of the integrand. For the fixed-point integration, either three or six points may be specified, and for the adaptive integration, Simpson's rule is used; the integral over a certain interval is compared with that over half the interval, and the process repeated with the interval further subdivided, until a chosen accuracy criterion is satisfied. The latter method is more time-consuming than the former, but does take into account the fact that when absorption is strong, the major contribution to the integral stems from points close to the point of integration.

Once the upward and downward fluxes at the midpoint of each atmospheric layer have been calculated, the heating rate dT/dt for the particular spectral interval being considered may be calculated. The processes detailed above are then executed for the remaining intervals, and the upward and downward fluxes at each level summed in order that the total band heating rate may be derived. The entire procedure is subsequently repeated for each of the latitudes for which there are valid temperature and water vapour profiles.

4. PLANS FOR FUTURE WORK

Plans for future work include the production of a validated transmittance model for both Doppler and Voigt lineshapes, and further study of the effect on the heating rate of altering the width and position of the bands used in calculating the transmittance. The temperature dependence of transmittance will also be taken into account, and the effects of line mixing for constituents such as CO_2. Minor constituents which have been largely neglected in past models will be included in the scheme, as well as minor bands of gases such as ozone, carbon dioxide, methane, nitrogen dioxide and CFCs. By linking the radiation scheme to existing circulation models, it will be possible to investigate other wide-ranging phenomena, such as the dehydration taking place within the Antarctic polar vortex, during the time of the ozone depletion.

REFERENCES

(1) **Barnett, J. J. and M. Corney (1985),** Middle atmosphere reference model derived from satellite data. Atmospheric structure and its variation in the region 20–120 km: Draft of a new reference middle atmosphere, ed. Labitzke, K., J. J. Barnett and B. Edward, *Handbook for MAP,* **16**, 47–48.

(2) **Edwards, D. P. (1988),** Atmospheric transmittance and radiance calculations using line-by-line computer models. *SPIE critical review of technology; modelling of the atmosphere,* Orlando, Florida.

(3) **Manabe, S. and R. T. Wetherald (1967),** Thermal equilibrium of the atmosphere with a given distribution of relative humidity. *J. Atmos. Sci.,* **24**, 241–259.

(4) **Murray, F. W. (1967),** On the computation of saturation vapour pressure. *J. Appl. Met.,* **6**, 203–204

(5) **Rodgers, C. D. and C. D. Walshaw (1966),** The computation of infrared cooling rate in planetary atmospheres. *Quart. J. Roy. Met. Soc,* **92**, 67–92.

(6) **Rothman, L. S., R. R. Gamache, A. Goldman, J. R. Gillis, L. R. Brown, R. A. Toth, H. M. Pickett, L. L. Poynter, J.-M. Flaud, C. Camy-Peyret, A. Barbe, N. Husson, C. P. Rinsland and M. A. H. Smith (1987),** The HITRAN database: 1986 edition. *Appl. Optics,* **26**, 4058.

(7) **Russell III, J. M., J. C. Gille, E. E. Remsberg, L. L. Gordley, P. L. Bailey, H. Fischer, A. Girard, S. R. Drayson, W. F. J. Evans and J. E. Harries (1984),** Validation of water vapour results measured by the Limb Infrared Monitor of the Stratosphere experiment on Nimbus 7. *J. Geophys. Res.,* **89**, 5115–5124.

MODELLING THE EURASIAN ICE SHEET OVER THE LAST 122,000 YEARS

I. MARSIAT, Th. FICHEFET, H. GALLEE, Ch. TRICOT and A. BERGER
Université Catholique de Louvain
Institut d'Astronomie et de Géophysique G. Lemaître
2 Chemin du Cyclotron
1348 Louvain-la-Neuve
Belgium

Summary

A seasonal bidimensional climate model is used in order to simulate the transient response of the climatic system to the astronomical forcing over the last glacial to interglacial cycle. The simulated evolution of the ice volume of the Eurasian ice sheet shows a general agreement with the most recent geological reconstructions.

A seasonal bidimensional climate model representing in a realistic way the internal physical processes of each climatic subsystem and their complex interactions is used in order to simulate long-term climatic variations. The Milankovitch experiment presented here has been carried out over the last 122,000 years and show a general agreement with the most recent geological reconstructions (1).

During the last glacial maximum (18,000 YBP) the major ice sheets of Eurasia were probably the Scandinavian, the British, the Barents, the Kara and the Putorana Ice Sheets. Because of the 2-D character of our model, these ice sheets will be represented by a single ice sheet which is allowed to extend north to 75°N, south to 40°N and over the whole width of the Eurasian continent. Two others ice sheets, representing the American northern ice sheets and the Greenland are also modelled.

The model does not take into account the sea level variation which is associated to the build up of these huge ice sheets (-130 m below the present sea level). Therefore, the model can not simulate the major part of the Barents and Kara Ice Sheets nor the Iceland Ice Sheet.

The observed ice sheet distribution during the Riss-Würm interglacial was taken as initial condition. At this time, the northern American and the Eurasian continents were free of ice sheets (2) and the south of the Greenland ice sheet was probably lower than today (3). Therefore a Greenland ice sheet volume equal to 2/3 of its present value was taken as initial condition 122 kyr BP.

Although there are still a great number of uncertainties about the growth and decay of each one of the ice sheets, it seems that the chronology and the amplitude of the fluctuations of the Eurasian ice sheets differ considerably from Europe to Asia. The older and more firmly established reconstruction (2) calls for several disconnected terrestrial ice caps and ice sheets : the largest was the Scandinavian Ice Sheet and

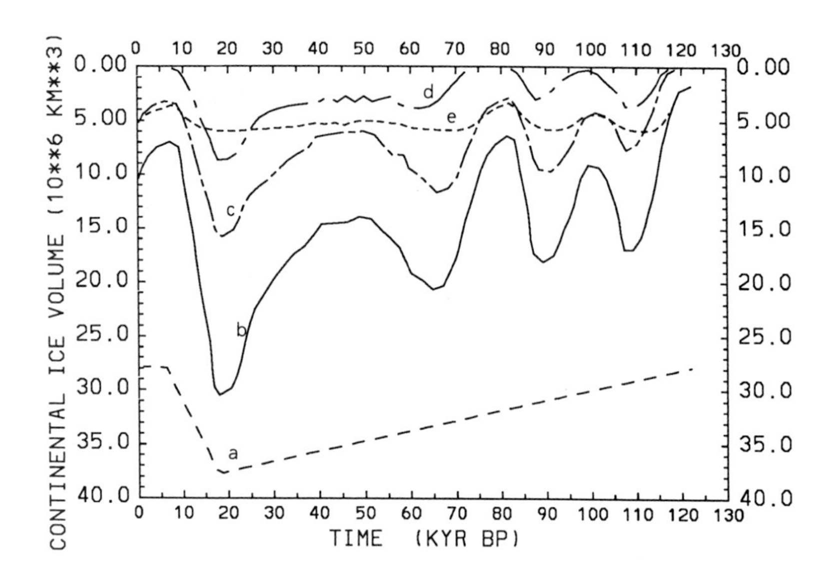

Figure 1. Long term variations over the last interglacial-glacial cycle of the ice volume reconstructed for Antarctica (a) and simulated by the Bimensional Climate Model for all the northern hemisphere ice sheets (b), the northern America Ice Sheets (c), the Eurasian Ice Sheet (d) and the Greenland Ice Sheet (e).

the Arctic Seas remained unglaciated. The simulation of the glaciations over northern Europe and Asia with a single ice sheet can not represent such a diversity of phenomena. However, our results (Figure 1) are in remarkable agreement with the chronology of the glacial events described by Mangerud (4) (Figure 2) which represents the behaviour of the Scandinavian Ice Sheet.

In our simulation, the Eurasian ice sheet disappears twice during stage 5 in response to rapid strong variations in the insolation (5). Real inception occurs at 74 kyr BP. This ice sheet reaches a first maximum around 60 kyr BP followed by small variations of the ice volume between 60 kyr BP and 29 kyr BP. This agrees very well with geological observations according to which, during stages 3 and 4, the most probable picture of the Scandinavian Ice Sheet was that of a relatively stable ice sheet interrupted only by minor fluctuations. After 29 kyr BP the ice volume shows a large increase until the maximum volume of 8.9 10^6 km^3 at 18 kyr BP which must be compared with estimates ranging from 8.2 to 14.3 10^6 km^3 following the kind of reconstruction accepted (2). After 18 kyr BP, the ice sheet shrinks rapidly and totally disappears around 8 kyr BP, 2,000 years earlier than the northern American Ice Sheet minimum, in total agreement with the observations.

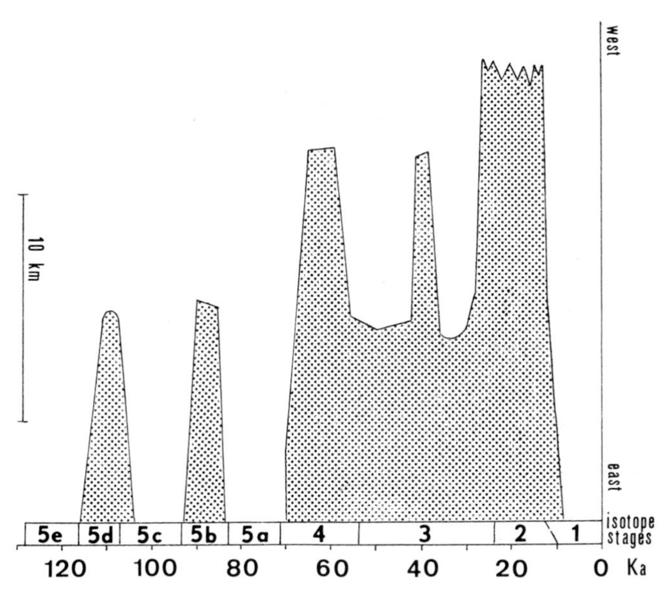

Figure 2. Glacial events for the Weichselian in western Norway. Redrawn from Mangerud (4).

This simulation has allowed us to emphasize some very important mechanisms :

(i) Between 80 kyr BP and 64 kyr BP the total ice volume increase is important. Although the insolation deficit around 70 kyr BP is not larger than around 116 kyr BP, the ice volume response is more pronounced. This is essentially due to the existence of ice sheets at 80 kyr BP, and therefore of higher albedo surfaces. Thus, the albedo-temperature feedback, together with the absence of any important positive anomaly in solar radiation after 70 kyr BP, allows a rapid increase of the ice volume everywhere. The three ice sheets developped extensively from 76 kyr BP up to the last glacial maximum.

(ii) When the ice sheets become large, their increasing altitude limits snow precipitations. This feature affects both snow accumulation and albedo which decreases with time when no snow precipitation occurs any more. Therefore, when early summer insolation increases after 18 kyr BP, a relatively low snow albedo of the ice sheets allows solar energy to input significantly the system which in turn restores a positive energy balance at the ice sheet surface. This leads to a surface temperature increase and melting begins, with an additional albedo decrease. Consequently, the ablation at the south of the ice sheets becomes so important that a depression in the ice forms there and ice flows rapidly into this depression. The altitude of the ice sheets decreases everywhere and the ablation zone becomes predominant.

Rapid deglaciation up to 6 kyr BP is thus a consequence of the feedback mechanism in the model initiated by a slight increase in insolation at a time where the maximum extent of ice makes the ice sheets most vulnerable.

ACKNOWLEDGMENTS

Th. Fichefet and I. Marsiat are supported by the Belgian Foundation for Scientific Research and H. Gallée by the Antarctic Programme of the Prime Minister's Office for Science Policy programming. This research was also partially sponsored by the Climate Programme of the Commission of the European Communities under Grants n° CLI-076-B(RS) and n° EV4C-0052-B(GDF).

REFERENCES

(1) MARSIAT, I., BERGER, A., FICHEFET, TH., GALLEE, H., TRICOT, C. (1987). Modeling the transient response of a coupled model to the astronomical forcing over the last glacial to interglacial cycle. In : The beginning of an inland glaciation-facts and problems of climate dynamics. Proceedings, Mainz, Nov. 87, B. Frenzel (Ed.), in press.

(2) HUGHES, T.J. (1981). Numerical reconstruction of paleo-ice sheets. In : The last Great Ice Sheet. Denton G.H. and Hughes T.J. (Eds), A Wiley Interscience Publ., USA, pp. 222-273.

(3) DANSGAARD, W., CLAUSEN, H.B., GUNDESTRUP, N., HAMMER, C.U., JOHNSEN, S.F., KRISTINDOTTIR, P.M., REEH, N. (1982). A new Greenland deep ice core. Science, 218, 1273-1277.

(4) MANGERUD, J. (1987). The last glacial history of Scandinavia between the last interglacial and the last glacial maximum. In : The beginning of an inland glaciation-facts and problems of climate dynamics. Proceedings, Mainz, Nov. 87, B. Frenzel (Ed.), in press.

(5) BERGER, A. (1979). Insolation signatures of Quaternary Climatic Changes. Il Nuovo Cimento, 2C(1), 63-87.

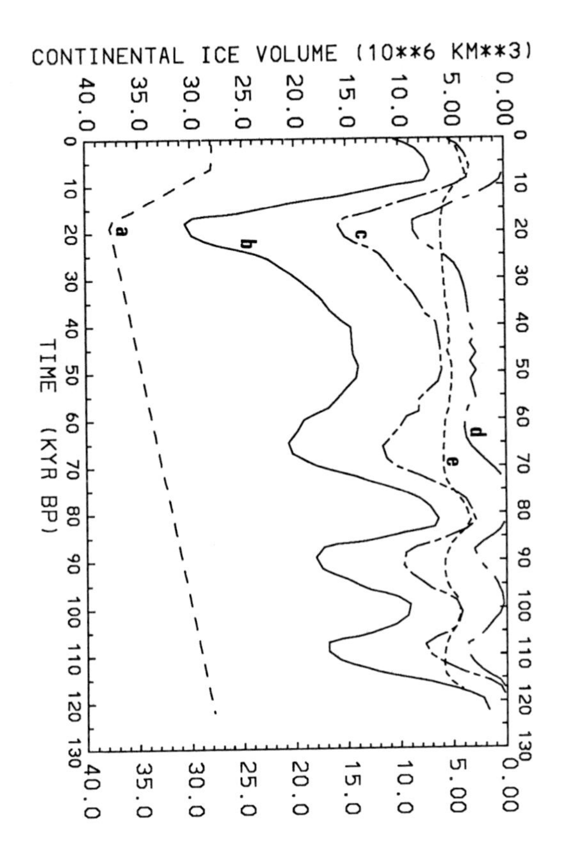

CONTINENTAL ICE VOLUME (10**6 KM**3)
TIME (KYR BP)
a
b
c
d
e

A COMPARISON OF TWO AGROCLIMATIC INDICES FOR USE IN INVESTIGATING THE EFFECTS OF CLIMATIC CHANGE ON AGRICULTURE IN EUROPE.

J.H. Porter
Atmospheric Impacts Research Group
School of Geography
University of Birmingham
P.O. Box 363
Birmingham B15 2TT
U.K.

Agroclimatic indices can be used as a measure of the agricultural potential of a given area with appropriate climatological data used as inputs. The Miami Model and Precipitation Effectiveness Index investigated in this study use temperature and precipitation alone, these variables were arbitrarily modified from a baseline climate to test the sensitivity of the indices to a changing climate. It would appear that both these indices are suitable for use in Europe, the Miami Model is particularly useful for identifying which of the two climate variables used, may limit potential, whereas the Precipitation Effectiveness Index can be used to examine combined changes in temperature and precipitation.

The potential for agricultural production

Although the pattern of agricultural development is mainly a function of anthropogenic factors, such as social, economic and political forces, the potential for agricultural production is largely governed by physical conditions (latitude, longitude, altitude, climate, soil characteristics etc.). A wide range of agricultural management practices have evolved to offset adverse environmental conditions in order to obtain the optimum yield possible in a given area. However, there is increasing evidence that radiatively active gases released by man into the atmosphere will alter the earths' atmosphere at a rate and magnitude unprecedented in human history. This phenomenon, known as the greenhouse effect, will have a great influence on the environment, agriculture and man.

Use of agroclimatic indices

A number of agroclimatic indices have been developed which combine the climatic factors most influencing plant growth into a single term. Appropriate climatological data from meteorological stations are required as "inputs" to these indices, and the computed index values are the "outputs". As they relate agricultural response to climatic variables these indices can be used to examine agricultural potential in Europe under different climatic conditions. This is done by perturbing the climatic variables input into the indices using scenarios of possible climatic changes obtained from either historical records of past changes (eg. 6) , general circulation models (eg. 1) or synthetic arbitrary methods (eg. 4, 8, 9). The new index values produced under the changed climate can then be compared with values obtained for the present-day climate.

The performance of two different agroclimatic indices - the Miami Model (2) and the Precipitation Effectiveness Index (12) - was tested using arbitary changes in climate to assess their potential for use in studying the effects of climatic change on agriculture in Europe.

The Miami Model

The Miami Model uses two correlation equations to predict, independently, the net primary productivity (NPP) for a given area from mean annual temperature and mean annual precipitation. Liebigs' law is assumed to apply whereby the minimum factor controls productivity, so the lower of these two values is taken to represent NPP. Primary productivity is organic matter created by photosynthetic plants using energy from sunlight, some of this matter is then used by the plants in respiration, the remainder represents net primary productivity. In this model NPP is measured in g/m^2/year and higher values are found in more favourable climates where productivity is not restricted by low annual temperatures or precipitation. Under baseline (1931-1960) conditions NPP in Europe ranges from 418 g/m^2/yr for Almeria in Southern Spain to 1477 g/m^2/yr for Plymouth in the U.K. Figure 1 shows the location of some of the stations used in this study.

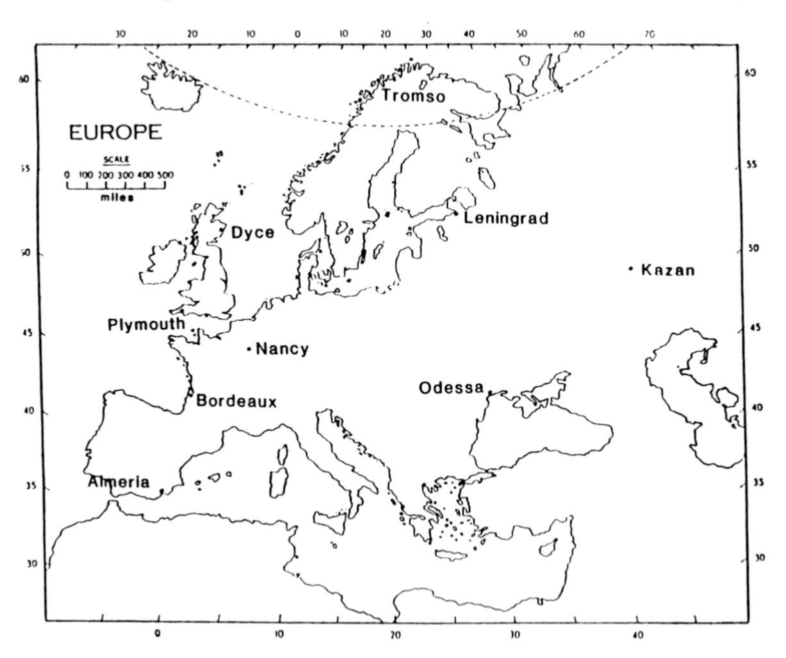

Figure 1. Map of Europe showing the location of some of the meteorological stations used in this study.

One advantage of the Miami Model is its' simplicity. It has been used in previous climate change experiments, both globally (4) and for individual countries Germany - (5), and Australia - (9). These authors and others found that the present day productivity results produced were in agreement with those obtained from other sources. The model also performed well with small arbitary changes in climate, producing consistent results in each of the studies.

In order to test the performance of the model under a range of temperature and precipitation regimes, and also give an indication of the range of index values to be found in Europe, meteorological data input into the models were taken from a number of stations in different parts of Europe. The 1931-1960 climatological normals data (11) were used as the baseline climate. By perturbing the climatic variables incrementally it is possible to assess the potential effects of a range of climate changes on the indices. Modification of temperature (from -1°C below to +4°C above baseline) and precipitation (-20% below to +20% above normal) simultaneously enabled examination of the combined effects of temperature and precipitation changes on the indices.

Preliminary results obtained using meteorological data from stations in Europe show a tendency for NPP predicted by the Miami Model under baseline conditions to be limited by temperature in northern Europe (>58°N) while productivity in lower latitude areas is restricted more by precipitation. The model is useful for identifying the most important of the two climatic variables when investigating climatic change, especially in higher latitudes which can be sensitive to both temperature and precipitation. For those stations where temperature is the limiting factor under baseline conditions, any increase in temperature leads to increased productivity, however, increases in temperature without a compensatory rise in precipitation levels means precipitation then begins to limit productivity.

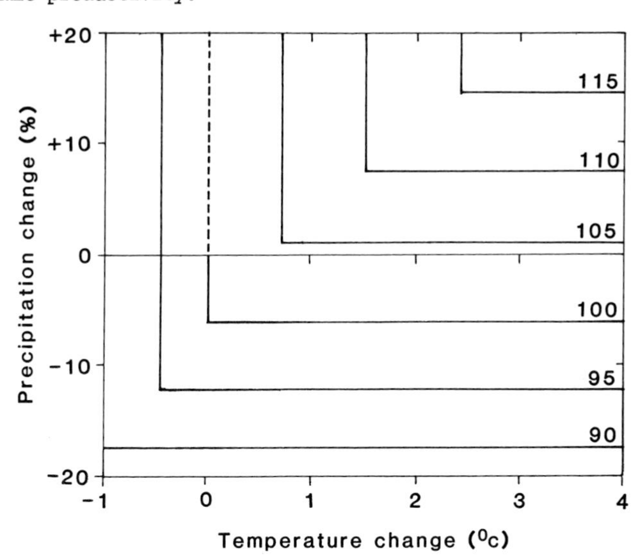

Figure 2. Influence of changes in temperature and precipitation on relative net primary productivity calculated using the Miami Model for Dyce, UK (57°N 02°W). Under baseline (1931-1960) conditions annual temperature = 7.9^{o}C, annual precipitation = 837mm and NPP = 1222.0 g/m^{2}/yr

Figure 2 shows the influence of changes in temperature and precipitation on NPP in Dyce, Scotland, UK. It can be seen that this location is sensitive to both temperature and precipitation, under baseline conditions productivity (1222.0 $g/m^2/yr$) is limited by temperature, given a 1^oC rise in temperature productivity could potentially reach 1309.1 g/m2/yr, 107% above present, yet at this temperature precipitation becomes limiting, so NPP reaches only 1279.1 $g/m^2/yr$, an increase of 104%.

Figure 3 shows the influence of changes on the NPP of Bordeaux, France where initial NPP is 1349.6 g/m2/yr. Bordeaux is 4^oC warmer and slightly (10%) wetter than the present climate of Dyce, but this could be analogous to Dyce under a changed climate, for example with a $2xCO_2$-induced warming of the climate. The current NPP in Bordeaux may also be of a similar magnitude to that attainable in Dyce under the changed climate (8).

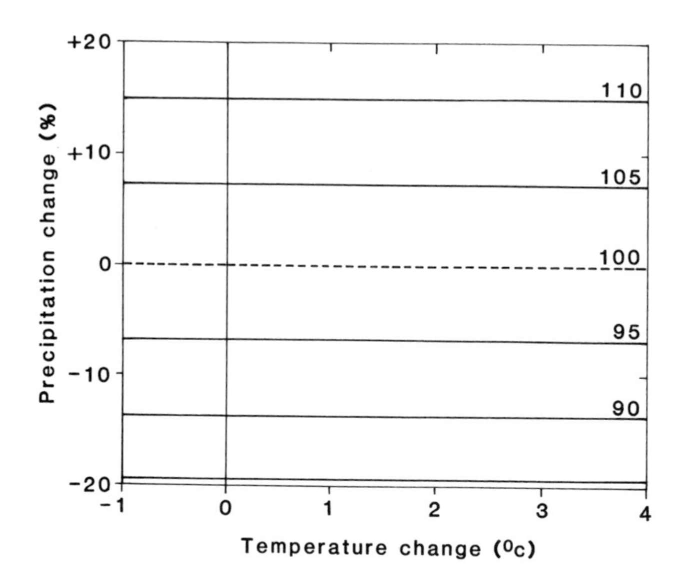

Figure 3. Influence of changes in temperature and precipitation on relative net primary productivity calculated using the Miami Model for Bordeux, France (44°N 00°W). Abcissa, change in temperature (°C), ordinate, change in precipitation (%). Under baseline (1931-1960) conditions annual temperature = 12.3°C, annual precipitation = 900mm and NPP = 1349.6 $g/m^2/yr$. The climate here could be analogous to Dyce under a changed climate with warmer and slightly wetter conditions.

Although the structure of the Miami Model is useful in enabling identification of limiting climate variables the major disadvantage of the model is that as the two variables are treated independently there is no direct interactive effect of one upon the other except when one becomes more limiting to productivity than the other, although precipitation does have an intrinsic dependent relationship on mean annual temperature (2). According to the model, under climates where NPP is always restricted by precipitation, an increase in temperature will have no impact on productivity levels. However, in reality there will be a change in evapotranspiration rates which may lead to changes in productivity levels.

By applying the adjoint approach (7) to the Miami Model results it is possible to produce an indication of the severity of the limiting factor on NPP. This involves looking at the increase in the limiting climate variable required to allow full potential NPP to be reached. For example, Odessa (southern USSR) would require a precipitation increase of 143% to reach the maximum potential productivity level, whereas Nancy (France) would require an increase of only 28%. Therefore, although both have similar annual temperatures, 9.9°C and 9.5°C respectively, precipitation restricts productivity in Odessa more severely than in Nancy. A similar adjoint approach can also be used with the Precipitation Effectiveness Index.

The Precipitation Effectiveness Index

The Precipitation Effectiveness Index combines monthly values of temperature and precipitation into a single index value which can be used to characterize the spatial pattern of moisture regimes. In Europe, the highest index values are to be found at more northerly latitudes with a general decline further south indicating drier conditions, values range from 167.7 for Tromso in northern Norway to as low as 14.4 for Almeria in southern Spain. Thornthwaite defined the following vegetation classes using the index:

A (wet) Rain forest, I = 128 or greater
B (humid) Forest, I = 64-127
C (subhumid) Grassland, I = 32-63
D (semi arid) Steppe, I = 16-31
E (arid) Desert, I = <16

Preliminary results using this index suggest that it can be used for mapping the climate of Europe in terms of spatial moisture patterns as successfully as it has been used for North America (12) and regions of Canada (10). Continentality may also influence the index as values tend to be lower in more central areas - Leningrad (59°N 30°E) has an index value of 60.7 compared with 41.1 at Kazan (55°N 49°E). Comparison of these results with the pattern obtained using an alternative moisture index would be a good way of evaluating the usefulness of the Precipitation Effectiveness Index for application in Europe. Under Canadian conditions, for example, results from Thornthwaites' index compare well with those from other indices (10).

Applying the incremental changes in temperature and precipitation to input data for all the stations produced a decline in the index value both with increasing temperature and with decreasing precipitation, this effect was intensified by a combination of the two. This result, along with the spatial variations in the index values, could be expected as there is a relationship between temperature and precipitation in terms of evaporation rate included implicitly within the model. As temperature increases so too does evaporation, hence there is less moisture available for use ("effective" precipitation), an effect analogous to a decrease in precipitation.

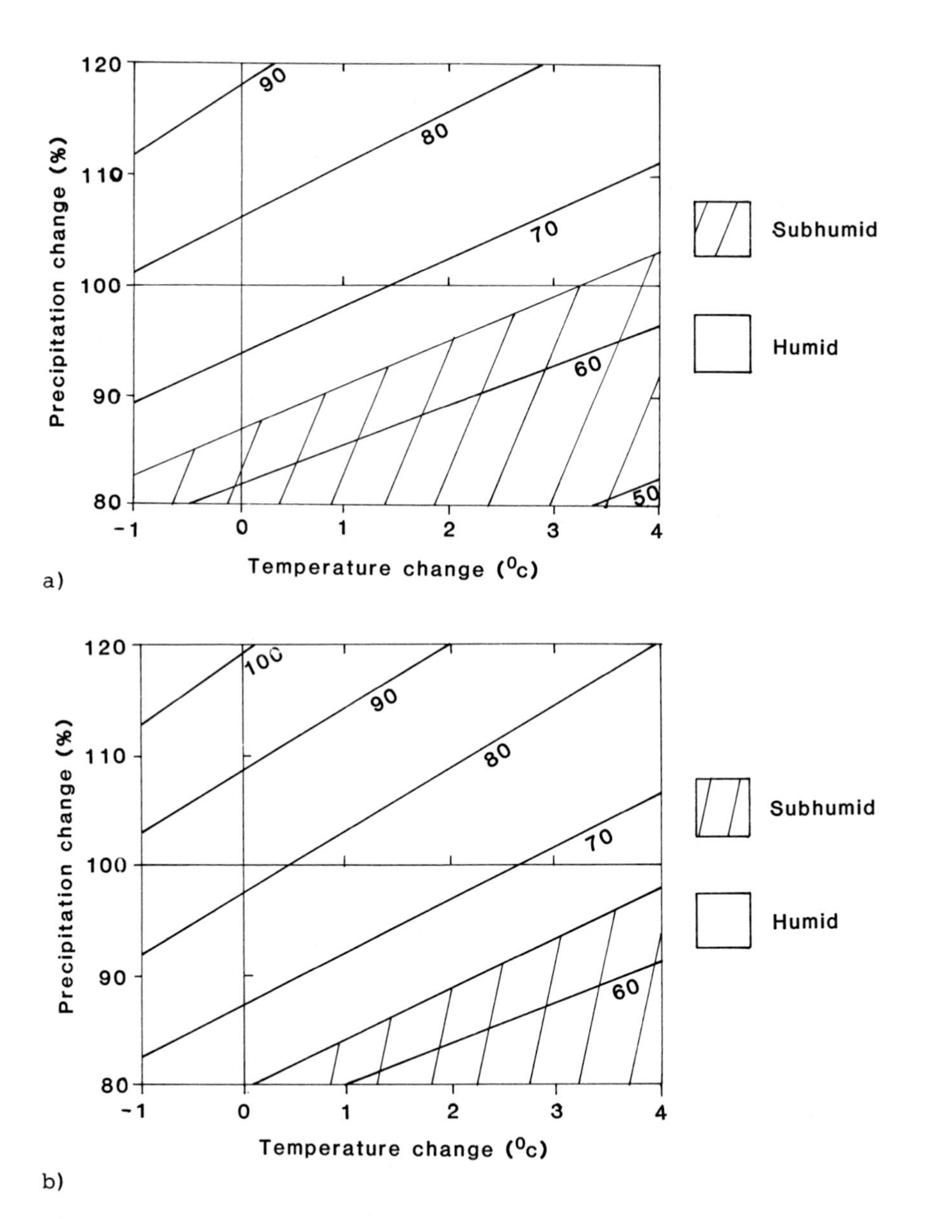

Figure 4. Influence of changes in temperature and precipitation on the Precipitation Effectiveness Index value for a) Dyce, UK b) Bordeaux, France. Abcissa, change in temperature (°C); ordinate, change in precipitation (%). Under the baseline climate for Dyce annual temperature = 7.9°C, annual precipitation = 837mm and Index value = 82.3, for Bordeaux annual temperature = 12.3°C, annual precipitation = 900mm and Index value = 75.2.

Figure 4 shows the influence of changes in temperature and precipitation on the index values for Dyce. Bordeaux, shown in figure 5, again provides an approximate analogue of conditions in Dyce under a warmer and slightly wetter climate. Although Bordeaux has marginally more rainfall than Dyce the increased temperature results in lower index values, indicating drier conditions, so, although like Dyce, Bordeaux is classified as humid, it is at the lower end of this division and would require only slightly drier conditions before becoming subhumid.

Conclusions

Both the Miami Model and Precipitation Effectiveness Index appear suitable for use under the wide range of climatic conditions found in Europe. While the Miami Model is useful for identifying the single climate variable limiting agricultural potential, the Precipitation Effectiveness Index is suitable for investigating the effect of combined changes in the two variables. A global agroclimatic index has not yet been developed, and it is unlikely that it ever will be as it would be impossible to take into account the many factors influencing plant growth and combine them into a simple formula. However, it should be possible to use a number of agroclimatic indices, such as the Miami Model and Precipitation Effectiveness Index, together with different climatic scenarios to identify those areas which are potentially sensitive to changes in the climate. By focusing on these areas, where the effects of climatic change are likely to be most readily detected, an improved understanding of agricultural sensitivity to climate can then be obtained through the use of more advanced analytical techniques.

References

(1) Bach, W. (1988). Development of climatic scenarios: A. From general circulaton models. In: M.L. Parry, T.R. Carter and N.T.Konijn (eds.), The Impact of Climatic Variations on Agriculture. Vol. 1: Assessments in cool Temperate and Cold Regions, Kluwer, The Netherlands, pp.125-157.

(2) Lieth, H. (1975). Modeling the primary productivity of the World. In: H. Lieth and R.H. Whittaker (eds.), Primary Productivity of the Biosphere. Springer-Verlag, New York, pp.237-263.

(3) Lieth, H. (1976). Possible effects of climate changes on natural vegetation. In: R.J. Kopec (ed.), Atmospheric Quality and Climatic Change, Studies in Geography, Number 9, University of North Carolina, Chapel Hill, pp. 150-159.

(4) Lieth, H. (1978). Vegetation and CO_2 changes. In: J. Williams (ed.), Carbon Dioxide, Climate and Society, Pergamon Press, Oxford, pp. 103-109.

(5) Lieth, H. and Aselmann, I. (1983). Comparing the primary productivity of natural and managed vegetation - an example from Germany. In: W. Holzner, M.J.A. Werger and I. Ikusima (eds.), Man's Impact on Vegetation, Junk Publishers, The Netherlands, pp. 25-44

(6) Jager, J. (1988). Development of climatic scenarios: B. Background to the instrumental record. In: M.L. Parry, T.R. Carter and N.T.Konijn (eds.), The Impact of Climatic Variations on Agriculture. Vol. 1: Assessments in cool Temperate and Cold Regions, Kluwer, The Netherlands, pp. 159-182.

(7) Parry, M.L. and Carter, T.R. (eds.), (1984). Assessing the Impact of Climatic Change in Cold Regions. Summary Report, SR-84-1, International Institute for Applied Systems Analysis, Laxenburg, Austria.

(8) Parry, M.L. and Carter, T.R. (1988). The assessment of effects of climatic variations on agriculture: aims, methods and summary of results. In: M.L. Parry, T.R. Carter and N.T.Konijn (eds.), The Impact of Climatic Variations on Agriculture. Vol. 1: Assessments in cool Temperate and Cold Regions, Kluwer, The Netherlands, pp. 11-95.

(9) Pittock, A.B. and Nix, H.A. (1986). The effect of changing climate on Australian biomass production - a preliminary study. Climatic Change, **8**, 243-255.

(10) Williams, G.D.V., Fautley, R.A., Jones, K.H. Stewart, R.B. and Wheaton, E.E. (1988). Estimating effects of climatic change on agriculture in Saskatchewan, Canada. In: M.L. Parry, T.R. Carter and N.T.Konijn (eds.), The Impact of Climatic Variations on Agriculture. Vol. 1: Assessments in cool Temperate and Cold Regions, Kluwer, The Netherlands, pp. 221-379.

(11) WMO (1971). Climatological normals (Clino) for Climat and Climat Ship Stations for the Period 1931-1960. WMO/OMM - No. 117.TP. 52, World Meteorological Organization, Geneva.

(12) Thornthwaite, C.W. (1931). The climates of North America according to a new classification. Geog. Rev., **21**, 633-655.

THE ROLE OF SEA-ICE IN THE SIMULATED ANTARCTIC WINTER CLIMATE

C. Senior
Dynamical Climatology Branch
UK Meteorological Office

A recent study of the climate sensitivity of a version of the UKMO atmospheric general circulation model (AGCM) to increased atmospheric CO_2 concentrations showed that at equilibrium winter antarctic sea-ice extents were reduced by up to 5° of latitude in response to the surface warming. This inspired several numerical simulations to be performed investigating the sensitivity of a higher resolution AGCM to changes in sea-ice extent during the antarctic winter. One such experiment, described here, consisted of three 120 day integrations in which all sea-ice equatorwards of 67.5° S was removed and replaced with water at 271.2 K . The three integrations were started from consecutive years of a control experiment commencing on the 1st June

One necessary factor to establish before using models for climate perturbation experiments is that they show realistic simulations of present day climate. The antarctic winter climatology of the model used here has been verified against operational analyses (2) and shows a reasonable climatology. In particular sea-ice extents were less extensive and therefore more realistic than in an earlier model used for a similar experiment (1).

The immediate response to removing sea-ice is to raise the surface temperature (Figure 1, full line). The greatest warming occurs at the latitude just to the north of the new sea-ice edge and is largely confined to within 10° latitude of this point. In a previous experiment (1) the warming in the lowest model layer spread across Antarctica where it was greater than 7 K, in this experiment the warming of the Antarctic plateaux is less than 2 K perhaps due to the improved simulation of katabatic winds blowing off the continent. The strong inversion present in southern high latitude winter confines the warming vertically to below about 700mb. The presence of a warm anomaly at about 65° S above the boundary layer produces increases in easterly wind to the north and in westerly wind to the south, consistent with the thermal wind equation.

The rise in surface temperature produces an increase in sensible heat of about 50 WM^{-2} at the region where the insulating effect of the sea-ice has been removed and open water exposed and a reduction to the north reflecting the poleward shift of the previous maximum. The change in latent heat shows a very similar pattern with the maximum again being of the order of 50 WM^{-2}.

A fall in zonally averaged surface pressure coincides with the latitude of the maximum in surface and atmospheric heating and there are also decreases over Antarctica (Figure 2, full line). The main falls in surface pressure are found near the new ice edge just downstream from the largest decrease in ice extent at around 65° S, 30° E and these have been shown to be statistically significant at the 95% confidence level.

These results are largely consistent with those found by other authors (1),(4),(5). However Simmonds and Simmonds and Dix found statistically significant <u>rises</u> in surface pressure over areas where they had removed sea-ice. In the present experiment and in Mitchell and Hills and Simmonds and Dix the removal of sea-ice was accompanied by a reduction in drag coefficient. To investigate the relative effects of the thermal and mechanical forcings on the change in surface pressure a further series of experiments was performed in which the surface roughness length was reduced from 10^{-1} m to 10^{-4} m over the area of sea-ice that was removed in the original experiment. As expected this produced very little thermal forcing (Figure 1, dashed line) however a fall of about 2mb in the zonally averaged surface pressure occurred around the same latitude as in the original experiments (Figure 2, dashed line) showing that part of the reduction in surface pressure noticed in the first experiment is due to the change in drag coefficient produced by the reduction of surface roughness length. Only when the two processes are considered together does a highly significant response occur.

Further differences between the results of the various authors mentioned are partly due to the formulations of the ice anomaly experiments and to the inherent variability of climate at high latitudes.

In summary, the results of the present experiments indicate that removing antarctic sea-ice produces a reduction in surface pressure over the area of largest change in sea-ice extent. The thermal forcing due to the increase in turbulent heating is likely to be exaggerated in the model as a constant depth of 2 m is used for all sea-ice points wheras in reality antarctic sea-ice is on average much thinner and contains a substantial fraction of open water. There is much observational evidence indicating that the change in drag coefficient when going from sea-ice to open water is relatively large (eg. (3)). These factors must be considered in models used to determine the effect of sea-ice anomalies.

References

(1) Mitchell, J.F.B. and Hills, T.S. (1986). Sea-ice and the antarctic winter circulation, a numerical experiment. Q.J.R. Meteorol. Soc. 113 953-969.

(2) Mitchell, J.F.B. and Senior,C.A. (1989). The antarctic winter; simulations with climatological and reduced sea-ice extents. Accepted for publication in Q.J.R. Meteorol. Soc.

(3) Overland, J.E. (1985). Atmospheric boundary layer structure and drag coefficients over sea-ice. J. Geophys. Res.,90, 9029-9050

(4) Simmonds I. (1981. The effect of sea-ice on a general circulation model of the southern hemisphere. In. Allison I. (Ed.) Sea level, Ice and Climatic Change, 193-206. IAHS pub. No. 131.

(5) Simmonds, I. and Dix, M. (1986). The circulation changes induced by the removal of antarctic sea-ice in a general circulation model. pp 107-110 in report of second international conference on southern hemisphere meteorology, December 1-5 Wellington, New Zealand. American Met. Soc., Boston.

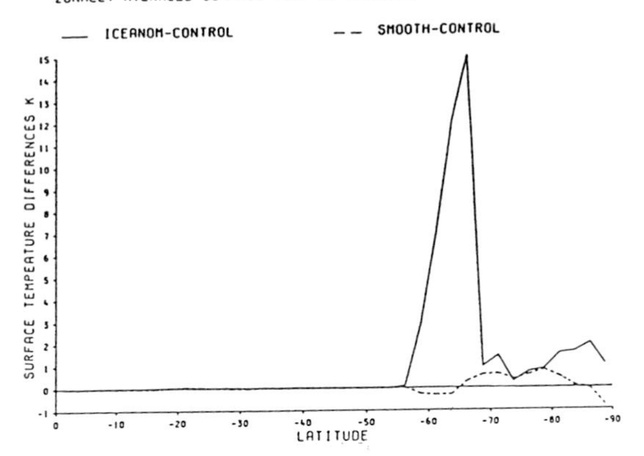

FIGURE 1. Change in zonally averaged surface temperature (k) for Experiments ICEANOM (Full line) and SMOOTH (dashed line), averaged over three years, July to September.

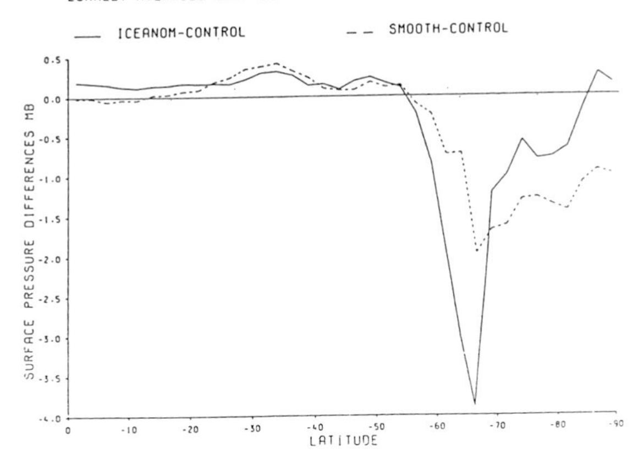

FIGURE 2. As Figure 1. but for zonally averaged surface pressure (mb).

THE EFFECTS OF CLIMATIC CHANGE AND INCREASING CO_2 ON EUROPEAN CROP PRODUCTION - INITIAL STUDIES.

ROSAMUND A. SPENCE
Department of Meteorology
University of Reading, U.K.

1 INTRODUCTION

The aim of the project is to deduce indices which relate crop yield to climatic factors by incorporating underlying physiological processes in quantitative relationships. These indices will then enable the evaluation of the output from crop growth models which incorporate the known physiological responses of crops to increased ambient CO_2 when run on current -climate data. Such simulation models are designed to be used for specific 'field level' studies and therefore require this intermediate evaluation to facilitate their extrapolation to wider regional yield predictions.

2 DATA

The agricultural and climatic data will be handled on a regional or county scale for Europe with more detailed data being available for the U.K.

Yield data comes from the European Communities Farm Management Survey (an annual, principally economic survey of European Agriculture based on surveys of individual farms performed by several designated centres within each member state). When summarised at the county scale this data has the advantage that it must include most of the sources of variation such as differing soil types, management practices and crop variety that otherwise obscure the crop-weather relationship at the field level.

Initially cereal crops are being studied, principally wheat, as cereals are grown throughout Europe.

Climatic data has been obtained from various sources provided by the U.K. Meteorological Office from their network of climatological reporting stations. Initially monthly figures from daily climatic observations have been used in attempting to match yield to climatic information.

Meteorological and agricultural data operate over different time scales with 12 monthly meteorological data running from January to December whereas for a crop such as winter wheat a more relevant 12 monthly total runs from September to August. Therefore in this study all 12 monthly totals have been extracted so that they run over this agricultural year.

3 INITIAL STUDIES

Preliminary work has involved the selection of detailed data at the farm scale (representing 116 different farms) for an area of $288km^2$ in Central-Southern Britain. Three climatically contrasting harvest years 1983, 1984 and 1985 were chosen. Encouragingly wheat yields also differed considerably between these three year. Figure 1 shows the average yield per county for each of the 3 years. Note that in all but one case the highest yield per county occured in 1984 and the lowest in 1985.

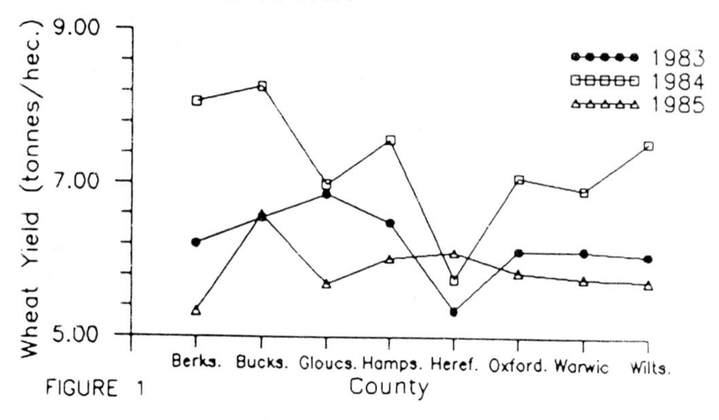

FIGURE 1

Figures 2 and 3 show that with very few exceptions the meteorological stations in the study area show 1984 to be a year of higher total hours of sunshine and low total rainfall with the opposite conditions in 1985. A climatological 'middle year' is provided by 1983 in all cases.

FIGURE 2 (September to August)

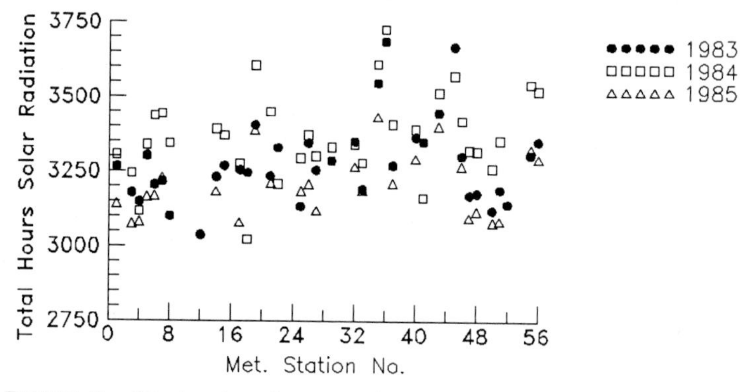

FIGURE 3 (September to August)

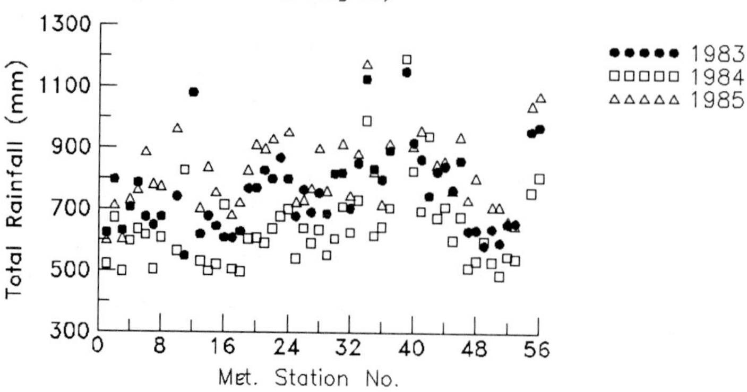

The spatial and temporal distribution of climatic features such as rainfall, solar radiation and potential soil moisture deficit are being studied. As the farms providing yield data are often situated at some distance from the meteorological stations it is important to describe the distribution of any weather variable accurately and adequately over the whole study area. There may also be different numbers of stations reporting in each of the 3 years studied and it is thus desirable to know how many meteorological data points are required to produce an adequate description of the weather variables.

The distribution pattern of rainfall for the twelve months from September 1982 to August 1983 over the study area is shown in Figures 4 and 5. To produce Figure 4, 48 data points were used whereas Figure 5 was produced using only 19 of the possible 48 points. The latter, smaller data set was obtained via a computerised easy-access data retrieval system which although making the data available more quickly only provides the records from a limited number of stations. Thus if the smaller data set could produce accurate results, calculation time could be reduced. As Figures 4 and 5 show, rainfall distribution and amount were adequately described using the smaller data set even if some of the finer detail was lost.

However, when total sunshine hours received from September 1984 to August 1985 were plotted with similar full and reduced data sets it was found that the 13 point 'rapid access' data set does not accurately reproduce the distribution pattern of sun hours as described by the 32 station data set. This is shown in Figures 6 and 7.

These results show that the accurate description of weather variables over even small areas with a temperate maritime climate such as that of the U.K. is not simple. Great care is required when only limited data is available as the amount of data required for accurate description of the various aspects of climate varies substantially

Isoline plots such as those shown in Figures 4 to 7 (plotted using 'Surfer Software',(Golden Software Inc.),run on an IBM compatible P.C.) allow the calculation of average weather variable values at the $100km^2$ scale which may then be compared with wheat yield values available at this scale. Meteorological data on any time scale may be handled in this way

4 SUMMARY

An understanding of the climatic features which determine crop yield and their relation to the plant processes forming the components of the yield response is being sought. These relationships are to be quantified without using a stepwise regression technique as this rarely gives realistic, comprehensible coefficients which are therefore of little use in assessing the output of physiological crop growth models.

Statistics only allow for interpolation between existing data points and not extrapolation beyond them. As climatic change and the consequent response of crops requires extrapolation from crop growth simulation models beyond existing knowledge, it is hoped to provide a more mechanistic understanding of crop-climate interactions permitting model evaluation prior to their use for prediction.

FIGURE 4

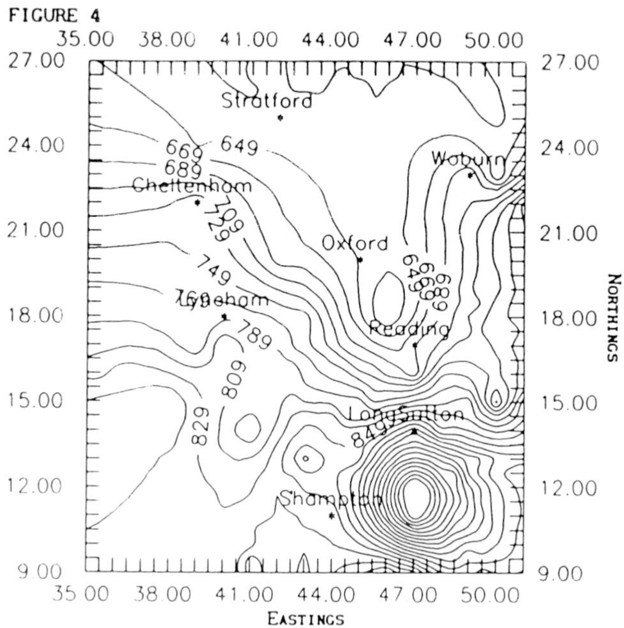

Rainfall Distribution (total mm) over Central-Southern England using 48 data points (September 1983-August 1984)

FIGURE 5

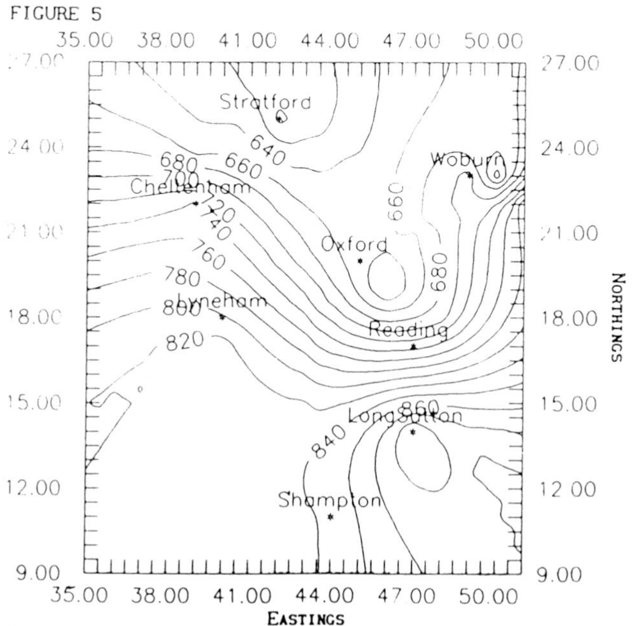

Rainfall Distribution (total mm) over Central-Southern England using 19 data points (September 1983-August 1984)

FIGURE 6

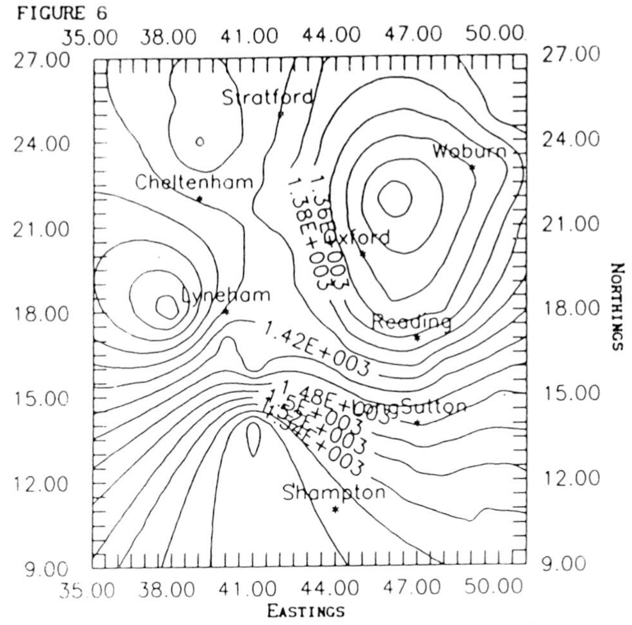

Total Sunshine Hours over Central-Southern England using 32 data points (September 1984-August 1985)

FIGURE 7

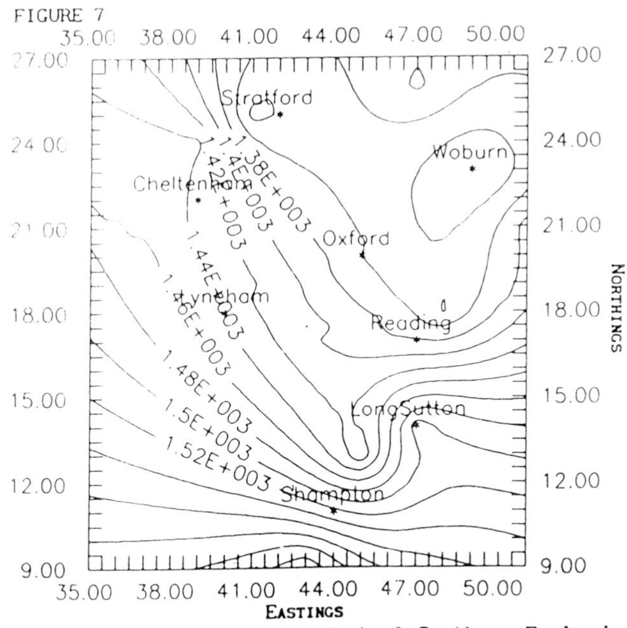

Total Sunshine Hours over Central-Southern England using 13 data points (September 1984-August 1985)

European Communities — Commission

EUR 11943 — Climatic change and impacts: A general introduction

Proceedings of the European School of Climatology and Natural Hazards course, held in Florence from 11 to 18 September 1988

Edited by: *R. Fantechi, G. Maracchi, M. E. Almeida-Teixeira*

Luxembourg: Office for Official Publications of the European Communities

1991 — XI, 454 pp., num. tab., fig. — 16.2 x 22.9 cm

Environment and quality of life series

ISBN 92-826-0564-7

Catalogue number: CD-NA-11943-EN-C

Price (excluding VAT) in Luxembourg: ECU 37.50

The European School of Climatology and Natural Hazards is part of the training and education activities of Epoch (European programme on climatology and natural hazards).

It annually organizes courses open to undergraduate, graduate or postgraduate students in appropriate fields of climatology, natural hazards and closely related fields.

The courses are organized in cooperation with European institutions involved in the carrying out of the Community's R&D programmes on climatology and natural hazards, and are aimed at allowing students to attend formal lectures and to participate in informal discussions with leading research workers. The opportunity for demonstrations, case-studies or presentations of posters is given to the students attending the courses.

The teachers are selected from among European scientists who are leading authorities in their respective fields.

The present volume contains the lectures delivered at the course held in Florence, Italy, from 11 to 18 September 1988, and is the first within the European School of Climatology and Natural Hazards.